Systems Biology and Biotechnology of *Escherichia coli*

Sang Yup Lee
Editor

Systems Biology and Biotechnology of *Escherichia coli*

Springer

Editor

Prof. Dr. Sang Yup Lee
Korea Advanced Institute of Science
and Technology (KAIST)
Dept. of Chemical and Biomolecular Engineering
335 Gwahangno
Yuseong-gu
Daejeon 305-701
Republic of Korea (South Korea)
leesy@kaist.ac.kr

ISBN 978-1-4020-9393-7 e-ISBN 978-1-4020-9394-4

DOI 10.1007/978-1-4020-9394-4

Library of Congress Control Number: 2008942030

Printed on acid-free paper

9 8 7 6 5 4 3 2 1

springer.com

To Hyejean and Gina, with love

Preface

Escherichia coli has been a workhorse not only in fundamental biological and microbiological studies, but also in various biotechnological applications. Recent advances in systems biology have been changing the way biological studies are performed – we are hoping to move towards system-wide understanding of the cell and organism; however, we are still far from truly doing so due to the lack of thorough understanding on complex metabolic, gene regulatory and signaling networks and their interactions. *E. coli*, being a much simpler organism compared with higher organisms, can be a good model system in systems biological studies. The strategies, methods, and tools developed through the systems biological studies on *E. coli* can be extended to other organisms, just like recombinant DNA techniques and other molecular biological tools did. Also, biotechnological applications developed using *E. coli* can be extended to other organisms, in particular other microorganisms, by taking similar metabolic and cellular engineering strategies. Significant advances have been made during the last several years on the systems biology and biotechnology of *E. coli*. In this book, the worldwide experts in the field provide us with the state-of-the-art reviews on the systems-level analyses and applications of *E. coli*.

In Chapter 1, Dr. Kim and his colleagues present the genome project of *E. coli* B. They performed comprehensive functional genomic studies after finishing the genome sequencing. This will be an important addition to the literature as we now have the complete picture of another *E. coli* workhorse, B strain, in addition to the K12 strains. In Chapter 2, Professor Kim and his colleagues present the state-of-the-art strategies and applications of genome minimization. Impressive genome engineering strategies are presented with the practical applications of *E. coli* strains with reduced genomes. In Chapter 3, Professor Tomita and his colleagues review the systems biology of *E. coli* based on multi-omics analyses. Strategies for the combined analyses of transcriptome, proteome, metabolome, and fluxome are presented. Cellular robustness observed based on this systems biological studies is discussed. Structural proteomics is, as we all know, as equally important as quantitative proteomics for better biological and biochemical understanding and many applications. Professor Cygler and his colleagues review the current status of structural proteomics of *E. coli* and methods and techniques available to perform the studies.

In Chapters 5 and 6, resources available for systems biological studies on *E. coli* are reviewed. Professor Mori and Professor Wanner review the important techniques, tools, and libraries available, including the famous gene knock-out system and the complete KEIO single gene knock-out *E. coli* library. Then, Dr. Karp presents an updated status of EcoCyc, the very important database for *E. coli* community and beyond. Through this chapter, he proves that EcoCyc is not just a database for genes and proteins, but is a comprehensive platform for performing systems biological studies.

In Chapter 7, Professor Jeong gives his insights on how to investigate the complex metabolic network of *E. coli* by applying the methods developed in other studies on networks. He applied these methods to metabolic network as well as protein-protein interaction network of *E. coli*. The global and local characteristics of the network structures along with the recent studies on the dynamic aspects of these networks are discussed. In addition to Professor Jeong's chapter on the physical aspects of *E. coli* networks, Dr. Galperin provides an in-depth review on signal transduction network of *E. coli* in Chapter 8. He reviews 6 classes of sensor proteins and 32 response regulators, chemotaxis proteins and several others. Several levels of responses ranging from gene expression to chemotaxis to biofilm formation are discussed.

Chapters 9 to 12 deal with the computational analyses of the genome-scale *E. coli* metabolic network. In Chapter 9, Professor Palsson, one of the pioneers in genome-scale metabolic modeling and simulation, together with his colleagues describe foundational concepts central to the reconstruction process and model formulation and the history of reconstruction of the *E. coli* metabolic network. They also describe the development of reconstruction technology, constraints-based modeling and simulation, and future insights. In Chapters 10 and 11, Professor Goryanin's group and Professor Reuss's group describe kinetic modeling and simulation of *E. coli* metabolism. Starting by describing the basic principles of kinetic description of enzymatic reactions using *in vitro* enzyme assays, Goryanin and his colleagues report detailed kinetic modeling of key enzymes in *E. coli* metabolism. They emphasize that these kinetic models are important to understand key regulatory properties of enzymes. Professor Reuss and his colleagues describe integration of the different networks of *E. coli* exposed to an increasing carbon limitation during the fed-batch process with constant feeding of glucose. They report the analysis and dynamic modeling of regulation phenomena in the catabolism based on the global observation of flux distribution and gene expression in the central metabolism. In Chapter 12, Palsson and Applebee describe the use of genome-scale metabolic model in the analysis of the functions of acquired adaptive mutations for understanding their system-wide effect on phenotype. The constraints-based flux analysis proves to be a powerful tool to study genome-wide characteristics of metabolism.

In Chapter 13, Professor Busby and his colleagues describe the promoters in *E. coli* and molecular characteristics of RNA polymerase recruitment. In addition, transcription factors and their roles in regulation are described. Importance of plasmids in basic biological and applied biotechnological studies does not need to be emphasized. Ow *et al.* report in Chapter 14 our current understanding on plasmid replication and the effects of plasmid presence on host cell physiology at systems-level including the results of *in silico* analysis.

In Chapter 15, the research team of Dr. Rinas and Professor Villaverde review various factors affecting recombinant protein production, especially in the form of inclusion bodies. Protein folding, the machinery of protein quality control, structure and composition of inclusion bodies, and how to control inclusion body formation are described. In Chapter 16, Professor Georgiou's group covers the protein secretion system in *E. coli*. Starting with the general protein transport in *E. coli*, expression and secretion of proteins as well as folding of exported proteins are described. They also discuss on the display of proteins using phage display system and *E. coli*-based cell surface display system.

Chapters 17 and 18 cover the key aspects of *E. coli* central metabolism. Professor Bennett and Professor San's group review a systems view of the central metabolic network and strategies for engineering the metabolic network for the production of various primary and secondary metabolites. Emphasis was also given to cofactor balance and cofactor engineering issues during the metabolic engineering of *E. coli*. Then, Dr. Shiloach and Dr. Rinas further describe the characteristics of central carbon metabolism with a focus on acetate production in *E. coli*. They describe the results of comparative analysis of acetate metabolism in *E. coli* K12 and B strains, and the effects of recombinant protein production on glucose catabolism. The bottlenecks in the primary metabolism and how to overcome these by metabolic engineering are also described.

Chapters 19 and 20 concern with synthetic biology and systems metabolic engineering of *E. coli*. Professor Voigt and his colleagues review the reprogramming of *E. coli* metabolic and regulatory circuits. They describe in detail three classes of genetic parts: sensors, circuits and actuators. Construction of genetic sensors and circuits, and genetic methods to provide perturbation are described. The final chapter contributed by my own team deals with metabolic engineering of *E. coli*, in particular a new paradigm shift towards systems metabolic engineering. Systems metabolic engineering allows genome-wide metabolic engineering based on the findings of systems biological studies including omics and computational analyses. Detailed strategies for systems metabolic engineering are described using *E. coli* as a model organism.

For many decades, *E. coli* has been the organism of choice in studying basic microbiology, genetics and molecular biology as well as in developing important biotechnological applications. As compiled in this book, *E. coli* is now serving as a platform system for systems biological studies as well. It is hoped that this book comprised of the state-of-the-art reviews provided by the worldwide experts would be a good starting point for the new comers in the field and also an important update for those who have been around in this field. It is hoped that this book will be of a long lasting value to the scientists and engineers working in the field. I wish to thank all the contributing authors who made this book possible. Special thanks go to the members of my lab, led by Dr. Tae Yong Kim, who exerted much effort to uniformly format the book. Last but not least, I want to thank Springer people for their help in the production of this book.

KAIST
Daejeon, Republic of Korea

Sang Yup Lee

Contents

Contributors

M. Kenyon Applebee Department of Chemistry & Biochemistry, University of California San Diego, La Jolla, CA 92093-0412, USA, mapplebe@ucsd.edu

George N. Bennett Department of Biochemistry and Cell Biology, Rice University, Houston, TX 77005-1892, USA, gbennett@rice.edu

Douglas F. Browning School of Biosciences, University of Birmingham, Birmingham, B15 2TT UK

Stephen J. W. Busby School of Biosciences, University of Birmingham, Birmingham, B15 2TT UK, s.j.w.busby@bham.ac.uk

Sue Lin-Chao Institute of Molecular Biology, Academia Sinica, 128 Sec. 2, Academia Rd, Nankang, Taipei 115, Taiwan, R.O.C.

Miroslaw Cygler Biotechnology Research Institute, National Research Council Canada, Department of Biochemistry, McGill University, 6100 Royalmount Ave., Montreal, QC H4P2R2 Canada, mirek.cygler@bri.nrc.ca

Kirill A. Datsenko Department of Biological Sciences, Purdue University, West Lafayette, IN 47907, USA

Oleg V. Demin A.N. Belozersky Institute of Physico-Chemical Biology, Moscow State University, Moscow, Russia; Institute for Systems Biology SPb, Sankt-Petersburg, Russia, demin@genebee.msu.su

Hitomi Dose Graduate School of Biological Sciences, Nara Institute of Science and Technology, Ikoma, Nara 630-0101, Japan

Irena Ekiel Biotechnology Research Institute, National Research Council Canada, Department of Chemistry and Biochemistry, Concordia University, 6100 Royalmount Ave., Montreal, QC H4P2R2, Canada

Adam M. Feist Department of Bioengineering, University of California San Diego, La Jolla, CA, USA, afeist@ucsd.edu

Michael Y. Galperin National Center for Biotechnology Information, National Library of Medicine, National Institutes of Health, Bethesda, MD, USA, galperin@ncbi.nlm.nih.gov

Elena García-Fruitós Institute for Biotechnology and Biomedicine and Department of Genetics and Microbiology, Autonomous University of Barcelona, and CIBER-BBN Network in Bioengineering, Biomaterials and Nanomedicine, Bellaterra, 08193 Barcelona, Spain

Kalle Gehring Department of Biochemistry, McGill University, Montreal, QC, Canada, kalle.gehring@mcgill.ca

George Georgiou Department of Chemical Engineering, Biomedical Engineering, Molecular Genetics and Microbiology and the Institute for Cellular and Molecular Biology, University of Texas at Austin, Austin, TX USA, gg@che.utexas.edu

Nuria González-Montalbán Institute for Biotechnology and Biomedicine and Department of Genetics and Microbiology, Autonomous University of Barcelona, and CIBER-BBN Network in Bioengineering, Biomaterials and Nanomedicine, Bellaterra, 08193 Barcelona, Spain

Igor I. Goryanin Centre for Systems Biology, University of Edinburgh, Darwin Building, The Kings Buildings, Mayfield Road, Edinburgh EH9 3JU, 0131 6519063, Scotland UK; Informatics Forum, University of Edinburgh, Crichton Street, Edinburgh EH8 9LE Scotland, UK, goryanin@inf.ed.ac.uk

David C. Grainger School of Biosciences, University of Birmingham, Birmingham B15 2TT, UK

Eli S. Groban Department of Pharmaceutical Chemistry, University of California, San Francisco, CA 94143-2280 USA; Biophysics Program, San Francisco, CA 94158 USA

Timo Hardiman Institute of Biochemical Engineering and Centre Systems Biology, University of Stuttgart, Allmandring 31, 70569, Stuttgart, Germany

Nobuyoshi Ishii Institute for Advanced Biosciences, Keio University, Tsuruoka, Yamagata 997-0017, Japan

Haeyoung Jeong Systems Microbiology Research Center, Korea Research Institute of Bioscience and Biotechnology (KRIBB), 111 Gwahangno, Yuseong-gu, Daejeon 305-806, Republic of Korea

Hawoong Jeong Department of Physics, Institute for the BioCentury, KAIST, Daejeon 305-701, Korea, hjeong@kaist.edu

Zongchao Jia Department of Biochemistry, Queen's University, Kingston, ON, Canada

Peter D. Karp SRI International, Menlo Park, CA 94025, USA, pkarp@ai.sri.com

Jihyun F. Kim Industrial Biotechnology and Bioenergy Research Center, Korea Research Institute of Bioscience and Biotechnology (KRIBB), 111 Gwahangno, Yuseong-gu, Daejeon 305-806, Republic of Korea; Field of Functional Genomics, School of Science, University of Science and Technology (UST), 113 Gwahangno, Yuseong, Daejeon 305-333, Republic of Korea, jfk@kribb.re.kr

Sun Chang Kim Department of Biological Sciences, Institute for the BioCentury, Korea Advanced Institute of Science and Technology (KAIST), Daejeon 305-701, Korea, sunkim@kaist.ac.kr

Soon-Kyeong Kwon Systems Microbiology Research Center, Korea Research Institute of Bioscience and Biotechnology (KRIBB), 111 Gwahangno, Yuseong-gu, Daejeon 305-806 Republic of Korea; Field of Functional Genomics, School of Science, University of Science and Technology (UST), 113 Gwahangno, Yuseong, Daejeon 305-333, Republic of Korea

Galina V. Lebedeva Centre for Systems Biology, University of Edinburgh, Darwin Building, The Kings Buildings, Mayfield Road, Edinburgh, EH9 3JU, 0131 6519063, UK, galina.lebedeva@ed.ac.uk

Dong-Yup Lee Bioprocessing Technology Institute, Agency for Science, Technology and Research (A*STAR), 20 Biopolis Way, #06-01 Centros, Singapore 138668; Department of Chemical and Biomolecular Engineering, National University of Singapore, 4 Engineering Drive 4, Singapore 117576

Jun Hyoung Lee Department of Biological Sciences, Korea Advanced Institute of Science and Technology (KAIST), Daejeon 305-701, Korea

Sang Yup Lee Metabolic and Biomolecular Engineering National Research Laboratory, Department of Chemical and Biomolecular Engineering (BK21 program), Department of Bio and Brain Engineering, Department of Biological Sciences, and Bioinformatics Research Center Center for Systems and Synthetic Biotechnology, Institute for the BioCentury, BioProcess Engineering Research Center, KAIST, 335 Gwahangno, Yuseong–gu, Daejeon 305-701, Korea, leesy@kaist.ac.kr

Karin Lemuth Institute of Biochemical Engineering and Centre Systems Biology, University of Stuttgart, Allmandring 31, 70569, Stuttgart, Germany

Allan Matte Biotechnology Research Institute, National Research Council Canada, 6100 Royalmount Ave., Montreal, Quebec H4P2R2, Canada

Mónica Martínez-Alonso Institute for Biotechnology and Biomedicine and Department of Genetics and Microbiology, Autonomous University of Barcelona, and CIBER-BBN Network in Bioengineering, Biomaterials and Nanomedicine, Bellaterra, 08193 Barcelona, Spain

Eugeniy A. Metelkin A.N. Belozersky Institute of Physico-Chemical Biology, Moscow State University, Moscow, Russia; Institute for Systems Biology SPb, Sankt-Petersburg, Russia, emetelkin@yandex.ru

Ekaterina A. Mogilevskaya A.N. Belozersky Institute of Physico-Chemical Biology, Moscow State University, Moscow, Russia; Institute for Systems Biology SPb, Sankt-Petersburg, Russia, zobova@genebee.msu.su

Hirotada Mori Graduate School of Biological Sciences, Nara Institute of Science and Technology, Ikoma, Nara 630-0101, Japan; Advanced Institute of Biosciences, Keio University, Tsuruoka, Yamagata 997-0017, Japan, hmori@gtc.naist.jp

Kenji Nakahigashi Advanced Institute of Biosciences, Keio University, Tsuruoka, Yamagata, Japan, knakahig@sfc.keio.ac.jp

Dave Siak-Wei Ow Bioprocessing Technology Institute, Agency for Science, Technology and Research (A*STAR), 20 Biopolis Way, #06-01 Centros, Singapore 138668, dave_ow@bti.a-star.edu.sg

Bernhard Ø. Palsson Department of Bioengineering, University of California San Diego, La Jolla, CA 92093-0412, USA, palsson@ucsd.edu

Jin Hwan Park Metabolic and Biomolecular Engineering National Research Laboratory, Department of Chemical and Biomolecular Engineering (BK21 program) Center for Systems and Synthetic Biotechnology, Institute for the BioCentury, BioProcess Engineering Research Center, KAIST, 335 Gwahangno, Yuseong–gu, Daejeon 305-701, Korea

Kirill V. Peskov A.N. Belozersky Institute of Physico-Chemical Biology, Moscow State University, Moscow, Russia, kirillpeskov@gmail.com

Tatiana Y. Plyusnina Institute for Systems Biology SPb, Sankt-Petersburg, Russia; Biophysics Department, Biological Faculty, Moscow State University, Moscow, Russia, plusn@yandex.ru

Matthias Reuss Institute of Biochemical Engineering and Centre Systems Biology, University of Stuttgart, Allmandring 31, 70569, Stuttgart, Germany, reuss@ibvt.uni-stuttgart.de

Ursula Rinas Helmholtz Center for Infection Research, Inhoffenstr. 7, 38124, Braunschweig, Germany, uri@helmholtz-hzi.de

Ka-Yiu San Department of Bioengineering, Rice University, Houston, TX 77005-1892, USA, ksan@rice.edu

Joseph Shiloach Biotechnology Unit, NIDDK, NIH, Bethesda, MD 20892, USA, ljs@helix.nih.gov

Martin Siemann-Herzberg Institute of Biochemical Engineering and Centre Systems Biology, University of Stuttgart, Allmandring 31, 70569, Stuttgart, Germany

Eva-Maria Strauch Department of Chemistry and Biochemistry, The Institute for Cellular and Molecular Biology, University of Texas at Austin, Austin, TX; Department of Biochemistry, University of Washington, Seattle, Washington, USA

Bong Hyun Sung Department of Biological Sciences, Korea Advanced Institute of Science and Technology (KAIST), Daejeon 305-701, Korea

Jeffrey J. Tabor Department of Pharmaceutical Chemistry, University of California, San Francisco, CA 94143-2280, USA

Ines Thiele Program in Bioinformatics, Department of Bioengineering, University of California San Diego, La Jolla, CA 92093-0412, USA, ithiele@ucsd.edu

Masaru Tomita Institute for Advanced Biosciences, Keio University, Tsuruoka, Yamagata 997-0017, Japan, tomita@ttck.keio.ac.jp

Hsiu-Hui Tung Institute of Molecular Biology, Academia Sinica, 128 Sec. 2, Academia Rd, Nankang, Taipei 115, Taiwan, R.O.C.

Antonio Villaverde Institut de Biotecnologia i de Biomedicina, Universitat Autònoma de Barcelona, Bellaterra, Barcelona, 08193 Spain, antoni.villaverde@uab.es

Christopher A. Voigt Department of Pharmaceutical Chemistry, University of California San Francisco, CA 94143-2280 USA; Biophysics Program, San Francisco, CA 94158, USA, cavoigt@picasso.ucsf.edu

Barry L. Wanner Department of Biological Sciences, Purdue University, West Lafayette, IN, USA, blwanner@purdue.edu

Natsuko Yamamoto Graduate School of Biological Sciences, Nara Institute of Science and Technology, Ikoma, Nara 630-0101, Japan

Sung Ho Yoon Systems Microbiology Research Center, Korea Research Institute of Bioscience and Biotechnology (KRIBB), 111 Gwahangno, Yuseong-gu, Daejeon 305-806, Republic of Korea

About the Editor

Dr. Sang Yup Lee is a Distinguished Professor and LG Chem Chair Professor at the Department of Chemical and Biomolecular Engineering at KAIST, Daejeon, Korea. He is currently Dean of the College of Life Science and Bioengineering, Director of BioProcess Engineering Research Center, Director of Bioinformatics Research Center, and Director of Center for Systems and Synthetic Biotechnology. His major research interests are on metabolic engineering of microorganisms for industrial and medical biotechnology by integrating systems biology and synthetic biology approaches. He has published more than 270 journal papers, 40 books/book chapters, and is holding 300 patents registered/applied. He received numerous awards including National Order of Merit from Korean government, the First Young Scientist's Award from the President of Korea, the Elmer Gaden Award from *Biotechnology and Bioengineering*, and many others, including the most recent Merck Metabolic Engineering Award. He is a Fellow of Korean Academy of Science and Technology, and Junior Fellow of National Academy of Engineering of Korea, Fellow of AAAS, and Fellow of American Academy of Microbiology. He is currently serving Editor-in-Chief of Biotechnology Journal, and Associate Editor and editorial board member of many journals in the field of biotechnology and biochemical engineering.

Chapter 1
Genomics, Biological Features, and Biotechnological Applications of *Escherichia coli* B: "Is B for better?!"

Sung Ho Yoon, Haeyoung Jeong, Soon-Kyeong Kwon, and Jihyun F. Kim

Contents

Abstract Strains of *Escherichia coli* B, especially BL21, have been widely used for overproducing recombinant proteins, ethanol, and other biomolecules. Almost all laboratory strains of *E. coli* are derivatives of non-pathogenic K-12 or B strains. While most genetic and metabolic studies have been performed with K-12 strains, little has been done on B strains. Recently, genome sequences of two *E. coli* strains of the B lineage, REL606 and BL21(DE3), have been determined, and results of multi-omics analyses were compared between B and K-12. As compared to K-12,

---

J.F. Kim (✉)
Industrial Biotechnology and Bioenergy Research Center, Korea Research Institute of Bioscience and Biotechnology (KRIBB), 111 Gwahangno, Yuseong, Daejeon 305-806, Republic of Korea;
Field of Functional Genomics, School of Science, University of Science and Technology (UST), 113 Gwahangno, Yuseong, Daejeon 305-333, Republic of Korea
e-mail: jfk@kribb.re.kr

S.Y. Lee (ed.), *Systems Biology and Biotechnology of Escherichia coli*,
DOI 10.1007/978-1-4020-9394-4_1,

B strains show a number of phenotypes such as faster growth in minimal media, lower acetate production, higher expression levels of recombinant proteins, and less degradation of such proteins during purification. In this review, we summarize the unique biological features of the B strains and overview their academic and industrial applications.

1.1 Introduction

Escherichia coli, a common inhabitant of the mammalian intestines, undoubtedly has been one of the best studied organisms and plays important roles in biological sciences, medicine, and industry. *E. coli* strain B was named by Delbrück and Luria in 1942 (Delbruck and Luria 1942), but the early history of B is less well established whereas the origin of *E. coli* K-12 is clear. In any case, derivatives of *E. coli* B have been serving not only as a research model for the study of phage sensitivity, restriction systems, and bacterial evolution in the laboratories, but as a major workhorse for protein expression in the biotechnological industry. However, genetic bases of the apparent superiority of B in many industrial setups have been restricted to a limited number of topics and the rest have been left largely undetermined.

The genomes of two derivative strains of B recently have been deciphered through an international collaboration among scientists in Korea (KRIBB; our group), the United States (Michigan State University and Brookhaven National Laboratory), and France (Genoscope). This has opened a new possibility of examining B through various omics technologies including DNA microarray for gene expression profiling and two dimensional gel electrophoresis followed by MALDI-TOF identification of proteins. In this chapter, we review the current understanding of the biological features of B strains in the context of genomic information and results of the multi-omics analyses together with their utility in scientific studies and biotechnical applications.

1.2 Genomics of *E. coli* B Strains

1.2.1 Genomic Comparison of E. coli *REL606 and MG1655*

The first complete genome sequence of a B strain was determined by an international consortium (Jeong et al. submitted). The strain of choice was REL606, an Ara$^-$ clone derived from chemical mutation of Bc251 (F$^-$ mal^+ λ^s) (Lederberg 1966). REL606 has been long used as a founder strain for long-term evolution experiments by Richard E. Lenski at Michigan State University (Cooper and Lenski 2000, Lenski et al. 1991). A Sanger chemistry-based, standard shotgun approach was exploited for the genome sequencing of REL606. KRIBB participated in the initial shotgun sequencing and final process for genome annotation, while Genoscope led genome sequencing to completion and automatic annotation based on MaGe (Vallenet 2006).

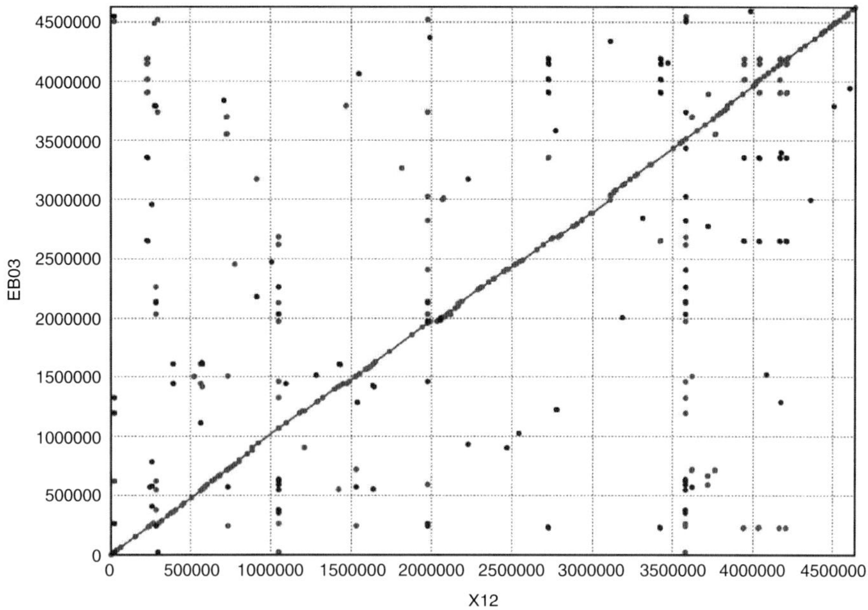

Fig. 1.1 Whole-genome alignment of *E. coli* K-12 MG1655 (*x*-axis) and B REL606 (*y*-axis). NUCMER script in MUMMER 3.0 was used for the generation of alignment with default parameters (http://mummer.sourcefourge.net/)

E. coli B REL606 has a single circular chromosome of 4,612,812 bp with no plasmid, which makes it the most compact genome among the completely sequenced strains of *E. coli*. Its chromosome size is most similar to that of K-12 MG1655 (4,639,675 bp), and as shown by the whole-genome alignment plot no chromosomal rearrangement was observed other than some insertions or deletions (Fig. 1.1). When MUMMER was used as an alignment generator, the total length of aligned regions between REL606 and MG1655 amounted to more than 96%. Average percent identity of the aligned regions is 97.5%, and it further increases up to 99.09% if it is length-weighted average.

Though overall genome organization is very similar between two strains, several prominent factors contribute to shaping peculiarities of each genome (Fig. 1.2, see below). First, highly divergent regions, occupying the equivalent positions on each genome, are readily identified by the broken lines appearing on the whole-genome alignment plot (Fig. 1.3). Most of them are related to genes involved in surface characteristics (e.g., LPS core oligosaccharide biosynthesis), which are probably ones under strong positive selection. Second, there are several horizontally transferred genomic segments that represent genome-specific regions. Fitness island encoding gene sets for the metabolism of aromatic hydrocarbon is a good example.

Lastly, distribution of mobile genetic elements such as prophages and insertion sequence (IS) elements are significantly different between the two strains. Specifically, IS seems to exert most dramatic effect to its host genome, since it can deacti-

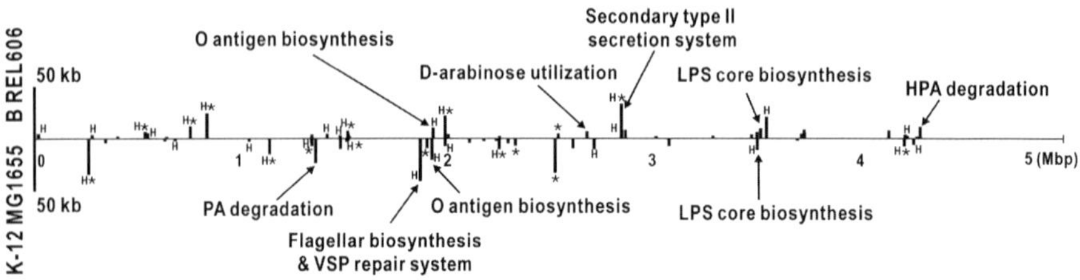

Fig. 1.2 Genomic regions unique to *E. coli* B REL606 or K-12 MG1655. The horizontal axis represents the map coordinates of backbone regions common to REL606 and MG1655; *vertical bars* denote locations and lengths of strain-specific regions larger than 2 kb. A H denotes the region overlapping predicted genomic islands (Yoon et al. 2005). Genomic regions containing prophage genes are marked by asterisks. Abbreviations: LPS, lipopolysaccharide; HPA, 3-hydroxyphenylacetic acid and 4-hydroxyphenylacetic acid; PA, phenylacetic acid; VSP, very short-patch

vate functional genes by insertion events, and can result in genomic rearrangement by homologous recombination as well. For example, REL606 has a 41-kb deletion ($\sim$0.8% of the entire chromosome), probably mediated by IS*1* (the most abundant IS in B), from *yecF* to *yedS*. The deleted segment, that includes *sdiA*, *amyA*, *rcsA*, *vsr*, *dcm*, and the *fli* cluster, produce relevant B-specific phenotypes that were previously known. Functions affected by IS transpositional inactivation include porin expression, restriction-modification, and Lon protease (saiSree et al. 2001, Schneider et al. 2002).

1.2.2 Comparative Genomics of E. coli *BL21(DE3)*

BL21(DE3) is a specifically engineered B descendant harboring the T7 RNA polymerase gene for high-level expression of recombinant proteins (Studier and Moffatt 1986). With other B-specific traits such as deficiency in proteases and amenability to high-density culture as mentioned above, BL21(DE3) has become the most widely used strain for biotechnological applications. Since BL21(DE3) and REL606 are very close to each other being diverged just dozens of years ago from their common ancestor, we applied a hybrid approach that combines hybridization-based genome resequencing by NimbleGen's CGS (Comparative Genome Sequencing) technology and 454 pyrosequencing by Roche GS 20 at about 10X coverage to assemble genome sequence of BL21(DE3) using the REL606 genome (Jeong et al. submitted). All potential mutagenic sites were confirmed by Sanger sequencing of PCR-amplified products, and genomic rearrangement that could be only identified by pairwise comparison between the reference sequence and *de novo*-assembled 454 contigs were also verified by the same method. Over 98% of the final BL21(DE3) sequence could be covered by 454 pyrosequencing contigs, and $\sim$166 kb were confirmed by conventional sequencing.

We confirmed 415 SNPs, 16 insertions, and 28 deletions in the genome of BL21(DE3) with respect to REL606. 79% of SNPs (326 out of 415) occur in a nar-

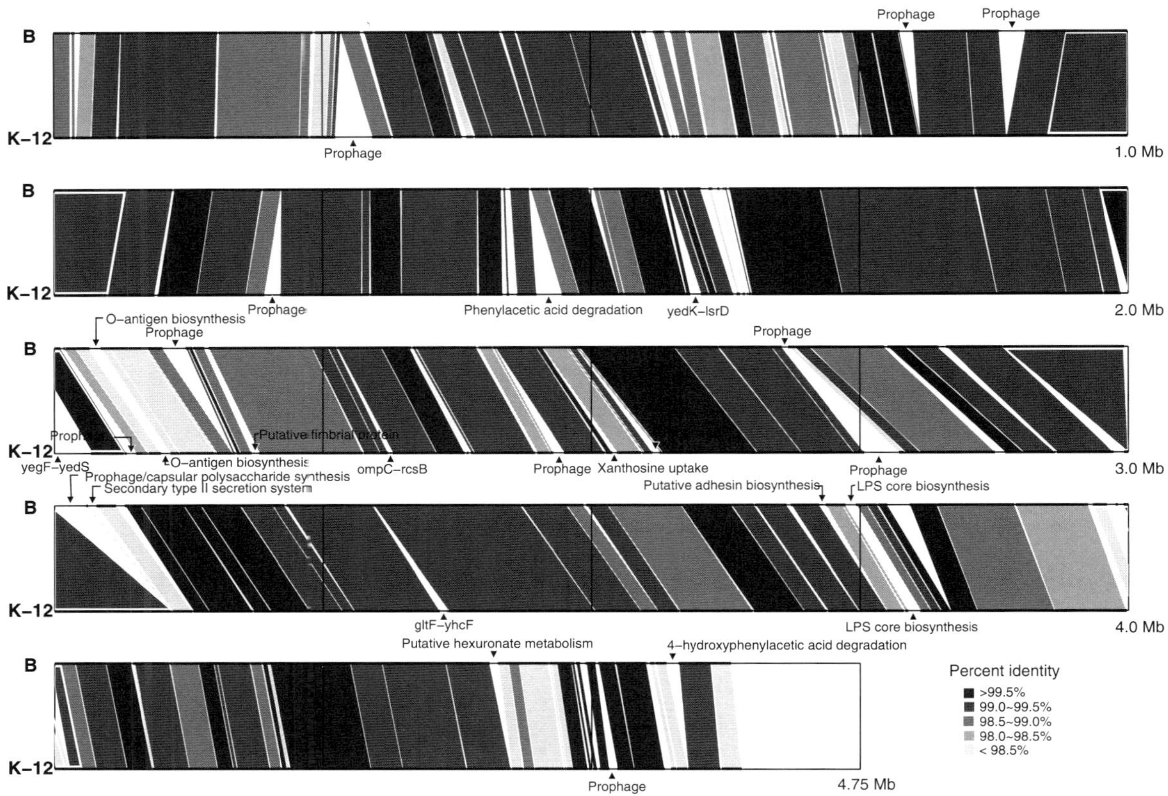

Fig. 1.3 A schematic diagram showing alignment of the blocks of homologous sequences (*thick lines*) between *E. coli* B REL606 (*top lines*) and K-12 MG1655 (*bottom lines*). Percent identity for each block is indicated as a grey parallelogram of a different shade. Texts denote genes or functions encoded by specific genomic segments

row region around 4.2 Mb-position, which occupies only 1.4% of the chromosome. Chemical mutagenesis applied to Bc251 to produce the progenitor to REL606 only cannot accounts for such a big differences. Close inspection revealed the highly divergent region was indeed transferred from W3110 by P1 transduction to produce Bc251. Another P1 transduction to create BL21 replaced again most of the W3110 DNA with B DNA, which resulted in K-12-derived sequence remaining only in REL606.

To obtain a complete genome sequence of BL21(DE3) based on 454 contigs, we recently constructed a fosmid library and produced paired-end sequences. Scaffolds have been made by a mixed assembly approach using fosmid end reads and verified 454 contigs, and gap closure is in progress (Jeong and Kim, unpublished data).

1.2.3 Comparative Omics Analysis of E. coli *B and K-12*

Recent advances in high-throughput omics technology are providing us with the possibility of deciphering an organism's genotype-to-phenotype relationships. Recently, we have carried out comparative and integrated analysis of the genome, transcriptome, proteome, and phenome data of B and K-12 strains that are closely related (Yoon et al. submitted). We also reconstructed an *in silico* metabolic network of B that accommodates the multidimensional omics data. From the study, we identified many important differences in cellular metabolism and physiology between B and K-12.

Lack of flagellar biosynthetic genes and low expression of motility-related genes make B non-motile. This is an important property of B when used as a cell factory because flagella biosynthesis is energy-intensive and is not necessary under an industrial setup of constant agitation and generous supply of nutrients (Posfai et al. 2006, Yu et al. 2002). Differences in the composition of the LPS core and expression of outer membrane proteins may influence the permeability and integrity of the cell envelope, which presumably result in alterations to screening barriers that control import and export of materials such as antibiotics, nutrients, and proteins. Importantly, the existence of the second T2S system and enhanced capability for protein release qualify B strains as the first choice for extracellular production of recombinant proteins. Information on naturally exported proteins can be useful in developing a strategy of excretory protein production, as exemplified by the use of OmpF fusion approach for the extracellular production of human proteins (Jeong and Lee 2002). B strains exhibited up-regulation of many amino acid biosynthetic genes, and showed lower expression of proteases. These characteristics are desirable for the enhanced production of recombinant proteins.

1.3 Biological Features of *E. coli* B

Elucidation of biological features of a strain is always the starting point for its biotechnological applications. B and its derivatives have been widely used for the production of recombinant proteins and biomolecules. B strains have also served as research models for studies of phage sensitivity, restriction systems, mutagenic assays, and bacterial evolution (Cooper and Lenski 2000, Herrera et al. 2002, Swartz 1996). Although the genome sequences of B and K-12 are highly similar based on comparison of IS elements (Schneider et al. 2002), B often shows phenotypes distinct from those of K-12 (Swartz 1996).

1.3.1 Catabolism and Acetate Metabolism

B strains have been widely used for overproduction of recombinant proteins because they offer faster cell growth in minimal media and lower production of acetate than K-12 derivatives. Such differences between B and K-12 strains can be attributed to their genetic backgrounds. However, from the analysis of genome sequences of two B strains, REL606 and BL21(DE3) (Jeong et al. submitted), we found that there was no genetic difference in genes involved in glycolysis, TCA cycle, pentose phosphate pathway, glyoxylate pathway, gluconeogenesis, and acetate production among genomes of B and K-12. Thus, it is possible that the metabolic genes are under different regulation of in these two groups of *E. coli*. Transcriptome analysis of BL21 and JM109 (B derivative) during batch fermentation with high initial glucose concentration (Phue et al. 2007) demonstrated that genes involved in glyoxylate shunt, TCA cycle, fatty acid, gluconeogenesis and anaplerotic pathways were expressed differently between the two strains, while no apparent differences were detected for those in glycolysis and pentose phosphate pathway.

Acetate accumulation is one of the major problems encountered during high cell density cultivation of *E. coli*, because it inhibits cell growth and production of foreign proteins (Eiteman and Altman. 2006). Generally, B strains accumulate less acetate than K-12 strains during high cell density cultivation with glucose as a carbon source. A common explanation for low acetate accumulation by B is the active glyoxylate shunt which is the main pathway for acetate utilization due to the high expression of acetate operon (*aceBAK*). Analyses of DNA microarray and Northern blot demonstrated that BL21 showed high activity in glyoxylate shunt, TCA cycle, gluconeogenesis pathway, conversion of acetate to acetyl CoA, and fatty acid degradation irrespective of glucose concentration in culture media (Phue et al. 2005). In case of JM109, a K-12 derivative, the trend was similar at low glucose culture conditions, while it was lowered at high glucose conditions. Phue et al. (2005) suggested that insensitivity of BL21 to glucose concentration can be attributed to absence of a regulatory mechanism and possibility of altered activity of FruR, a transcriptional regulator of the control of carbon and energy metabolism, in BL21. Much effort should be made to fully understand the metabolism of glucose and acetate of the B and K-12.

1.3.2 Anabolism

Sequence differences in amino acids biosynthetic pathways have been found in genes for L-arginine and branched-chain amino acids biosyntheses. In K-12, formation of the enzymes involved in arginine biosynthetic pathway is under feedback inhibition by arginine, while those enzyme levels are not affected by arginine in B. The different regulation mode is due to the differences in a single amino acid in the arginine repressor protein (ArgR), changing from the proline residue in K-12 to leucine in B (Tian et al. 1994). K-12 contains two genes encoding ornithine carbamoyltransferase in arginine biosynthesis, *argI* and *argF*, while B, *E. coli* W, and other species in *Enterobacteriaceae* have only *argI* or its equivalents (Legrain et al. 1976).

B is insensitive to extracellular valine while K-12 cannot grow in the presence of valine (Yoon et al. submitted). The first reaction in the biosynthesis of branched-chain amino acids (leucine, isoleucine and valine) is catalysed by three isozymes, acetohydroxy acid synthase (AHAS) I, II, and III encoded by *ilvBN*, *ilvGM*, and *ilvIH* respectively. It is known that valine exerts feedback inhibition on isozymes I and III (Umbarger 1996). Thus, exogenous valine can inhibit cell growth because the valine makes the cell unable to synthesize leucine and isoleucine, which is known as valine toxicity. It is thus essential that the functional isozyme II is expressed for cell growth in the presence of exogeneous valine. K-12 has a frameshift mutation in the *ilvG* gene, but B has an intact *ilvG* mediating valine resistance.

1.3.3 Utilization of Carbon Sources

Ability to utilize a variety of substrates is quite different between B and K-12, which can be attributed largely to genetic discrepancy in nutrient uptake systems. Most enteric bacteria cannot grow on D-arabinose which is uncommon in the natural environments. As enzymes for L-fucose utilization can degrade D-arabinose to dihydroxyacetone phosphate and glycoaldehyde, regulatory mutations of the L-fucose pathway in K-12 led to growth on D-arabinose (LeBlanc and Mortlock 1971). In contrast to K-12, B strains cannot grow on L-fucose because of the lack of L-fuculose-1-phosphate aldolase (Boulter et al. 1974). Interestingly, B strains can degrade D-arabinose without mutation. This is due to the possession of gene cluster for converting D-arabinose to D-xylulose 5-phosphate, which appears to have acquired through horizontal gene transfer (Elsinghorst and Mortlock 1994).

Eliminating environmental pollutants such as aromatic compounds by microorganisms is a competitive alternative to the commonly used chemical processes (Pieper and Reineke 2000). Aromatic compounds are highly abundant in soil and water, and *Pseudomonas* strains and other soil bacteria can catabolize a wide range of aromatic compounds. Unexpectedly, some *E. coli* strains and other enteric bacteria are reported to be able to degrade aromatic amino acids (Diaz et al. 2001). *E. coli* B and C can grow on 3- and 4-hydroxyphenylacetic acid (HPA) but not on phenylacetic acid (PA), while K-12 grow on PA but not on 3-HPA and 4-HPA (Diaz et al. 2001). Recent studies that compared the genomes of B and K-12 (Yoon et al. submitted) demonstrated that each has a different gene cluster for the catabolism of aromatic compounds – the *paa* cluster for the catabolism of phenylacetic acid in K-12 and the *hpa* cluster for the degradation of 3- and 4-HPA in B.

1.3.4 Cell Surface Features

Cell envelope is the principal stress-bearing and shape-maintaining element in *E. coli*, and its integrity is of critical importance to cell viability. B strains have been widely used for mutagenic assays and toxicological studies because they show higher membrane permeability than does K-12 (Herrera et al. 2002). Structural studies on the LPS core oligosaccharides have revealed that K-12 is devoid of the O antigen while B lacks the O antigen plus the distal part of the polysaccharide core of the outer membrane (Jansson et al. 1981). Sequence comparison has revealed that the outer membrane structure of B is quite different from that of K-12 (Jeong et al. submitted). IS elements were found to be inserted at the gene clusters for O antigen biosynthesis: at *wbbL* for K-12 and between *manC* and *wbbD* for B strain. In the B genome, the core part of LPS was further disrupted by the insertion of IS*1* at *waaT* encoding the UDP-galatose:(glucosyl) LPS α1,2-galactosyltransferase.

Importantly, flagellar biosynthesis genes are missing in B (Jeong et al. submitted). A 38-kb region of K-12 from *yecF* to *yedS*, containing *fliYZACDSTEFGHIJKLMNOPQR* genes was deleted in the genome of B. Thus, B cannot form the flagella and thus is non-motile.

Porin proteins control the permeability of polar solutes across the outer membrane and play important roles in the nutrient uptake process (Nikaido 1996). In K-12, though the total amount OmpC and OmpF is constant, their relative proportion changes depending on the culture medium osmolality, which is controlled by the EnvZ-OmpR regulatory system. By contrast, B strains express only OmpF in large quantity (Pugsley and Rosenbusch 1983). This is attributed to the fact that IS insertion in the B genome results in the deletion of the first 114 bp of *ompC* and the upstream region containing *micF* which posttranscriptionally prevents the production of OmpF (Schneider et al. 2002). Noxious agents such as antibiotics and bile acids diffuse far better through OmpF because OmpF produces a larger channel than OmpC (Nikaido 2003). Thus, *ompF* mutants became highly resistant to β-lactam compounds (Harder et al. 1981). In the phenotype microarray test, we discovered that B displayed sensitivity to various stress conditions of osmolarity, pH stress, and antibiotics much higher than K-12 (Yoon et al. submitted).

1.3.5 Heat Shock Proteins

Heat shock proteins (HSPs) including molecular chaperones and proteases make sure cellular proteins being in the right shape and in the right place at the right time (Gross 1996). Thus, they are required both during stress and normal growth conditions. Among the ATP-dependent proteases, B strains are naturally deficient in the major protease Lon which degrades abnormally folded proteins. This is due to the insertion of IS*186* in the promoter region of *lon* (saiSree et al. 2001). Additionally, the BL21 cells lack the OmpT outer membrane protease. Besides its major role in protein quality control, Lon is involved in many biological processes such as cell differentiation, pathogenicity, motility, stringent response to amino acid starvation, and regulation of the toxin-antitoxin module (Tsilibaris et al. 2006). Lon mutants are viable, but display sensitivity to ultraviolet light and overproduce capsular polysaccharide, which are the result of the elevated levels of regulatory proteins (SulA and RcsA) that are normally degraded by Lon.

As HSPs are up-regulated by the heat-shock sigma factor σ^{32} encoded by *rpoH* when cells are exposed to stress condition such as temperature upshift and production of recombinant proteins, they can be used as a stress probe for monitoring cellular stress (Cha et al. 1999, Vostiar et al. 2004). When cellular stress levels of BL21 and K-12 strains (JM105, HB101, and TOP10) were measured by fusing promoters of heat-shock genes (*rpoH*, *dnaK* or *clpB*) to the reporter gene (*gfp*), BL21 exhibited the lowest cellular stress level and expressed the highest foreign protein (Seo et al. 2003). Possibly, lower cellular stress level of B strain is one of the reasons for high capacity in foreign protein production.

1.3.6 Cell Cycle and Growth

Bacterial cell growth is closely coordinated with DNA replication and chromosome segregation (Haeusser and Levin 2008). The cell cycle of slowly growing bacteria can be divided into three time periods: (i) period B, cell division to the initiation of chromosome replication, (ii) period C, chromosome replication, and (iii) period D, termination of replication to cell division. From the cytometry data that measured periods C and D of *E. coli*, the D period in B/r is much shorter than in K-12 strains (Michelsen et al. 2003).

Normally, B strains grow faster than K-12 in minimal media. The widely used K-12 strains, MG1655 and W3110, grow slowly in a pyrimidine-free medium than in a medium containing uracil. In the *rph-pyrE* operon involved in *de novo* pyrimidine biosynthesis of these strains, *rph* is frame-shifted to produce truncated RNase PH, and the premature translation stop leads to decreased expression of *pyrE* encoding orotate phosphoribosyltransferase (Jensen 1993). However, the *rph* gene is intact in the B strains (Yoon et al. submitted).

1.3.7 Secretion Capacity

Bacterial extracellular proteins perform important biological processes such as assembly of flagella and fimbriae, nutrient acquisition, cell-to-cell communication, and pathogenesis. In Gram-negative bacteria, excretory proteins are much less than in Gram-positive species because they should cross the two membranes of the cell envelope. Laboratory *E. coli* strains normally does not secrete extracellular proteins because the genes encoding type II secretion (T2S) pathway operon (*gsp*) are silenced by H-NS (Francetic et al. 2000). B strains released more proteins according to analyses of the extracellular proteomes of B and K-12 during flask culture (Yoon et al. submitted) and high cell density cultivation (Xia et al. 2008). This could be at least partly due to an additional gene cluster for T2S in the B strains (Yoon et al. submitted). Phylogenetic analysis revealed that the T2S system commonly found in REL606 and MG1655 were clustered into the clade of *E. coli* and other genera in the *Enterobacteriaceae* family, whereas the sequence of the additional T2S system in REL606 was grouped into the branches of *E. coli* strains having multiple T2S systems and families other than *Enterobacteriaceae*. This implies that the two T2S systems of the B strains have evolved independently or more specifically the latter might have been introduced.

1.4 Usage of B Strains

1.4.1 As an Academic Lab Rat

Since 1940s, K-12 and B strains have been widely used as a laboratory strain and have had significant impact on biological sciences, medicine, and industry (Daegelen et al. submitted, Lederberg 2004). While K-12 strains have been mainly used for developing recombinant DNA techniques, B strains have been the subject of physiological studies (Swartz 1996). B strains also served as hosts for the historical studies of T1-T7 bacteriophages (Delbruck 1946), which led to the construction of BL21(DE3) (Studier and Moffatt 1986). Due to the rapid growth in minimal media and enhanced membrane permeability, B strains were favored by physiologists.

A radiation-resistant mutant, *E. coli* B/r, was isolated after UV-irradiation (Witkin 1946), and has been used for determining cell cycle-related parameters (Helmstetter 1968, Michelsen et al. 2003). Chemical composition measurements of B/r was measured (Neidhardt and Umbarger 1996), which is essential to estimate growth requirements such as energy distribution, reducing power (Neijssel et al. 1996) and metabolic fluxes in a genome-scale metabolic model (Feist et al. 2007).

Due to the increased membrane permeability, B strains have been used widely for mutagenic assays and toxicological studies. Mutants of B strain WP2 (e.g. WP2 *uvrA* and WP2 *uvrA*/pKM101) have been used as a tester strain in mutagenic assays (Gatehouse et al. 1994) and officially included in the OECD guideline for bacterial reverse mutation test (OECD guideline for testing of chemicals: bacterial reverse mutation test, 1998). These mutants are sensitive to oxidizing mutagens, cross-linking agents and hydrazines (Blanco et al. 1998, Herrera et al. 1993, Wilcox et al. 1990). The higher permeability can make B a primary choice for functional studies by flow cytometry and fluorescence microscopy. When B and K-12 strains were stained with several fluorochromes, B strain showed higher uptake of fluorescent dyes and higher fluorescent intensity (Herrera et al. 2002).

Evolution experiments with microorganisms are of a great interest because they allow one to investigate genetic and phenotypic evolution in action under the controlled environment (Elena and Lenski 2003, Philippe et al. 2007). For decades, B has served as a research model for long-term bacterial evolution. Twelve populations derived from a common ancestor have propagated by daily serial transfer in a glucose-limited minimal medium for more than 40,000 generations (Cooper et al. 2003). In the experiment, all the populations have adapted to the growth environment via beneficial mutations. Critical issues in evolution are being addressed using laboratory populations of bacteria, e.g. the dynamics of evolutionary adaptation, the genetic bases of adaptation, interactions between different genotypes in a population, and between interacting microbial species (Elena and Lenski 2003). Application of evolutionary principle to strain development and process optimization, which is so called evolutionary engineering, is becoming an important strategy in the field of metabolic engineering (Sauer 2001). Recently, the efficiency of natural selection using long-term evolution experiment is successfully exploited to improve industrial strains (de Crecy et al. 2007, Fong et al. 2005).

1.4.2 As an Industrial Workhorse

A primary goal of the bioprocess development is the cost-effective production of desired products such as therapeutic and industrial proteins on a large scale. B and K-12 are preferred production hosts because of fast growth, facility in genetic modification and cultivation, and high yields for many recombinant proteins. As mentioned above, B and its derivatives have salient features desirable for high cell density culture such as low acetate production even when grown on excess glucose, faster growth in minimal media, protease deficiency, and simple cell surfaces that enhance permeability. Thus, they have been widely used for the overproduction of recombinant proteins, ethanol, and other biomolecules on a large scale (Choi et al. 2006).

The most popular strains, BL21 and its derivatives (Studier and Moffatt 1986), are derived from B, thus are naturally deficient in the major protease Lon. Additionally, their chromosomes are deleted from the gene for the outer membrane protease OmpT. The absence of these proteases can lead to higher expression levels of recombinant proteins and less degradation of such proteins during purification. A derivative of BL21, BL21(DE3), was constructed to harbor a recombinant phage λ carrying the T7 RNA polymerase gene under the control of the *lacUV5* promoter (Studier and Moffatt 1986). Addition of isopropyl β-D-thiogalactopyranoside (IPTG) into growth media is required to express the T7 RNA polymerase, which then transcribes target genes located in a plasmid under the control of the T7 promoter. Due to its high selectivity and activity, BL21(DE3) is extremely popular for mass-production of recombinant proteins which are toxic to the host cells (Choi et al. 2006). Various versions of the T7 RNA polymerase-based expression system have been developed to use cheap and nontoxic inducers instead of IPTG or to minimize basal expression of the cloned gene (Sorensen and Mortensen 2005).

Membrane proteins (MPs) account for more than 50% of all drug targets and are of major pharmaceutical and biotechnological interests. Generally, a large amount of MPs is required for their functional and structural studies. However, in many cases, over-expression of MPs is lethal to host cells. To overcome this difficulty, mutant hosts, C41(DE3) and C43(DE3) were derived from BL21(DE3) over-producing some membrane proteins (Miroux and Walker 1996). These two mutant hosts, especially C43(DE3), showed reduced toxicity of expressed MPs and are widely used for a variety of MPs. Although the genetic mutation(s) responsible for the changes have not yet been identified, comparative genome sequence analyses of C41(DE3), C43(DE3), and their parental BL21(DE3) revealed several interesting genetic changes (Kwon and Kim, unpublished data). Comparative analysis revealed that there are six SNPs in C41(DE3) and seven in C43(DE3) as compared to BL21(DE3). Interestingly and perplexingly, only two of them overlaps between the two strains. Also, there are two IS-mediated deletions that have been observed – one in both C41(DE3) and C43(DE3) and the other only in C43(DE3). It is reported that C41(DE3) and C43(DE3) are also superior to BL21(DE3) in the production of some cytoplasmic proteins and in the stability of their cloning plasmid (Dumon-Seignovert et al. 2004). However, it is hard to predict an expression host and system working best for a target protein, and so, screening process is required to some extent.

1.5 Future Prospects

Until now, most genetic and metabolic studies of *E. coli* have been performed with K-12 or its derivatives. Also, a variety of omics analyses (Choi et al. 2003, Franchini and Egli 2006, Han and Lee 2006, Ishii et al. 2007, Nandakumar et al. 2006, Yoon et al. 2003) and *in silico* metabolic modeling of K-12 (Covert et al. 2004, Feist et al. 2007) have been accelerated by the availability of the complete genome sequences of the MG1655 and W3110 strains (Blattner et al. 1997, Hayashi et al. 2006). In contrast to K-12, little studies have been performed for B strains. This can be attributed to the fact that K-12 strains are the best fit in the current recombinant DNA techniques and a wealth of safety information makes them preferred recombinant organisms by the biosafety communities (Swartz 1996). Additionally, many B derivatives as industrial hosts have been developed in private companies, which can make some difficulties in academic and public research. However, various features of B strains are beneficial for the overexpression of foreign proteins and studies of *E. coli* physiology. B strain is now in its early stages of global omics studies and systems biology (Xia et al. 2008, Yoon et al. submitted). With the availability of genome sequences of B strains (Jeong et al. submitted), the omics information on the cellular metabolism and physiology should be pivotal in better understanding the underlying biological networks and is invaluable for designing strains having customized genomes as well as establishing rational fermentation strategies.

Acknowledgments This work was supported by the 21C Frontier Microbial Genomics and Applications Center Program, Ministry of Science and Technology, Republic of Korea.

References

Blanco, M., A. Urios, and A. Martinez. 1998. New *Escherichia coli* WP2 tester strains highly sensitive to reversion by oxidative mutagens. Mutat. Res. **413**:95–101.

Blattner, F. R., G. Plunkett, 3rd, C. A. Bloch, N. T. Perna, V. Burland, M. Riley, J. Collado-Vides, J. D. Glasner, C. K. Rode, G. F. Mayhew, J. Gregor, N. W. Davis, H. A. Kirkpatrick, M. A. Goeden, D. J. Rose, B. Mau, and Y. Shao. 1997. The complete genome sequence of *Escherichia coli* K-12. Science **277**:1453–74.

Boulter, J., B. Gielow, M. McFarland, and N. Lee. 1974. Metabolism of D-arabinose by *Escherichia coli* B/r. J. Bacteriol. **117**:920–3.

Cha, H. J., R. Srivastava, V. N. Vakharia, G. Rao, and W. E. Bentley. 1999. Green fluorescent protein as a noninvasive stress probe in resting *Escherichia coli* cells. Appl. Environ. Microbiol. **65**:409–14.

Choi, J. H., K. C. Keum, and S. Y. Lee. 2006. Production of recombinant proteins by high cell density culture of *Escherichia coli*. Chem. Eng. Sci. **61**:876–85.

Choi, J. H., S. J. Lee, and S. Y. Lee. 2003. Enhanced production of insulin-like growth factor I fusion protein in *Escherichia coli* by coexpression of the down-regulated genes identified by transcriptome profiling. Appl. Environ. Microbiol. **69**:4737–42.

Cooper, T. F., D. E. Rozen, and R. E. Lenski. 2003. Parallel changes in gene expression after 20,000 generations of evolution in *Escherichia coli*. Proc. Natl. Acad. Sci. USA **100**:1072–7.

Cooper, V. S., and R. E. Lenski. 2000. The population genetics of ecological specialization in evolving *Escherichia coli* populations. Nature **407**:736–9.

Covert, M. W., E. M. Knight, J. L. Reed, M. J. Herrgard, and B. O. Palsson. 2004. Integrating high-throughput and computational data elucidates bacterial networks. Nature **429**:92–6.

Daegelen, P., F. W. Studier, R. E. Lenski, S. Cure, and J. F. Kim. Tracing ancestors and relatives of *Escherichia coli* B, and the derivation of B strains REL606 and BL21(DE3). submitted.

de Crecy, E., D. Metzgar, C. Allen, M. Penicaud, B. Lyons, C. J. Hansen, and V. de Crecy-Lagard. 2007. Development of a novel continuous culture device for experimental evolution of bacterial populations. Appl. Microbiol. Biotechnol. **77**:489–96.

Delbruck, M. 1946. Bacterial viruses or bacteriophages. Biol. Rev. **21**:30–40.

Delbruck, M., and S. E. Luria. 1942. Interference between bacterial viruses. I. Interference between two bacterial viruses acting upon the same host, and the mechanism of virus growth. Arch. Biochem. **1**:111–41.

Diaz, E., A. Ferrandez, M. A. Prieto, and J. L. Garcia. 2001. Biodegradation of aromatic compounds by *Escherichia coli*. Microbiol. Mol. Biol. Rev. **65**:523–69.

Dumon-Seignovert, L., G. Cariot, and L. Vuillard. 2004. The toxicity of recombinant proteins in *Escherichia coli*: a comparison of overexpression in BL21(DE3), C41(DE3), and C43(DE3). Protein Expr. Purif. **37**:203–6.

Eiteman, M. A., and E. Altman. 2006. Overcoming acetate in *Escherichia coli* recombinant protein fermentations. Trends Biotechnol. **24**:530–6.

Elena, S. F., and R. E. Lenski. 2003. Evolution experiments with microorganisms: the dynamics and genetic bases of adaptation. Nat. Rev. Genet. **4**:457–69.

Elsinghorst, E. A., and R. P. Mortlock. 1994. Molecular cloning of the *Escherichia coli* B L-fucose-D-arabinose gene cluster. J. Bacteriol. **176**:7223–32.

Feist, A. M., C. S. Henry, J. L. Reed, M. Krummenacker, A. R. Joyce, P. D. Karp, L. J. Broadbelt, V. Hatzimanikatis, and B. O. Palsson. 2007. A genome-scale metabolic reconstruction for *Escherichia coli* K-12 MG1655 that accounts for 1260 ORFs and thermodynamic information. Mol. Syst. Biol. **3**:121.

Fong, S. S., A. P. Burgard, C. D. Herring, E. M. Knight, F. R. Blattner, C. D. Maranas, and B. O. Palsson. 2005. In silico design and adaptive evolution of *Escherichia coli* for production of lactic acid. Biotechnol. Bioeng. **91**:643–8.

Francetic, O., D. Belin, C. Badaut, and A. P. Pugsley. 2000. Expression of the endogenous type II secretion pathway in *Escherichia coli* leads to chitinase secretion. Embo J. **19**: 6697–703.

Franchini, A. G., and T. Egli. 2006. Global gene expression in *Escherichia coli* K-12 during short-term and long-term adaptation to glucose-limited continuous culture conditions. Microbiology **152**:2111–27.

Gatehouse, D., S. Haworth, T. Cebula, E. Gocke, L. Kier, T. Matsushima, C. Melcion, T. Nohmi, T. Ohta, S. Venitt, et al. 1994. Recommendations for the performance of bacterial mutation assays. Mutat. Res. **312**:217–33.

Gross, C. A. 1996. Function and regulation of the heat shock proteins. *In* F. C. Neidhardt, R. Curtiss III, J. L. Ingraham, E. C. C. Lin, K. B. Low, B. Magasanik, W. S. Reznikoff, M. Riley, D. Schneider, and H. E. Umbarger (eds.), *Escherichia coli* and *Salmonella typhimurium*: Cellular and Molecular Biology, 2nd ed. ASM Press, Washington, DC, pp. 1382–99.

Haeusser, D. P., and P. A. Levin. 2008. The great divide: coordinating cell cycle events during bacterial growth and division. Curr. Opin. Microbiol. **11**:94–9.

Han, M. J., and S. Y. Lee. 2006. The *Escherichia coli* proteome: past, present, and future prospects. Microbiol. Mol. Biol. Rev. **70**:362–439.

Harder, K. J., H. Nikaido, and M. Matsuhashi. 1981. Mutants of *Escherichia coli* that are resistant to certain beta-lactam compounds lack the *ompF* porin. Antimicrob. Agents Chemother. **20**:549–52.

Hayashi, K., N. Morooka, Y. Yamamoto, K. Fujita, K. Isono, S. Choi, E. Ohtsubo, T. Baba, B. L. Wanner, H. Mori, and T. Horiuchi. 2006. Highly accurate genome sequences of *Escherichia coli* K-12 strains MG1655 and W3110. Mol. Syst. Biol. **2**:2006 0007.

Helmstetter, C. E. 1968. DNA synthesis during the division cycle of rapidly growing *Escherichia coli* B/r. J. Mol. Biol. **31**:507–18.

Herrera, G., A. Martinez, M. Blanco, and J. E. O'Connor. 2002. Assessment of *Escherichia coli* B with enhanced permeability to fluorochromes for flow cytometric assays of bacterial cell function. Cytometry **49**:62–9.

Herrera, G., A. Urios, V. Aleixandre, and M. Blanco. 1993. Mutability by polycyclic hydrocarbons is improved in derivatives of *Escherichia coli* WP2 *uvrA* with increased permeability. Mutat. Res. **301**:1–5.

Ishii, N., K. Nakahigashi, T. Baba, M. Robert, T. Soga, A. Kanai, T. Hirasawa, M. Naba, K. Hirai, A. Hoque, P. Y. Ho, Y. Kakazu, K. Sugawara, S. Igarashi, S. Harada, T. Masuda, N. Sugiyama, T. Togashi, M. Hasegawa, Y. Takai, K. Yugi, K. Arakawa, N. Iwata, Y. Toya, Y. Nakayama, T. Nishioka, K. Shimizu, H. Mori, and M. Tomita. 2007. Multiple high-throughput analyses monitor the response of *E. coli* to perturbations. Science **316**:593–7.

Jansson, P. E., A. A. Lindberg, B. Lindberg, and R. Wollin. 1981. Structural studies on the hexose region of the core in lipopolysaccharides from Enterobacteriaceae. Eur. J. Biochem. **115**:571–7.

Jensen, K. F. 1993. The *Escherichia coli* K-12 "wild types" W3110 and MG1655 have an *rph* frameshift mutation that leads to pyrimidine starvation due to low *pyrE* expression levels. J. Bacteriol. **175**:3401–7.

Jeong, H., V. Barbe, D. Vallenet, S.-H. Choi, C. H. Lee, S.-W. Lee, B. Vacherie, S. H. Yoon, D.-S. Yu, L. Cattolico, C.-G. Hur, H.-S. Park, B. Ségurens, M. Blot, D. Schneider, F. W. Studier, S. C. Kim, T. K. Oh, R. E. Lenski, P. Daegelen, and J. F. Kim. Genome sequencing and comparative analysis of *Escherichia coli* B REL606 and BL21(DE3). submitted.

Jeong, K. J., and S. Y. Lee. 2002. Excretion of human beta-endorphin into culture medium by using outer membrane protein F as a fusion partner in recombinant *Escherichia coli*. Appl. Environ. Microbiol. **68**:4979–85.

LeBlanc, D. J., and R. P. Mortlock. 1971. Metabolism of D-arabinose: a new pathway in *Escherichia coli*. J. Bacteriol. **106**:90–6.

Lederberg, J. 2004. *E. coli* K-12. Microbiol. Today **31**:116.

Lederberg, S. 1966. Genetics of host-controlled restriction and modification of deoxyribonucleic acid in *Escherichia coli*. J. Bacteriol. **91**:1029–36.

Legrain, C., V. Stalon, and N. Glansdorff. 1976. *Escherichia coli* ornithine carbamolytransferase isoenzymes: evolutionary significance and the isolation of $\lambda argF$ and $\lambda argI$ transducing bacteriophages. J. Bacteriol. **128**:35–8.

Lenski, R. E., M. R. Rose, S. C. Simpson, and S. C. Tadler. 1991. Long-term experimental evolution in *Escherichia coli*. I. Adaptation and divergence during 2,000 generations. Am. Nat. **138**.

Michelsen, O., M. J. Teixeira de Mattos, P. R. Jensen, and F. G. Hansen. 2003. Precise determinations of C and D periods by flow cytometry in *Escherichia coli* K-12 and B/r. Microbiology **149**:1001–10.

Miroux, B., and J. E. Walker. 1996. Over-production of proteins in *Escherichia coli*: mutant hosts that allow synthesis of some membrane proteins and globular proteins at high levels. J. Mol. Biol. **260**:289–98.

Nandakumar, M. P., A. Cheung, and M. R. Marten. 2006. Proteomic analysis of extracellular proteins from *Escherichia coli* W3110. J. Proteome Res. **5**:1155–61.

Neidhardt, F. C., and H. E. Umbarger. 1996. Chemical composition of *Escherichia coli*. *In* F. C. Neidhardt, R. Curtiss III, J. L. Ingraham, E. C. C. Lin, K. B. Low, B. Magasanik, W. S. Reznikoff, M. Riley, D. Schneider, and H. E. Umbarger (eds.), *Escherichia coli* and *Salmonella typhimurium*: Cellular and Molecular Biology, 2nd ed. ASM Press, Washington, DC, pp. 13–16.

Neijssel, O. M., M. J. Teixeira de Mattos, and D. W. Tempest. 1996. Growth yield and energy distribution. *In* F. C. Neidhardt, R. Curtiss III, J. L. Ingraham, E. C. C. Lin, K. B. Low, B. Magasanik, W. S. Reznikoff, M. Riley, D. Schneider, and H. E. Umbarger (eds.), *Escherichia coli* and *Salmonella typhimurium*: Cellular and Molecular Biology, 2nd ed. ASM Press, Washington, DC, pp. 1683–92.

Nikaido, H. 2003. Molecular basis of bacterial outer membrane permeability revisited. Microbiol. Mol. Biol. Rev. **67**:593–656.

Nikaido, H. 1996. Outer membrane. *In* F. C. Neidhardt, R. Curtiss III, J. L. Ingraham, E. C. C. Lin, K. B. Low, B. Magasanik, W. S. Reznikoff, M. Riley, D. Schneider, and H. E. Umbarger (eds.), *Escherichia coli* and *Salmonella typhimurium*: Cellular and Molecular Biology, 2nd ed. ASM Press, Washington, DC, pp. 29–47.

Philippe, N., E. Crozat, R. E. Lenski, and D. Schneider. 2007. Evolution of global regulatory networks during a long-term experiment with *Escherichia coli*. Bioessays **29**:846–60.

Phue, J. N., B. Kedem, P. Jaluria, and J. Shiloach. 2007. Evaluating microarrays using a semiparametric approach: application to the central carbon metabolism of *Escherichia coli* BL21 and JM109. Genomics **89**:300–5.

Phue, J. N., S. B. Noronha, R. Hattacharyya, A. J. Wolfe, and J. Shiloach. 2005. Glucose metabolism at high density growth of *E. coli* B and *E. coli* K: differences in metabolic pathways are responsible for efficient glucose utilization in *E. coli* B as determined by microarrays and Northern blot analyses. Biotechnol. Bioeng. **90**:805–20.

Pieper, D. H., and W. Reineke. 2000. Engineering bacteria for bioremediation. Curr. Opin. Biotechnol. **11**:262–70.

Posfai, G., G. Plunkett, 3rd, T. Feher, D. Frisch, G. M. Keil, K. Umenhoffer, V. Kolisnychenko, B. Stahl, S. S. Sharma, M. de Arruda, V. Burland, S. W. Harcum, and F. R. Blattner. 2006. Emergent properties of reduced-genome *Escherichia coli*. Science **312**:1044–6.

Pugsley, A. P., and J. P. Rosenbusch. 1983. OmpF porin synthesis in *Escherichia coli* strains B and K-12 carrying heterologous *ompB* and/or *ompF* loci. FEMS Microbiol. Lett. **16**:143–47.

saiSree, L., M. Reddy, and J. Gowrishankar. 2001. IS186 insertion at a hot spot in the *lon* promoter as a basis for Lon protease deficiency of *Escherichia coli* B: identification of a consensus target sequence for IS186 transposition. J. Bacteriol. **183**:6943–6.

Sauer, U. 2001. Evolutionary engineering of industrially important microbial phenotypes. Adv. Biochem. Eng. Biotechnol. **73**:129–69.

Schneider, D., E. Duperchy, J. Depeyrot, E. Coursange, R. Lenski, and M. Blot. 2002. Genomic comparisons among *Escherichia coli* strains B, K-12, and O157:H7 using IS elements as molecular markers. BMC Microbiol. **2**:18.

Seo, J. H., D. G. Kang, and H. J. Cha. 2003. Comparison of cellular stress levels and green-fluorescent-protein expression in several *Escherichia coli* strains. Biotechnol. Appl. Biochem. **37**:103–7.

Sorensen, H. P., and K. K. Mortensen. 2005. Advanced genetic strategies for recombinant protein expression in *Escherichia coli*. J. Biotechnol. **115**:113–28.

Studier, F. W., and B. A. Moffatt. 1986. Use of bacteriophage T7 RNA polymerase to direct selective high-level expression of cloned genes. J. Mol. Biol. **189**:113–30.

Swartz, J. R. 1996. *Escherichia coli* recombinant DNA technology. *In* F. C. Neidhardt, R. Curtiss III, J. L. Ingraham, E. C. C. Lin, K. B. Low, B. Magasanik, W. S. Reznikoff, M. Riley, M. Shaechter, and H. E. Umbarger (eds.), *Escherichia coli* and *Salmonella*: cellular and molecular biology, 2nd ed. ASM Press, Washington, DC, pp. 1693–1712.

Tian, G., D. Lim, J. D. Oppenheim, and W. K. Maas. 1994. Explanation for different types of regulation of arginine biosynthesis in *Escherichia coli* B and *Escherichia coli* K12 caused by a difference between their arginine repressors. J. Mol. Biol. **235**:221–30.

Tsilibaris, V., G. Maenhaut-Michel, and L. Van Melderen. 2006. Biological roles of the Lon ATP-dependent protease. Res. Microbiol. **157**:701–13.

Umbarger, H. E. 1996. Biosynthesis of the branched-chain amino acids. *In* F. C. Neidhardt, R. Curtiss III, J. L. Ingraham, E. C. C. Lin, K. B. Low, B. Magasanik, W. S. Reznikoff, M. Riley, D. Schneider, and H. E. Umbarger (eds.), *Escherichia coli* and *Salmonella typhimurium*: Cellular and Molecular Biology, 2nd ed. ASM Press, Washington, DC, pp. 442–57.

Vallenet, D., L. Labarre, Z. Rouy, V. Barbe, S. Bocs, S. Cruveiller, A. Lajus, G. Pascal, C. Scarpelli, and C. Medigue. 2006. MaGe: a microbial genome annotation system supported by synteny results. Nucl. Acids Res. **34**:53–65.

Vostiar, I., J. Tkac, and C. F. Mandenius. 2004. Off-line monitoring of bacterial stress response during recombinant protein production using an optical biosensor. J. Biotechnol. **111**:191–201.

Wilcox, P., A. Naidoo, D. J. Wedd, and D. G. Gatehouse. 1990. Comparison of *Salmonella typhimurium* TA102 with *Escherichia coli* WP2 tester strains. Mutagenesis **5**:285–91.

Witkin, E. M. 1946. Inherited differences in sensitivity to radiation in *Escherichia coli*. Proc. Natl. Acad. Sci. USA **32**:59–68.

Xia, X. X., M. J. Han, S. Y. Lee, and J. S. Yoo. 2008. Comparison of the extracellular proteomes of *Escherichia coli* B and K-12 strains during high cell density cultivation. Proteomics **8**:2089–103.

Yoon, S. H., M. J. Han, H. Jeong, C. H. Lee, X. X. Xia, J. H. Shim, S. Y. Lee, T. K. Oh, and J. F. Kim. Comparative multi-omics systems analysis of closely related *Escherichia coli* strains. submitted.

Yoon, S. H., M. J. Han, S. Y. Lee, K. J. Jeong, and J. S. Yoo. 2003. Combined transcriptome and proteome analysis of *Escherichia coli* during high cell density culture. Biotechnol. Bioeng. **81**:753–67.

Yoon, S. H., C. G. Hur, H. Y. Kang, Y. H. Kim, T. K. Oh, and J. F. Kim. 2005. A computational approach for identifying pathogenicity islands in prokaryotic genomes. BMC Bioinformatics **6**:184.

Yu, B. J., B. H. Sung, M. D. Koob, C. H. Lee, J. H. Lee, W. S. Lee, M. S. Kim, and S. C. Kim. 2002. Minimization of the *Escherichia coli* genome using a Tn5-targeted Cre/loxP excision system. Nat. Biotechnol. **20**:1018–23.

Chapter 2
Escherichia coli Genome Engineering and Minimization for the Construction of a Bioengine

Bong Hyun Sung, Jun Hyoung Lee, and Sun Chang Kim

Contents

Abstract A profusion of diverse genome-related information has been obtained by the sequencing of genomes from many microorganisms, functional analyses of these genomes, and the application of bioinformatics techniques to genomics, proteomics, and systems biology. The resulting barrage of data coupled with large scale gene inactivation studies have allowed researchers to produce a genetic blueprint for a streamline, custom-designed microbe that carries the minimal gene set required for the organism to replicate in a given environment. On the basis of this minimal genome information, several research groups have generated minimal-genome *Escherichia coli* strains using sophisticated genome engineering techniques, such as the dual transposition, site-specific recombinations, and markerless genome recombination. These minimal genomes display various desirable traits for biological researches, such as improved genome stability, increased transformation efficacy, and higher production of biological materials. Therefore, the generation of

S.C. Kim (✉)
Department of Biological Sciences, Institute for the BioCentury, Korea Advanced Institute of Science and Technology (KAIST), Daejeon 305-701, Korea
e-mail: sunkim@kaist.ac.kr

S.Y. Lee (ed.), *Systems Biology and Biotechnology of Escherichia coli*,
DOI 10.1007/978-1-4020-9394-4_2, © Springer Science+Business Media B.V. 2009

a large number of deletion mutants of the minimal *E. coli* genomes coupled with restructuring of regulatory circuits may lead to facilitate the construction of a variety of custom-designed bacterial strains (also called a "bioengine") with myriad practical and commercial applications.

2.1 Introduction

Nucleotide sequencing and comparative analysis of multiple diverse genomes is revolutionizing contemporary biology by providing a framework for interpreting and predicting the physiological properties of an organism. A variety of emerging postgenomic techniques, such as genome-wide gene expression profiling and monitoring of interactions among macromolecules, are helping to define the molecular compositions of cells. Scientists have developed, and continue to refine, sophisticated new computational approaches that allow one to explore the inherent organization of cellular networks as well as the mode and dynamics of interactions among cellular constituents (Hasty et al. 2001, Herring et al. 2006, Ishii et al. 2007, Kitano 2002, Koonin et al. 2002). These intricate tools and techniques have introduced a new paradigm in cell biology research: the construction of custom-designed, minimal-genome microbes (bioengines) that perform functions that raise the quality of life for human beings.

A minimal genome can be defined as a one that contains the smallest set of genes that allows the organism to replicate in a given environment (Mushegian 1999). The creation and study of minimal microbial genomes can help to increase our understanding of complex genetic material and provide a basis for the design of custom bacterial strains. Drawing on the complete genome sequences of more than 800 microorganisms (June 2008, Genomes OnLine Database http://www.genomesonline.org/) as well as extensive functional analyses of their gene products, researchers have proposed two different approaches for the construction of minimal genomes (Cho et al. 1999, Luisi 2002, Maniloff 1996). The first is a "top-down" approach, which involves trimming the genome of sequences that appear to be unnecessary on the basis of functional genomic studies of microorganisms. The second is a "bottom-up" approach, which entails synthesizing the proposed minimal genome from basic chemical building blocks and inserting it into an environment that allows metabolic activity and replication (Forster and Church 2006, 2007, Szostak et al. 2001). Although simple biological constructs can be synthesized artificially (Cello et al. 2002, Gibson et al. 2008, Itaya et al. 2008, Smith et al. 2003, Tumpey et al. 2005), the bottom-up approach is technically challenging and the actual synthesis of an artificial minimal genome from chemical building blocks is not possible if one lacks a complete functional analysis of the genes needed for life. However, the top-down approach, which starts with the intact genome of a well-characterized microorganism, is more technically feasible; this is because the top-down approach can be initiated in parallel with rapidly progressing functional genomics research in microorganisms.

Back in 1984, H. Morowitz first raised the idea of using mycoplasmas as models for the construction of a minimal genome in a living cell (Razin 1997a, b). The complete genomic sequence of the human pathogen *Mycoplasma genitalium* (Fraser et al. 1995) revealed that the genome is only 580 kb in length and contains only about 470 predicted protein coding genes, as compared with 1,703 and 4,288 in the *Haemophilus influenzae* (Fleischmann et al. 1995) and *Escherichia coli* (Blattner et al. 1997) genomes, respectively. However, mycoplasmas are parasites that evolved, by degenerative or reductive evolution, from Gram-positive bacteria and they have not been well characterized, because they are hard to grow and their genomes are difficult to manipulate. Therefore, scientists chose to attempt minimal genome construction in free-living microorganisms (Koob et al. 1994).

Among the various free-living microorganisms, *E. coli* is the most logical choice for minimal genome construction experiments, as it is more fully defined at the molecular level than any other microorganism. Furthermore, because of its nearly ubiquitous use as a research tool and its favorable growth characteristics, this bacterium has been the organism of choice in the development of tools for sophisticated genetic engineering. Finally, *E. coli* is one of the best commercially applicable hosts for the pharmaceutical and fermentation industries (Blattner et al. 1997, Riley et al. 2006).

Functional analysis of the *E. coli* genome has revealed that bacteria that are grown under a given condition use only a fraction of their genes for growth, replication, and production of important biological materials (Richmond et al. 1999, Tao et al. 1999). This is because the *E. coli* genome contains nonessential genes, transposable elements, bacteriophage DNA, cryptic prophages, pseudogenes, gene remnants, damaged operons, and virulence factors, some of which yield unnecessary or unwanted products that interfere with rational strain improvement and the production of desired biological substances (Blattner et al. 1997, Hayashi et al. 2006). Therefore, the identification and deletion of nonessential genes and other dispensable sequences in the microbial genome is necessary for the construction of a custom-designed bioengine, in which cellular metabolites and energy sources are efficiently optimized and directed toward the production of both essential and desired gene products (Edwards and Palsson 2000, Park et al. 2007). Concomitantly, the bioengine's metabolic waste and bio-pollution can be minimized, and the quality and stability of its products can be maximized (Cho et al. 1999, Kolisnychenko et al. 2002, Westers et al. 2003).

In this chapter, we describe approaches for minimizing the *E. coli* genome by eliminating unnecessary genes, to create a self-sustaining, self-replicating, artificial bioengine.

2.2 Estimating the Size and Gene Content of Minimal Genomes

Identification of the regulatory and protein-coding DNA sequences that are most essential for maintaining and replicating a free-living cellular organism is a logical first step in the construction of a custom-designed bioengine. Several approaches have been used to identify essential and nonessential genes in microorganisms under given conditions, with the goal of defining the minimal set of genes necessary for cell survival and self-replication.

2.2.1 Identification of Minimal Essential Gene Sets by Sequence Comparison in silico

The sequenced genome of the parasitic bacterium *M. genitalium* contains only about 470 identified protein-coding genes, and these have been dubbed a minimal gene complement (Fraser et al. 1995). Although the complete gene complement of *M. genitalium* is nearly the smallest one among cellular life forms with sequenced genomes, there is no evidence that this collection of 470 genes represents a minimal self-sufficient gene set. To derive such a set, Mushegian and Koonin (1996) compared the 468 predicted *M. genitalium* protein sequences with the 1,703 predicted protein sequences encoded by the other completely sequenced microbial genome, that of *H. influenzae* (Fleischmann et al. 1995). Because these microorganisms belong to two distinct ancient bacterial lineages [that is, Gram-positive (*M. genitalium*) and Gram-negative (*H. influenzae*) bacteria], genes that are conserved in these two organisms are almost certainly essential for growth and replication. This genomic comparison suggested that the minimal number of genes necessary and sufficient to sustain the existence of a modern-type cell is closer to 256 (Koonin et al. 2002, Mushegian 1999). On the basis of these 256 genes, Mushegian and Koonin suggested the following key features that must be specified by a minimal gene set: rudimentary systems for gene transcription, protein translation, DNA replication, recombination, and repair; chaperone-like proteins and machinery for protein export and metabolite transport; and nucleotide salvage pathways.

Since then, comparisons of protein-coding regions in complete genome sequences from diverse organisms have revealed clusters of orthologous groups (COGs), (Tatusov et al. 1997); because orthologous proteins likely have similar functions, the COGs have been used to define a minimal gene set of life (Mushegian 1999). Using a similar approach, researchers compared genome sequences from uropathogenic *E. coli* CFT073, enterohemorrhagic DEL933, and the *E. coli* laboratory strain MG1655 and defined a combined set of nonredundant protein-encoding genes. Of these genes, only 39.2% (2,996 genes) are common to all three strains (Welch et al. 2002). However, when such an analysis was carried out with about 100 sequenced genomes, only 63 genes were found to be ubiquitous; most of these genes encode translational components, and a few specify basic components of the transcription machinery (Koonin 2003).

The challenge of this comparative genomic approach to the identification of a minimal gene set is that certain function-related complexities are not taken into account. For example, the minimal number of essential functions may be larger than that predicted by genome sequence comparison, because not all proteins that perform the same function share detectable sequence similarity (Riley and Serres 2000). Therefore, because it may substantially underestimate the size of the minimal gene set, one cannot rely exclusively on the comparative approach (Feher et al. 2007).

2.2.2 Identification of Essential Genes by Large–Scale Gene Inactivation

Global transposition mutagenesis has been used to identify nonessential genes in an effort to learn whether the naturally occurring gene complements behave as true minimal genomes under laboratory conditions.

The positions of 2,209 transposon insertions in the completely sequenced genomes of *M. genitalium* and its close relative *Mycoplasma peumoniae* were determined by sequencing across the junctions of the transposons and the genomic DNA (Hutchison et al. 1999). These junctions defined 1,354 distinct sites in which transposon insertion did not lead to lethality. This analysis suggests that 265–350 of the 480 protein-coding genes of *M. genitalium* are essential under laboratory conditions, including about 100 genes of unknown function (Hutchison et al. 1999).

In *Bacillus subtilis*, Itaya (1995) introduced mutations in a small set of randomly selected genetic loci, examined the growth properties of the mutants, and determined the percentage of genes that could be disrupted without loss of viability. This led to the hypothesis that the minimal *B. subtilis* genome may comprise about 318–562 kb which, given the average size of $\sim$1 kb for a bacterial protein-coding gene, corresponds to 300–500 genes (Kunst et al. 1997). And with a systematic approach that employed single-gene disruptions that covered the complete *Bacillus* genome, Kobayashi et al. (2003) identified about 270 genes that are indispensable for growth of *B. subtilis* in a rich medium at 37 °C.

In *Pseudomonas aeruginosa*, *H. influenzae*, *Corynebacterium glutamicum*, and *E. coli*, global transposition mutagenesis has identified 678, 670, 658, and 620 genes, respectively, those essential for growth under laboratory conditions (Akerley et al. 2002, Gerdes et al. 2003, Jacobs et al. 2003, Suzuki et al. 2006). However, when Baba et al. (2006) used Red recombination to generate a set of precisely defined, single-gene deletions of all putative protein-coding genes in *E. coli* K-12, of the 4,288 genes targeted, only 303 genes, including 37 of unknown function, were unable to be disrupted in Lulia-Bertani (LB) medium. When these 303 essential genes were divided into functional groups, the major subsets contained members of COGs that play roles in protein translation, ribosomal structure, cell division, lipid metabolism, transcription, and cell envelope biogenesis. And only 67% (205 genes) of the 303 essential genes overlap with those in the essential gene set predicted by global transposition (Baba et al. 2006, Gerdes et al. 2003). These differences can be attributed to the use of different mutagenesis strategies and different growth conditions. However, because the global transposition system measures the effect of mutations on cell populations, a mutation that causes very slow growth can appear to be lethal and hence be falsely classified as essential. Furthermore, of the 3,988 single-gene deletion mutants made by Baba et al., 119 gave rise to mutant *E. coli* strains that were unable to grow on glycerol-supplemented M9 minimal medium

Table 2.1 Number of estimated essential genes in various microorganisms

Method	Strain	No. of essential genes/total ORFs (%)	Ref.
Comparative genomics	*M. genitalium and H. influenzae*	256	Mushegian and Koonin 1999
	M. genitalium	265–350/468 (79%)	Hutchison et al. 1999
	H. influenzae RD	670/1703 (38%)	Akerley et al. 2002
Global transposition	*E. coli* K12	620/4296 (14%)	Gerdes et al. 2003
	P. aeruginosa	678/5500 (12%)	Jacobs et al. 2003
	H. pyroli	255/1590 (16%)	Salama et al. 2004
	C. glutamicum R	658/2990 (22%)	Suzuki et al. 2006
Single-gene deletion or knockout	*S. cerevisiae*	1105/5916 (19%)	Giaever et al. 2002
	B. subtilis 168	271/4099 (6.8%)	Kobayashi et al. 2003
	S. typhimurium LT2	490/4597 (11%)	Knuth et al. 2004
	E. coli K12	303/4296 (7%)	Baba et al. 2006

(Joyce et al. 2006). Information about the variously defined minimal gene sets is summarized in Table 2.1.

2.3 Techniques for Experimental Genome Minimization

For some bacteria, strains that carry spontaneous deletion mutations can be obtained by long-term serial passage of the cells under laboratory culture conditions (Cooper et al. 2001). However, the spontaneous deleterious mutation rate for the *E. coli* genome is too low ($\sim 2 \times 10^{-4}$ per generation) to allow deletion mutants to be efficiently created using this technique (Elena and Lenski 2003, Kibota and Lynch 1996). Therefore, for the rapid construction of minimal genomes by removing protein-coding genes and other genomic sequences that previously were shown to be nonessential, researchers have developed a variety of deletion methods, including homologous recombination using suicide plasmids, linear DNA recombination using the phage Red system, site-specific recombination system, and random deletion by double transposition.

2.3.1 Suicide Plasmid–Mediated Genomic Deletion

Suicide plasmids are convenient vehicles for the delivery of DNA into the *E. coli* chromosome. Link et al. (1997) have described a method for gene replacement in *E. coli* that uses a homologous recombination between the bacterial chromosome and a suicide plasmid, carrying cloned chromosomal fragments (homologous arms), whose replication ability is temperature sensitive (Fig. 2.1A). At the non-permissive temperature (42 °C), cells maintain drug resistance only if the plasmid integrates into the chromosome by homologous recombination between the cloned fragment and the bacterial chromosome (Hamilton et al. 1989, Posfai et al. 1999). Alternative suicide plasmids that contain plasmid R6K-origin of replication (*ori*) can be made in an *E. coli* cell that supports synthesis of the replication initiation protein and

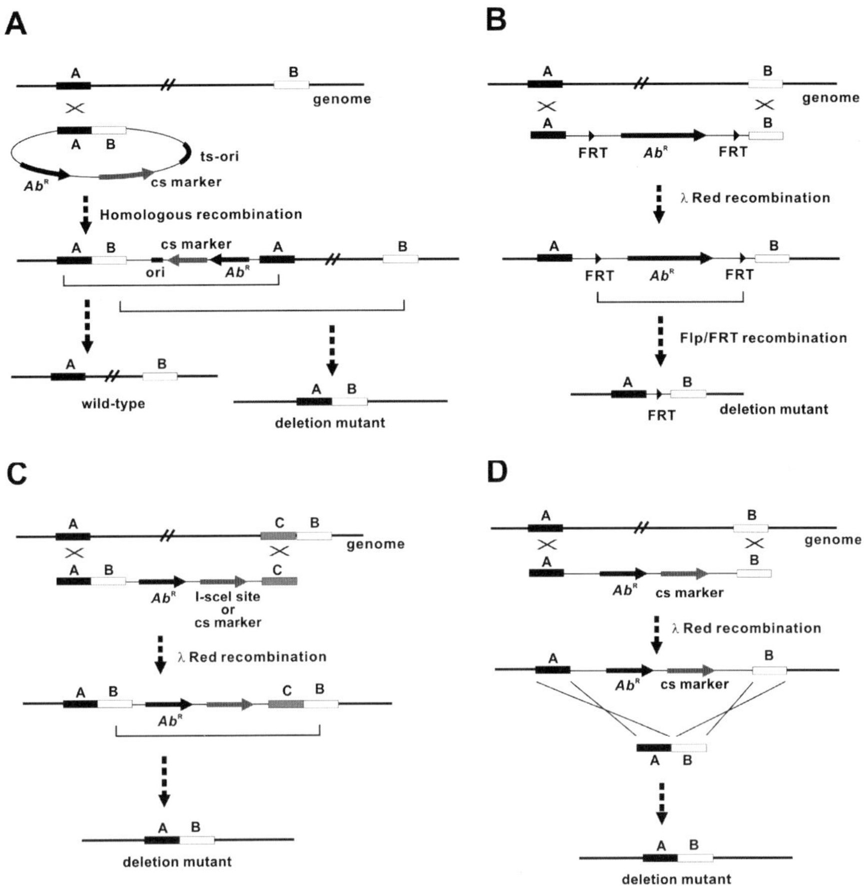

Fig. 2.1 General schemes of the deletion procedures. In all schemes, the chromosomal DNA is shown at the top of the figure, and the targeted plasmid or DNA fragment is shown below the chromosomal DNA schematic. The sequences of the **A** and **B** regions in the chromosome and the targeted plasmid or DNA fragment are the same thus can undergo homologous recombination. (**A**) Protocol for use of the suicide plasmid-mediated deletion system. The plasmid, which carries two PCR-amplified DNA fragments (**A** and **B**, >500 bp) that flank a target genomic region to be deleted, is transformed into *E. coli,* and cells are plated at the nonpermissive temperature (42 °C) of the plasmid replicon. When shifted to the permissive temperature (30 °C), the plasmid is excised from the chromosome by recombination at either A or B. The resistance colonies against the counterselection marker are screened for deletion of the target region (Link et al. 1997, Posfai et al. 1999). (**B**) Overview of the Red-mediated linear DNA deletion method. An *FRT*-flanked antibiotic resistance gene is amplified by PCR and transformed into *E. coli* expressing the Red recombinase. After selection of the antibiotic-resistant transformants, the inserted antibiotic resistance gene is eliminated by the Flp/*FRT* recombination system (Datsenko and Wanner 2000). (**C**) The scarless deletion system. The targeted DNA fragment, which carries the actual post-deletion sequence joints, was amplified by PCR and integrated into the genome by Red recombination. Integration creates a duplication of the segment that flanks the deletion target region. Cleavage of the inserted DNA by meganuclease I-*Sce*I introduces a DSB between the duplicated segments and stimulates their intramolecular recombination, resulting in a scarless genome deletion (Kolisnychenko et al. 2002). (**D**) Deletion of a genomic region by two serial linear DNA recombination events. A DNA

then transferred to a target cell that lacks the replication protein, which permits the plasmid to integrate into the *E. coli* genome (Koob et al. 1994, Posfai et al. 1994, Yoon et al. 1998).

After the integration of a suicide plasmid that contains a homologous arm into the chromosome, the suicide plasmid can be excised out by the second homologous recombination. Depending on the position of the second recombination event that excises the plasmid, the chromosome either retains the wild-type sequence or has a deletion between the target sites (Fig. 2.1A). For easy identification of the resolved products, a counterselection marker, such as *sacB* (sucrose sensitivity) (Dedonder 1966, Gay et al. 1985, Link et al. 1997) or *rpsL* (streptomycin sensitivity) (Hashimoto et al. 2005, Russell et al. 1989, Wang et al. 1993), is integrated in the suicide plasmids. The resolution of the plasmids is thus screened on media supplemented with sucrose or streptomycin, respectively. Another efficient mode of enhancing the plasmid excision step is the introduction of an 18-base pair (bp) meganuclease I-*Sce*I cleavage site in the suicide plasmid (Posfai et al. 1999). Cleavage of the genome at this unique site creates a double-strand break, which simultaneously stimulates recombination and selects for resolution of the integrated plasmid.

Although methods that employ suicide plasmids can be used to delete genomic segments, for each deletion experiment, these procedures require the creation of specific targeting vectors before recombining them into the chromosome.

2.3.2 Linear DNA–Mediated Genomic Deletion

To avoid the inconvenience of constructing the targeting vectors in the suicide plasmid-mediated genomic deletion, linear DNA-mediated genomic deletion approach has been introduced. In *Saccharomyces cerevisiae* and a few naturally competent bacteria, genes or genomic regions can be directly disrupted by transformation with double-stranded DNA (dsDNA) fragments, created with the polymerase chain reaction (PCR), that encode a selectable marker and have only about 50 bp of flanking DNA (called homology arms) that are homologous to the chromosome region of interest. Through homology arm-directed homologous recombination between the DNA fragments and the chromosome, this procedure facilitates the generation of specific chromosomal mutations and thus functional analysis of the genome (Baudin et al. 1993, Oliver et al. 1998, Wilson et al. 1999). In *E. coli*, however, intracellular exonucleases, such as RecBCD, degrade linear DNA (Lorenz and Wackernagel

Fig. 2.1 (continued) fragment that contains an antibiotic resistance gene and a counterselection marker cassette flanked by DNA segments that are homologous to the chromosomal target region was amplified by PCR and inserted into the chromosomal DNA by Red–mediated recombination. The inserted markers are replaced with a linear DNA fragment consisting of only the chromosomal fragment by a second Red recombination (Hashimoto et al. 2005). Ab^R stands for antibiotic resistance marker gene; ori indicates an origin of replication that functions only under permissive conditions; and cs marker indicates a counterselection marker

1994) and inhibit recombination with the PCR products. Linear DNA recombination systems in *E. coli* were developed by finding ways to inhibit the intracellular exonucleases. For example, one such system uses the RecET recombinase to disrupt plasmid-borne genes with linear DNA fragments (Zhang et al. 1998). Other methods make use of the fact that the phage lambda Red (*gam*, *bet*, *exo*) function promotes a greatly enhanced rate of recombination when using linear DNA (Datsenko and Wanner 2000, Murphy 1998, Yu et al. 2000). In the Red system, the Gam protein inhibits the RecBCD nuclease and prevents it from attacking the linear DNA fragments, and Exo (a 5′-to-3′ exonuclease) and Bet (a single-strand DNA binding protein) generate recombination activity for the linear DNA. Because Bet binds to linear DNA strands that are 36 bases in length or longer, it is recommended that, for efficient recombination, the DNA have homology arms of more than 40 bp (Yu et al. 2000).

One method that uses Red-mediated recombination for efficient deletion of specific genomic segments in *E. coli* is that of Datsenko and Wanner (2000). Their basic strategy (Fig. 2.1B) is to replace a chromosomal target with a linear DNA fragment that carries a selectable antibiotic resistance gene flanked by 50-nucleotide (nt) extensions that are homologous to selected sequences in the bacterial chromosome (Fig. 2.1B, segments labeled A and B). This is accomplished by Red-mediated recombination between the *E. coli* genome and these flanking homologies. After selection, the resistance gene can be eliminated by using a helper plasmid that expresses the Flp recombinase, a *S. cerevisiae* enzyme that acts on the directly repeated *FRT* sites that flank the resistance gene (Broach and Hicks 1980). However, because this method is dependent on yeast Flp/*FRT* site-specific recombination to eliminate the selection marker gene, each deletion event leaves behind one *FRT* site in the bacterial chromosome, which interferes with subsequent rounds of chromosomal deletions.

To delete genomic segments without leaving behind remnants of the selection marker, researchers developed a scarless deletion method that combines Red–mediated recombination and double strand break (DSB)–stimulated recombination (Kolisnychenko et al. 2002). For this method, the PCR generated, linear DNA fragments must be constructed so that they carry the actual post-deletion sequence joints (Fig. 2.1C). Thus, integration of such a DNA fragment by Red recombination creates a duplication of the segment that flanks the deletion target region. Cleavage of the inserted DNA by meganuclease I-*Sce*I introduces a DSB between the duplicated segments and stimulates their intramolecular recombination. Eventually, repair of the DSB by this recombination event results in a scarless genome deletion. In another system for markerless deletion, a counterselection marker, *sacB*, and an I-*Sce*I recognition site were used simultaneously to increase the efficiency of resolution of the inserted markers (Sung et al. 2006).

Hashimoto et al. (2005) developed yet another deletion method that goes through two serial Red–mediated recombination events (Fig. 2.1D). In the first recombination, the targeted genomic region is replaced with the Cm^R-*rpsL*-*sacB* (CRS) cassette flanked by DNA fragments that are homologous to the chromosomal target. In this step, the chloramphenicol-resistance gene (Cm^R) is used as a positive selection marker for the deletion mutants. In the second step, the inserted CRS cassette is replaced with a linear DNA fragment that consists of only the chromosomal sequences, which produces a markerless deletion. For this second round of recombination, *rpsL* and *sacB* are used as the two counterselection markers, and sensitivity to chloramphenicol is assessed in the selected transformants.

In addition to the protocols described above, an improved method for the rapid markerless deletion with linear DNA was developed recently by Yu et al. (2008). In this method, the deletion process is mediated by a single helper plasmid that carries genes that encode the Red recombination proteins and the I-*Sce*I nuclease under the control of inducible arabinose and rhamnose promoters, respectively. Genomic deletions are performed by first growing the bacteria in a medium that contains arabinose as the carbon source (to spur synthesis of the Red recombination proteins, which introduces linear DNAs into the bacterial chromosome) and then changing the carbon source in the growth medium from arabinose to rhamnose (to stimulate production of the I-*Sce*I nuclease, which introduces a DSB that stimulates intramolecular recombination). Only two days are required for the deletion of a genomic segment without remaining a selection marker.

Finally, a DNA recombination system mediated by single-strand DNAs (ssDNAs) also has been developed. The ssDNA-binding protein Bet of phage lambda stimulates recombination in chromosomal DNA by using synthetic ssDNAs as short as 30 bases in length. This ssDNA recombination can be used to mutagenize or repair the chromosome with efficiencies that generate up to 6% recombinants among treated cells (Ellis et al. 2001).

2.3.3 Site-Specific Recombination–Mediated Genomic Deletion

Site-specific recombination is a useful genetic tool for deleting undesired DNA sequences and modifying chromosome architecture. The frequently used Cre/*loxP* and Flp/*FRT* site-specific recombination systems share many features. Cre (Flp) is a site-specific recombinase that mediates the recombination of a DNA sequence flanked by two 34-bp *loxP* (*FRT*) sites. A *loxP* (*FRT*) site consists of two 13-bp inverted repeats that flank an 8-bp core region. Intramolecular recombination between the two uni-directionally oriented *loxP* (*FRT*) sites that flank a genomic region of interest results in deletion of the intervening DNA fragment (Broach and Hicks 1980, Sternberg and Hamilton 1981).

To integrate a *loxP* (*FRT*) site and an antibiotic resistance marker into a predetermined chromosomal site, PCR-amplified segments of the selected chromosomal region are cloned into a suicide plasmid that contains the R6K-*ori* and an antibiotic-resistance marker. Independently, another PCR-amplified genomic segment is cloned into another suicide plasmid that carries a different antibiotic-resistance marker. The suicide plasmids are inserted into the genome by homologous recombination at predetermined sites. One of the inserted chromosomal sites is then transferred by P1 transduction into a cell that carries the other insertion (Miller 1992). The predetermined genomic segments, which are flanked by the *loxP* (*FRT*) sequences, are then excised by the Cre (Flp) recombinase (Posfai et al. 1994, Yoon et al. 1998). For the rapid integration of *loxP* sites into the genome, Fukiya et al. (2004) also reported a method that involves the integration of *loxP*-containing DNA fragments into the two ends of the target genomic sequence using the Red system.

Yu et al. (2002) established a transposon-coupled, site-specific excision system (Fig. 2.2A). Using modified Tn*5* transposons, the authors constructed two large pools of independent transposon-generated mutants and then mapped precisely the chromosomal locations of 800 of these transposons, which carry a *loxP* site and either the chloramphenicol- or kanamycin-resistance markers (Cm^R or Km^R). Yu et al. then combined selected mutants in a single cell by P1 transduction, and large genomic target regions (57–117 kb) flanked by *loxP* sites were excised by Cre/*loxP* recombination. The advantage of this method is the rapid and easy deletion of almost any desired segment of the *E. coli* genome using the transposon-generated mutant libraries without having to go through the time-consuming process of generating

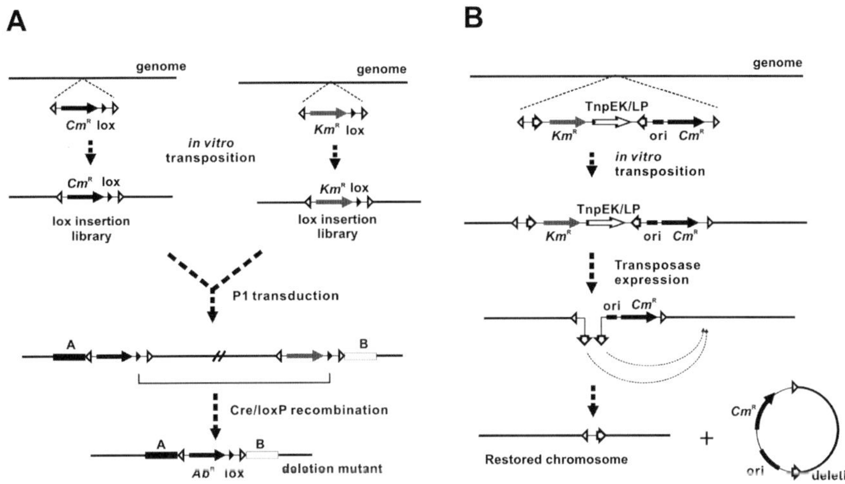

Fig. 2.2 General scheme of the Tn-coupled Cre/*loxP* excision system and random deletion by double transposition. (**A**) Tn*5*-mediated insertions and deletion of the target regions using the Cre/*loxP* excision system. Two modified Tn*5*-transposons are introduced into the *E. coli* genome randomly by *in vitro* transposition. Two mutant strains with a *loxP* site in the same orientation are selected, one from each mutant library, depending on the target region to be deleted. The two selected *loxP* sites are brought, in parallel, into a single strain by P1 transduction. The target region between the two *loxP* sites is deleted by the action of Cre recombinase (Yu et al. 2002). (**B**) The recursive deletion system. This strategy for deletion can be employed after integration of the transposon into the host genome using external transposon ends (*white triangles*). The internal transposon ends (*white arrows*) are used in the second transposition event. The intramolecular transposition leads to the removal of the internal part of the transposon and the deletion of a portion of the chromosome. The addition of a conditional origin of replication allows for capture of the deleted chromosomal material into a complementary self-replicating plasmid (Goryshin et al. 2003). Ab^R stands for antibiotic resistance marker gene; ori indicates an origin of replication; Cm^R *and* Km^R indicates a chloramphenicol resistance and a kanmycin resistance gene, respectively

either customized deletion constructs or PCR products. However, one *loxP* site and an antibiotic resistance marker are retained within the genome after a segment is deleted, which necessitates the use of further engineering techniques to remove the remnants if serial accumulation of deletions is desired. Nevertheless the Cre/*loxP* recombination system can be a useful engineering tool, especially for high-throughput construction of extremely large deletions.

2.3.4 Transposon–Mediated Random Deletion

Goryshin et al. (2003) have described a Tn*5*-based deletion technology (Fig. 2.2B) that uses a composite, linear Tn*5* derivative that carries a replication origin, Km^R and Cm^R selectable markers, Tn*5* transposon end sequences at its 5′ and 3′ termini (external ends), and an internal pair of end sequences from a different transposon (internal ends). In this method, the engineered transposon is introduced into the bacterial host genome by the electroporation of preformed transposome complexes into the cell (Goryshin et al. 2000). The external ends drive integration of this transposon into the host chromosome. A mutant transposase (TnpEK/LP) is expressed from the integrated transposon and then binds to and carries out blunt-end cleavage at the internal ends, which results in (i) the loss of a fragment of the integrated transposon that houses the Km^R gene and (ii) the facilitation of intramolecular transposition. The intramolecular transposition event can create host genome inversions or deletions that begin at the internal ends and extend for varying distances along the host chromosome. Deletions result in loss of the transposon DNA, with the exception of a linker sequence.

Repeated use of this procedure in the same cell, yields a series of random deletions. The average deletion size per round is about 11 kb. The addition of a conditionally active *ori* (one that is induced by IPTG) in the transposon allows for capture of the deleted chromosomal material into a self-replicating plasmid that is complementary to the host chromosome (Fig. 2.2B). Because the transposon integration sites and genomic deletions are random, screening of mutants obtained by this strategy requires significant time and effort. However, this approach allows the deletion of genomic segments in the absence of complete genome sequence information and without prior knowledge of which genes are dispensable for viability.

2.4 Genome Minimization of Microorganisms

E. coli minimal genomes have been constructed by the diverse genomic deletion methods described above, generating bacterial strains that house genomes that are 5–30% smaller than that of a wild-type *E. coli* strain. The genomes of other microorganisms, such as *B. subtilis*, *C. glutamicum*, and the yeasts also have been reduced for the construction of minimal-genome factories (Fujio 2007). The deletion sizes and characteristics of these minimal genomes are summarized in Table 2.2.

Table 2.2 Deleted functions and characteristics of minimal genomes

Strain	Deletion Size	Deleted Functions (D) and Characteristics (C)	Ref.
E. coli			
Δ20-4	218.7 kb (5.6%)	(D) Random genomic regions	Goryshin et al. 2003
CDΔ3456	313.1 kb (6.7%)	(D) Nonspecific target regions (C) Presentation of mutually exclusive regions	Yu et al. 2002
MDS12	376.1 kb (8.1%)	(D) K-strain–specific islands (C) No significant difference to the parent cell	Kolisnychenko et al. 2002
MDS43	708.3 kb (15.3%)	(D) K-islands, mobile elements (C) Increased genome stability and electroporation efficiency	Posfai et al. 2006
MGF-01	1.03 Mb (22%)	(D) Nonessential regions without growth deficiency (C) Increased threonine production (2-fold)	Mizoguchi et al. 2007
Δ16	1.38 Mb (29.7%)	(D) Nonessential genes in the literature (C) Growth deficiency, abnormal nucleoid location	Hashimoto et al. 2005
B. subtilis			
Δ6	320 kb (7.7%)	(D) Prophage, polyketide synthesis	Westers et al. 2003
MG1M	991 kb (24%)	(D) Prophage, polyketide synthesis, secondary metabolites (C) Growth deficiency	Ara et al. 2007
MGB874	873.5 kb (20.7%)	(C) Increased productivity of extracellular cellulase (1.7-fold) and protease (2.5-fold)	Morimoto et al. 2008
C. glutamicum	190 kb (5.7%)	(D) R-strain–specific regions	Suzuki et al. 2005b
S. pombe	~500 kb (4%)	(C) Growth at a lower rate	Giga-Hama et al. 2007
S. cerevisiae	531.5 kb (5%)	(C) Increased production of ethanol (1.8-fold) and glycerol (2-fold)	Murakami et al. 2007

2.4.1 E. coli

2.4.1.1 Random Genome Minimization by Transposition

Using the composite Tn*5* derivative (Fig. 2.2B), Goryshin et al. (2003) developed a unique method for random and recursive deletion of genomic segments that can be applied to gene essentiality studies and minimal genome construction. The authors repeated the random integration/deletion process 20 times per cell to reduce the size of the *E. coli* MG1655 genome, generating several different multi-deletion strains (Fig. 2.3). For 4 of these minimized strains, pulsed-field gel electrophoresis (PFGE)

29.7% smaller than that of the parent strain (MG1655, 4.64 Mb). For the construction of Delta16, each deletion was generated through two serial lambda Red–mediated recombination events (Fig. 2.1D). The deletions were then accumulated, one at a time, in a single strain, by serial P1 transductions using the original single-deletion strains (Fig. 2.3). Phenotypic analysis revealed that the various large deletion-containing strains that gave rise to Delta16 grew more slowly than did the parental cell. In addition, mutant cells that harbored 13 or more deletions were longer in size and wider than the wild type. Hashimoto et al. (2005) also noticed that, while the parental cells contained one or two nucleoids localized at the midcell, or at the $^{1}/_{4}$ and $^{3}/_{4}$ positions, most of the mutant cells had four or more nucleoids that were localized at the periphery of the cell, near the envelope.

Recently, Kato and Hashimoto (2007) used consecutive genome deletion to discover that the origin of DNA replication is the only unique *cis*-acting DNA sequence in the *E. coli* genome that is necessary for survival.

To construct the minimal *E. coli* genome for industrial applications, Mizoguchi et al. (2007a) selected for deletion regions that were not expected to affect the growth or basic metabolism of the bacteria. Ninety-five regions of the *E. coli* genome, with a total size of 1.8 Mb, were deleted independently using markerless deletion mediated by the *sacB-cat* cassette (Mizoguchi et al. 2007b), and the individual deletions were transferred to a single chromosome by P1 transduction. Throughout the genome-size reduction process, Mizoguchi et al. assessed the growth properties (in minimal medium) of each intermediate strain and selected only those strains that displayed no growth-deficiency for subsequent deletions. In the final mutant strain, called MGF-01, the genome size was reduced by 22% (1.03 Mb). During the exponential growth phase, MGF-01 displayed doubling times in M9 minimal medium that were as fast as that of the wild-type strain, and the final cell density reached by MGF-01 was 1.5 times higher than that of the wild-type strain. When the genetic circuit for threonine production ($\Delta met::thrABC\text{-}Cm^{R}$) was integrated into the chromosomes of MGF-01 and its wild-type parental strain, the resulting wild-type- and MGF-01-based strains produced 5 and 10 g/l of threonine in 48 h, respectively.

2.4.2 Other Microorganisms

B. subtilis *B. subtilis*, one of the most extensively studied model microorganisms, displays a superior ability to produce various secretory enzymes. Many industrial scientists have exploited this capability in the production of a variety of useful enzymes (Westers et al. 2004). The 4.2-Mb *B. subtilis* genome contains 10 horizontally acquired prophage (SPβ and PBSX) and prophage-like (pro1-7 and skin) sequences (Kunst et al. 1997). In addition, 2.8% of the genome encompasses two large operons that produce secondary metabolites (pks and pps).

Using a suicide plasmid-based chromosomal integration-excision system (Leenhouts et al. 1996), Westers et al. (2003) have produced a *B. subtilis* Delta6 mutant strain, with a 7.7% reduction in the genome (0.53 Mb), by deleting two prophage (SPβ, PBSX) and three prophage-like sequences (pro1, pro6, skin) as well as one of the secondary metabolite operon (pks) from wild-type *B. subtilis* 168. However, phenotypic characterization of the Delta6 cells disclosed no unique properties, relative to wild-type 168 cells. Recently, Ara et al. (2007) deleted, from the *B. subtilis* genome, all prophage and prophage-like sequences, with the exception of pro7, as well as the pks and pps operons, which resulted in a *B. subtilis* strain (MG1M) that lacked 0.99 Mb of the wild-type genome. However, MG1M strain displays unstable phenotypes with respect to growth rate, cell morphology, and recombinant protein production after successive culture, making it inappropriate for further study.

Another *B. subtilis* minimal genome strain, MGB874, was created recently by introducing deletions, step-by-step, into 28 regions in which single deletions do not affect cell growth (Morimoto et al. 2008). A total of 873.5 kb of DNA (20.7% of the genome) was deleted from wild-type *B. subtilis*. In order to assess the ability of MGB874 to synthesize and secrete proteins, wild-type *B. subtilis* and MGB874 were transformed with plasmids that encoded extracellular cellulase and protease enzymes, and protein production was measured. Cellulase and protease enzyme production was enhanced 1.7- and 2.5-fold, respectively, in MGB874, relative to the wild-type strain.

C. glutamicum The bacterium *C. glutamicum* is used widely for the industrial production of amino acid and organic acids. Therefore, Suzuki et al. (2005a, b, c, and d) used modified Cre/*loxP* recombination to generate a minimal-genome *C. glutamicum* strain that lacked 190 kb of the wild-type genome that encodes a total of 188 ORFs. This deletion mutant exhibits normal growth under standard laboratory conditions. In addition, Tsuge et al. (2007) have generated 42 *C. glutamicum* mutants (with deletions of 0.2–186 kb) using a deletion method similar to that described by Yu et al. (2002), which combines an transposon and the Cre/*loxP* excision system. Tsuge et al. showed that a total of 393.6 kb (11.9%) of the *C. glutamicum* R genome is nonessential for growth under standard laboratory conditions.

Schizosaccharomyces pombe In the fission yeast *S. pombe*, Hirashima et al. (2006) reported a method for the deletion of a large genomic region using homologous recombination between the chromosome and a fragment of linear DNA. Giga-Hama et al. (2007) used this method to create an *S. pombe* mutant dedicated to heterologous protein production. The authors deleted a total of more than 500 kb from a wild-type *S. pombe* strain by repeating the deletion procedure multiple times. Although the authors succeeded in developing a viable strain of *S. pombe* with a minimal genome, the phenotypic characteristics of this organism have not yet been reported.

S. cerevisiae To decipher the number of genes required for growth and the genome organization responsible for ethanol production, various segments of the *S. cerevisiae* genome were deleted, and a viable mutant was created that had lost about 5% (531.5 kb) of the wild-type genome (Murakami et al. 2007). This mutant displays an increase in ethanol (1.8-fold) and glycerol (2-fold) production, relative to the wild type, while also exhibiting levels of resistance to various stresses (heat-shock, acidic or alkaline growth conditions, the presence of 7.5% ethanol, 1 M NaCl, or 1.5 M sorbitol in the growth medium) that are nearly equivalent to those of the parental strain.

2.5 Conclusions and Perspectives

In order to realize our dream of creating ideal, robust host organisms for novel uses that benefit humankind, scientists must first unravel microbial genomes to determine the minimal components that are sufficient for life in specific controlled environments. Researchers are tackling this task by using a comprehensive approach based on computational, experimental, and literature-based studies. For some organisms, the removal of all genomic regions save those that are part of the defined core element has been accomplished, without any deleterious effect on basic cell physiology, by adapting well-characterized recombination systems to the deletion of large genomic segments. Indeed, some of the minimal *E. coli* genomes constructed to date even show improved genome stability and/or increased production of industrial products, relative to wild-type strains.

With the aid of systems biology and synthetic biology approaches, the minimal *E. coli* genomes now in existence may be reduced further to produce a genome that houses the absolute minimal number of genes essential for life. This would represent an important step toward acquiring the ability to genetically engineer organisms (novel and existing) knowing only the sequences of their genomes. Minimal genome research also may provide insights into the origins of life; bacterial evolution; regulation of microbial metabolism; and the genomes of more complex modern organisms. Finally, minimized *E. coli* genomes can lead to the construction of elite, custom-designed bioengines with a plethora of practical and commercial applications.

Bacteria are now commonly engineered to produce useful products, ranging from industrial chemicals to pharmaceutical proteins. Therefore, the first benefits of minimal-genome research likely will be in microbial engineering. A minimal-genome organism might require less energy to produce the same amount of a given protein or produce fewer waste products that contaminate the desired product. A minimal-genome organism also may be used as the basis for novel bioengines created to perform specific tasks, such as the breakdown of environmental toxins, the neutralization of toxic spills, or the creation of renewable energy.

References

Akerley BJ, Rubin EJ, Novick VL et al. (2002) A genome-scale analysis for identification of genes required for growth or survival of *Haemophilus influenzae*. Proc Natl Acad Sci U S A 99(2): 966–71

Ara K, Ozaki K, Nakamura K et al. (2007) *Bacillus* minimum genome factory: effective utilization of microbial genome information. Biotechnol Appl Biochem 46(Pt 3):169–78

Baba T, Ara T, Hasegawa M et al. (2006) Construction of *Escherichia coli* K-12 in-frame, single-gene knockout mutants: the Keio collection. Mol Syst Biol 2:2006 0008

Baudin A, Ozier-Kalogeropoulos O, Denouel A et al. (1993) A simple and efficient method for direct gene deletion in *Saccharomyces cerevisiae*. Nucleic Acids Res 21(14):3329–30

Blattner FR, Plunkett G, 3rd, Bloch CA et al. (1997) The complete genome sequence of *Escherichia coli* K-12. Science 277(5331):1453–74

Broach JR, Hicks JB (1980) Replication and recombination functions associated with the yeast plasmid, 2 mu circle. Cell 21(2):501–8

Cello J, Paul AV, Wimmer E (2002) Chemical synthesis of poliovirus cDNA: generation of infectious virus in the absence of natural template. Science 297(5583):1016–8

Cho MK, Magnus D, Caplan AL et al. (1999) Policy forum: genetics. Ethical considerations in synthesizing a minimal genome. Science 286(5447):2087, 2089–90

Cooper VS, Schneider D, Blot M et al. (2001) Mechanisms causing rapid and parallel losses of ribose catabolism in evolving populations of *Escherichia coli* B. J Bacteriol 183(9): 2834–41

Datsenko KA, Wanner BL (2000) One-step inactivation of chromosomal genes in *Escherichia coli* K-12 using PCR products. Proc Natl Acad Sci U S A 97(12):6640–5

Dedonder (1966) Levansucrase from *Bacillus subtilis*. Methods Enzymol 8:500–5

Edwards JS, Palsson BO (2000) Metabolic flux balance analysis and the in silico analysis of *Escherichia coli* K-12 gene deletions. BMC Bioinformatics 1:1

Elena SF, Lenski RE (2003) Evolution experiments with microorganisms: the dynamics and genetic bases of adaptation. Nat Rev Genet 4(6):457–69

Ellis HM, Yu D, DiTizio T et al. (2001) High efficiency mutagenesis, repair, and engineering of chromosomal DNA using single-stranded oligonucleotides. Proc Natl Acad Sci U S A 98(12):6742–6

Feher T, Papp B, Pal C et al. (2007) Systematic genome reductions: theoretical and experimental approaches. Chem Rev 107(8):3498–513

Fleischmann RD, Adams MD, White O et al. (1995) Whole-genome random sequencing and assembly of *Haemophilus influenzae* Rd. Science 269(5223):496–512

Forster AC, Church GM (2006) Towards synthesis of a minimal cell. Mol Syst Biol 2:45

Forster AC, Church GM (2007) Synthetic biology projects in vitro. Genome Res 17(1):1–6

Fraser CM, Gocayne JD, White O et al. (1995) The minimal gene complement of *Mycoplasma genitalium*. Science 270(5235):397–403

Fujio T (2007) Minimum genome factory: innovation in bioprocesses through genome science. Biotechnol Appl Biochem 46(Pt 3):145–6

Fukiya S, Mizoguchi H, Mori H (2004) An improved method for deleting large regions of *Escherichia coli* K-12 chromosome using a combination of Cre/loxP and lambda Red. FEMS Microbiol Lett 234(2):325–31

Gay P, Le Coq D, Steinmetz M et al. (1985) Positive selection procedure for entrapment of insertion sequence elements in gram-negative bacteria. J Bacteriol 164(2):918–21

Gerdes SY, Scholle MD, Campbell JW et al. (2003) Experimental determination and system level analysis of essential genes in *Escherichia coli* MG1655. J Bacteriol 185(19):5673–84

Giaever G, Chu AM, Ni L et al. (2002) Functional profiling of the *Saccharomyces cerevisiae* genome. Nature 418(6896):387–91

Gibson DG, Benders GA, Andrews-Pfannkoch C et al. (2008) Complete chemical synthesis, assembly, and cloning of a *Mycoplasma genitalium* genome. Science 319(5867): 1215–20

Giga-Hama Y, Tohda H, Takegawa K et al. (2007) *Schizosaccharomyces pombe* minimum genome factory. Biotechnol Appl Biochem 46(Pt 3):147–55

Goryshin IY, Jendrisak J, Hoffman LM et al. (2000) Insertional transposon mutagenesis by electroporation of released Tn5 transposition complexes. Nat Biotechnol 18(1):97–100

Goryshin IY, Naumann TA, Apodaca J et al. (2003) Chromosomal deletion formation system based on Tn5 double transposition: use for making minimal genomes and essential gene analysis. Genome Res 13(4):644–53

Hamilton CM, Aldea M, Washburn BK et al. (1989) New method for generating deletions and gene replacements in *Escherichia coli*. J Bacteriol 171(9):4617–22

Hartman JL, 4th Garvik B, Hartwell L (2001) Principles for the buffering of genetic variation. Science 291(5506):1001–4

Hashimoto M, Ichimura T, Mizoguchi H et al. (2005) Cell size and nucleoid organization of engineered *Escherichia coli* cells with a reduced genome. Mol Microbiol 55(1):137–49

Hasty J, McMillen D, Isaacs F et al. (2001) Computational studies of gene regulatory networks: in numero molecular biology. Nat Rev Genet 2(4):268–79

Hayashi K, Morooka N, Yamamoto Y et al. (2006) Highly accurate genome sequences of *Escherichia coli* K-12 strains MG1655 and W3110. Mol Syst Biol 2:7

Herring CD, Raghunathan A, Honisch C et al. (2006) Comparative genome sequencing of *Escherichia coli* allows observation of bacterial evolution on a laboratory timescale. Nat Genet 38(12):1406–12

Hirashima K, Iwaki T, Takegawa K et al. (2006) A simple and effective chromosome modification method for large-scale deletion of genome sequences and identification of essential genes in fission yeast. Nucleic Acid Res 34(2)11

Hutchison CA, Peterson SN, Gill SR et al. (1999) Global transposon mutagenesis and a minimal Mycoplasma genome. Science 286(5447):2165–9

Ishii N, Nakahigashi K, Baba T et al. (2007) Multiple high-throughput analyses monitor the response of *E. coli* to perturbations. Science 316(5824):593–7

Itaya M (1995) An estimation of minimal genome size required for life. FEBS Lett 362(3): 257–60

Itaya M, Fujita K, Kuroki A et al. (2008) Bottom-up genome assembly using the *Bacillus subtilis* genome vector. Nat Methods 5(1):41–3

Jacobs MA, Alwood A, Thaipisuttikul I et al. (2003) Comprehensive transposon mutant library of *Pseudomonas aeruginosa*. Proc Natl Acad Sci U S A 100(24):14339–44

Joyce AR, Reed JL, White A et al. (2006) Experimental and computational assessment of conditionally essential genes in *Escherichia coli*. J Bacteriol 188(23):8259–71

Kato J, Hashimoto M (2007) Construction of consecutive deletions of the *Escherichia coli* chromosome. Mol Syst Biol 3:132

Kibota TT, Lynch M (1996) Estimate of the genomic mutation rate deleterious to overall fitness in *E. coli*. Nature 381(6584):694–6

Kitano H (2002) Computational systems biology. Nature 420(6912):206–10

Knuth K, Niesalla H, Hueck CJ et al. (2004) Large-scale identification of essential Salmonella genes by trapping lethal insertions. Mol Microbiol 51(6):1729–44

Kobayashi K, Ehrlich SD, Albertini A et al. (2003) Essential *Bacillus subtilis* genes. Proc Natl Acad Sci U S A 100(8):4678–83

Kolisnychenko V, Plunkett G, 3rd, Herring CD et al. (2002) Engineering a reduced *Escherichia coli* genome. Genome Res 12(4):640–7

Koob MD, Shaw AJ, Cameron DC (1994) Minimizing the genome of *Escherichia coli*. Motivation and strategy. Ann NY Acad Sci 745:1–3

Koonin EV (2003) Comparative genomics, minimal gene-sets and the last universal common ancestor. Nat Rev Microbiol 1(2):127–36

Koonin EV, Wolf YI, Karev GP (2002) The structure of the protein universe and genome evolution. Nature 420(6912):218–23

Kunst F, Ogasawara N, Moszer I et al. (1997) The complete genome sequence of the gram-positive bacterium *Bacillus subtilis*. Nature 390(6657):249–56

Leenhouts K, Buist G, Bolhuis A et al. (1996) A general system for generating unlabelled gene replacements in bacterial chromosomes. Mol Gen Genet 253(1–2):217–24

Link AJ, Phillips D, Church GM (1997) Methods for generating precise deletions and insertions in the genome of wild-type *Escherichia coli*: application to open reading frame characterization. J Bacteriol 179(20):6228–37

Lorenz MG, Wackernagel W (1994) Bacterial gene transfer by natural genetic transformation in the environment. Microbiol Rev 58(3):563–602

Luisi PL (2002) Toward the engineering of minimal living cells. Anat Rec 268(3):208–14

Maniloff J (1996) The minimal cell genome: "on being the right size". Proc Natl Acad Sci U S A 93(19):10004–6

Miller JH (1992) A short course in bacterial genetics: A laboratory manual and handbook for *Escherichia coli* and related bacteria. Cold Spring Harbor Laboratory Press, New York

Mizoguchi H, Mori H, Fujio T (2007a) *Escherichia coli* minimum genome factory. Biotechnol Appl Biochem 46(Pt 3):157–67
Mizoguchi H, Tanaka-Masuda K, Mori H (2007b) A simple method for multiple modification of the *Escherichia coli* K-12 chromosome. Biosci Biotechnol Biochem 71(12): 2905–11
Morimoto T, Kadoya R, Endo K et al. (2008) Enhanced recombinant protein productivity by genome reduction in *Bacillus subtilis*. DNA Res 15(2):73–81
Murakami K, Tao E, Ito Y et al. (2007) Large scale deletions in the Saccharomyces cerevisiae genome create strains with altered regulation of carbon metabolism. Appl Microbiol Biotechnol 75(3):589–97
Murphy KC (1998) Use of bacteriophage lambda recombination functions to promote gene replacement in *Escherichia coli*. J Bacteriol 180(8):2063–71
Mushegian A (1999) The minimal genome concept. Curr Opin Genet Dev 9(6):709–14
Mushegian AR, Koonin EV (1996) A minimal gene set for cellular life derived by comparison of complete bacterial genomes. Proc Natl Acad Sci U S A 93(14):10268–73
Oliver SG, Winson MK, Kell DB et al. (1998) Systematic functional analysis of the yeast genome. Trends Biotechnol 16(9):373–8
Park JH, Lee KH, Kim TY et al. (2007) Metabolic engineering of *Escherichia coli* for the production of L-valine based on transcriptome analysis and *in silico* gene knockout simulation. Proc Natl Acad Sci U S A 104(19):7797–802
Perna NT, Plunkett G, 3rd, Burland V et al. (2001) Genome sequence of enterohaemorrhagic *Escherichia coli* O157:H7. Nature 409(6819):529–33
Posfai G, Kolisnychenko V, Bereczki Z et al. (1999) Markerless gene replacement in *Escherichia coli* stimulated by a double-strand break in the chromosome. Nucleic Acids Res 27(22): 4409–15
Posfai G, Koob M, Hradecna Z et al. (1994) In vivo excision and amplification of large segments of the *Escherichia coli* genome. Nucleic Acids Res 22(12):2392–8
Posfai G, Plunkett G, 3rd, Feher T et al. (2006) Emergent properties of reduced-genome *Escherichia coli*. Science 312(5776):1044–6
Razin S (1997a) Comparative genomics of mycoplasmas. Wien Klin Wochenschr 109(14–15): 551–6
Razin S (1997b) The minimal cellular genome of mycoplasma. Indian J Biochem Biophys 34(1–2):124–30
Richmond CS, Glasner JD, Mau R et al. (1999) Genome-wide expression profiling in *Escherichia coli* K-12. Nucleic Acids Res 27(19):3821–35
Riley M, Abe T, Arnaud MB et al. (2006) *Escherichia coli* K-12: a cooperatively developed annotation snapshot – 2005. Nucleic Acids Res 34(1):1–9
Riley M, Serres MH (2000) Interim report on genomics of *Escherichia coli*. Annu Rev Microbiol 54:341–411
Russell CB, Dahlquist FW (1989) Exchange of chromosomal and plasmid alleles in *Escherichia coli* by selection for loss of a dominant antibiotic sensitivity marker. J Bacteriol 171(5):2614–8
Salama NR, Shepherd B, Falkow S (2004) Global transposon mutagenesis and essential gene analysis of Helicobacter pylori. J Bacteriol 186(23):7926–35
Smalley DJ, Whiteley M, Conway T (2003) In search of the minimal *Escherichia coli* genome. Trends Microbiol 11(1):6–8
Smith HO, Hutchison CA, 3rd, Pfannkoch C et al. (2003) Generating a synthetic genome by whole genome assembly: phiX174 bacteriophage from synthetic oligonucleotides. Proc Natl Acad Sci U S A 100(26):15440–5
Sternberg N, Hamilton D (1981) Bacteriophage P1 site-specific recombination. I. Recombination between loxP sites. J Mol Biol 150(4):467–86
Sung BH, Lee CH, Yu BJ et al. (2006) Development of a biofilm production-deficient *Escherichia coli* strain as a host for biotechnological applications. Appl Environ Microbiol 72(5): 3336–42

Suzuki N, Nonaka H, Tsuge Y et al. (2005a) Multiple large segment deletion method for *Corynebacterium glutamicum*. Appl Microbiol Biotechnol 69(2):151–161

Suzuki N, Nonaka H, Tsuge Y et al. (2005b) New multiple-deletion method for the *Corynebacterium glutamicum* genome, using a mutant lox sequence. Appl Environ Microbiol 71(12):8472–80

Suzuki N, Okayama S, Nonaka H et al. (2005c) Large-scale engineering of the *Corynebacterium glutamicum* genome. Appl Environ Microbiol 71(6):3369–72

Suzuki N, Tsuge Y, Inui M et al. (2005d) Cre/loxP-mediated deletion system for large genome rearrangements in *Corinebacterium glutamicum*. Appl Microbiol Biotechnol 67(2):225–33

Suzuki N, Okai N, Nonaka H et al. (2006) High-throughput transposon mutagenesis of *Corynebacterium glutamicum* and construction of a single-gene disruptant mutant library. Appl Environ Microbiol 72(5):3750–5

Szostak JW, Bartel DP, Luisi PL (2001) Synthesizing life. Nature 409(6818):387–90

Tao H, Bausch C, Richmond C et al. (1999) Functional genomics: expression analysis of *Escherichia coli* growing on minimal and rich media. J Bacteriol 181(20):6425–40

Tatusov RL, Koonin EV, Lipman DJ (1997) A genomic perspective on protein families. Science 278(5338):631–7

Tsuge Y, Suzuki N, Inui M et al. (2007) Random segment deletion based on IS31831 and Cre/loxP excision system in *Corynebacterium glutamicum*. Appl Microbiol Biotechnol 74(6):1333–41

Tumpey TM, Basler CF, Aguilar PV et al. (2005) Characterization of the reconstructed 1918 Spanish influenza pandemic virus. Science 310(5745):77–80

Wang G, Blakesley RW, Berg DE et al. (1993) pDUAL: a transposon-based cosmid cloning vector for generating nested deletions and DNA sequencing templates in vivo. Proc Natl Acad Sci U S A 90(16):7874–8

Welch RA, Burland V, Plunkett G, 3rd et al. (2002) Extensive mosaic structure revealed by the complete genome sequence of uropathogenic *Escherichia coli*. Proc Natl Acad Sci U S A 99(26):17020–4

Westers H, Dorenbos R, van Dijl JM et al. (2003) Genome engineering reveals large dispensable regions in *Bacillus subtilis*. Mol Biol Evol 20(12):2076–90

Westers L, Westers H, Quax WJ (2004) *Bacillus subtilis* as cell factory for pharmaceutical proteins: a biotechnological approach to optimize the host organism. Biochim Biophys Acta 1694(1–3):299–310

Wilson RB, Davis D, Mitchell AP (1999) Rapid hypothesis testing with *Candida albicans* through gene disruption with short homology regions. J Bacteriol 181(6):1868–74

Yoon YG, Cho JH, Kim SC (1998) Cre/loxP-mediated excision and amplification of large segments of the *Escherichia coli* genome. Genet Anal 14(3):89–95

Yu BJ, KK, Lee JH (2008) Rapid and efficient construction of markerless deletions in the *Escherichia coli* genome. Nucleic Acids Research 36:84

Yu BJ, Sung BH, Koob MD et al. (2002) Minimization of the *Escherichia coli* genome using a Tn5-targeted Cre/loxP excision system. Nat Biotechnol 20(10):1018–23

Yu BJ, Sung BH, Lee JY et al. (2006) sucAB and sucCD are mutually essential genes in *Escherichia coli*. FEMS Microbiol Lett 254(2):245–50

Yu D, Ellis HM, Lee EC et al. (2000) An efficient recombination system for chromosome engineering in *Escherichia coli*. Proc Natl Acad Sci U S A 97(11):5978–83

Zhang Y, Buchholz F, Muyrers JP et al. (1998) A new logic for DNA engineering using recombination in *Escherichia coli*. Nat Genet 20(2):123–8

Chapter 3
Multi-Omics Data-Driven Systems Biology of *E. coli*

Nobuyoshi Ishii and Masaru Tomita

Contents

Abstract The omics, which means comprehensive analysis of a specific layer in a cellular system, are emerging as essential methodological approaches for molecular biology and systems biology. However, single omics analysis does not always provide enough information to understand the behaviors of a cellular system. Therefore, a combination of multiple omics analyses, the multi-omics approach, is required to acquire a precise picture of living organisms. In this chapter, basic concepts of omics studies, and recent technologies in the omics of metabolism and published multi-omics analyses of *Escherichia coli*, are reviewed. Subsequently, a large-scale multi-omics analysis of *E. coli* K-12, including transcriptomics, proteomics, metabolomics and fluxomics, is presented. This study uncovered the complementary strategies of *E. coli* that result in a metabolic network robust against various types of perturbations, therefore demonstrating the power of a multi-omics, data-driven approach for understanding the functional principles of total cellular systems.

M. Tomita (✉)
Institute for Advanced Biosciences, Keio University, Tsuruoka, Yamagata 997-0017, Japan
e-mail: tomita@ttck.keio.ac.jp

S.Y. Lee (ed.), *Systems Biology and Biotechnology of Escherichia coli*,
DOI 10.1007/978-1-4020-9394-4_3, © Springer Science+Business Media B.V. 2009

3.1 Overview of Omics and Multi-omics Analysis

3.1.1 A New Approach in Molecular Biology – Omics

The history of molecular biology is defined by a number of innovations. Currently, a new innovative breakthrough is joining the world of the molecular biology – the so-called "omics" (Lee et al. 2005, Yadav 2007). The basic methods of modern molecular biology have been, to simplify the situation somewhat, hypothesis-driven, reductionist, and bottom-up. In most cases, only a few biochemical species are focused on in any one study based on hypotheses formed by the researcher(s) prior to the start of the study. After that, exhaustive investigation is directed towards understanding the properties of the target molecules.

Although such an approach is still valuable for obtaining detailed and precise knowledge of the target molecular species, some inherent problems exist in how the conventional research flows. At the beginning stage of such classical research schemes, the selection of the target strongly depends on the personal experience and intuition of individual researchers. Moreover, information about limited numbers of molecular species does not always provide insight into a biological "system" that consists of networks formed by a number of interacting molecular species (Bruggeman and Westerhoff 2007).

To overcome these weak points in traditional molecular biology, a novel research area, the "omics", is emerging. Omics means a comprehensive analysis of biochemical molecular species or interactions of molecules belonging to a specific layer in a cellular system. For example, "genomics" is defined as the study of whole DNA sequences and the information contained therein. Many different words having the suffix of "omics" have been proposed - transcriptomics, proteomics, lipidomics, glycomics, interactomics, phenomics, and so on. However, all omics approaches can be considered to share two major features in contrast to traditional procedures.

One feature is changing the direction of the flow of analysis. Unlike traditional methods, in omics approaches massive data is first collected with no prior hypothesis, and meaningful targets are searched for within the obtained data set. The second feature of omics is the attempt to understand the target as a total "system" by using information of the relationships between many measured molecular species. From this point of view, the omics can be expected to contribute to the progress of systems biology.

In brief, omics can be said to be a data-driven, holistic, and top-down approach, as opposed to traditional approaches. Rapid advances in the development of high-throughput measuring instruments are inducing dramatic growth in the omics research area. The extreme progression of information technologies, including enhancements of public web-databases of biological knowledge (Caspi et al. 2008, Kanehisa et al. 2008, Teufel et al. 2006, Wittig and De Beuckelaer 2001), also support the expansion of omics studies, which require the handling of hundreds or thousands of measured values.

3.1.2 Omics for Metabolic Systems – Metabolomics and Fluxomics

While many technologies are included in the "omics" family, most activity is found in the following three areas: genomics, transcriptomics, and proteomics. Currently, whole genome sequences of many species have been deciphered by high-performance DNA sequencers. Using the information of a complete genome sequence, most of the products (mRNAs and proteins) coded in the genome can be predicted; nonetheless, post-transcriptional modifications cannot be ignored. Thus, with transcriptomic or proteomic experiments, the near complete detection of biochemical species, i.e., true omics analysis, is possible in principle. However, for other omics studies, it is impossible to define an explicit number of targets.

Although difficult to comprehensively measure, omics analysis of metabolism in cellular systems is highly important (Fiehn 2002). The phenotype of a strain is strongly connected to the profile of metabolite concentrations in the cell. In many cases, adaptations of living cells to environmental changes can be achieved by reconfiguration of enzymatic reaction rates in some metabolic pathways. Therefore, metabolomics (Dettmer et al. 2007, Kell 2004, Mashego et al. 2007, Oldiges et al. 2007, Rabinowitz 2007, Wang et al. 2006), which is the omics study for metabolic compounds (low molecular weight, typically less than 1 kDa), is desired to obtain a more precise overview of life.

Traditionally, large-scale metabolite analysis has been performed by gas chromatography mass spectrometry (GC-MS) (Fiehn et al. 2000), and GC-MS is frequently used in plant metabolomics studies (Sanchez et al. 2008). Other instruments, including liquid chromatography mass spectrometry (LC-MS) (Chen et al. 2007, Tolstikov et al. 2007) and nuclear magnetic resonance (NMR) (Grivet et al. 2003, Jordan and Cheng 2007, Ward et al. 2007), have also been successfully applied to metabolome analyses.

Capillary electrophoresis mass spectrometry (CE-MS) has emerged as a powerful new tool, and various CE-MS methods have been developed for the analysis of charged metabolites (Gaspar et al. 2008, Monton and Soga 2007, Sniehotta et al. 2007, Song et al. 2008). The advantages of CE-MS compared to other separation technologies are that this method exhibits extremely high resolution and that almost any charged species can be infused into MS. (Soga et al. 2003) developed a metabolome analysis method by CE-MS whereby metabolites were first separated by CE based on charge and size and then selectively detected using MS by monitoring over a large range of m/z values. Since hundreds of metabolites can be detected simultaneously by CE-MS, our understanding of the metabolic layer in cellular systems is being greatly expanded. More recently, (Soga et al. 2006) also constructed a coupling of CE and time-of-flight MS (TOFMS), and their CE-TOFMS analysis revealed that serum ophthalmate is a sensitive indicator of hepatic glutathione depletion in mice.

Another new methodology, called fluxomics (Sanford et al. 2002, Sauer et al. 1999, Wiechert et al. 2007), which means detailed metabolic flux analysis (MFA)

(Stephanopoulos et al. 1998) of large-scale metabolic pathways, has joined the omics family for investigating metabolic systems. MFA includes mathematical procedures for the estimation of unmeasurable reaction rates in a metabolic pathway by using measurable data, such as specific consumption rates of substrate and specific production rates of byproduct. MFA itself has relatively a long history – the first MFA is believed to have been conducted by Aiba and Matsuoka in 1979 (Aiba and Matsuoka 1979, Stephanopoulos et al. 1998). However, after the 1990s, the use of stable-isotope labeled substrates has become a common technique, and some advanced algorithms to handle the information of labeled metabolites for calculating metabolic fluxes have been developed (Noh et al. 2006, Sauer 2006, Shimizu 2004, Wiechert 2001). Accordingly, metabolic pathways that have complex topologies can be treated by current MFA technologies, i.e., metabolic fluxes distributed in a wide network can now be estimated. Therefore, fluxomics can be considered as one of the omics methodologies. Figure 3.1 shows a bibliographic search containing the words "metabolomics or metabolome" or "fluxomics or fluxome" using PubMed (http://www.pubmed.gov/). An exponential increase in the number of metabolomics studies and the genesis of fluxomics research can be observed.

A combination of metabolomics and fluxomics has been established by Toya et al. (Toya et al. 2007). They used CE-TOFMS to measure mass distributions of intermediate metabolites in cells cultured by isotope-labeled glucose, and performed flux analysis with the measured mass distribution patterns. Since the pool sizes of intermediate metabolites are generally so small, isotopic pseudo-steady states of intermediate metabolites are immediately achieved (Wiechert and Noh 2005). Accordingly, MFA using CE-TOFMS can be applied to metabolic systems under drastic dynamical change, such as in a batch culture, which is practically important in fermentation industries. Other methods of MFA using LC-MS to determine labeling patterns of intermediate metabolites have also been reported (Costenoble et al. 2007, Noh et al. 2007, Schaub et al. 2008, van Winden et al. 2005). Further

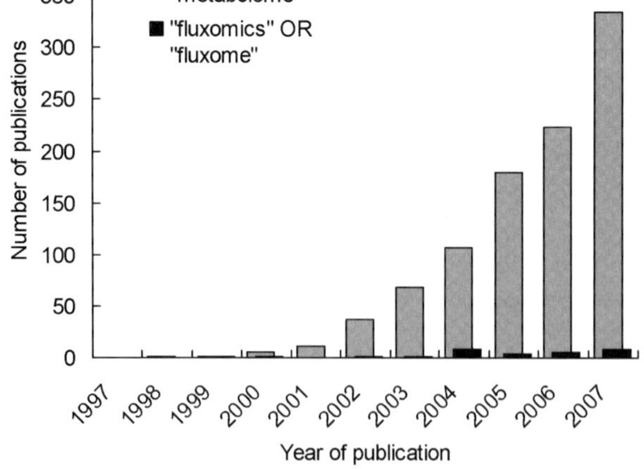

Fig. 3.1 Bibliographic searches in PubMed (http://www.pubmed.gov/) containing the keywords "metabolome OR metabolomics" and "fluxome OR fluxomics" (as of May 20, 2008)

collaborations of metabolomics and fluxomics are expected to be developed and to be employed in investigations of complex and large-scale metabolic systems.

3.1.3 Integration of Various Omics Analyses – Multi-omics

These days various omics analyses are frequently employed in many experimental studies. However, it has been gradually realized that obtaining useful biological knowledge from a single type of omics data (for example, DNA microarray only) is no easy task. One reason is that single omics analysis provides us with information about only one layer of a cellular system. Obviously, multiple functional cellular layers, including the mRNA, protein, and metabolite layers, are interacting with each other; thus the response of a total cellular system to given perturbations cannot be fully captured from a single layer. Figure 3.2 shows a schematic diagram of the functional layers and their interactions in a cellular system.

In conclusion, not just one omics analysis, but multiple omics analyses are required for deep and precise understanding of a cellular system. This recognition seems to be shared by many researchers (Andersen et al. 2008, De Keersmaecker et al. 2006, Joyce and Palsson 2006, Steinfath et al. 2007). Toyoda et al. proposed the concept of the "omic space", which consists of multi-layered state variables, and suggested a data integration framework and graphic presentation method of multiple omics data (Toyoda et al. 2007, Toyoda and Wada 2004). Figure 3.3 displays a conceptual diagram of the "omic space". (Lee et al. 2005) indicated the essen-

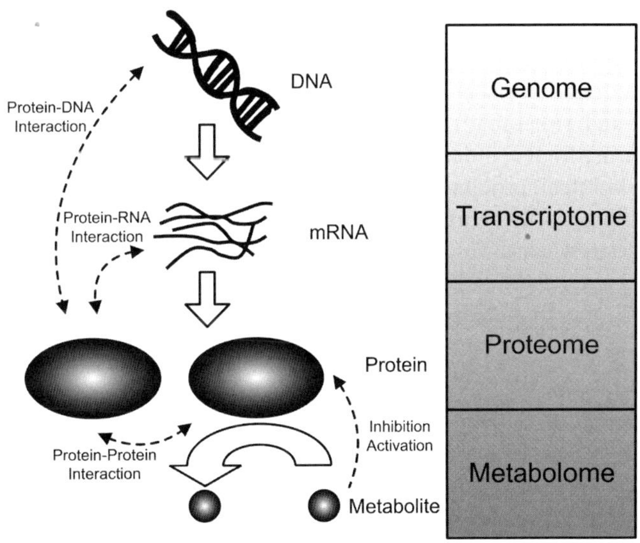

Fig. 3.2 Schematic diagram of interactions among various functional layers in a cellular system. *Blank arrow*, flow of biological information; *dashed line*, possible interaction between various biomolecular species

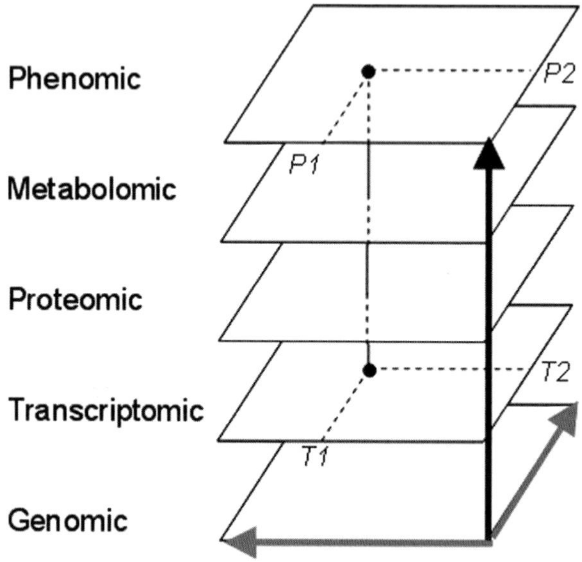

Fig. 3.3 Conceptual projection view of omic space (Toyoda and Wada 2004). *Black* arrow, direction of ascending order in transcriptomic, proteomic, metabolomic and phenomic planes. *Gray* arrow genomic-coordinate axes. The epistatic P1–P2 interaction on the phenomic plane corresponds to T1 and T2 genes interacting on the transcriptomic planes

tiality of the combination of multiple omics analyses for strain improvements in fermentation industries. (Paley and Karp 2006) developed the "Omics viewer" that can show different types of data sets (for example, measurements of gene expression and metabolite concentrations) simultaneously on a metabolic pathway map. (Arakawa et al. 2005) also developed a mapping tool to display complex omics data together.

Excellent studies using a combination of multiple omics methods have begun to be reported. Confining examples to studies of *Escherichia coli*, the following works can be found: Yoon et al. (2003) carried out combined transcriptomic (DNA microarray) and proteomic (two-dimensional gel electrophoresis; 2-DE) analyses of *E. coli* during high cell density cultivation, which is required for higher productivity of recombinant proteins. They showed that patterns of gene expression were mostly similar to patterns of protein expression, except for several discrepancies observed for a few genes (Fig. 3.4). Fong et al. (2006) investigated transcriptomics (DNA microarray) and fluxomics (^{13}C-labeled glucose was used as a substrate, and label patterns of amino acids of hydrolyzed cells were measured by GC-MS) of *E. coli* to reveal the mechanisms of adaptive mutations of some gene-disrupted strains. They found that activation of latent pathways and flux changes in the tricarboxylic acid (TCA) cycle in the adaptive evolved strains correlate well with changes in the transcriptome. Bore et al. (2007) performed transcriptomics (quantitative reverse-transcription polymerase chain reaction; qRT-PCR) and proteomics (peptide mass fingerprinting) to study *E. coli* adaptation to benzalkonium chloride, which is a commonly used disinfectant and preservative. Their analysis indicated that benzalkonium chloride treatment might result in superoxide stress in *E. coli*. Wittmann et al. (2007) studied the fluxome (GC-MS analysis of labeled proteinogenic amino acids) and metabolome (enzymatic analyses) of *E. coli* during temperature-induced

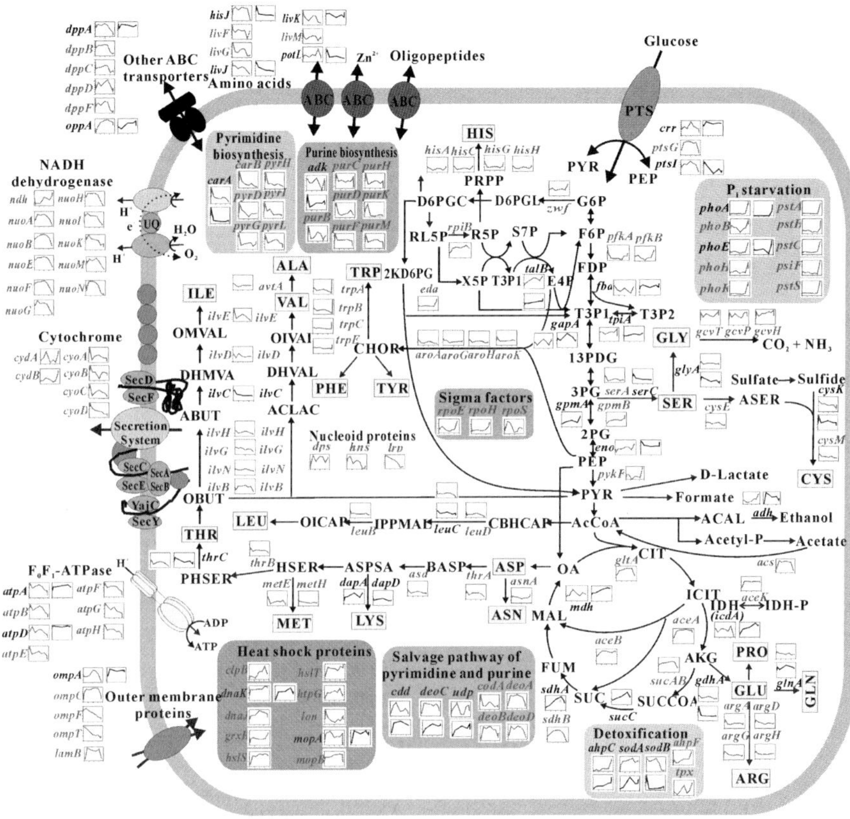

Fig. 3.4 Transcriptome and proteome analysis of *E. coli* during high cell density culture. (From Yoon SH, Han MJ, Lee SY, Jeong KJ, Yoo JS (2003) Combined transcriptome and proteome analysis of *Escherichia coli* during high cell density culture. Biotechnol Bioeng. 81(7):753–767. Copyright © 2003 by Wiley Periodicals, Inc. Reprinted with permission of Wiley-Liss, Inc., a subsidiary of John Wiley & Sons, Inc.). X axis, cell concentration (g DCW/L); Y axis, expression level in log2 scale for transcriptome (*gray*) and in absolute value of volume % for proteome (*black*); *gray-colored* gene name, only mRNA level was detected; *black-colored* gene name, both mRNA and protein level were detected

recombinant production of human fibroblast growth factor. Their analysis showed a relationship between the adenylate energy charge drop and an increase in the glycolytic flux. Other regulations in central carbon metabolism were also estimated. (Durrschmid et al. 2008) performed transcriptomics (DNA microarray) and proteomics (two-dimensional difference gel electrophoresis (Marouga et al. 2005); 2D-DIGE) analyses of *E. coli* stress response mechanisms towards recombinant protein expression. Their investigation of the expression of two model proteins demonstrated that there is a distinct impact of recombinant proteins, particularly on levels of known stress regulatory genes and proteins, as well as on the response

Table 3.1 Reported multi-omics analyses of *E. coli*

	Transcriptome	Proteome	Metabolome	Fluxome
Yoon et al. 2003	o	o		
Fong et al. 2006	o			o
Bore et al. 2007	o	o		
Wittmann et al. 2007			o	o
Durrschmid et al. 2008	o	o		

associated with ArcA and *psp*. Table 3.1 summarizes these multi-omics research studies targeting *E. coli*.

In 2001, the Institute for Advanced Biosciences (IAB) of Keio University was founded at Tsuruoka City, Yamagata, Japan. The purpose of the IAB is to actualize the crossover association of different research fields, including genomics, proteomics, metabolomics and informatics, for the establishment of "integrative systems biology" to obtain a more complete picture of living organisms. *E. coli* was selected as the primary target of the IAB, and a multi-omics approach was applied to reveal the basic principles of cellular responses of *E. coli* to genetic or environmental perturbations (Ishii et al. 2007). In the following section, a large-scale multi-omics study performed in the IAB is presented.

3.2 Multi-omics Analysis of *E. coli*

3.2.1 Chemostat Cultures of the Keio Collection

The Keio collection (Baba et al. 2006), which is the complete collection of all single-gene disruptants of *E. coli* K-12, was used for this study. From the Keio collection, 24 single-gene disrupted strains were selected. These strains are disruptants of genes in glycolysis or pentose phosphate pathway metabolism. These metabolic pathways are parts of the "central carbon metabolism", which functions to supply energy and synthesize essential precursors used for cellular components. Since central carbon metabolism is crucial for living cells, the disruption of genes involved in this metabolism was expected to result in dramatic changes in the cellular system. A uniform dilution rate of $0.2\,h^{-1}$ was applied to the chemostat cultures of the Keio collection strains.

Gene disruption can be thought of as an "internal" perturbation to the cell. On the other hand, "external" perturbation can be added by changing environmental factors. In this study, we chose substrate concentration change as the external perturbation. This was carried out by changing the dilution rate of the chemostat culture. Table 3.2 summarizes the strains and culture conditions used, and Fig. 3.5 shows the pathway map of central carbon metabolism of *E. coli* and the positions of disrupted single genes examined in this study.

Table 3.2 Strain and culture conditions

Strain	*E. coli* BW25113
	galM, glk, pgm, pgi, pfkA, pfkB, fbp, fbaB, gapC, gpmA, gpmB, pykA, pykF, ppsA, zwf, pgl, gnd, rpe, rpiA, rpiB, tktA, tktB, talA, talB
Medium	Modified M9
Carbon source	Glucose
Oxygen supply	Aerobic
Temperature	37 °C
pH	7.0
Dilution rate	0.2 h^{-1} (for single-gene disruptants)
	0.1, 0.2, 0.4, 0.5, 0.7 h^{-1} (for wild-type)

3.2.2 Transcriptome, Proteome, Metabolome, and Fluxome Analysis

The performed multi-omics analysis included layers closest to the genome and those closest to the phenotype, i.e., including transcriptomics, proteomics, metabolomics and fluxomics. Both cell-wide semi-quantitative analysis and targeted quantitative methods were employed in the transcriptome and proteome analyses. The transcriptome analysis was performed by DNA microarray for 4213 genes and qRT-PCR for 85 genes involved in central carbon metabolism. The proteome analysis was carried out with 2D-DIGE (approximately 2000 proteins were detected) and quantitative methods using liquid chromatography-mass spectrometry/mass spectrometry (LC-MS/MS) for 57 proteins involved in central carbon metabolism. The metabolome analysis was performed by CE-TOFMS for 579 metabolites. To perform the fluxome analysis, ^{13}C-labeled glucose was used as a substrate and mass distributions of proteinogenic amino acids of cultured cells were measured by GC-MS. The metabolic fluxes were calculated from the information of the obtained mass distributions. Table 3.3 summarizes the omics technologies employed in this study. All measurement data is published on our website (http://ecoli.keio.ac.jp/).

The obtained data set was used to analyze the response of the cellular system to the perturbations. For this purpose, two-step normalizations were applied to the measurement values (Ishii et al. 2007), and final converted values are named as "expression index" (EI).

3.2.3 Observed Robustness in E. coli Metabolic System

Figure 3.6 shows EIs of quantitative measurements (qRT-PCR for mRNAs, LC-MS/MS applied methods for proteins, and CE-TOFMS for metabolites) for all samples. Upon first glance of this figure, mRNAs and proteins seem to vary with the change of specific growth rate (equal to the dilution rate in a chemostat culture). Surprisingly, no clear changes of mRNAs and proteins were found for most single-gene disruptants, even when the disrupted gene concerns crucial central carbon

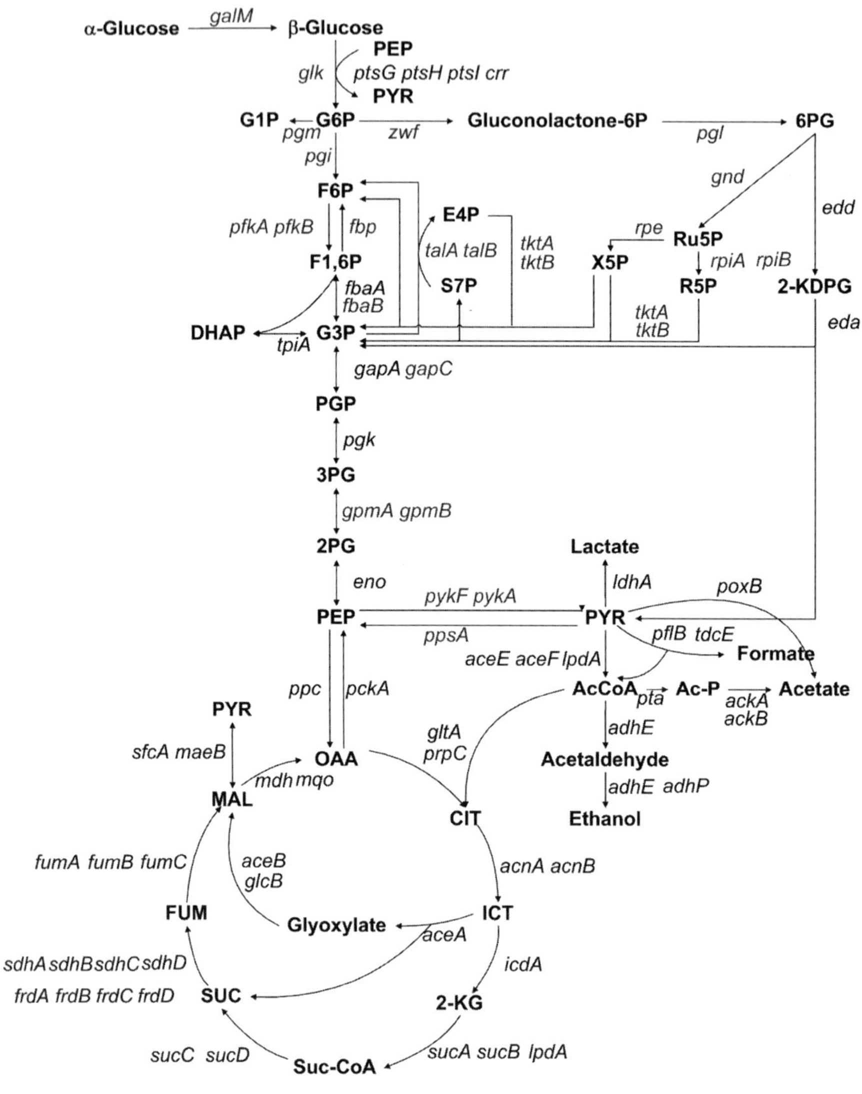

Fig. 3.5 Map of *E. coli* K-12 central carbon metabolism. (Modified from Ishii et al. 2007). Bold font, metabolites; italics, genes. *Gray* character genes are examined single-gene disruptions

metabolism. Moreover, no significant or regular change was observed for metabolites in both growth rate changed samples and single-gene disrupted samples. Some nucleotides in single-gene disruptants showed relatively large variances, but this is probably because of instability and/or low extraction efficiency of the nucleotide compounds. To authenticate these findings, averages of absolute values of the EI included in a specific category (i.e., mRNAs, proteins, or metabolites) were calculated and referred to as the average expression index (AEI). Figure 3.7 shows the AEIs of each category.

Table 3.3 Employed technologies in the multi-omics study of *E. coli*

	Used technology	Number of measured chemical species
Transcriptomics	DNA microarray	4213
	qRT-PCR	85
Proteomics	2D-DIGE	2000 (approximately)
	LC-MS/MS	57
Metabolomics	CE-TOFMS	579
Fluxomics	GC-MS	104 (isotopomers of fragment from proteinogenic amino acids)

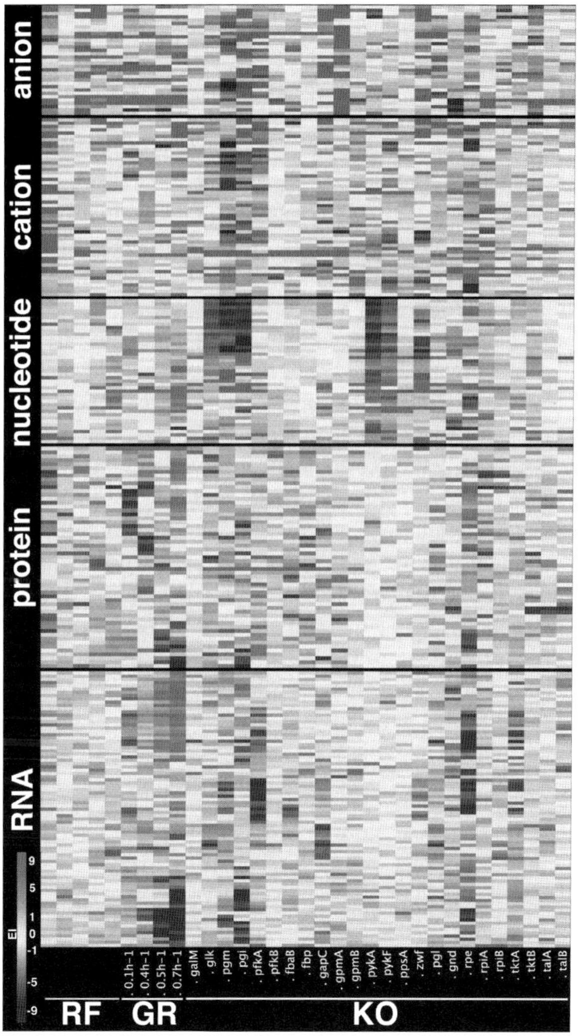

Fig. 3.6 Heatmap of the EI values of intracellular components. (Modified from Ishii et al. 2007). The heatmap shows the EI values of intracellular components that were detected in more than half the samples. RF, reference sample (wild-type cells cultured at a specific growth rate of $0.2\,h^{-1}$); GR, wild-type cells cultured at the indicated specific growth rates; KO, single-gene knockout mutants cultured at a specific growth rate of $0.2\,h^{-1}$

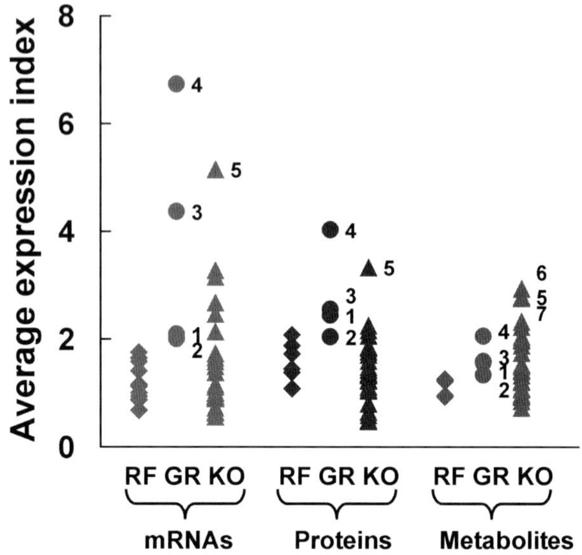

Fig. 3.7 AEI values for quantitative measurements obtained by targeted analysis. (Modified from Ishii et al. 2007). RF, reference sample (wild-type cells cultured at a specific growth rate of $0.2\,h^{-1}$); GR, wild-type cells cultured at the indicated specific growth rates; KO, single gene knockout mutants cultured at a specific growth rate of $0.2\,h^{-1}$. Numbers 1, 2, 3 and 4, correspond to specific growth rates of 0.1, 0.4, 0.5 and $0.7\,h^{-1}$, and numbers 5, 6 and 7 correspond to *rpe*, *pgi* and *pgm* disruptants, respectively

The AEIs for mRNAs and proteins gradually increased at higher growth rates. This suggests that *E. coli* actively regulates global gene and protein levels to meet increasing metabolic demands. Meanwhile, the AEI values for metabolites did not change significantly with the growth rate. This relative stability in metabolite level may be a consequence of the active regulation of enzyme expression. Focusing on local pathways, large changes of expression levels of proteins related to energy supply under aerobic condition were observed accompanying an increase in the specific growth rate (Ishii et al. 2007).

In contrast to the changes observed in wild-type cells cultured at various growth rates, the AEIs for both mRNAs and proteins in most gene-disruptants showed small changes, which fell within the range of variation observed in wild-type samples at the same specific growth rate (i.e., reference samples). In comparisons of targeted analyses of mRNAs (qRT-PCR) and proteins (LC-MS/MS), the AEI values in all disruptants were smaller than the AEI values observed for wild-type cells at a specific growth rate of $0.7\,h^{-1}$. Similar results were obtained for the AEI values representing the global analysis of expression of mRNAs (DNA microarray) and proteins (2D-DIGE) (Ishii et al. 2007). An overview of the changes in AEIs explained above is displayed in Table 3.4.

Table 3.4 Changes in AEIs

	Growth rate change (wild-type)	Most of examined single-gene disruptants
mRNAs	+	−
Proteins	+	−
Metabolites	−	−

+: Variation among samples was large. −: Variation among samples was small.

These findings suggest that *E. coli* does not appreciably respond to the loss of a single enzyme in central carbon metabolism by regulating the abundance of other compensatory enzymes. Actually, in most single-gene disruptants, the expression level of isozyme coding genes was almost same as the level in the wild-type strain (Ishii et al. 2007). In single-gene disruptants, a stable metabolic state is maintained by using remaining isozymes or by rerouting metabolic fluxes. For example, in the *zwf*-disruptant, some fluxes flow in a countercurrent direction compared to the wild-type, as reported in a previous study (Zhao et al. 2004)

Two exceptions were the *pfkA*-disruptant and *rpiA*-disruptant (Ishii et al. 2007). In these strains, potential mutations in genes other than the disrupted gene were checked, and various mutations enhancing the expression level of compensatory isozymes of the disrupted gene (*pfkB* in *pfkA*-disruptant and *rpiB* in *rpiA*-disruptant) were found, as reported in previous studies (Daldal 1983, Skinner and Cooper 1974).

3.3 Concluding Remarks

Changes in the dilution rate of a chemostat culture correspond to changes in the concentration of growth-limiting substrates, and thus various settings of the dilution rate can be regarded as an environmental perturbation for *E. coli*. On the other hand, the disruption of a gene can be thought of as an intracellular perturbation. Our multi-omics analysis demonstrates that the metabolic network of *E. coli* is markedly robust against both types of perturbations. *E. coli* can actively respond to changes in the concentration of growth-limiting substrates by regulating the level of enzyme expression to maximize growth rate, which is reflected in the observed stability of metabolite levels. However, this strategy may come at a high cost, because the cell must prepare additional systems (such as sensor proteins, signal mediators, and transcriptional regulators) to detect and react appropriately to each specific perturbation. This strategy contrasts with the finding that *E. coli* does not appear to reconfigure mRNA or protein levels actively when most single metabolic genes are disrupted. In this case, structural redundancy in the metabolic network itself provides the necessary robustness. As a result, the levels of most metabolites remain at wild-type levels, although some localized perturbations may occur. This strategy seems to save more energy than the active regulation of mRNA or proteins, because it requires no specific molecular machinery for detecting each mutation. Even if this strategy appears insufficient in the face of some mutations, *E. coli* may survive by accumulating additional mutations, as observed for *pfkA* and *rpiA* disruptants. Using multiple strategies may thus enable *E. coli* to maintain a stable metabolic state when exposed to various types of perturbations.

Biological robustness is one of the central subjects in systems biology (Kitano 2004), and conceptual descriptions or analyses with mathematical models have been attempted to explain how robustness is achieved (Kitano 2007). Furthermore, some omics or multi-omics analyses to study robustness in real cells have also been reported. For example, (Becker et al. 2006) performed a proteomics analysis of *Salmonella*, and found extensive metabolic redundancies and access to diverse host nutrients. (Gibon et al. 2006) measured the transcriptomes, proteomes, and metabolomes of *Arabidopsis rosettes* wild-type and *pgm*-disruptant, and demonstrated that the amplitudes of diurnal changes in metabolite levels in *pgm* were similar or smaller than those in the wild-type. The above mentioned multi-omics analysis of *E. coli* also supports the existence of robustness as a common principle to ensure survival in the face of countless accidents.

The next challenge of the multi-omics data-driven systems biology of *E. coli* is to construct a mathematical model incorporating the obtained multi-omics data to elucidate a tangible mechanism of metabolic robustness in *E. coli*. Analyses using a mathematical model will suggest methods for breaking cellular robustness to enhance the productivity of useful metabolic compounds. Finally, and needless to say, the integrative multi-omics approach can be applied to many organisms, not just microorganisms, and thus expanding applications of this approach can be expected in the future.

Acknowledgments This study was carried out with the collaboration of following members of the IAB: Ishii N*, Nakahigashi K*, Baba T*, Robert M*, Soga T*, Kanai A*, Hirasawa T*, Naba M, Hirai K, Hoque A, Ho PY, Kakazu Y, Sugawara K, Igarashi S, Harada S, Masuda T, Sugiyama N, Togashi T, Hasegawa M, Takai Y, Yugi K, Arakawa K, Iwata N, Toya Y, Nakayama Y, Nishioka T, Shimizu K, Mori H and Tomita M. (*These members contributed equally to this work.) This study was supported by grants from the New Energy and Industrial Technology Development Organization (NEDO) of the Ministry of Economy, Trade, and Industry of Japan (METI) (Development of a Technological Infrastructure for Industrial Bioprocess Project), the Core Research for Evolutional Science and Technology (CREST) of the Japan Science and Technology Agency (JST) (Development of Modeling/Simulation Environment for Systems Biology), the Japan Society for the Promotion of Science (JSPS), and a "Grant-in-Aid" from the Ministry of Education, Culture, Sports, Science and Technology (MEXT) for the 21st Century Center of Excellence (COE) Program (Understanding and Control of Life's Function via Systems Biology), Scientific Research on Priority Areas "Systems Genomics," and Scientific Research on Priority Areas "Lifesurveyor", as well as funds from the Yamagata Prefectural Government and Tsuruoka City. We thank the faculty and students of the Systems Biology Program at Keio University for useful discussions, as well as members of the International *E. coli* Alliance (IECA).

References

Aiba S, Matsuoka M (1979) Identification of metabolic model: Citrate production from glucose by *Candida lipolytica*. Biotechnol Bioeng 21(8):1373–86

Andersen MR, Nielsen ML, Nielsen J (2008) Metabolic model integration of the bibliome, genome, metabolome and reactome of *Aspergillus niger*. Mol Syst Biol 4:178

Arakawa K, Kono N, Yamada Y et al. (2005) KEGG-based pathway visualization tool for complex omics data. In Silico Biol 5(4):419–23

Baba T, Ara T, Hasegawa M et al. (2006) Construction of *Escherichia coli* K-12 in-frame, single-gene knockout mutants: the Keio collection. Mol Syst Biol 2:2006 0008 doi:10.1038/msb4100050

Becker D, Selbach M, Rollenhagen C et al. (2006) Robust *Salmonella* metabolism limits possibilities for new antimicrobials. Nature 440(7082):303–7

Bore E, Hebraud M, Chafsey I et al. (2007) Adapted tolerance to benzalkonium chloride in *Escherichia coli* K-12 studied by transcriptome and proteome analyses. Microbiology 153(Pt 4):935–46

Bruggeman FJ, Westerhoff HV (2007) The nature of systems biology. Trends Microbiol 15(1): 45–50

Caspi R, Foerster H, Fulcher CA et al. (2008) The MetaCyc Database of metabolic pathways and enzymes and the BioCyc collection of Pathway/Genome Databases. Nucleic Acids Res 36(Database issue):D623–31

Chen C, Gonzalez FJ, Idle JR (2007) LC-MS-based metabolomics in drug metabolism. Drug Metab Rev 39(2–3):581–97

Costenoble R, Muller D, Barl T et al. (2007) ^{13}C-Labeled metabolic flux analysis of a fed-batch culture of elutriated *Saccharomyces cerevisiae*. FEMS Yeast Res 7(4):511–26

Daldal F (1983) Molecular cloning of the gene for phosphofructokinase-2 of *Escherichia coli* and the nature of a mutation, *pfkB1*, causing a high level of the enzyme. J Mol Biol 168(2):285–305

De Keersmaecker SC, Thijs IM, Vanderleyden J et al. (2006) Integration of omics data: how well does it work for bacteria? Mol Microbiol 62(5):1239–50

Dettmer K, Aronov PA, Hammock BD (2007) Mass spectrometry-based metabolomics. Mass Spectrom Rev 26(1):51–78

Durrschmid K, Reischer H, Schmidt-Heck W et al. (2008) Monitoring of transcriptome and proteome profiles to investigate the cellular response of *E. coli* towards recombinant protein expression under defined chemostat conditions. J Biotechnol 135(1):34–44

Fiehn O (2002) Metabolomics–the link between genotypes and phenotypes. Plant Mol Biol 48(1–2):155–71

Fiehn O, Kopka J, Dormann P et al. (2000) Metabolite profiling for plant functional genomics. Nat Biotechnol 18(11):1157–61

Fong SS, Nanchen A, Palsson BO et al. (2006) Latent pathway activation and increased pathway capacity enable *Escherichia coli* adaptation to loss of key metabolic enzymes. J Biol Chem 281(12):8024–33

Gaspar A, Englmann M, Fekete A et al. (2008) Trends in CE-MS 2005–2006. Electrophoresis 29(1):66–79

Gibon Y, Usadel B, Blaesing OE et al. (2006) Integration of metabolite with transcript and enzyme activity profiling during diurnal cycles in *Arabidopsis rosettes*. Genome Biol 7(8):R76

Grivet JP, Delort AM, Portais JC (2003) NMR and microbiology: from physiology to metabolomics. Biochimie 85(9):823–40

Ishii N, Nakahigashi K, Baba T et al. (2007) Multiple high-throughput analyses monitor the response of *E. coli* to perturbations. Science 316(5824):593–97

Jordan KW, Cheng LL (2007) NMR-based metabolomics approach to target biomarkers for human prostate cancer. Expert Rev Proteomics 4(3):389–400

Joyce AR, Palsson BO (2006) The model organism as a system: integrating 'omics' data sets. Nat Rev Mol Cell Biol 7(3):198–210

Kanehisa M, Araki M, Goto S et al. (2008) KEGG for linking genomes to life and the environment. Nucleic Acids Res 36(Database issue):D480-84

Kell DB (2004) Metabolomics and systems biology: making sense of the soup. Curr Opin Microbiol 7(3):296–307

Kitano H (2004) Biological robustness. Nat Rev Genet 5(11):826–37

Kitano H (2007) Towards a theory of biological robustness. Mol Syst Biol 3:137

Lee SY, Lee DY, Kim TY (2005) Systems biotechnology for strain improvement. Trends Biotechnol 23(7):349–58

Marouga R, David S, Hawkins E (2005) The development of the DIGE system: 2D fluorescence difference gel analysis technology. Anal Bioanal Chem 382(3):669–78

Mashego MR, Rumbold K, De Mey M et al. (2007) Microbial metabolomics: past, present and future methodologies. Biotechnol Lett 29(1):1–16

Monton MR, Soga T (2007) Metabolome analysis by capillary electrophoresis-mass spectrometry. J Chromatogr A 1168(1–2):237–46; discussion 236

Noh K, Gronke K, Luo B et al. (2007) Metabolic flux analysis at ultra short time scale: Isotopically non-stationary ^{13}C labeling experiments. J Biotechnol 129(2):249–67

Noh K, Wahl A, Wiechert W (2006) Computational tools for isotopically instationary ^{13}C labeling experiments under metabolic steady state conditions. Metab Eng 8(6):554–77

Oldiges M, Lutz S, Pflug S et al. (2007) Metabolomics: current state and evolving methodologies and tools. Appl Microbiol Biotechnol 76(3):495–511

Paley SM, Karp PD (2006) The Pathway Tools cellular overview diagram and Omics Viewer. Nucleic Acids Res 34(13):3771–8

Rabinowitz JD (2007) Cellular metabolomics of *Escherchia coli*. Expert Rev Proteomics 4(2): 187–98

Sanchez DH, Siahpoosh MR, Roessner U et al. (2008) Plant metabolomics reveals conserved and divergent metabolic responses to salinity. Physiol Plant 132(2):209–19

Sanford K, Soucaille P, Whited G et al. (2002) Genomics to fluxomics and physiomics - pathway engineering. Curr Opin Microbiol 5(3):318–22

Sauer U (2006) Metabolic networks in motion: ^{13}C-based flux analysis. Mol Syst Biol 2:62

Sauer U, Lasko DR, Fiaux J et al. (1999) Metabolic flux ratio analysis of genetic and environmental modulations of *Escherichia coli* central carbon metabolism. J Bacteriol 181(21):6679–88

Schaub J, Mauch K, Reuss M (2008) Metabolic flux analysis in *Escherichia coli* by integrating isotopic dynamic and isotopic stationary ^{13}C labeling data. Biotechnol Bioeng 99(5):1170–85

Shimizu K (2004) Metabolic flux analysis based on ^{13}C-labeling experiments and integration of the information with gene and protein expression patterns. Adv Biochem Eng Biotechnol 91:1–49

Skinner AJ, Cooper RA (1974) Genetic studies on ribose 5-phosphate isomerase mutants of *Escherichia coli* K-12. J Bacteriol 118(3):1183–85

Sniehotta M, Schiffer E, Zurbig P et al. (2007) CE - a multifunctional application for clinical diagnosis. Electrophoresis 28(9):1407–17

Soga T, Baran R, Suematsu M et al. (2006) Differential metabolomics reveals ophthalmic acid as an oxidative stress biomarker indicating hepatic glutathione consumption. J Biol Chem 281(24):16768–76

Soga T, Ohashi Y, Ueno Y et al. (2003) Quantitative metabolome analysis using capillary electrophoresis mass spectrometry. J Proteome Res 2(5):488–94

Song EJ, Babar SM, Oh E et al. (2008) CE at the omics level: towards systems biology–an update. Electrophoresis 29(1):129–42

Steinfath M, Repsilber D, Scholz M et al. (2007) Integrated data analysis for genome-wide research. Exs 97:309–29

Stephanopoulos GN, Nielsen J, Aristidou A (1998) Metabolic Engineering: Principles and Methodologies. Academic Press, San Diego

Teufel A, Krupp M, Weinmann A et al. (2006) Current bioinformatics tools in genomic biomedical research (Review). Int J Mol Med 17(6):967–73

Tolstikov VV, Fiehn O, Tanaka N (2007) Application of liquid chromatography-mass spectrometry analysis in metabolomics: reversed-phase monolithic capillary chromatography and hydrophilic chromatography coupled to electrospray ionization-mass spectrometry. Methods Mol Biol 358:141–55

Toya Y, Ishii N, Hirasawa T et al. (2007) Direct measurement of isotopomer of intracellular metabolites using capillary electrophoresis time-of-flight mass spectrometry for efficient metabolic flux analysis. J Chromatogr A 1159(1–2):134–41

Toyoda T, Mochizuki Y, Player K et al. (2007) OmicBrowse: a browser of multidimensional omics annotations. Bioinformatics 23(4):524–6

Toyoda T, Wada A (2004) Omic space: coordinate-based integration and analysis of genomic phenomic interactions. Bioinformatics 20(11):1759–65

van Winden WA, van Dam JC, Ras C et al. (2005) Metabolic-flux analysis of *Saccharomyces cerevisiae* CEN.PK113–7D based on mass isotopomer measurements of ^{13}C-labeled primary metabolites. FEMS Yeast Res 5(6–7):559–68

Wang QZ, Wu CY, Chen T et al. (2006) Integrating metabolomics into a systems biology framework to exploit metabolic complexity: strategies and applications in microorganisms. Appl Microbiol Biotechnol 70(2):151–61

Ward JL, Baker JM, Beale MH (2007) Recent applications of NMR spectroscopy in plant metabolomics. Febs J 274(5):1126–31

Wiechert W (2001) ^{13}C metabolic flux analysis. Metab Eng 3(3):195–206

Wiechert W, Noh K (2005) From stationary to instationary metabolic flux analysis. Adv Biochem Eng Biotechnol 92:145–72

Wiechert W, Schweissgut O, Takanaga H et al. (2007) Fluxomics: mass spectrometry versus quantitative imaging. Curr Opin Plant Biol 10(3):323–30

Wittig U, De Beuckelaer A (2001) Analysis and comparison of metabolic pathway databases. Brief Bioinform 2(2):126–42

Wittmann C, Weber J, Betiku E et al. (2007) Response of fluxome and metabolome to temperature-induced recombinant protein synthesis in *Escherichia coli*. J Biotechnol 132(4):375–84

Yadav SP (2007) The wholeness in suffix -omics, -omes, and the word om. J Biomol Tech 18(5):277

Yoon SH, Han MJ, Lee SY et al. (2003) Combined transcriptome and proteome analysis of *Escherichia coli* during high cell density culture. Biotechnol Bioeng 81(7):753–67

Zhao J, Baba T, Mori H et al. (2004) Effect of *zwf* gene knockout on the metabolism of *Escherichia coli* grown on glucose or acetate. Metab Eng 6(2):164–74

Chapter 4
A Medium-Throughput Structural Proteomics Approach Applied to the Genome of *E. coli*

Allan Matte, Irena Ekiel, Zongchao Jia, Kalle Gehring, and Miroslaw Cygler

Contents

Abstract Structural information of the protein complement of *E. coli* represents an important component in our quest for a more complete understanding of this organism at the molecular level. Structural proteomics, the application of technologies to enhance the rate of protein structure determination at the genome level, has significantly increased the structural coverage of the *E. coli* proteome. The Bacterial Structural Genomics Initiative (BSGI) has focused on the structure determination of *E. coli* proteins of both known and unknown function, using a combination of NMR and X-ray crystallography. This program has resulted in the implementation of several technologies, including robotics platforms, in a coordinated manner in order to streamline the steps involved in protein structure determination. Here, we describe our experimental approaches as well as some examples high-lighting new structural and functional insights of specific targets.

M. Cygler (✉)
Biotechnology Research Institute, National Research Council Canada,
Department of Biochemistry, McGill University, Montreal, QC H4P2R2 Canada
e-mail: mirek.cygler@bri.nrc.ca.

S.Y. Lee (ed.), *Systems Biology and Biotechnology of Escherichia coli*,
DOI 10.1007/978-1-4020-9394-4_4, © Springer Science+Business Media B.V. 2009

4.1 Introduction

Structural genomics and proteomics can be defined in various ways, but a common theme is the application of high or medium-throughput methods to determining protein structures with little or no sequence similarity to known structures. This goal of increasing the coverage of structure space with respect to protein sequence was a formative concept in initiating structural genomics programs world-wide nearly ten years ago (Shapiro and Lima 1998). A second reason for these large-scale protein structure determination efforts has been to utilize the power of structural information to infer protein function, especially in those cases where similarity at the level of the protein sequence has been lost (Watson et al. 2007). A number of computational approaches have been developed over the past several years in order to accomplish this task (Lee et al. 2007), and lay the groundwork for specific experiments to validate or refute these predictions.

While the field of structural genomics is still relatively young, it has immensely influenced modern biology, through the provision of many thousands of new protein structures which are made quickly available to biologists through deposition in the Protein Data Bank (Berman et al. 2000), as well as through the development and dissemination of a number of technologies and tools that can be used to speed up the various steps in the structure determination pipeline (Manjasetty et al. 2008). This wealth of new protein structure information has, in turn, permitted generation of comparative protein structure models for a large number of related sequences, which can be used to guide further experiments (Ginalski 2006). Indeed, the greatest legacy of structural genomics may not be the structural information that ultimately results from its practice, but rather the change in thinking that these technological advances have caused, making protein structure information available at a lower cost, for more targets, and to more scientists more quickly than was formerly conceivable. An indication of the impact of structural genomics is seen through the more than 2000 hits resulting from a search using PubMed. Importantly for the biological community at large, the quality of the resulting crystal structures from structural genomics initiatives are at least on par, if not better, on average, than those from investigator-driven laboratories (Brown and Ramaswamy 2007). One of the "Holy Grails" of structural biology generally is the hope to eventually predict, with some level of accuracy, those proteins that are likely to crystallize, and those which will not. The flood of structures from structural genomics projects as well as the target status information contained within TargetDB are slowly shedding light on this matter (Slabinski et al. 2007a,b, Smialowski et al. 2006).

Many early structural genomics initiatives focused on microbial genomes due to the relative ease in cloning and expression, the large number of potential microbial ortholgue sequences available as backups, and the fact that a proportion of these proteins are sequence-conserved in eukaryotes. We chose *Escherichia coli* as a suitable system, in part due to its status as a "model" organism for many physiological and biochemical processes. Many parallel genomics and functional proteomics studies had been initiated on *E. coli* concurrently, all with the shared goal of increasing the breadth and depth of understanding of this bacterium so that ultimately, its metabolism could be modeled at the level of an intact cell (Feist et al. 2007). In addition to the BSGI (Matte et al. 2003, 2007), several other centers invested heavily in the determination of *E. coli* protein structures, including the Midwest Center for Structural Genomics (MWCSG), the North East Center for Structural genomics (NESG) as well as groups in France (Abergel et al. 2003) and Japan (Yokoyama et al. 2000).

4.2 Genomics and Proteomics Studies of *E. coli*

The complete genome sequences of *E. coli* K-12 (Blattner et al. 1997), O157:H7 (Hayashi et al. 2001, Perna et al. 2001), CFT073 (Welch et al. 2002) and DH10B (Durfee et al. 2008) set the stage for large-scale proteomics studies on this organism, with most of these studies focusing on the well-annotated K-12 bacterium (Karp et al. 2007, Riley et al. 2006). Based on data within the *E. coli* genome and proteome database GenProtEC (http://genprotec.mbl.edu/), as of Aug 2007, a total of 4485 ORFs have been identified within *E. coli* K-12, with ~58% having an experimentally-verified function, ~24% having an imputed function based on sequence similarity, and ~10% remain of unknown function. The *E. coli* K-12 annotation available through EcoCyc is similar, with 66% of genes having an experimentally-verified function and 76% an assigned function (Karp et al. 2007). Structural information for proteins of unknown function can contribute significantly to narrowing down possible functions, and can lead to the design of targeted experiments to test the resulting predictions (Adams et al. 2007, Kim et al. 2003, Watson et al. 2007).

Structural studies on *E. coli* proteins at the level of its genome are only one of many "omics" approaches that are being utilized to understand this bacterium at the cellular level. Other approaches include analysis of the *E. coli* phosphoproteome (Macek et al. 2008), combined 2D gel and mass-spectrometry analysis of *E. coli* proteins (Maillet et al. 2007), protein-protein interactions using pull-downs (Arifuzzaman et al. 2006) or TAP/SPA tagging (Butland et al. 2005) and systematic knockouts of all non-essential *E. coli* genes (Baba et al. 2008), to name but a few. These experimental approaches are supplemented by a variety of databases and bioinformatics tools that serve to organize and make data on *E. coli* available to the larger community. Examples of such tools include the encyclopedia of *Escherichia coli* K-12 genes and metabolism, EcoCyc (http://ecocyc.org/), the *E. coli* genome and proteome database, GenProtEC (http://genprotec.mbl.edu/), the Bacteriome.org database (http://www.bacteriome.org), cataloging *E. coli* protein-protein interactions (Su et al. 2008) and EchoBASE, a database dedicated to the annotation of *E. coli* proteins of unknown function (Misra et al. 2005). Many of these and other resources are available through the central, EcoliHub web site (http://www.ecolicommunity.org/). Together this information will allow modeling of the physiology, metabolism and overall behavior of the *E. coli* cell. Such a systems-biology approach has many potential outcomes, ranging from a more thorough understanding of how bacteria grow and function, to the genetic manipulation of *E. coli* for industrial-scale fermentation (Chou 2007, Herrgard et al. 2006), and a better understanding of *E. coli*-host interactions resulting in pathogenesis (Fogg et al. 2006) as well as novel antibiotic discovery (Abergel et al. 2003).

4.3 Methodology

A key element of structural genomics initiatives is the leverage of medium to high-throughput, parallelized methods in conjunction with automated and robotics platforms, used together to increase sample throughput. At the BSGI, several methodologies have been implemented to enhance various stages in the pipeline (Fig. 4.1).

This process is not linear, however, as it is frequently required to revisit one or more previous steps in the process in order to optimize the "product", either a protein sample for NMR analysis or protein crystals for X-ray diffraction. It should be noted that in order for such projects to succeed, it is necessary to maximize efficiency at every step, as the overall process is very much like a funnel, with attrition at each point in the pipeline having a cumulative influence on the outcome. For those inter-

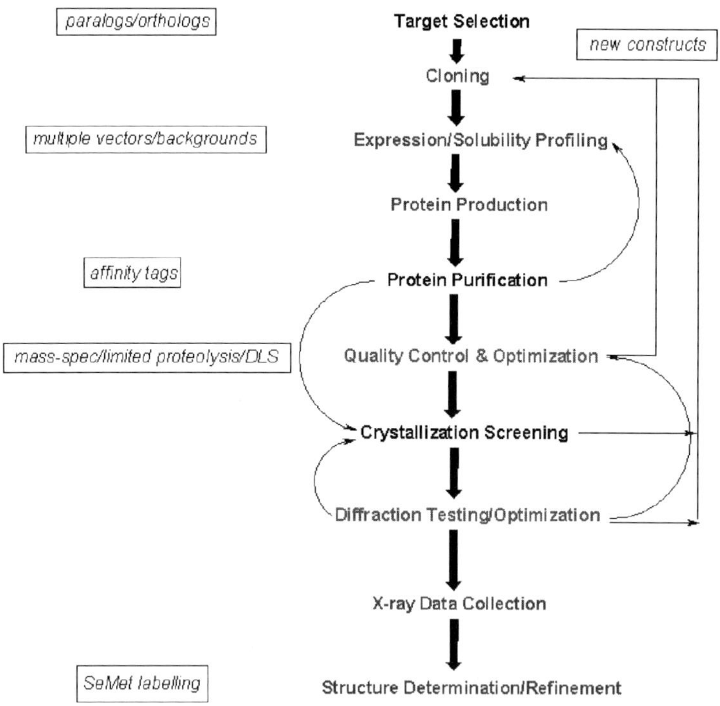

Fig. 4.1 Schematic representation of steps associated with the structural genomics pipeline. In several instances, it is necessary to revisit previous steps in the pipeline in order to futher optimize purity, behavior or the specific construct in order to proceed to structure determination

ested, more detailed protocols relating to the various steps within the experimental pipeline are available (Cygler et al. 2008)

4.3.1 SPeX-DB Database and Target Management

A key element of all structural proteomics efforts is information management, both in terms of protein targets, but also experimental information and the accumulated materials that result from each step in the pipeline. A laboratory information management system (LIMS) is a key component, used to tie these different types of information together in a single, coherent manner which is accessible to a variety of users within the project. While various LIMS have been developed for structural genomics projects (Albeck et al. 2005, Goh et al. 2003, Prilusky et al. 2005), each have specific strengths and weaknesses, and some level of customization for a specific project is inevitable. Within the BSGI, the SPeX-DB database and its associated web-based interface have been developed over several years, and have served as a one-stop center for data management for the project (Raymond et al. 2004).

There are several aspects to target management and selection where a LIMS system plays a key role in improving functionality. A broad variety of information is required on proteins within an organism's genome, such as *E. coli*, in order to determine whether or not it is a suitable target. The extent to which a particular sequence is found in other prokaryotic and eukaryotic genomes is an important criterion for its potential value as a structure target, and requires access to such information stored within databases such as PFAM (Finn et al. 2008) or InterPro (Mulder and Apweiler 2008). These tools also give a glimpse of potential, more distant structural relationships for the chosen sequence, and may indicate possible functional domains. Analysis for potential signal sequences using SignalP (Bendtsen et al. 2004) or transmembrane-spanning regions identified using TMHMM (Krogh et al. 2001) is essential in order to identify these elements in order to select appropriate positions for fusion tags or in the case of TM regions, design alternative constructs where these are removed. Of course, the resulting sequences must be checked against those for structures within the Protein Data Bank (Berman et al. 2000) using tools such as BLAST as well as against the central registry of structural genomics targets, TargetDB (Chen et al. 2004). Such a centralized registry is necessary in order to avoid unnecessary duplication of effort in those cases where a different center is highly advanced on a specific target sequence. Links to NCBI DNA and protein sequence records for specific targets, as well as to the ExPASy server (Gasteiger et al. 2003) allow quick access to a variety of information and bioinformatics tools that expedite target selection.

4.3.2 Cloning and Expression of E. coli *Proteins*

Once targets for cloning are selected, the corresponding DNA sequences for each ORF are obtained from NCBI and analyzed for restriction sites using in-house developed software. This allows different, compatible restriction enzymes to be selected, with a minimal number of targets "lost" due to the presence of internal sites. The same software also generates the sequences of the forward and reverse primers, and organizes them into an Excel table which can be used to order directly from an appropriate vendor. A set of three compatible vectors have been engineered which can accept the same digested PCR products, and which express proteins as in-frame fusions with a N-terminal His_6-tag and no cleavage site (designated pFO4), an N-terminal His_8-tag followed by a tobacco-etch virus protease (TEV) cleavage site (pJW234) and a third vector yielding a N-terminal GST fusion protein with a TEV cleavage site (pRL652). Cleavage sites compatible for TEV protease were selected due to the high specificity of this protease, with virtually no secondary cleavage of the protein of interest, the ready expression and purification of TEV protease from an appropriate expression vector, and the presence of a His-tag fused to the protease, permitting easy removal from the purified protein sample (Kapust et al. 2001, Kapust and Waugh 2000). Most aspects of cloning, from arraying of primers for PCR, to digestion and purification of PCR products and plasmid DNA mini-preps, have been automated using a Beckman-Coulter dual bridge liquid handling robot. This robot is equipped with a vacuum manifold, span-8 and 96-pipetting units, as well an orbital mixer (Fig. 4.2).

As a result of this process, a significant quantity of information and materials are generated that must be recorded and stored, including primers and primer sequences, PCR-products, plasmids, and the transformed expression strains as glycerol stocks. Here again, SPeX-DB has been designed to handle these tasks, with the ability to search and locate individual plasmids and glycerol stocks *via* the database. Glycerol stocks are of particular importance as they represent the link between those who generate the clones and the various users which will express and purify proteins from them. We have adopted the use of 2D bar-coded capped tubes sold by Matrix Technologies (http://www.matrixtechcorp.com/) for their storage, as the bar-code represents a convenient way of identifying the particular tube with respect to its database entry.

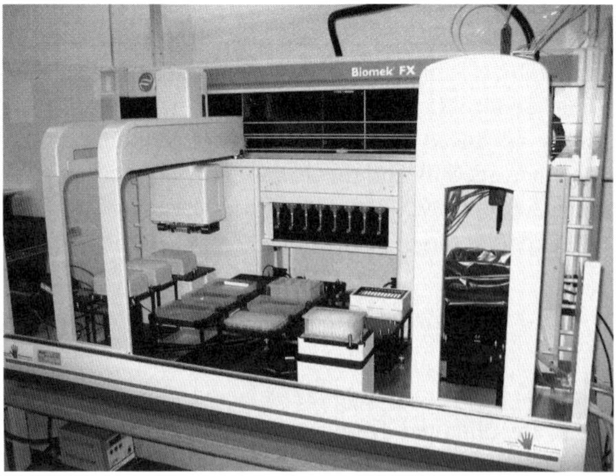

Fig. 4.2 Beckman-Coulter dual-bridge robot equipped with span-8 and 96-pipettor units, a vacuum manifold used for filtering 96-well plates, and an orbital mixing unit

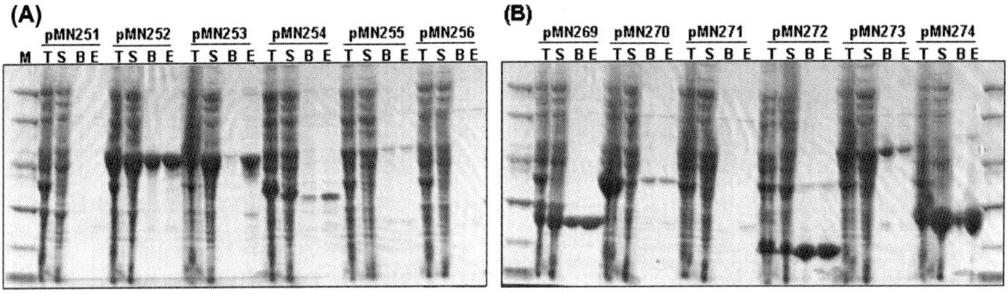

Fig. 4.3 Representative SDS-PAGE gels from small-scale protein purification experiments. Each set of four lanes corresponds to a single expression clone containing a His-tag (marked pMNxxx), with lanes corresponding to low-molecular weight markers (M), total protein (T), soluble protein (S), Ni-beads after washing and before elution (B) and the protein eluting using 250 mM imidazole (E)

Not all expression clones express soluble protein at a sufficient level suitable to move forward in the pipeline. Protein expression is evaluated using small-scale expression tests, initially using total and soluble fraction as analyzed by SDS-PAGE (Fig. 4.3), and later using dot blots or small-scale, semi-automated purification of His-tagged or GST-tagged fusion proteins using a Beckman dual-bridge robot (Fig. 4.4).

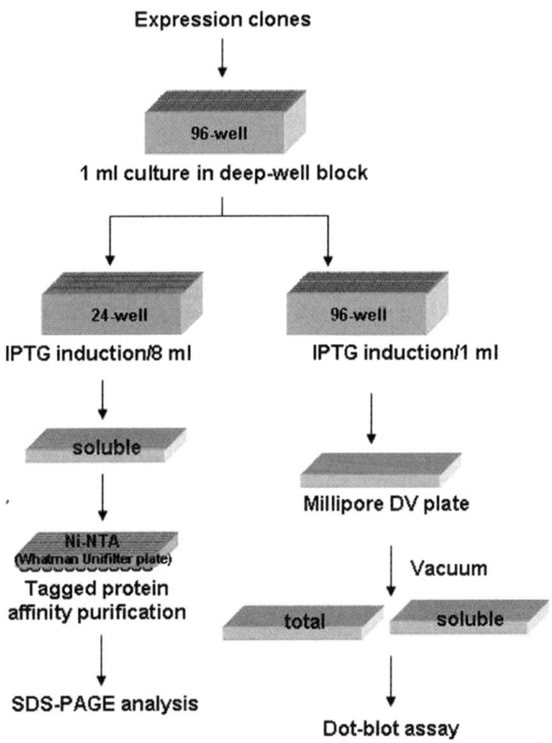

Fig. 4.4 Strategy for semi-automated expression testing using the Beckman dual-bridge robot. Cells are cultured in 24- or 96-well blocks, induced with IPTG and cells lysed using BugBuster bacterial cell lysis detergent (Novagen). Proteins purified using Ni-NTA resin (Qiagen) are then eluted and analyzed using in a dot-blot detected using an anti-His-tag antibody

4.3.3 Protein Production, Purification and Quality Assessment

We adopted the strategy of triaging expression clones into three groups, representing proteins that (a) expressed at a high level and are soluble, (b) expressed at a low level and were soluble, or (c) did not express or were insoluble. Proteins from groups (a) and (b) were then scheduled for production and purification, although using different protocols. Those in group (a) could be cultured in a smaller volume (500 ml) and induced for a shorter time (3–6 h) compared to group (b), which required larger culture volumes (1–2 l) and typically longer induction times (12–18 h). Cells are cultured using 2.8 l Fernback flasks in either terrific broth (TB) or 2YT media.

While it is conceivable to purify proteins in medium-throughput using conventional column chromatography, the process is greatly expedited by the use of affinity tags. These tags also can also be used to readily detect proteins using antibodies where necessary. We adopted the use of Ni-NTA resin from Qiagen (www.qiagen.com) as the best overall affinity resin for capturing His-tagged proteins. In some instances Ni-Sepharose 6 Fast Flow resin (GE Healthcare) or Talon superflow metal affinity resin (Clonetech Laboratories) is used, depending on the specific protein. In order to minimize non-specific binding of contaminating proteins, it is important to utilize a low concentration of imidazole (20–40 mM) in the binding buffer and wash buffers, as well as to appropriately "match" the quantity of Ni-NTA resin to the amount of expected fusion protein. Our experiences suggest that Ni-NTA has a much larger binding capacity for many His-tagged proteins than suggested by the manufacturer (20 mg/ml).

Secondary purification following the affinity step utilizes either anion exchange chromatography alone or in combination with gel filtration using an Äkta Purifier or Äkta Explorer FPLC system (GE Healthcare). In many cases, the gel filtration step is useful not only for the removal of contaminating proteins and other biomolecules, but is also used to remove aggregated protein species that may impede crystallization as well as to exchange the sample into the final buffer to be used for crystallization screening. The typical drawbacks to gel filtration (low flow rate, small sample load) can be partially compensated by the use of prep-grade columns such as Superdex-75 and Superdex-200 which have high flow rates and loading capacities.

Due to the large attrition of purified protein samples at the level of structural analysis, it is important to ensure homogeneity so that the sample has the best chance of crystallizing or yielding an interpretable NMR spectrum. We think of protein homogeneity in two ways; one is chemical purity, most often assessed by SDS-PAGE and electron-spray mass spectroscopy (ESI-MS) analysis following the final purification step. The second is solution homogeneity, or how many different macromolecular forms are represented in solution. This second criterion for homogeneity is more subtle, but can be equally important with respect to protein crystallization (Valente et al. 2005). Frequently, a sample which looks good by SDS-PAGE will have problems related to solubility, stability and aggregation, especially at the relatively high protein concentrations required for crystallization. Protein solution behavior can be evaluated using native PAGE, dynamic light scattering (DLS), and analytical gel filtration. We have adopted a number of approaches, including the use of optimal solubility screening, in which buffer pH, composition and various additives are utilized in improving protein solution behavior by DLS (Collins et al. 2005, Jancarik et al. 2004), the use of enzyme substrates, products or inhibitors to stabilize the conformation of enzymes for crystallization, the use of fluorescence melting analysis to find one or more ligands, salts or buffers that enhance the thermal melting point (T_m) of a protein sample (Ericsson et al. 2006, Vedadi et al. 2006), and finally, the use of size exclusion chromatography to remove aggregates prior to crystallization screening. This last method has proven very useful, in conjunction with DLS analysis, to prepare monodisperse protein samples suitable for crystallization screening (Matte and Cygler 2007).

4.3.4 Crystallization and Structure Determination

All crystallization screening is performed in 96-well sitting drop vapor diffusion or microbatch (under oil) plates using a Hydra II^{+one} crystallization robot (Fig. 4.5a; Thermo Fisher Scientific, Hudson, NH). Typically, drops consisting of 0.2–0.3 μl protein in buffer and 0.3–0.4 μl reservoir solution are set at either 20 °C or 4 °C. Plates set at 20 °C are monitored using a CrystalFarm™ imaging system (Fig. 4.5b; Bruker AXS, Madison, WI), consisting of a plate hotel and CCD camera to record images of drops, which are accessible to users through a web-based interface. The contents of the drops (clear solution, precipitate, etc) can then be scored and recorded using the web-based software. Visualization of plates occurs on a predetermined schedule, with all images stored within a centralized database.

We have developed an in-house set of ~400 conditions that serve as the primary crystallization screen for each new protein sample. There are two advantages to this screen – (a) the solutions are organized according to precipitant type, so

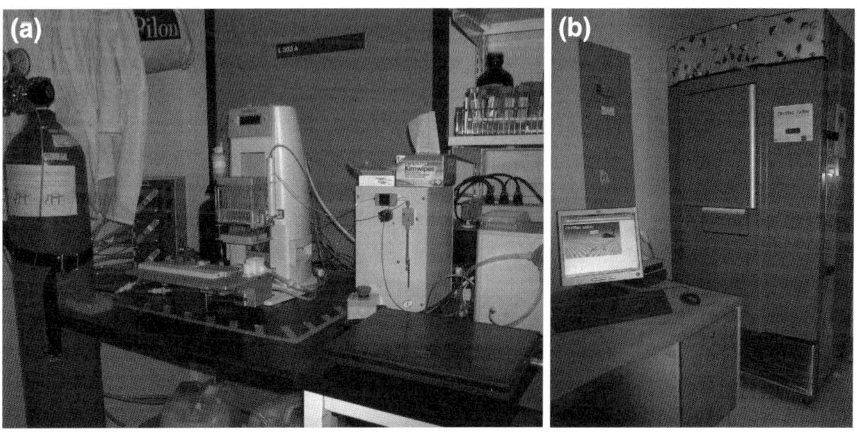

Fig. 4.5 (**a**) Hydra II^{+one} Protein Crystallization robot (Thermo Fisher Scientific). (**b**) CrystalFarm™ system for imaging protein crystallization trays (Bruker AXS)

that a protein can be screened against conditions containing polyethylene glycol or salts, as desired and (b) stock solutions of the screening solutions are kept available so that crystallization hits can be quickly reproduced using identical solutions to those contained within the screens. In addition to these, the Classic and Classic II screens from Qiagen (www.qiagen.com) and the Index screen from Hampton Research (www.hamptonresearch.com) are used. The precipitation behavior of the sample is first checked using solutions of polyethylene glycol (PEG) with an average molecular weight of 8000 and saturated ammonium sulfate to determine if the protein concentration is within a suitable range; it should neither be too concentrated nor too dilute. Crystals from the initial screens are checked for diffraction properties using a Rigaku 007 rotating anode source with a HTC imaging plate detector. Initial crystallization hits are reproduced and optimized if necessary in 24-well Limbro plates. All of the usual strategies for crystal optimization are employed, including varying precipitant concentrations, inclusion of additives, varying of drop ratios and temperatures and the use of both micro- and macro-seeding methods.

For solving structures of proteins having low sequence similarity to structures in the PDB, single wavelength anomalous diffraction using L-selenomethionine (SeMet)-substituted proteins in the method of choice (Hendrickson et al. 1990). To accomplish labeling of proteins, an *E. coli metA* auxotroph is transformed with the plasmid, and grown in semi-defined LeMaster media containing 25 mg L^{-1} L-SeMet. Frequently, the expression, purification and crystallization behavior of the SeMet-substituted protein are similar to that of the unlabelled protein, although some changes (lower solubility) can result if the methionine content of the protein is unusually high. X-ray diffraction data are collected for these crystals at X-ray beamlines within the Protein Crystallography X-ray Resource (PXRR; http://www.px.nsls.bnl.gov/) at the National Synchrotron Light Source, Brookhaven National Laboratory, or at the SGX-CAT beamline at the Advanced Photon Source (http://www.sgxpharma.com/pipeline/beamline/beamline.php).

The most common methods for flash-cooling of crystals for X-ray data collection are either (a) addition of glycerol at a concentration of 20–30% to the reservoir solution or (b) increasing the concentration of precipitant contained within the reservoir such that no ice forms upon flash cooling to 100K. Diffraction data resulting from these experiments are integrated and scaled using HKL2000(Otwinowski and Minor 1997). Our preferred method for structure solution is the collection of the Se anomalous signal at the Se peak, the Se-SAD method. The location of Se sites and the consequent calculation of phases are then determined using any of the standard packages, including SOLVE (Terwilliger and Berendzen 1999), SHELX (Sheldrick 2008) using the HKL2MAP interface (Pape and Schneider 2004) or autoSHARP (Vonrhein et al. 2007). Density modification is performed either using the programs RESOLVE (Terwilliger 2000) or DM (Cowtan and Main 1998). If diffraction data extend to 2.5 Å resolution or better then model building is performed using ARP/wARP (Cohen et al. 2004), otherwise it is done using RESOLVE (Terwilliger 2003). Further alternating cycles of refinement using Refmac (Murshudov et al. 1999) and fitting using COOT (Emsley and Cowtan 2004) result in the final structure.

4.3.5 Structure Determination by NMR Spectroscopy

For structural studies, NMR spectroscopy offers the advantage that it can be used both as a tool for protein characterization and as a method of structure determination. We typically acquire NMR spectra of all proteins whether they are strictly speaking NMR structure targets or not. While ^{15}N-correlation spectra (HSQC) are the most useful (requiring ^{15}N-labeled protein), even one-dimensional spectra of unlabeled proteins are useful for assessment of the homogeneity, stability and solubility of protein samples. The amide signal of the indole ring of tryptophan has a characteristic chemical shift downfield of most other protein signals. A comparison of the observed and expected number of indole signals is a quick check on the solution homogeneity of the sample. In addition, the line width (sharpness) of the NMR signals gives a qualitative assessment of the protein rotational diffusion rate and hence it's aggregation state. Proteins that aggregate slightly show broader, less well-resolved spectra than highly soluble, monomeric proteins. Conversely, unfolded proteins show overly narrow peaks with a characteristic pattern of chemical shifts. Time-dependent changes in the behavior of the protein are also easily detected by NMR.

A distinct advantage of NMR for protein characterization is the ability to obtain residue-specific information. This generally requires ^{15}N-labeled or ^{15}N,^{13}C doubly-labeled protein obtained from bacteria grown in minimal media using $(^{15}NH_4)_2SO_4$ and ^{13}C-glucose. ^{15}N labeling is inexpensive and allows very rapid acquisition (< 15 minutes for a typical protein at 10 mg/ml) of two-dimensional ^{15}N-HSQC spectra in which each amino acid residue gives rise to a distinct signal (peak) in the spectrum. Proline residues are an exception and do not have a signal, while amino acid side chain amides such as tryptophan, glutamine or asparagine give rise to extra peaks. The assignment of each peak to the corresponding amino acid residue in the protein is obtained via triple-resonance NMR experiments using ^{15}N,^{13}C-labeled protein. At the BSGI, NMR spectra are typically collected at 303K on a Bruker Avance DRX600 MHz spectrometer with a triple resonance cryoprobe. Data are processed using NMRPipe (Delaglio et al. 1995), GIFA (Pons et al. 1996) or XWIN-NMR (Bruker Biospin). Data analysis is carried out using NmrDraw (Delaglio et al. 1995) or XEASY (Bartels et al. 1995).

An advantage of initial screening using ^{15}N-HSQC spectra is that flexible regions of proteins are easily identified as they give rise to strong, easily detected signals. This contrasts with X-ray crystallography where the flexible portions are invisible and generally impede protein crystallization. Identification of the disordered regions is very useful for the optimization of protein crystallization since it allows new constructs expressing different regions of the protein to be made and the disordered regions removed. In functional studies, ^{15}N-HSQC spectra are also used for detecting intermolecular interactions either between proteins or, most frequently, between an ^{15}N-labeled protein and a low molecular weight compound such as a peptide. Mapping the peaks that shift upon the binding of a ligand allows identification of the ligand binding site and plotting the size of the shift as a function of the free ligand concentration gives the binding affinity (K_d). NMR is one of the few techniques that can measure millimolar binding affinities, and somewhat surprisingly, the analysis is easier and accuracy better for weak interactions. For modeling complexes using NMR chemical shift changes, we use the program HADDOCK (Dominguez et al. 2003).

NMR structural information comes in several forms and, in contrast to crystallography, is obtained piecewise. This has the advantage that it can be used to generate low resolution models as stepping stones toward higher resolution structures but the diverse nature of the information ultimately makes NMR structure determination more labor intensive and slower than X-ray crystallography. The workhorse of NMR structure determination is NOE restraints which reflect short (<5 Å) distances between hydrogen atoms. These are obtained from ^{15}N and ^{13}C-edited multidimensional NOESY experiments for proteins and 2D homonuclear NOESY experiments for peptides and small molecules. The assignment of NOEs is the most time-consuming part of NMR structure determination and several semi-automated and fully-automated programs have been developed by research groups in the field. Most NMR structures at the BSGI have used either ARIA (Nilges et al. 1997) or CYANA 2.0 (Guntert et al. 1997) to complete NOE assignments based on a preliminary structural model.

Other structural information comes in the form of values for ϕ and ψ torsion angles that are determined from NMR coupling constants and chemical shifts (Cornilescu et al. 1999). Hydrogen bond constraints based on deuterium exchange rates are often added at the end of the structure determination once NOEs have identified hydrogen bond acceptors. The process of NMR structure determination is highly iterative and involves several rounds of refinement of the structural restraints and structural models.

An important innovation in NMR structure determination was the development of residual dipolar couplings (RDC) as a routine technique (Bax and Grishaev 2005). RDCs give information about the relative orientation of internuclear vectors (typically backbone ^{1}H-^{15}N amides) and complement the local, geometric information of NOEs, torsion angles and hydrogen bonds (Trempe and Gehring 2003). For measuring RDCs, we try different media (typically Pf1 bacteriophage, alkyl poly(ethylene glycol)/n-alcohol, or strained polyacrylamide) until a satisfactory set of spectra are obtained. The resulting RDCs are analyzed using the MODULE software (Dosset et al. 2001) and incorporated into CNS (Brunger et al. 1998) for final structure calculations using the full complement of NOE, torsion angle, and other NMR information. Chemical shift data determined at the BSGI are deposited at the BioMagResBank (http://www.bmrb.wisc.edu).

4.4 Structures of Selected Targets

As discussed earlier, one of the strategies for target selection was to choose sequences for relatively large protein families for which structural information was currently unavailable. Using the methods described herein, over 80 protein structures, mostly from *E. coli*, have been determined within the project using either X-ray crystallography or NMR spectroscopy. In the following, we will provide examples of some of the interesting structural features that resulted from these studies, as well describe important functional insights that were obtained.

4.4.1 Proteins of Previously Known Function

4.4.1.1 Enzymes of the Histidine Biosynthetic Pathway

Unlike eukaryotes, bacteria such as *E. coli* contain all of the enzymes necessary to synthesize histidine, in a total of ten enzymatic steps, beginning with condensation of ATP with 5-phosphoribosyl 1-pyrophosphate (Alifano et al. 1996). At the beginning of our project, no structural information was available for most of these enzymes, and due to their broad distribution in other organisms, several were selected for structural analysis. We have determined the crystal structures of *E. coli* histidinol phosphate phosphatase, HisB (Rangarajan et al. 2006a), histidinol phosphate aminotransferase, HisC (Sivaraman et al. 2001), histidinol dehydrogenase, HisD (Barbosa et al. 2002) as well as the HisI paralog from *Methanobacterium thermoautotrophicum* (Sivaraman et al. 2005).

E. coli and *Salmonella typhimurium* histidinol phosphatase is a bi-functional enzyme which catalyzes the sixth and eighth steps in histidine biosynthesis (Loper 1961). The imidazole glycerol phosphate dehydratase activity (E.C. 4.2.1.19) and the histidinol phosphate phosphatase activity (E.C. 3.1.3.15) are carried out independently by separate domains of the enzyme. The N-terminal domain of *E. coli* HisB, residues 1–167, contains the histidinol phosphate phosphatase activity. This domain has been classified as a member of the haloacid dehalogenase-like hydrolase (HAD) family of enzymes, based on the presence of four conserved aspartate residues (Thaller et al. 1998). This family of enzymes utilizes a single metal ion in catalysis and proceeds through a phosphoaspartate intermediate (Allen and Dunaway-Mariano 2004). The crystal structure of *E. coli* N HisB was determined as complexes with several combinations of ligands, including Mg^{2+}, Mg^{2+}/L-histidinol and Ca^{2+}, a known inhibitor (Houston and Graham 1974), alone and in the presence of the phosphoaspartate intermediate. All of these structures were determined and refined at resolution ranges between 1.7–2.2 Å (Rangarajan et al. 2006a). The crystal structure of N-HisB indicates it is a dimer, consistent with solution data, with each monomer adopting a Rossmann fold. Different crystal structures were found to contain different combinations of metal ions. *E. coli* N-HisB contains a structural Zn^{2+} site, as revealed both in the crystal structure and by EXAFS measurements. Unlike other HAD enzymes, two distinct metal-binding sites, a primary site with high affinity and a secondary site having lower affinity, were identified in the active site region. This conclusion is based on both X-ray crystallographic evidence as well as data from isothermal titration calorimetry (ITC). The proposed catalytic mechanism involves nucleophilic attack of the substrate by Asp10, resulting in formation of a phosphoaspartate intermediate and inversion of configuration about the phosphorous atom. The metal at the high-affinity site, Mg^{2+} under physiological conditions, would play two roles, to neutralize the negative charge of the substrate phosphoryl moiety, as well as to stabilize the phosphor-aspartate intermediate.

The transfer of an amino group to histidinol phosphate, the seventh step in histidine biosynthesis, is performed by L-histidinol phosphate aminotransferase (Hsu et al. 1989). Like many aminotransferases, HisC is a PLP-dependent enzyme, with the protype member of this family represented by aspartate aminotransferase. The crystal structure revealed a dimeric α/β protein, with each monomer consisting of two domains; a larger PLP-binding domain, and a smaller domain (Sivaraman et al. 2001). The N-terminal region of the enzyme is associated with dimer formation. We successfully trapped the external aldimine form of the enzyme, forming a covalent complex with PLP, as well as the cognate complexes with PMP, and with both PLP and L-histidinol phosphate. This last complex represents structurally the *gem*-diamine intermediate formed during conversion of the internal aldimine to the external aldimine forms of the enzyme during the catalytic cycle. Several residues interacting with PLP, including Tyr55, Asn157, Asp184, Tyr187, Ser213, Lys214 and Arg222 were identified as conserved in other, related aminotransferases. The residue Tyr10 was identified as important for interaction with the imidazole ring of the histidinol phosphate substrate *via* a hydrogen bond. The structure of the *gem*-diamine intermediate, in particular, is unusual, and we suggest was trapped through non-productive binding of L-histidinol phosphate at the HisC active site, preventing conversion to the external aldimine (Sivaraman et al. 2001).

The final two steps in histidine biosynthesis, the NAD^+-dependent oxidations of L-histidinol to L-histidinaldehyde and then to L-histidine, are performed by the bifunctional enzyme L-histidinol dehydrogenase, HisD. The catalytic mechanism for HisD involves retention of the histidinaldehyde (histidinal) intermediate at the HisD active site (Adams 1955). HisD had previously been characterized biochemically as a Zn-metallo-enzyme, in which the Zn^{2+} ion performed a structural or catalytic function (Grubmeyer et al. 1989). Overall, 2 molecules of NAD^+ are reduced to NADH with concomitant transfer of four electrons to the substrate.

The HisD molecule was found to be dimeric, with each subunit consisting of two large and two small domains, with the two large domains adopting similar, incomplete Rossmann folds, arguing in favor of an ancient gene duplication event (Barbosa et al. 2002; Fig. 4.6a). Domains 3 and 4 together form a tail-like structure which forms domain-swapping interactions with the other subunit of the dimer. An octahedrally-coordinated Zn^{2+} site was observed, coordinated by four side chains of the enzyme as well as the ND1 and N atoms of L-histidinol (Fig. 4.6b). The observation of the amino moiety of L-histidinol coordinating Zn^{2+} was unexpected based on previous NMR studies (Kanaori et al. 1996). The NAD^+ molecule is bound within a cleft formed at the C-terminal ends of the β-sheet within domain 1, making contacts with only one subunit of the dimer. Mechanistically, the overall reaction is expected to proceed *via* four proton abstraction steps (Teng and Grubmeyer 1999), which we propose are performed by Glu326 (step 2) and His327 (steps 1, 3 and 4).

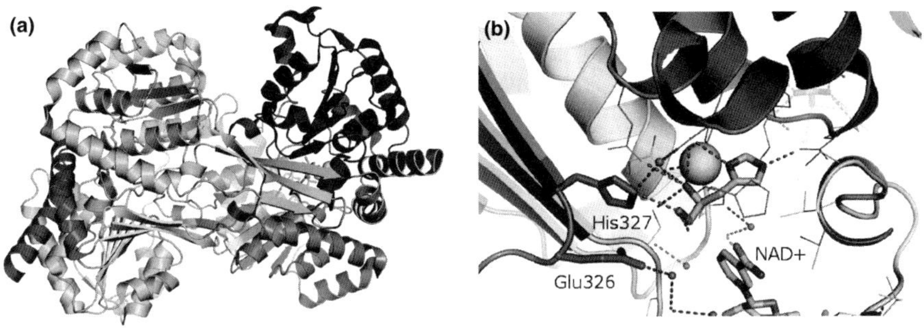

Fig. 4.6 Crystal structure of *E. coli* L-histidinol dehydrogenase (HisD; PDB 1KAE). (**a**) Structure of the HisD dimer, with one subunit shown in light gray and the second subunit colored in dark gray. The swapping of domains 3 and 4 with the other subunit of the dimer occurs within the dimer. (**b**) Interactions between HisD, the cofactor NAD+ and the L-histidinol product. The Zn^{2+} ion is shown as a sphere as well as the key active site residues Glu326 and His327. This and subsequent images of protein structures were prepared using the program PyMol (http://sourceforge.pymol.net)

4.4.1.2 Other Metabolic Enzymes

Several of the structures determined as part of this project include metabolic enzymes. Two examples of these are *N*-succinylarginine dihydrolase, AstB, of the arginine succinyltransferase (AST) pathway (Tocilj et al. 2005) and the exopolyphosphatase, PpX, which degrades polyphosphate to inorganic phosphate (Rangarajan et al. 2006b).

In *E. coli*, the AST pathway is the dominant route for arginine catabolism, leading to glutamate, ammonia and formation of CoA and succinate from succinyl-CoA (Schneider et al. 1998). This pathway consists of five enzymes with AstB catalyzing the second step, the conversion of *N*-succinylarginine to *N*-succinylornithine with the concomitant release of ammonia and carbon dioxide. The structure of AstB revealed a dimeric enzyme, with each monomer consisting of repeating β-β-α-β units generating a propeller structure having pseudo-5-fold symmetry. Characterization of the site-specific mutant enzyme, Cys365Ser, resulted in a reduction of specific activity of ~2 orders of magnitude, allowing the co-crystal structure with the *N*-succinylarginine substrate to be determined (Tocilj et al. 2005). The substrate-binding site is within a ~15 Å deep tunnel, binding such that the guanidinium group is at the bottom of the tunnel and the succinate carboxylate closest to the surface. The substrate is anchored through extensive H-bonding interactions involving a number of highly sequence-conserved residues. The structure of the complex supports the prediction of Cys365, His248 and Glu174 as being the key catalytic residues (Shirai and Mizuguchi 2003), with Cys365 being the key nucleophile, attacking the guanidinium group of the substrate.

Polyphosphate, a high-energy, linear polymer made up of hundreds of phosphate units, is required for stationary-phase survival of *E. coli* (Crooke et al. 1994) and plays a role in protecting cells against stress resulting from heat or oxidation. Polyphosphate is degraded by both endo- and exo-polyphosphatases, with

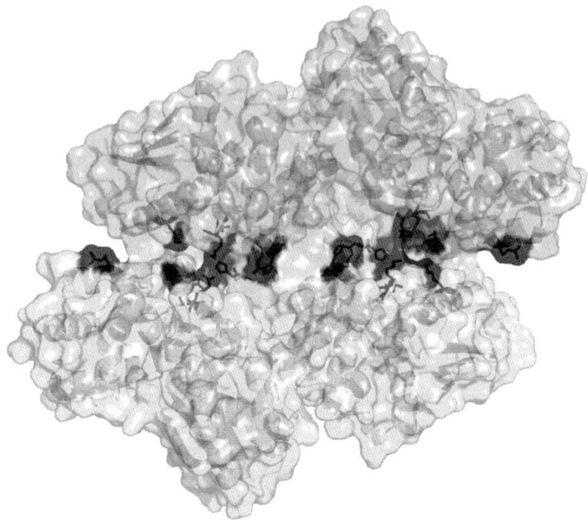

Fig. 4.7 Crystal structure of the exopolyphosphatase PPX from *E. coli* (PDB 2FL0) showing the inter-subunit cleft, the putative polyphosphate-binding site, containing a number of basic residues

the exopolyphosphatase from *E. coli* being a dimer made up of 58 kDa subunits (Bolesch and Keasling 2000). Each PpX monomer was found to consist of four structural domains, with domains 1 and 2 being structurally similar to one another (Rangarajan et al. 2006b). The dimer is formed through head-to-tail association of the two monomers, resulting in a long, deep S-layered cleft at the subunit interface. Many basic and polar residues are located within this cleft, which is the most likely location for polyphosphate binding (Fig. 4.7). A number of conserved residues were identified and found to cluster near the interface between domains 1 and 2, which is near one end of the cleft and represents the putative PpX active site. Active site features include a glycine-rich phosphate-binding loop (P-loop, Gly145-Ser148) for anchoring phosphoryl groups or functioning as an oxy-anion hole, as well as Asp143 and Glu150, which most likely function to coordinate a catalytically-important Mg^{2+} ion.

4.4.1.3 Pseudouridine Synthases

Pseudouridine (5-β-D-ribofuranosyluracil, ψ), is amongst the most abundant modifications found in RNA molecules, and has been characterized extensively in structural RNAs including rRNA, tRNA and small nuclear and nucleolar sn(o)RNA (Charette and Gray 2000). This modified base plays a variety of roles within RNA, including assembly of the catalytic RNA active site required for pre-mRNA splicing (Lin and Kielkopf 2008), pH-induced structural changes in 23S rRNA (Abeysirigunawardena and Chow 2008), ribosome-mediated translational termination (Ejby et al. 2007) and the structural stability of some tRNAs (Cabello-Villegas and Nikonowicz 2005) and 23S rRNA.

In bacteria, site-specific formation of pseudouridine by isomerization of uridine is catalyzed by ψ-synthases (Ferre-D'Amare 2003, Ofengand 2002). Based on amino acid sequences, these enzymes are divided into 5 subgroups, with little overall sequence similarity between the groups (Kaya and Ofengand 2003, Koonin 1996). Pseudouridine synthases have been extensively characterized from *E. coli*, where a total of 11 distinct enzymes have been identified (Del Campo et al. 2001, Kaya and Ofengand 2003). While we now have some insight into key catalytic residues (Del Campo et al. 2001, Hamilton et al. 2005, Phannachet et al. 2005) and the likely chemical mechanism of catalysis (Gu et al. 1999, Hamilton et al. 2006), much remains to be learned about the exquisite specificity of these remarkable enzymes for their target uridine residues.

As part of the BSGI project, the crystal structures of three *E. coli* ψ–synthases have been determined, RsuA and its complexes with uracil and UMP (Sivaraman et al. 2002), RluD (Sivaraman et al. 2004) and RluF (Sunita et al. 2006). All three enzymes possess a domain organization, with RsuA and RluD both having two domains and RluF three domains. The crystal structure of RsuA revealed a small, N-terminal $\alpha_3\beta_4$ domain with structural similarity to ribosomal protein S4, while in RluD, this domain had to be removed in order to obtain diffraction-quality crystals. All three enzymes contain a catalytic α/β domain made up from a central, mixed β-sheet that contains a cleft with the key catalytic Asp residue, Asp102 in RsuA, Asp139 in RluD and Asp107 in RluF (Fig. 4.8a,b). This central β-sheet is the most structurally-conserved feature of this enzyme family. All three enzymes possess a cleft within the catalytic domain that represents the predicted RNA-binding site, with many structurally-equivalent residues from bacterial Ψ-synthase structures

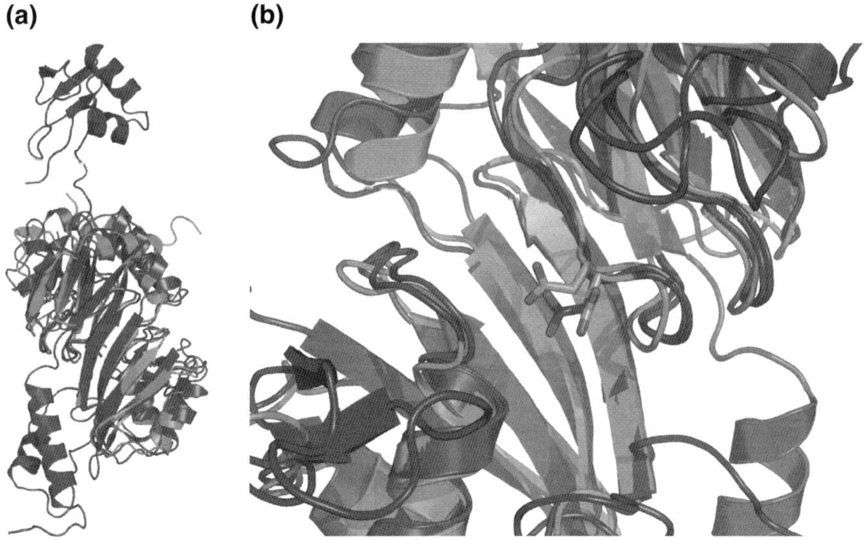

Fig. 4.8 (**a**) Superposition of the structures of the *E. coli* pseudouridine synthases RsuA (PDB 1KSK), RluD (PDB 1PRZ) and RluF (PDB 2GML) showing the common fold of the catalytic domain. (**b**) Structural conservation of the key catalytic Asp residue in each structure, Asp102 in RsuA, Asp139 in RluD and Asp42 in RluF, are shown in stick representation

mapping to this region. In RluF, the domain organization appears more complex, with the enzyme having N-terminal and catalytic domains as well as a C-terminal domain. It is unclear the exact role of these auxiliary domains, although they are evidently flexible with respect to the catalytic domain and a role in RNA recognition is possible (Matte et al. 2005, Sivaraman et al. 2002).

4.4.2 Hypothetical ("Y") Proteins

4.4.2.1 Heme Oxygenase – ChuS

Pathogenic bacteria, including *E. coli*, make use of a variety of systems to sequester iron, an essential nutrient, from their environment. One of these systems involves sequestering and uptake of heme. Following transport into the cell, a series of enzymatic reactions are required to breakdown the heme and release free iron within the bacterium. ChuS, a protein contained within the heme-degradation operon in *E. coli*, had not been previously structurally or functionally characterized.

The crystal structure of ChuS revealed a protein consisting of two domains joined by a flexible linker, with each domain having a central, nine-stranded β-sheet flanked by α-helices (Suits et al. 2005). These domains bear high structural similarity to one another, and can be superposed with a root mean squares deviation of 2.1 Å, indicative of a structural duplication and possibly duplication of function. Several side-chains of the two domains are also found to superpose in this comparison. Interestingly, spectral analysis showed both the full-length protein as well as independent N- and C-terminal domains were found to possess heme-binding activity. Enzymatic analysis showed that ChuS and its isolated domains each have heme oxygenase activity, based on their ability to form biliverdin and release CO_2 from heme.

The co-crystal structure of ChuS with heme revealed a different heme binding mode than for other heme oxygenases, with the C-terminal domain of the protein playing an important role (Suits et al. 2006). A key residue involved in axial coordination of heme was identified to be His193. This residue was also shown by directed mutagenesis to asparagine to be the key player in the heme-degrading activity of ChuS.

4.4.2.2 Shikimate/Quinate Dehydrogenases

Aromatic amino acids and other aromatic metabolites are synthesized in *E. coli* using the shikimate pathway, starting with phosphoenolpyruvate and D-erythrose-4-phosphate as precursors (Herrmann and Weaver 1999). As this pathway is absent in mammals, it represents a potential anti-microbial target (Coggins et al. 2003), and has been targeted in plants for the development of the herbicide glyphosate. This pathway consists of seven enzymatic steps, resulting in the formation of chorismate, with the fourth step, the NADP-dependent reduction of 3-dehydroshiimate to shikimate catalyzed by shikimate dehydrogenase, AroE (Anton and Coggins 1988). A sequence-related enzyme, YdiB, was characterized and was shown to have

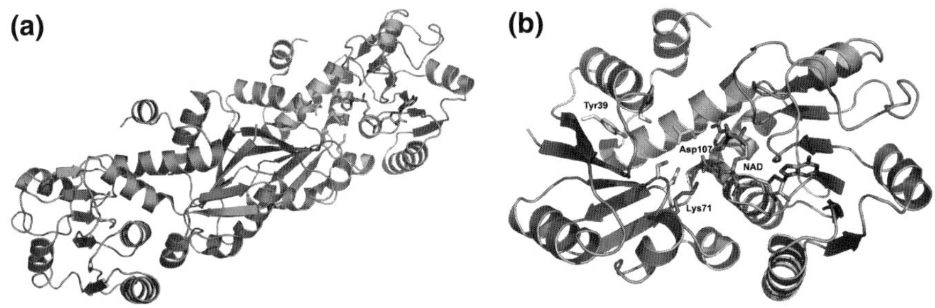

Fig. 4.9 (**a**) Structure of the *E. coli* YdiB dimer (PDB 1O9B), showing the NAD cofactor bound at each active site in stick representation. (**b**) The YdiB active site, with residues selected for site-directed mutagenesis and subsequent kinetic analysis shown in stick representation. Of these residues, Lys71 and Asp107 were found to be most important for substrate binding

quinate/shikimate dehydrogenase activity that, unlike AroE, could use either NADP or NAD as cofactor (Michel et al. 2003). Both AroE and YdiB display a similar fold, with two α/β domains separated by a cleft suitable for binding the cofactor. The C-terminal, NAD(P)-binding domain adopts a nearly canonical Rossmann fold. Interestingly, despite their relatedness in sequence, AroE is a monomeric enzyme while YdiB has been shown to be a dimer (Fig. 4.9a). While both AroE and YdiB recognize the nicotinamide and pyrophosphate moieties of the cofactors in a similar manner, there are important structural differences in interaction with the adenosine portion. The residues Arg150 and Arg154 of AroE were found to form an "electrostatic clamp" ideally suited for binding the phosphate moiety found in NADP but absent from NAD. Kinetic analysis of active site mutants of YdiB revealed roles for Lys71 and Asp107 in substrate binding, with no specific role for a residue participating in general acid-base catalysis (Fig. 4.9b; Lidner et al. 2005). This later study allowed us to differentiate between two previously proposed models of substrate binding to these enzymes (Benach et al. 2003, Michel et al. 2003), with the data in support of the model proposed by Michel et al. (2003), in which residues making interactions with 3-dehydroshikimate are expected to include Lys71, Asp107, Gln262, Tyr234, Ser20 and Ser22.

4.4.2.3 Protein Structures Determined by NMR

A significant number of structures of smaller proteins of little or previously-unknown function have been determined using NMR methods as part of the BSGI project. Examples include the solution structures of *E. coli* YbeD, a conserved protein which shows structural similarity to regulatory domain of 3-phosphoglycerate dehydrogenase (Kozlov et al. 2004), the oxidative stress-related protein YggX (Osborne et al. 2005), CsrA, a member of a new class of RNA-binding regulatory proteins (Gutierrez et al. 2005), the Rho-specific transcription factor YaeO (Gutierrez et al. 2007) and YcgL, a conserved protein representing the DUF709 sequence family (Minailiuc et al. 2007).

The *ybeD* gene is located between *dacA* and genes of the *lip* operon, required for lipoic acid biosynthesis. Lipoic acid is found as a prosthetic group used by a number of metabolic enzymes, including pyruvate and 2-oxoglutarate dehydrogenases and branched-chain keto acid dehydrogenases. The structure of YbeD revealed a β-α-β-β-α-β fold, with the two α–helices located on one side of a four-stranded, antiparallel β-sheet. A patch of conserved hydrophobic residues are found on the β-sheet surface, suggesting a role for this region in protein-protein interactions. Most intriguingly, YbeD shows high structural similarity to the regulatory domain of 3-phosphoglycerate dehydrogenase, possibly indicating an allosteric role in regulation of lipoic acid biosynthesis (Kozlov et al. 2004).

YggX has been characterized as playing a role in diminishing the effects of oxidative damage, partly through protection of DNA from iron-mediated oxidative damage (Gralnick and Downs 2003). In *E. coli*, the *yggX* gene is part of the SoxRS regulon used as an antioxidant defense system. The structure of YggX reveals a single domain protein containing two antiparallel β-sheets and three α-helices. While the YggX sequence is found in a number of other gram-negative bacteria, only a single structurally-related protein, of unknown function, could be found in *Pseudomonas aeruginosa*. Unexpectedly, YggX was found unable to bind iron salts *in vitro*, suggesting that other cofactors may be involved in mediating iron-dependent oxidative damage (Osborne et al. 2005).

Carbon storage regulator A (CsrA) is a founding member of a family of bacterial regulatory proteins that function by controlling the level of mRNA translation. In *E. coli*, CsrA is responsible for the repression of a variety of stationary-phase genes and carbon metabolism (Babitzke and Romeo 2007). The protein binds specific mRNAs to repress the initiation of protein synthesis. Derepression occurs by an unusual mechanism in which non-coding, regulatory RNAs bind to CsrA and displace it from the mRNA. The solution structure of CsrA was the first in the Csr/Rsm family and revealed a novel fold consisting of a symmetric dimer composed of five beta-strands and a short alpha-helix in each subunit. NMR titration experiments identified elements involved in RNA binding and the mechanism for the recognition of mRNAs regulated by CsrA (Gutierrez et al. 2005).

Termination of transcription in bacteria is either dependent on a hexameric helicase, Rho, or can occur independently of the helicase. The protein YaeO binds tightly to Rho, inhibiting Rho-dependent transcriptional termination (Pichoff et al. 1998). The structure of YaeO revealed an N-terminal α-helix and a seven-stranded β-sandwich. NMR titration experiments designed to probe the interaction between Rho and YaeO revealed that the binding site on Rho for YaeO overlapped with its binding site for RNA, revealing that YaeO is a competitive inhibitor with respect to RNA. These results were supported by gel-shift experiments, revealing the loss of nucleic acid-binding activity by Rho upon binding YaeO. Together, these data and computational molecular docking resulted in a model of the Rho-YaeO complex (Gutierrez et al. 2007).

YcgL is a conserved protein of unknown function, representing the DUF709 sequence family. The NMR structure of this 108-residue protein was determined, revealing a protein with the topology $\beta 1$-$\beta 2$-$\alpha 1$-$\beta 3$-$\alpha 2$-$\beta 4$, forming a three-layered

$\alpha/\beta/\alpha$ sandwich (Minailiuc et al. 2007). The structure of YcgL is differs from those of proteins available within the PDB, indicating it represents a novel fold.

4.5 Summary and Conclusions

Structural proteomics has and continues to contribute new structural information for *E. coli* in order to further our understanding of the proteome of this organism. This structural information in turn will be exploited by biologists in many different ways, ranging from the relationships between sequence and structure and comparative protein modeling to details of enzyme function and of protein-protein recognition. Methods continue to be refined, mainly with the goal of improving the success in protein sample preparation and protein crystallization. The next challenges involve tackling the membrane protein complement and protein-protein complexes of *E. coli*.

Acknowledgments The authors wish to recognize the critical contributions made by many members of their laboratories, past and present, who have contributed to the work described here. X-ray data collection was performed at X-ray beamlines at the National Synchrotron Light Source, Brookhaven National Laboratory and at the Advanced Photon Source, Argonne National Laboratory. This work was funded by an operating grant from the Canadian Institutes of Health Research. This is National Research Council of Canada publication No. 49581.

References

Abergel, C., Coutard, B., Byrne, D., Chenivesse, S., Claude, J.B., Deregnaucourt, C., Fricaux, T., Gianesini-Boutreux, C., Jeudy, S., Lebrun, R., Maza, C., Notredame, C., Poirot, O., Suhre, K., Varagnol, M. and Claverie, J.M. (2003). Structural genomics of highly conserved microbial genes of unknown function in search of new antibacterial targets. *J. Struct. Funct. Genomics* **4**, 141–157.

Abeysirigunawardena, S.C. and Chow, C.S. (2008). pH-dependent structural changes of helix 69 from *Escherichia coli* 23S ribosomal RNA. *RNA* **14**, 782–792.

Adams, E. (1955). L-histidinal, a biosynthetic precursor of histidine. *J. Biol. Chem.* **217**, 325–344.

Adams, M.A., Suits, M.D., Zheng, J. and Jia, Z. (2007). Piecing together the structure-function puzzle: experiences in structure-based functional annotation of hypothetical proteins. *Proteomics* **7**, 2920–2932.

Albeck, S., Burstein, Y., Dym, O., Jacobovitch, Y., Levi, N., Meged, R., Michael, Y., Peleg, Y., Prilusky, J., Schreiber, G., Silman, I., Unger, T. and Sussman, J.L. (2005). Three-dimensional structure determination of proteins related to human health in their functional context at the Israel structural proteomics center (ISPC). *Acta Cryst.* **D61**, 1364–1372.

Alifano, P., Fani, R., Lio, P., Lazcano, A., Bazzicalupo, M., Carlomagno, M.S. and Bruni, C.B. (1996). Histidine biosynthetic pathway and genes: structure, regulation and evolution. *Microbiol. Rev.* **60**, 44–69.

Allen, K.N. and Dunaway-Mariano, D. (2004). Phosphoryl group transfer: evolution of a catalytic scaffold. *Trends Biochem. Sci.* **29**, 495–503.

Anton, I.A. and Coggins, J.R. (1988). Sequencing and overexpression of the *Escherichia coli aroE* gene encoding shikimate dehydrogenase. *Biochem. J.* **249**, 319–326.

Arifuzzaman, M., Maeda, M., Itoh, A., Nishikata, K., Takita, C., Saito, R., Ara, T., Nakahigashi, K., Huang, H.C., Hirai, A., Tsuzuki, K., Nakamura, S., Altaf-Ul-Amin, M., Oshima, T., Baba, T., Yamamoto, N., Kawamura, T., Ioka-Nakamichi, T., Kitagawa, M., Tomita, M., Kanaya, S., Wada, C. and Mori, H. (2006). Large-scale identification of protein-protein interaction of *Escherichia coli* K-12. *Genome Res.* **16**, 686–691.

Baba, T., Huan, H.C., Datsenko, K., Wanner, B.L. and Mori, H. (2008). The applications of systematic in-frame, single-gene knockout mutant collection of *Escherichia coli* K-12. *Methods Mol. Biol.* **416**, 183–194.

Babitzke, P. and Romeo, T. (2007). CsrB sRNA family: sequestration of RNA-binding regulatory proteins. *Curr. Opin. Microbiol.* **10**(2):156–163.

Barbosa, J.A., Sivaraman, J., Li, Y., larocque, R., Matte, A., Schrag, J.D. and Cygler, M. (2002). Mechanism of action and NAD+-binding mode revealed by the crystal structure of L-histidinol dehydrogenase. *Proc. Natl. Acad. Sci. USA* **99**, 1859–1864.

Bartels, C., Xia, T.-H., Billeter, M., Guntert, P. and Wuthrich, K. (1995). The program XEASY for computer-supported NMR spectral analysis of biological macromolecules. *J. Biomol. NMR* **6**, 1–10.

Bax, A. and Grishaev, A. (2005). Weak alignment NMR: a hawk-eyed view of biomolecular structure. *Curr. Opin. Struct. Biol.* **15**(5), 563–570.

Benach, J., Lee, I., Edstrom, W., Kuzin, A.P., Chiang, Y., Acton, T.B., Montelione, G.T. and Hunt, J.F. (2003). The 2.3 Å crystal structure of the shikimate 5-dehydrogenase orthologue YdiB from *Escherichia coli* suggests a novel catalytic environment for an NAD-dependent dehydrogenase. *J. Biol. Chem.* **278**, 19176–19182.

Bendtsen, J.D., Nielsen, H., von Heijne, G. and Brunak, S. (2004). Improved prediction of signal peptides: SignalP 3.0. *J. Mol. Biol.* **340**, 783–795.

Berman, H.M., Westbrook, J., Feng, Z., Gilliland, G., Bhat, T.N., Weissig, H., Shindyalov, I.N. and Bourne, P.E. (2000). The Protein Data Bank. *Nucleic Acids Res.* **28**, 235–242.

Blattner, F.R., Plunkett, G. 3rd, Bloch, C.A., Perna, N.T., Burland, V., Riley, M., Collado-Vides, J., Glasner, J.D., Rode, C.K., Mayhew, G.F., Gregor, J., Davis, N.W., Kirkpatrick, H.A., Goeden, M.A., Rose, D.J., Mau, B. and Shao, Y. (1997). The complete genome sequence of *Escherichia coli* K-12. *Science* **277**, 1453–1474.

Bolesch, D.G. and Keasling, J.D. (2000). Polyphosphate binding and chain length recognition by *Escherichia coli* exopolyphosphatase. *J. Biol. Chem.* **275**, 33814–33819.

Brown, E.N. and Ramaswamy, S. (2007). Quality of protein crystal structures. *Acta Cryst.* **D63**, 941–950.

Brunger, A.T., Adams, P.D., Clore, G.M., DeLano, W.L., Gros, P., Grosse-Kunstleve, R.W., Jiang, J.S., Kuszewski, J., Nilges, M., Pannu, N.S. et al. (1998). Crystallography and NMR system: A new software suite for macromolecular structure determination. *Acta Cryst.* **D54**, 905–921.

Butland, G., Peregrin-Alvarez, J.M., Li, J., Yang, W., Yang, X., Canadien, V., Starostine, A., Richards, D., Beattie, B., Krogan, N., Davey, M., Parkinson, J., Greenblatt, J. and Emili, A. (2005). Interaction network containing conserved and essential protein complexes in *Escherichia coli*. *Nature* **433**, 531–537.

Cabello-Villegas, J. and Nikonowicz, E.P. (2005). Solution structure of psi32-modified anticodon stem-loop of *Escherichia coli* tRNAPhe. *Nucleic Acids Res.* **33**, 6961–6971.

Charette, M. and Gray, M.W. (2000). Pseudouridine in RNA: What, where, how and why. *IUBMB Life* **49**, 341–351.

Chen, L., Oughtred, R., Berman, H.M. and Westbrook, J. (2004). targetDB: a target registration database for structural genomics projects. *Bioinformatics* **20**, 2860–2862.

Chou, C.P. (2007). Engineering cell physiology to enhance recombinant protein production in *Escherichia coli*. *Appl. Microbiol. Biotechnol.* **76**, 521–532.

Coggins, J.R., Abell, C., Evans, L.B., Frederickson, M., Robinson, D.A., Roszak, A.W. and Lapthorn, A.P. (2003). Experiences with the shikimate-pathway enzymes as targets for rational drug design. *Biochem. Soc. Trans.* **31**, 548–552.

Cohen, S.X., Morris, R.J., Fernandez, F.J., Ben Jelloul, M., Kakaris, M., Partasarathy, V., Lamzin, V.S., Kleywegt, G.J. and Perrakis, A. (2004). Towards complete validated models in the next generation of ARP/wARP. *Acta Cryst.* **D60**, 2222–2229.

Collins, B., Stevens, R.C. and Page, R. (2005). Crystallization optimum solubility screening: using crystallization results to identify the optimal buffer for protein crystal formation. *Acta Cryst* **F61**, 1035–1038.

Cornilescu, G., Delaglio, F. and Bax, A. (1999). Protein backbone angle restraints from searching a database for chemical shift and sequence homology. *NMR* **13**, 289–302.

Cowtan, K. and Main, P. (1998). Miscellaneous algorithms for density modification. *Acta Cryst.* **D54**, 487–493.

Crooke, E., Akiyama, M., Rao, N.N. and Kornberg, A. (1994). Genetically altered levels of inorganic polyphosphate in *Escherichia coli. J. Biol. Chem.* **269**, 6290–6295.

Cygler, M., Hung, M., Wagner, J. and Matte, A. (2008). Bacterial genomics initiative: Overview of methods and technologies applied to the process of structure determination. In: *Methods in Molecular Biology, volume 426, Structural proteomics – High-throughput methods*, B. Kobe, M. Guss and T. Huber, eds. Humana Press, Australia, pp 537–559.

Delaglio, F., Grzesiek, S., Vuister, D.W., Zhu, G., Pfeifer, J. and Bax, A. (1995). NMR-Pipe: a multidimensional spectral processing system based on UNIX pipes. *J. Biomol. NMR* **6**, 277–293.

Del Campo, M., Kaya, Y. and Ofengand, J. (2001). Identification and site of action of the remaining four putative pseudouridine synthases in *Escherichia coli. RNA* **7**, 1603–1615.

Dominguez, C., Boelens, R. and Bonvin, A.M. (2003). HADDOCK: a protein-protein docking approach based on biochemical or biophysical information. *J. Am. Chem. Soc.* **125**, 1731–1737.

Dosset, P., Hus, J.C., Marion, D. and Blackledge, M. (2001). A novel interactive tool for rigid-body modeling of multi-domain macromolecules using residual dipolar couplings. *J. Biomol. NMR*, **20**, 223–231.

Durfee, T., nelson, R., Baldwin, S., Plunkett, G. 3rd, Burland, V., Mau, B., Petrosine, J.F., Qin, X., Muzny, D.M., Ayele, M., Gibbs, R.A., Csorge, B., Posfai, G., Weinstock, G.M. and Blattner, F.R. (2008). The complete genome sequence of *Escherichia coli* DH10B: insights into the biology of a laboratory workhorse. *J. Bacteriol.* **190**, 2597–2606.

Ejby, M., Sørensen, M.A. and Pedersen, S. (2007). Pseudouridylation of helix 69 of 23S rRNA is necessary for an effective translation termination. *Proc. Natl. Acad. Sci. USA* **104**, 19410–19415.

Emsley, P. and Cowtan, K. (2004). Coot: model-building tools for molecular graphics. *Acta Cryst.* **D60**, 2126–2132.

Ericsson, U.B., Hallberg, B.M., Detitta, G.T., Dekker, N. and Nordlund, P. (2006). Thermofluor-based high-throughput stability optimization of proteins for structural studies. *Anal. Biochem.* **357**, 289–298.

Feist, A.M., Henry, C.S., Reed, J.L., Krummenacker, M., Joyce, A.R., Karp, P.D., Broadbelt, L.J., Hatzimanikatis, V. and Palsson, B.Ø. (2007). A genome-scale metabolic reconstruction for *Escherichia coli* K-12 MG1655 that accounts for 1260 ORFs and thermodynamic information. *Mol. Syst. Biol.* **3**, 121.

Ferré-D'Amaré, A.R. (2003). RNA-modifying enzymes. *Curr. Op. Struct. Biol.* **13**, 49–55.

Finn, R.D., Tate, J., Mistry, J., Coggill, P.C., Sammut, S.J., Hotz, H.R., Ceric, G., Forslund, K., Eddy, S.R., Sonnhammer, E.L. and Bateman, A. (2008). The Pfam protein families database. *Nucleic Acids Res.* **36**, D281–D288.

Fogg, M.J., Alzari, P., Bahar, M., Bertini, I., Betton, J.M., Burmeister, W.P., Cambillau, C., Canard, B., Corrondo, M.A., Coll, M., Daenke, S., Dym, O., Egloff, M.P., Enquita, F.J., Geerlof, A., Haouz, A., Jones, T.A., Ma, Q., Manicka, S.N., Migliardi, M., Nordlund, P., Owens, R.J., Peleg, Y., Schneider, G., Schnell, R., Stuart, D.I., Tarbouriech, N., Unge, T., Wilkinson, A.J., Wilmanns, M., Wilson, K.S., Zimhony, O. and Grimes, J.M. (2006). Application of the use of high-throughput technologies to the determination of protein structures of bacterial and viral pathogens. *Acta Cryst.* **D62**, 1196–1207.

Gasteiger, E., Gattiker, A., Hoogland, C., Ivanyi, I., Appel, R.D. and Bairoch, A. (2003). ExPASY: the proteomics server for in-depth protein knowledge and analysis. *Nucleic Acids Res.* **31**, 3784–3788.

Ginalski, K. (2006). Comparative modeling for protein structure prediction. *Curr. Opin. Struct. Biol.* **16**, 172–177.

Goh, C.S., Lan, N., Echols, N., Douglas, S.M., Milburn, D., Bertone, P., Xiao, R., Ma, L.C., Zheng, D., Wunderlich, Z., Acton, T., Montelione, G.T. and Gerstein, M. (2003). SPINE 2: a system for collaborative structural proteomics within a federated database framework. *Nucleic Acids Res.* **31**, 2833–2838.

Gralnick, J. and Downs, D. (2003). The YggX protein of *Salmonella enterica* is involved in Fe(II) trafficking and minimizes the DNA damage caused by hydroxyl radicals: Residue CYS-7 is essential for YggX function. *J. Biol. Chem.* **278**, 20708–20715.

Grubmeyer, C., Skiadopoulos, M. and Senior, A.E. (1989). L-histidinol dehydrogenase, a Zn^{2+}-metalloenzyme. *Arch. Biochem. Biophys.* **272**, 311–317.

Gu, X., Liu, Y. and Santi, D.V. (1999). The mechanism of pseudouridine synthase I as deduced from its interaction with 5-fluorouracil-tRNA. *Proc. Natl. Acad. Sci. USA* **96**, 14270–14275.

Guntert, P., Mumenthaler, C. and Wuthrich, K. (1997). Torsion angle dynamics for NMR structure calculation with the new program DYANA. *J. Mol. Biol.* **273**, 283–298.

Gutiérrez. P., Li, Y., Osborne, M.J., Pomerantseva, E., Liu, Q., Gehring, K. (2005). Solution structure of the carbon storage regulator protein CsrA from *Escherichia coli*. *J. Bacteriol.* **187**(10), 3496–3501.

Gutiérrez, P., Kozlov, G., Gabrielli, L., Elias, D., Osborne, M.J., Gallouzi, I.E. and Gehring, K. (2007). Solution structure of YaeO, a Rho-specific inhibitor of transcription termination. *J. Biol. Chem.* **282**(32), 23348–23353.

Hamilton, C.S., Spedaliere, C.J., Ginter, J.M., Johnston, M.V. and Mueller, E.G. (2005). The roles of the essential Asp-48 and highly conserved His-43 elucidated by the pH dependence of the pseudouridine synthase TruB. *Arch. Biochem. Biophys.* **433**, 322–334.

Hamilton, C.S., Greco, T.M., Vizthum, C.A., Ginter, J.M., Johnston, M.V. and Mueller, E.G. (2006). Mechanistic investigation of the pseudouridine synthase RluA using RNA containing 5-fluorouridine. *Biochemistry* **45**, 12029–12038.

Hayashi, T., Makino, K., Ohnishi, M., Kurokawa, K., Ishii, K., Yokayama, K., Han, C.G., Ohtsubo, E., Nakayama, K., Murata, T., Tanaka, M., Tobe, T., Iida, T., Takami, H., Honda, T., Sasakawa, C., Ogasawara, N., Yasunaga, T., Kuhara, S., Shiba, T., Hattori, M. and Shinagawa, H. (2001). Complete genome sequence of enterohemorrhagic *Escherichia coli* O157:H7 and genomic comparison with a laboratory strain K-12. *DNA Res.* **8**, 11–22.

Hendrickson, W.A., Horton, J.R. and LeMaster, D.M. (1990). Selenomethionyl proteins produced for analysis by multiwavelength anomalous diffraction (MAD): a vehicle for direct determination of three-dimensional structure. *EMBO J.* **9**, 1665–1672.

Herrgard, M.J., Fong, S.S. and Palsson, B.Ø. (2006). Identification of genome-scale metabolic network models using experimentally measured flux profiles. *PLoS Comput. Biol.* **2**,e72.

Hermann, K.M. and Weaver, L.M. (1999). The shikimate pathway. *Annu. Rev. Plant Physiol. Plant Mol. Biol.* **50**, 473–503.

Houston, L.L. and Graham, M.E. (1974). Divalent metal ion effects on a mutant histidinol phosphate phosphatase from *Salmonella typhimurium*. *Arch. Biochem. Biophys.* **162**, 513–522.

Hsu, L.C., Okamoto, M. and Snell, E.E. (1989). L-histidinol phosphate aminotransferase from *Salmonella typhimurium*: kinetic behavior and sequence at the pyridoxal-P binding site. *Biochimie* **71**, 477–489.

Jancarik, J., Pufan, R., Hong, C., Kim, S.H. and Kim, R. (2004). Optimum solubility (OS) screening: an efficient method to optimize buffer conditions for homogeneity and crystallization of proteins. *Acta Cryst.* **D60**, 1670–1673.

Kanaori, K., Uodome, N., Nagai, A., Ohta, D., Ogawa, A., Iwasaki, G. and Nosaka, A.Y. (1996) 113Cd nuclear magnetic resonance studies of cabbage histidinol dehydrogenase. *Biochemistry* **35**, 5949–5954.

Kapust, R.B. and Waugh, D.S. (2000). Controlled intracellular processing of fusion proteins by TEV protease. *Protein Expr. Purif.* **19**, 312–318.

Kapust, R.B., Tözsér, J., Fox, J.D., Anderson, D.E., Cherry, S., Copeland, T.D. and Waugh, D.S. (2001). Tobacco etch virus protease: mechanism of autolysis and rational design of stable mutants with wild-type catalytic proficiency. *Protein Eng.* **14**, 993–1000.

Karp, P.D., Keseler, I.M., Shearer, A., Latendresse, M., Krummenacker, M., Paley, S.M., Paulsen, I., Collado-Vides, J., Gama-Castro, S., Peralta-Gil, M., Santos-Zavaleta, A., Penaloza-Spinola, M.I., Bonavides-Martinez, C. and Ingraham, J. (2007). Multidimensional annotation of the *Escherichia coli* K-12 genome. *Nucleic Acids Res.* **35**, 7577–7590.

Kaya, Y. and Ofengand, J. (2003). A novel unanticipated type of pseudouridine synthase with homologs in bacteria, archaea and eukarya. *RNA* **9**, 711–721.

Kim, S.H., Shin, D.H., Choi, I.G., Schulze-Gahmen, U., Chen, S. and Kim, R. (2003). Structure-based functional inference in structural genomics. *J. Struct. Funct. Genomics* **4**, 129–135.

Koonin, E.V. (1996). Pseudouridine synthases: four families of enzymes containing a putative uridine-binding motif also conserved in dUTPases and dCTP deaminases. *Nucleic Acids Res.* **24**, 2411–2415.

Kozlov, G., Elias, D., Semesi, A., Yee, A., Cygler, M. and Gehring, K. (2004). Structural similarity of YbeD protein from *Escherichia coli* to allosteric regulatory domains. *J. Bacteriol.* **186**, 8083–8088.

Krogh, A., Larsson, B., von Heijne, G. and Sonnhammer, E.L. (2001). Predicting transmembrane protein topology with a hidden Markov model: application to complete genomes. *J. Mol. Biol.* **305**, 567–580.

Lee, D., Redfern, O. and Orengo, C. (2007). Predicting protein function from sequence and structure. *Nat. Rev. Mol. Cell Biol.* **8**, 995–1005.

Lidner, H.A., Nadeau, G., Matte, A., Michel, G., Ménard, R. and Cygler, M. (2005). Site-directed mutagenesis of the active site region in the quinate/shikimate 5-dehydrogenase YdiB of *Escherichia coli*. *J. Biol. Chem.* **280**, 7162–7169.

Lin, Y. and Kielkopf, C.L. (2008). X-ray structures of U2 snRNA-branchpoint duplexes containing conserved pseudouridines. *Biochemistry* **47**, 5503–5514.

Loper, J.C. (1961). Enzyme complementation in mixed extracts of mutants from the Salmonella histidine B locus. *Proc. Natl. Acad. Sci. USA* **47**, 1440–1450.

Macek, B., Gnad, E., Soufi, B., Kumar, C., Olsen, J.V., Mijakovic, I. and Mann, M. (2008). Phosphoproteome analysis of *E. coli* reveals evolutionary conservation of bacterial Ser/Thr/Tyr phosphorylation. *Mol. Cell. Proteomics* **7**, 299–307.

Maillet, I., Berndt, P., Malo, C., Rodriguez, S., Brunisholz, R.A., Pragai, Z., Arnold, S., Langen, H. and Wyss, M. (2007). From the genome sequence to the proteome and back: Evaluation of *E. coli* genome annotation with a 2-D gel-based proteomics approach. *Proteomics* **7**, 1097–1106.

Manjasetty, B.A., Turnbull, A.P., Panjikar, S., Bussow, K. and Chance, M.R. (2008). Automated technologies and novel techniques to accelerate protein crystallography for structural genomics. *Proteomics* **8**, 612–625.

Matte, A., Sivaraman, J., Ekiel, I., Gehring, K., Jia, Z. and Cygler, M. (2003). Contribution of structural genomics to understanding the biology of *Escherichia coli*. *J. Bacteriol.* **185**, 3994–4002.

Matte, A., Louie, G.V., Sivaraman, J., Cygler, M. and Burley, S.K. (2005). Structure of the pseudouridine synthase RsuA from *Haemophilus influenzae*. *Acta Cryst.* **F61**, 350–354.

Matte, A., Jia, Z., Sunita, S., Sivaraman, J. and Cygler, M. (2007). Insights into the biology of *Escherichia coli* through structural proteomics. *J. Struct. Funct. Genomics* **8**, 44–55.

Matte, A. and Cygler, M. (2007). Using dynamic light scattering to improve protein solution behavior for crystallization. *Am. Biotechnol. Lab.* **25**, 14–16.

Meroueh, M., Grohar, P.J., Qiu, J., SantaLucia, Jr. J., Scaringe, S.A. and Chow, C.S. (2000). Unique structural and stabilizing roles for the individual pseudouridine residues in the 1920 region of *Escherichia coli* 23S rRNA. *Nucleic Acids Res.* **28**, 2075–2083.

Michel, G., Roszak, A.W., Sauvé, V., Maclean, J., Matte, A., Coggins, J.R., Cygler, M. and Lapthorn, A.J. (2003). Structures of shikimate dehydrogenase AroE and its paralog YdiB. A common structural framework for different activities. *J. Biol. Chem.* **278**, 19463–19472.

Minailiuc, O.M., Vavelyuk, O., Gandhi, S., Hung, M-N., Cygler, M. and Ekiel, I. (2007). NMR structure of YcgL, a conserved protein from *Escherichia coli* representing the DUF709 family, with a novel $\alpha/\beta/\alpha$ sandwich fold. *Proteins* **66**, 1004–1007.
Misra, R.V., Horler, R.S., Reindl, W., Goryanin, I.I. and Thomas, G.H. (2005). EchoBASE: an integrated post-genomic database for *Escherichia coli*. *Nucleic Acids Res.* **33**, D329–D333.
Mulder, N.J. and Apweiler, R. (2008). The InterPro database and tools for protein domain analysis. *Curr. Protoc. Bioinformatics*, Chapter 2, Unit 2.7.
Murshudov, G.N., Vagin, A.A., Lebedev, A., Wilson, K.S. and Dodson, E.J. (1999). Efficient anisotropic refinement of macromolecular structures using FFT. *Acta Cryst.* **D55**, 247–255.
Nilges, M., Macias, M.J., O'Donoghue, S.I. and Oschkinat, H. (1997). Automated NOESY interpretation with ambiguous distance restraints: the refined NMR solution structure of the pleckstrin homology domain from beta-spectrin. *J. Mol. Biol.* **269**(3), 408–422.
Ofengand, J. (2002). Ribosomal RNA pseudouridines and pseudouridine synthases. *FEBS Lett.* **514**, 17–25.
Osborne, M.J., Siddiqui, N., Landgraf, D., Pomposiello, P.J. and Gehring, K. (2005). The solution structure of the oxidative stress-related protein YggX from *Escherichia coli*. *Prot. Sci.* **14**, 1673–1678.
Otwinowski, Z. and Minor, W. (1997). Processing of X-ray diffraction data collected in oscillation mode. *Methods Enzymol.* **276**, 307–326.
Pape, T. and Schneider, T.R. (2004). HKL2MAP: a graphical user interface for phasing with SHELX programs. *J. Appl. Cryst.* **37**, 843–844.
Phannachet, K., Elias, Y. and Huang, R.H. (2005). Dissecting the roles of a strictly conserved tyrosine in substrate revognition and catalysis by pseudouridine 55 synthase. *Biochemistry* **44**, 15488–15494.
Perna, N.T., Plunkett, G. 3rd., Burland, V., Mau, B., Glasner, J.D., Rose, D.J., Mayhew, G.F., Evans, P.S., Gregor, J., Kirkpatrick, H.A., Pósfai, G., Hackett, J., Klink, S., Boutin, A., Shao, Y., Miller, L., Grotbeck, E.J., Davis, N.W., Lim, A., Dimalanta, E.T., Potamousis, K.D., Apodaca, J., Anantharaman, T.S., Lin, J., Yen, G., Schwartz, D.C., Welch, R.A. and Blattner, F.R. (2001). Genome sequence of enterohaemorrhagic *Escherichia coli* O157:H7. *Nature* **409**, 529–533.
Pichoff, S., Alibaud, L., Guédant, A., Castanié, M.P. and Bouché, J.P. (1998). An *Escherichia coli* gene (*yaeO*) suppresses temperature-sensitive mutations in essential genes by modulating Rho-dependent transcriptional termination. *Mol. Microbiol.* **29**, 859–869.
Pons, J.L., Malliavin, T.E. and Delsuc, M.A. (1996). Gifa V.4: a complete package for NMR data set processing. *J. Biomol. NMR* **8**, 445–452.
Prilusky, J., Oueillet, E., Ulryck, N., Pajon, A., Bernauer, J., Krimm, I., Quevillon-Cheruel, S., Leulliot, N., Graille, M., Liger, D., Tresaugues, L., Sussman, J.L., Janin, J., van Tilbeurgh, H. and Poupon, A. (2005). HalX: an open-source LIMS (Laboratory Information Management System) for small- to large-scale laboratories. *Acta Cryst.* **D61**, 671–678.
Rangarajan, E.S., Proteau, A., Wagner, J., Hung, M-N., Matte, A. and Cygler, M. (2006a). Structural snapshots of *Escherichia coli* histidinol phosphate phosphatase along the reaction pathway. *J. Biol. Chem.* **281**, 37930–37941.
Rangarajan, E.S., Nadeau, G., Li, Y., Wagner, J., Hung, M.N., Schrag, J.D., Cygler, M. and Matte, A. (2006b). The structure of the exopolyphosphatase (PPX) from *Escherichia coli* O157:H7 suggests a binding mode for long polyphosphate chains. *J. Mol. Biol.* **359**, 1249–1260.
Raymond, S., O'Toole, N. and Cygler, M. (2004). A data management system for structural genomics. *Proteome Sci.* **2**, 4.
Riley, M., Abe, T., Arnaud, M.B., Berlyn, M.K., Blattner, F.R., Chaudhuri, R.R., Glasner, J.D., Horiuchi, T., Keseler, I.M., Kosuge, T., Mori, H., Perna, N.T., Plunkett, G. 3rd, Rudd, K.E., Serres, M.H., Thomas, G.H., Thomson, N.R., Wishart, D. and Wanner, B.L. (2006). *Escherichia coli* K-12: a cooperatively developed annotation snapshot – 2005. *Nucleic Acids Res.* **34**, 1–9.
Sali, A. (1998). 100,000 protein structures for the biologist. *Nat. Struct. Biol.* **5**, 1019–1020.
Schneider, B.L., Kiupakis, A.K. and Reitzer, L.J. (1998). Arginine catabolism and the arginine succinyltransferase pathway in *Escherichia coli*. *J. Bacteriol.* **180**, 4278–4286.

Shapiro, L. and Lima, C.D. (1998). The argonne structural genomics workshop: Lamaze class for the birth of a new science. *Structure* **6**, 265–267.

Sheldrick, G.M., (2008). A short history of SHELX. *Acta Cryst.* **A64**, 112–122.

Shirai, H. and Mizuguchi, K. (2003). Prediction of the structure and function of AstA and AstB, the first two enzymes of the arginine succinyltransferase pathway of arginine catabolism. *FEBS Lett.* **555**, 505–510.

Sivaraman, J., Li, Y., Larocque, R., Schrag, J.D., Cygler, M. and Matte, A. (2001). Crystal structure of histidinol phosphate aminotransferase (HisC) from *Escherichia coli* and its covalent complex with pyridoxal-5'-phosphate and L-histidinol phosphate. *J. Mol. Biol.* **311**, 761–776.

Sivaraman, J., Sauvé, V., Larocque, R., Stura, E.A., Schrag, J.D., Cygler, M. and Matte, A. (2002). Structure of the 16S rRNA pseudouridine synthase RsuA bound to uracil and UMP. *Nat. Struct. Biol.* **9**, 353–358.

Sivaraman, J., Iannuzzi, P., Cygler, M. and Matte, A. (2004). Crystal structure of the RluD pseudouridine synthase catalytic module, an enzyme that modifies 23 S rRNA and is essential for normal cell growth of *Escherichia coli*. *J. Mol. Biol.* **335**, 87–101.

Sivaraman, J., Myers, R.S., Boju, L., Sulea, T., Cygler, M., Jo Davisson, V. and Schrag, J.D. (2005). Crystal structure of *Methanobacterium thermoautotrophicum* phosphoribosyl-AMP cyclohydrolase, HisI. *Biochemistry* **44**, 10071–10080.

Slabinski, L., Jaroszewski, L., Rodrigues, A.P., Rychlewski, L., Wilson, I.A., Lesley, S.A. and Godzik, A. (2007a). The challenge of protein structure determination – lessons from structural genomics. *Prot. Sci.* **16**, 2472–2482.

Slabinski, L., Jaroszewski, L., Rychlewski, L., Wilson, I.A., Lesley, S.A. and Godzik, A. (2007b). XtalPred: a web server for prediction of protein crystallizability. *Bioinformatics* **23**, 3403–3405.

Smialowski, P., Schmidt, T., Cox, J., Kirschner, A. and Frishman, D. (2006). Will my protein crystallize? A sequence-based predictor. *Proteins* **62**, 343–355.

Su, C., Peregrin-Alvarez, J.M., Butland, G., Phandase, S., Fong, V., Emili, A. and Parkinson, J. (2008). Bacteriome.org – an integrated protein interaction database for *E. coli*. *Nucleic Acids Res.* D632–D636.

Suits, M.D., Pal, G.P., Nakatsu, K., Matte, A., Cygler, M. and Jia, Z. (2005). Identification of an *Escherichia coli* O157:H7 heme oxygenase with tandem functional repeats. *Proc. Natl. Acad. Sci. USA* **102**, 16955–16960.

Suits, M.D.L., Jaffer, N. and Jia, Z. (2006). Structure of the Escherichia coli O157:H7 heme oxygenase ChuS in complex with heme and enzymatic inactivation by mutation of the heme coordinating residue His-193. *J. Biol. Chem.* **281**, 36776–36782.

Sunita, S., Zhenxing, H., Swaathi, J., Cygler, M., Matte, A. and Sivaraman, J. (2006). Domain organization and crystal structure of the catalytic domain of *E. coli* RluF, a pseudouridine synthase that acts on 23S rRNA. *J. Mol. Biol.* **359**, 998–1009.

Teng, H. and Grubmeyer, C. (1999). Mutagenesis of histidinol dehydrogenase reveals roles for conserved histidine residues. *Biochemistry* **38**, 7363–7371.

Terwilliger, T.C. and Berendzen, J. (1999). Automated MAD and MIR structure solution. *Acta Cryst.* **D55**, 849–861.

Terwilliger, T.C. (2000). Maximum likelihood density modification. *Acta Cryst.* **D56**, 965–972.

Terwilliger, T.C. (2003). Automated main-chain model building by template matching and iterative fragment extension. *Acta Cryst.* **D59**, 38–44.

Thaller, M.C., Schippa, S. and Rossolini, G.M. (1998). Conserved sequence motifs among bacterial, eukaryotic and archaeal phosphatases that define a new phosphohydrolase superfamily. *Prot. Sci.* **7**, 1647–1652.

Tocilj, A., Schrag, J.D., Li, Y., Schneider, B.L., Reitzer, L., Matte, A. and Cygler, M. (2005). Crystal structure of *N*-succinylarginine dihydrolase AstB, bound to substrate and product, an enzyme from the arginine catabolic pathway of *Escherichia coli*. *J. Biol. Chem.* **280**, 15800–15808.

Trempe, J.F. and Gehring, K. (2003). Observation and interpretation of residual dipolar couplings in biomolecules. In: *NMR of Orientationally Ordered Liquids*, E.E. Burnell, and C.A. de Lange, eds. Kluwer Academic Publishers B.V., Dordrecht.

Valente, J.J., Payne, R.W., Manning, M.C., Wilson, W.W. and Henry, C.S. (2005). Colloidal behavior of proteins: effects of the second virial coefficient on solubility, crystallization and aggregation of proteins in aqueous solution. *Curr. Pharm. Biotechnol.* **6**, 427–436.

Vedadi, M., Niesen, F.H., Allali-Hassani, A., Fedorov, O.Y., Finerty, P.J. Jr., Wasney, G.A., Yeung, R., Arrowsmith, C., Ball, L.J., Berglund, H., Hui, R., Marsden, B.D., Nordlund, P., Sundstrom, M., Weigelt, J. and Edwards, A.M. (2006). Chemical screening methods to identify ligands that promote protein stability, protein crystallization and structure determination. *Proc. Natl. Acad. Sci. USA* **103**, 15835–15840.

Vornhein, C., Blanc, E., Roversi, P. and Bricogne, G. (2007). Automated structure solution with autoSHARP. *Methods Mol. Biol.* **364**, 215–230.

Watson, J.D., Sanderson, S., Ezersky, A., Savchenko, A., Edwards, A., Orengo, C., Joachimiak, A., Laskowski, R.A. and Thornton, J.M. (2007). Towards fully automated structure-based function prediction in structural genomics: a case study. *J. Mol. Biol.* **367**, 1511–1522.

Welch, R.A., Burland, V., Plunkett, G. 3rd, Redford, P., Roesch, P., Rasko, D., Liou, S.R., Boutin, A., Hackett, J., Stroud, D., Mayhew, G.F., Rose, D.J., Zhou, S., Schwartz, D.C., Perna, N.T., Mobley, H.L., Donnenberg, M.S. and Blattner, F.R. (2002). Extensive mosaic structure revealed by the complete genome sequence of uropathogenic *Escherichia coli*. *Proc. Natl. Acad. Sci. USA* **99**, 17020–17024.

Yokoyama, S., Hirota, H., Kigawa, T., Yabuki, T., Shirouzu, M., Terada, T., Ito, Y., Matsuo, Y., Kuroda, Y., Nishimura, Y., Kyogoku, Y., Miki, K., Masui, R. and Kuramitsu, S. (2000). Structural genomics projects in Japan. *Nat. Struct. Biol.* **Suppl**, 943–945.

Chapter 5
Resources for *Escherichia coli* Systems Biology

Hirotada Mori, Natsuko Yamamoto, Hitomi Dose, Kenji Nakahigashi, Kirill A. Datsenko, and Barry L. Wanner

Contents

Abstract Since genomic sequencing project launched, during in 1990s, biological research environments has been dramatically changed by developments of interdisciplinary fields between biology and others such as chemistry, physics, information science, mathematics and engineering. Many high-throughput systems to obtain comprehensive analysis results have become available. As the result, accumulation of experimental data is now growing exponentially like sequence data in public databases. Experimental resources, such as plasmid clone and deletion mutant libraries, are the products of such high-throughput systems, and at the same time, motive force to generate further comprehensive information from experimental analyses. In this manuscript, we summarize the situation about the experimental

H. Mori (✉)
Graduate School of Biological Sciences, Nara Institute of Science and Technology, Ikoma, Nara 630-0101, Japan; Advanced Institute of Biosciences, Keio University, Tsuruoka, Yamagata 997-0017, Japan
e-mail: hmori@gtc.naist.jp

S.Y. Lee (ed.), *Systems Biology and Biotechnology of Escherichia coli*,
DOI 10.1007/978-1-4020-9394-4_5, © Springer Science+Business Media B.V. 2009

resources and how they have contributed in biology fields, especially in the 21st new generation of biology, such as systems biology.

5.1 Introduction

The research in genetics starts careful observation of phenotype and then looks for the cause in most case mutation on its genetic information. On the other hands, molecular biology has developed reverse genetic, which makes it possible by reverse direction, i.e. from the cause (gene) to the phenotype (function). In the last decade after the completion of genomic sequencing of the target organism, information about the approximate total number of genes coded on the chromosome and their predicted coding regions are available. Using this information with technology developments, the preparation of entire set of ORF clone and deletion mutant libraries has become possible. The importance of such comprehensive resources is not only for high-throughput comprehensive experiments but also for direct comparison across the strains on the same genetic background.

5.2 Information Resources

5.2.1 Sequence Information

Genomic sequence is one of the fundamental information of the organism. Fortunately, *Escherichia coli* K-12 genomic sequence has been determined using two sub-strains, MG1655 (Blattner et al. 1997) and W3110 (Aiba et al. 1996, Itoh et al. 1996, Oshima et al. 1996, Yamamoto et al. 1997), which were constructed from the same ancestral strain (Hayashi et al. 2006). We carefully compared genomic sequences of these two strains and confirmation by direct sequencing of conflict regions using PCR has been performed. Finally just 9 bases in 8 ORFs had been revealed as true nucleotide level differences. Larger differences were distribution of ISs and large inversion between *rrnD* and *rrnE* in W3110. Surprisingly, there were no nucleotide level changes in the inter-genic regions (Hayashi et al. 2006).

5.2.2 Biological Information (Annotation)

Finally, we obtained the most accurate genomic sequences both of two sub-strains and that was a good starting point to compile the latest functional information of the genes. To perform this, the annotation meetings were held by the international *E. coli* scientists and build up the first gene annotation Table with known knowledge and functional predictions (Riley et al. 2006). It is, still at present, so difficult to predict the true starting position of genes, however, it was meaningful to share gene Table with good agreements by the members of international researchers. This Table was then to be a starting point for further functional analyses and also be a blue print of construction of resources. The efforts to improve annotation are still kept and currently about 150 genes have been revised or added their annotation to the starting one, which we fixed in 2006 (Riley et al. 2006) (Refseq: NC000913, http://www.ncbi.nlm.nih.gov/).

5.3 Experimental Resources

5.3.1 ASKA and Mobile Plasmids Libraries

We first tried to construct ORF plasmid clone library, ASKA ORF library, of all of the predicted coding regions (Kitagawa et al. 2005). Each of the ORF regions was amplified by PCR from the second to the last amino acid coding region. The construction method and the structure of clone are shown schematically in Fig. 5.1. Currently, two libraries were established, fusion and non-fusion with GFP protein at C terminus of the target ORFs. All clones have His-tag at N terminus. Their expression is regulated by IPTG inducible promoter and synthetic SD signal. Without IPTG, the expression is tightly repressed by $LacI^q$ repressor protein in cis. Once such libraries have been established, many comprehensive analyses could be available, such as protein localization by monitoring GFP fluorescence (Niki in preparation), protein-protein interaction by pull-down assay etc (Arifuzzaman et al. 2006), etc.

The second construction was comprehensive mobile plasmid library using ColE1 derived vector. The features of this library are that, self-transmittable from male type cell to female and relatively low copy number. The ORF fragments were transferred and cloned from ASKA library into mobile plasmid vector using *Sfi*I restriction enzyme (Saka et al. 2005).

5.3.2 Keio Collection

This is a single gene deletion library of predicted ORFs of *E. coli* K-12 except essential genes (Baba et al. 2006). Before 2000, *E. coli* was thought to be a difficult organism to construct mutant by homologous recombination (Datsenko and Wanner 2000). This was one of the reasons for *E. coli* to be behind *Saccharomices serevisiae* (Giaever et al. 2002) and *Bacillus subtilis* (Kobayashi et al. 2003) in construction of comprehensive deletion mutant library. In 2000, however, the efficient method using lambda RED recombinase had been developed (Datsenko and Wanner 2000). Then, we started to construct comprehensive deletion (replacement with Km resistant gene cassette) mutant library by this technology (Baba et al. 2006). We observed that 303 ORFS had been repeatedly failed to be constructed as deletion mutant and defined as essential gene candidates. However, once established as deletion mutant, there might be possibility to keep another copy of the target gene, which is called partial duplication. This was the reason why we stored independent two isolates of each of the target genes. Currently, identification of such gene duplication is underway and the results might open to the public in near future. The construction and the structure are shown schematically in Fig. 5.2.

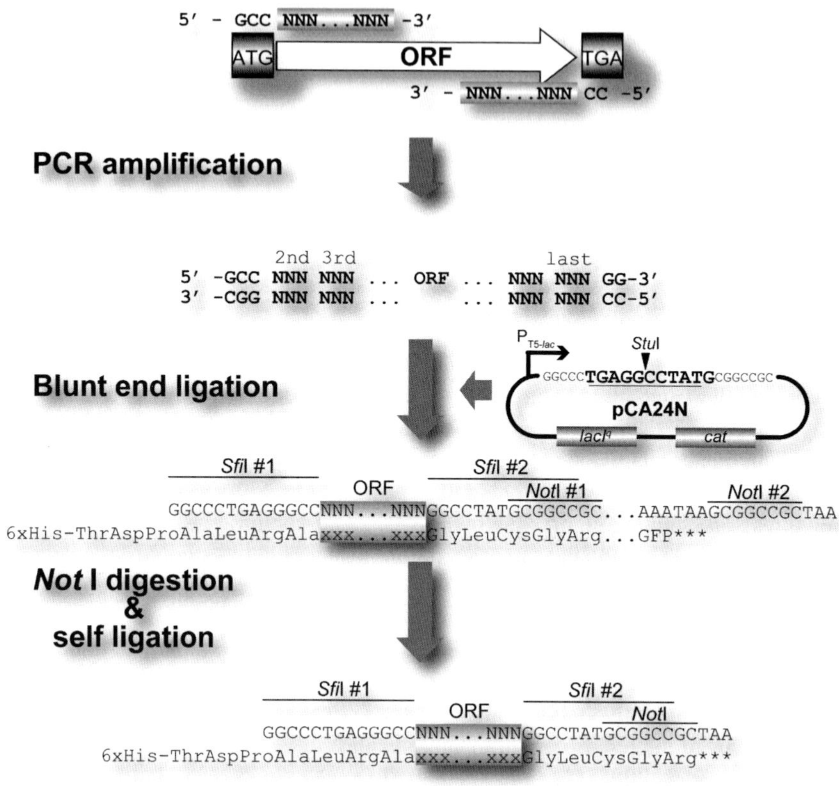

Fig. 5.1 Construction of AKSA ORF plasmid library. Predicted amino acid sequences at around the N- and C-terminal regions of the cloned ORF. The arrow indicates a target ORF and a pair of primers used for PCR amplification. The bottom nucleotide and amino acid sequences indicate the predicted final structure around the site of cloned ORF. The product proteins should contain 6 Histidine, 7 and 5 amino acids at the N- and C-terminal ends of the target gene, respectively, followed by GFP fragment: 6xHisThrAspProAlaLeuArgAlaXXX...XXXGlyLeuCysGlyArg...GFP, where XXX...XXX indicates second to the last amino acid codon of target gene. Removal of GFP fragment by NotI digestion. Each ORF clone has two NotI sites, one at the C-terminal spacer region (within SfiI #2) and region directly next to the termination codon of GFP gene. GFP can therefore be removed by NotI digestion followed by self-ligation

5.3.3 Other Comprehensive Experimental Resources

5.3.3.1 Random Transposon Insertion Mutants

To prepare disruption mutants of *E. coli*, random insertion mutagenesis by transposon has a long history in this organism. Miki and his colleagues started to construct a comprehensive random insertion mutant library of *E. coli* K-12. They used Tn10 derivative transposon and mutagenized lambda Kohara clones (Kohara et al. 1987). The mutagenized phages were then subjected to infection to wild type W3110 host

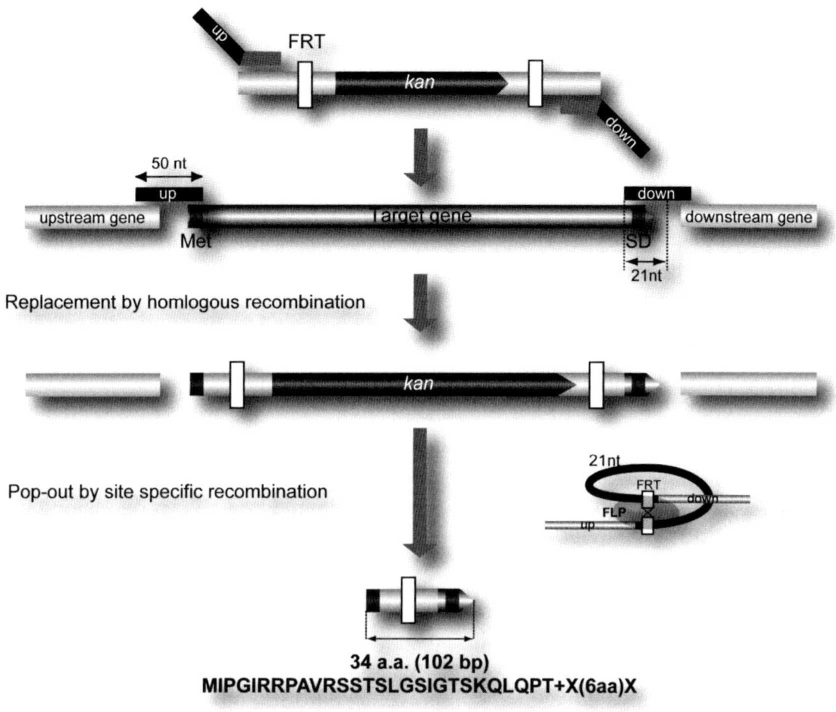

Fig. 5.2 Construction of Keio collection. Primer design and construction of single-gene deletion mutants. Gene knockout primers have 20-nt for priming upstream and downstream of the FRT sites flanking the kanamycin resistance gene in pKD13 and 50-nt homologous to upstream and downstream chromosomal sequences for the target gene deleted. The upstream primer includes the initiation codon of the target. The downstream primer includes codons for the six C-terminal residues, the stop codon, and 29-nt downstream. After establishment of deletion (replaced with kan cassette) mutant, over-supply of FLP recombinase leads elimination the kan cassette by site-specific recombination between two FRT sites. The final structure is designed as in-frame deletion translated using authentic SD, the initiation and termination codons. SD, Shine–Dalgarno ribosome binding sequence

strain. The chromosomal fragment of each of the mutagenized Kohara clone phages was integrated into the host chromosome by homologous recombination to make the host strain partial diploid. After selection of haploid type of mutant strain by antibiotic resistance, the library was established. 6404 insertion strains were established and are now distributed from the National Institute of Genetics, Mishima, Japan (http://www.shigen.nig.ac.jp/ecoli/strain/top/top.jsp).

Sun Chang Kim and his colleagues developed two series of random transposon insertion mutant libraries which carried *cre/loxP* excision system (Yu et al. 2002). They developed insertion mutants more than 400 each with transposons carrying *loxP* and chloramphenicol or kanamycin resistance genes. Insertion location were determined and established as Cmf and Kmf libraries. These libraries provide the system to remove large fragment from the chromosome. Two insertion mutations with appropriate direction and location were selected and combined by P1 transduction, then the region between two *loxP* sites was removed by Cre protein. The purpose of this resource is to remove large fragments from the chromosome and eliminate the size of its genome for providing improved *E. coli* both as a better model organism and as a better engineering tool. Other efforts to construct minimal genome of *E. coli* were performed (Kato and Hashimoto 2007, Kolisnychenko et al. 2002, Posfai et al. 2006).

Random insertion technology has been used to identify essential genes, however, mutants were not stored as a library (Gerdes et al. 2003, Goryshin et al. 2003).

5.3.3.2 Comprehensive Promoter Clone Library

To analyze gene expression profiling using any other technology than DNA microarray or chip, Robert A. Larossa and his colleagues developed genome-wide, genome-registered collection of *E. coli* bioluminescence reporter gene fusion (Van Dyk et al. 2001). They adopted random fusion of *E. coli* chromosomal DNA fragments to *lux* operon. Using these resources, gene expression profiling analysis comparing with DNA microarray technology for stress responses had been performed.

Recently, more designed based comprehensive fusion library was developed. Uri Alon and his colleagues developed comprehensive library of transcriptional fusion of *gfp* to measure expression dynamics in individual living cells (Zaslaver et al. 2006). Until then, there were no comprehensive tools that could provide quantitative blueprints of gene circuitry. To address this, they designed primers to flank intergenic region longer than 40 bp and cloned into pSC101 derived relatively low copy number vector to generate transcriptional fusion with *gfp*.

5.4 On going Project of New Resources

All resources developed have their original research purposes to be solved for their construction and expansion or new construction are needed for new research idea towards the next step. We developed single gene deletion library initially for functional analyses of all of the gene coded on the *E. coli* chromosome and make it possible to compare on the same platform both function known and unknown target genes. During construction of the library, as we expected, about 300 target genes had been failed to be isolated as deletion strain and we defined those as essential gene candidates (Baba et al. 2006, 2008, Baba and Mori 2008). For analyses of the essential genes from the global aspect on the same platform with high throughput way, new resources were required, for example single gene deletion mutant library of essential genes under the condition of in trans complementation from the expression plasmid clone.

Also we showed comprehensive protein-protein interaction by pull-down assay using His-tagged bait proteins expressed from our ASKA plasmid library (Arifuzzaman et al. 2006). Most of the membrane proteins were failed to be analyzed because of their solubility. To solve this problem, another system is also required. For this purpose, we are now designing of Gateway entry clone library and its destination plasmids.

In *E. coli*, comprehensive network analyses in transcriptome, proteome, interactome, metabolome etc., have been performed and are still continuing towards complete understandings of physiological networks in a cell. The remained important target network to be solved is genetic interaction. For this new target network, synthetic lethal/sickness analysis using double knockout strains is one of the appropriate analyses methods. To perform this, construction of another entire set of deletion library and the new system to combine two deletions into the same chromosome are required.

Needless to say that construction of resources is almost endless efforts to put science into the next step.

5.4.1 Gateway Entry Clone Library

Gateway technology is a site-specific recombination-assisted cloning method (http://www.invitrogen.com/) and its advantage is flexibility of shuttling insert DNA from an entry clone to other variety of vectors without any restriction enzyme cloning method. ASKA ORF library has *Sfi*I sites at the both ends of the target ORF fragment and these two sites have the different cohesive ends. This makes possible to uni-directional cloning to another vector. We developed the new vector, function as entry clone of Gateway technology and having *Sfi*I sites at both ends between *att* recombination sites to make entry clone by transferring *Sfi*I fragment from ASKA plasmid clone (Kitagawa et al. 2005). Pilot test of our new entry vector was finished, and systematic construction is now underway (Yamamoto et al. 2008a).

5.4.2 Low Copy Expression Plasmid Library

For physiological functional analysis, approaches in genetics, such as complementation, are powerful way. It is inevitable for this purpose to develop low copy expression plasmid vector, which provides precise expression regulation for genetical complementation. Some genes, especially physiologically important genes such as essential genes, are very difficult to clone into multi copy plasmid vector probably because of tight requirement for their cellular expression level. To perform this, we have developed F plasmid derived low copy vector systems and construction of essential genes has been almost done. Physiological analyses using these plasmid clone library are now underway and the expansion to the entire gene library is now under consideration.

5.4.3 The Second Set of Deletion Library

As mentioned above, comprehensive genetic network analysis is important remained target of *E. coli*. To perform this, the second set of deletion library having different antibiotics resistance is needed to combine single gene deletions to make double knockout mutant. We designed the new deletion strain with chloramphenicol resistance and 20 nt bar code sequence with similar idea of Yeast deletion library (Giaever et al. 2002). Construction of new deletion collection has been done and evaluation is now underway. Development of the system to combine two single deletion mutations by conjugation has also been finished. Towards the comprehensive genetic network analysis, the first tests have been done (Butland et al. 2008, Typas et al. 2008).

5.4.4 Essential Gene Deletion Library

The essential genes are the indispensable genes for cellular survival in a certain growth condition, generally defined in a rich medium condition such as LB. In *E. coli*, a several systematic studies to identify the essential genes have been endeavoured (Baba et al. 2006, Gerdes et al. 2003, Goryshin et al. 2003, Hashimoto et al. 2005, Miki et al. 2008). Although the extensive studies about the individual essential genes have been made, they have examined with the strains with the various genetic background and under the various growth conditions. For the systematic and genome-wide comparison, we are now constructing the new deletion library of essential genes under the condition of in trans complementation from the low copy plasmid clone library described above. Elimination of essential genes from the chromosome was performed by the same method as construction of Keio collection (Yamamoto et al. 2008b).

5.4.5 Chromosomal Fusion with GFP Protein

Fusion protein with fluorescence generating reporter gene might provide flexible purposes for its use in research. Monitoring protein localization, movement and quantification are small examples of the major uses. To quantify the protein expression level in living cells, we have constructed in frame chromosomal fusion of the target genes related to the central metabolic pathway with modified GFP (Nagai et al. 2002) to monitor their cellular dynamics (Dose et al. 2008). The comparison of the quantification between Western analysis and by fluorescence is now underway. Antibodies used for Western analysis were produced using purified proteins as antigen from ASKA ORF collection. Not only the chromosomal fusion of GFP will provide very efficient monitoring system of target protein concentration but also analysis system in a single cell level. Stochastic expression of those target genes is currently important research target.

5.5 Conclusion and Perspective

Table 5.1 summarize the current *E. coli* resources. These resources described above have opened new research fields especially in the researches of global aspects. And particularly in genetics, they provide the platform to test entire set of genes to be compared on the same genetic background. To analyze cells as systems level, these uniformly designed resources are now essential tools and the preparation of such

Table 5.1 Experimental comprehensive resources of *Escherichia coli* K-12

Plasmid Clone library				
ASKA ORF collection (GFP+)	published	entire genes	designed	(Kitagawa et al. 2005)
ASKA ORF collection (GFP−)	published	Entire grene	designed	(Kitagawa et al. 2005)
ASKA gateway entry collection	on going	entire genes	designed	(Yamamoto et al. 2008a).
Lux fusion library	published	~ 600	random fusion	(Van Dyk et al. 2001)
Promoter fusion	published	entire promoters (~ 2, 000 promoters)	designed	(Zaslaver et al. 2006)
ASKA low copy collection	on going	entire genes	designed	(Yamamoto et al. 2008a).
Chromosomal modification library				
Keio collection	published	entire genes (except essential genes)	designed	(Baba et al. 2006)
ASKA bar code deletion library	on going	entire genes (except essential genes)	designed	(Yamamoto et al. 2008b)
Cmf library	published	~ 400 insertion strains	random insertion	(Yu et al. 2002)
Kmf library	published	~ 400 insertion strains	random insertion	(Yu et al. 2002)
TAP tag fusion	published	~ 1, 000 fusion strains	designed	(Butland et al. 2005)
ASKA essential gene deletion collection	on going	essential genes	designed	(Yamamoto et al. 2008b)
Random transposon insertion library	published	~ 7000 insertion strains	random insertion	(Miki et al. 2008)
ASKA chromosomal fusion collection	unpublished	~ 100 in frame chromosomal fusion	designed	(Dose et al. 2008)
Large deletion strains				
large deletion library	published	30% elimination of the genome	large scale deletion	(Kato and Hashimoto 2007)
minimal genome	published	Elimination of non essential genes, IS, cryptic phaget etc. ~ 15% reduction	large scale deletion	(Posfai et al. 2006)

tools accelerate systems approaches using *E. coli*. For the last 50 years, contribution of *E. coli* in basic biology was huge to build up the concepts of genes. For the next 50 years, *E. coli* will also be an important organism for contribution of building up the concepts of cells. To make this reality, we need seriously to consider the community level activities rather than individual competitive researches. We hope more productive era sharing everything from cultivation to harvest (from construction of resources to getting analyses' results) will become reality in near future in *E. coli* research community.

Acknowledgments We thank T. Nakamichi, Y. Okumura, M. Hasegawa, Y. Takai, M. Yoshino, S. Kinjo, M. Baba, A. Furubayashi, A. Hirai and Y. Touda for technical assistance and more importantly distribution of more than 2 million of our strains the scientists in the world. Most of the construction done by our group was supported by a grant from Core Research for Evolutional Science and Technology (CREST), Japan Science and Technology Agency (JST), a grant from the New Energy and Industrial Technology Development Organization (NEDO) of the Ministry of Economy, Trade, and Industry of Japan (METI) and "grant-in-aid" research grants from the Ministry of Education, Culture, Sports, Science, and Technology (MEXT). We thank the faculty and students of the Systems Biology Program at Keio University for useful discussions, as well as members of the International *E. coli* Alliance (IECA).

References

Aiba H, Baba T, Hayashi K et al. (1996) A 570-kb DNA sequence of the *Escherichia coli* K-12 genome corresponding to the 28.0–40.1 min region on the linkage map. DNA Res 3(6):363–77

Arifuzzaman M, Maeda M, Itoh A et al. (2006) Large-scale identification of protein-protein interaction of *Escherichia coli* K-12. Genome Res 16(5):686–91

Baba T, Ara T, Hasegawa M et al. (2006) Construction of *Escherichia coli* K-12 in-frame, single-gene knockout mutants: the Keio collection. Mol Syst Biol 2:2006 0008

Baba T, Huan HC, Datsenko K et al. (2008) The applications of systematic in-frame, single-gene knockout mutant collection of *Escherichia coli* K-12. Methods Mol Biol 416:183–94

Baba T, Mori H (2008) The construction of systematic in-frame, single-gene knockout mutant collection in *Escherichia coli* K-12. Methods Mol Biol 416:171–81

Blattner FR, Plunkett G, 3rd, Bloch CA et al. (1997) The complete genome sequence of *Escherichia coli* K-12. Science 277(5331):1453–74

Butland G, Babu M, Diaz-Mejia JJ et al. (2008) eSGA: *E. coli* synthetic genetic array analysis. Nat Methods

Butland G, Peregrin-Alvarez JM, Li J et al. (2005) Interaction network containing conserved and essential protein complexes in *Escherichia coli*. Nature 433(7025):531–7

Datsenko KA, Wanner BL (2000) One-step inactivation of chromosomal genes in *Escherichia coli* K-12 using PCR products. Proc Natl Acad Sci USA 97(12):6640–5

Dose H, Nakamichi T, Yoshino M et al. (2008) in preparation

Gerdes SY, Scholle MD, Campbell JW et al. (2003) Experimental determination and system level analysis of essential genes in *Escherichia coli* MG1655. J Bacteriol 185(19):5673–84

Giaever G, Chu AM, Ni L et al. (2002) Functional profiling of the *Saccharomyces cerevisiae* genome. Nature 418(6896):387–91

Goryshin IY, Naumann TA, Apodaca J et al. (2003) Chromosomal deletion formation system based on Tn5 double transposition: use for making minimal genomes and essential gene analysis. Genome Res 13(4):644–53

Hashimoto M, Ichimura T, Mizoguchi H et al. (2005) Cell size and nucleoid organization of engineered *Escherichia coli* cells with a reduced genome. Mol Microbiol 55(1):137–49

Hayashi K, Morooka N, Yamamoto Y et al. (2006) Highly accurate genome sequences of *Escherichia coli* K-12 strains MG1655 and W3110. Mol Syst Biol 2:2006 0007, http://www.invitrogen.com/http://www.shigen.nig.ac.jp/ecoli/strain/top/top.jsp
Itoh T, Aiba H, Baba T et al. (1996) A 460-kb DNA sequence of the *Escherichia coli* K-12 genome corresponding to the 40.1–50.0 min region on the linkage map. DNA Res 3(6):379–92
Kato J, Hashimoto M (2007) Construction of consecutive deletions of the *Escherichia coli* chromosome. Mol Syst Biol 3:132
Kitagawa M, Ara T, Arifuzzaman M et al. (2005) Complete set of ORF clones of *Escherichia coli* ASKA library (A Complete Set of *E. coli* K-12 ORF Archive): Unique Resources for Biological Research. DNA Res 12:291–99
Kobayashi K, Ehrlich SD, Albertini A et al. (2003) Essential *Bacillus subtilis* genes. Proc Natl Acad Sci U S A 100(8):4678–83
Kohara Y, Akiyama K, Isono K (1987) The physical map of the whole *E. coli* chromosome: application of a new strategy for rapid analysis and sorting of a large genomic library. Cell 50(3):495–508
Kolisnychenko V, Plunkett G, 3rd, Herring CD et al. (2002) Engineering a reduced *Escherichia coli* genome. Genome Res 12(4):640–7
Miki T, Yamamoto Y, Matsuda H (2008) A novel, simple, high-throughput method for isolation of genome-wide transposon insertion mutants of *Escherichia coli* K-12. Methods Mol Biol 416:195–204
Nagai T, Ibata K, Park ES et al. (2002) A variant of yellow fluorescent protein with fast and efficient maturation for cell-biological applications. Nat Biotechnol 20(1):87–90
Niki H (in preparation)
Oshima T, Aiba H, Baba T et al. (1996) A 718-kb DNA sequence of the *Escherichia coli* K-12 genome corresponding to the 12.7–28.0 min region on the linkage map. DNA Res 3(3):137–55
Posfai G, Plunkett G, 3rd, Feher T et al. (2006) Emergent properties of reduced-genome *Escherichia coli*. Science 312(5776):1044–6
Riley M, Abe T, Arnaud MB et al. (2006) *Escherichia coli* K-12: a cooperatively developed annotation snapshot–2005. Nucleic Acids Res 34(1):1–9
Saka K, Tadenuma M, Nakade S et al. (2005) A complete set of *Escherichia coli* open reading frames in mobile plasmids facilitating genetic studies. DNA Res 12(1):63–8
Typas A, Nichols RJ, Siegele DA et al. (2008) High-throughput, quantitative analyses of genetic interactions in *E. coli*. Nat Methods
Van Dyk TK, Wei Y, Hanafey MK et al. (2001) A genomic approach to gene fusion technology. Proc Natl Acad Sci U S A 98(5):2555–60
Yamamoto N, Nakamichi T, Yoshino M et al. (2008a) unpublished
Yamamoto N, Nakamichi T, Yoshino M et al. (2008b) in preparation
Yamamoto Y, Aiba H, Baba T et al. (1997) Construction of a contiguous 874-kb sequence of the *Escherichia coli*-K12 genome corresponding to 50.0–68.8 min on the linkage map and analysis of its sequence features. DNA Res 4(2):91–113
Yu BJ, Sung BH, Koob MD et al. (2002) Minimization of the *Escherichia coli* genome using a Tn5-targeted Cre/loxP excision system. Nat Biotechnol 20(10):1018–23
Zaslaver A, Bren A, Ronen M et al. (2006) A comprehensive library of fluorescent transcriptional reporters for *Escherichia coli*. Nat Methods 3(8):623–8

Chapter 6
The Multiple Scientific Disciplines Served by EcoCyc

Peter D. Karp

Contents

Abstract The EcoCyc database integrates information about the *E. coli* genome, its metabolic pathways, and its regulatory network. EcoCyc is in use by scientists from a variety of disciplines. Experimental biologists use it as a reference source on *E. coli*, and to leverage information about *E. coli* to the study of other microbes. Because the *E. coli* genome has the largest number of experimentally characterized genes of any organism, EcoCyc is used in the annotation of other microbial genomes by sequence similarity. EcoCyc has also been used in a number of global biological studies by computational biologists, and to provide training and validation datasets for the development of new bioinformatics algorithms. EcoCyc serves as a reference source for metabolic engineers, and it is used in microbiology education. The software behind EcoCyc, called Pathway Tools, has been used to develop EcoCyc-like databases for many other organisms. Pathway Tools provides powerful query and visualization capabilities, including tools to analyze high-throughput datasets by painting those datasets onto genome-scale diagrams of the metabolic network, the transcriptional regulatory network, and the complete genome map.

P.D. Karp (✉)
SRI International, 333 Ravenswood Ave AE 206, Menlo Park, CA 94025 USA
e-mail: pkarp@ai.sri.com

S.Y. Lee (ed.), *Systems Biology and Biotechnology of Escherichia coli*,
DOI 10.1007/978-1-4020-9394-4_6,

6.1 Introduction

The EcoCyc database has been under development since 1992 with the goal of serving several different scientific communities that require knowledge of the molecular parts of *E. coli*. EcoCyc has evolved from an initial focus on the metabolic pathways of *E. coli* to describe its complete genome and proteome, its metabolism and transport capabilities, and its regulatory network.

This chapter surveys the scientific disciplines served by EcoCyc. It discusses how these scientists use EcoCyc, and how the information and software tools provided by EcoCyc have been designed to serve their needs. The article also describes recently released software tools within EcoCyc, such as its Omics Viewers, the new graph tracks for visualization of ChIP-chip datasets, and the comparative analysis tools that support comparisons between any of the 370 organisms (including *E. coli*) that are supported within the BioCyc collection.

Our knowledge of who uses EcoCyc comes from a survey of EcoCyc users that we conducted in the spring of 2005, and from a citation analysis we performed for EcoCyc. To date, publications about EcoCyc (and the associated database RegulonDB (Gama-Castro et al. 2008, Salgado et al. 2004), which draws most of its content from EcoCyc) have been cited more than 500 times according to the ISI Web of Knowledge (http://www.isiwebofknowledge.com/). Scientists who use EcoCyc fall into the following groups:

- Experimental biologists who work with *E. coli*, other microbes, and higher organisms
- Computational biologists
- Bioinformaticists
- Metabolic engineers
- Educators

6.2 EcoCyc as a Reference for Experimental Biologists

EcoCyc is a knowledge resource for experimentalists who work with *E. coli*, other microbes, and higher organisms. Over the last 50 years a tremendous amount of information has been gathered on the genetics, biochemistry, and cell biology of *E. coli*, and continues to be amassed at a rapid pace. The pertinent literature is spread among a large number of scientific journals and many books.

Our 2005 Web-based survey asked what responders use EcoCyc for. Wet-lab biology usage indicated by the survey was as follows (each sentence contains responses from one survey question): study the biology of *E. coli* (30%); use *E. coli* as a model organism to study a particular aspect of biology (41%); use *E. coli* as a tool (e.g., for protein expression) (23%); other microbial research (31%); and other biological research (13%).[1] In the survey, 67% of responders said they use EcoCyc

[1] For questions in our survey that allow responders to select multiple choices, percentages refer to percent of responders who selected that answer, and do not add up to 100.

as a general *E. coli* reference tool; 19% use it as a tool for understanding other nonpathogenic bacterial species; 27% use it as a tool for understanding pathogenic bacterial species; and 28% use it for hypothesis generation (developing ideas for new experiments).

EcoCyc can be thought of as an online review article. In EcoCyc version 12.0, released in April 2008, 3,444 of the gene products described in EcoCyc (out of 4,472 total genes) contain mini-reviews authored by EcoCyc curators who summarize and cite the experimental literature for that gene product. The majority of these summaries are from 50 to 2,000 words in length. EcoCyc version 12.0 cites more than 16,000 peer-reviewed publications that have formed the basis for curation. Summaries are also found in other EcoCyc pages, including pathway and transcription unit pages. EcoCyc evidence codes describe the types of experimental evidence that support assigned gene functions. In version 12.0, functional assignments for 2,853 gene products are supported by experimental evidence, which is the highest in both relative and absolute terms of any model organism (Karp et al. 2007).

6.2.1 EcoCyc Analysis of Functional-Genomics Experiments

The use of DNA microarrays within the *E. coli* community has expanded tremendously. Proteomics and metabolomics work in *E. coli* is also increasing steadily. These "omics" methods yield large quantities of data that are difficult to analyze, but promise to produce new insights into cell function; 44% of our survey responders said they use EcoCyc for analysis of the *E. coli* regulatory network. EcoCyc facilitates analysis of functional-genomics data in two unique respects.

First, the extensive catalog of transcriptional regulatory circuits within EcoCyc puts known mechanisms of gene regulation at the fingertips of experimentalists, allowing them to focus on discovering new regulatory mechanisms rather than rediscovering known mechanisms. EcoCyc describes the regulation by 183 transcription factors of 1,492 promoters through regulatory interactions with 1,982 transcription factor binding sites. The majority of these regulatory interactions are based on experimental assays reported in the literature.

A new effort within the EcoCyc project aims to expand the types of cellular regulation encoded within EcoCyc. In 2007, the Pathway Tools software underlying EcoCyc was expanded to be able to capture, display, and edit six subtypes of regulation by attenuation, and curation of attenuation began. In 2008 we will be extending Pathway Tools to accommodate regulation by small RNAs, and translational regulation, and curation of these types of regulation will begin.

The second way in which EcoCyc facilitates analysis of functional-genomics data is via unique bioinformatics analysis capabilities, namely, three Omics Viewers that paint omics data onto global diagrams of *E. coli* cellular networks and of the *E. coli* genome. The same omics dataset can be viewed on all three diagrams so that it may be interpreted from different biological perspectives. Omics measurements are mapped to the same color scale on all three diagrams. Animation can be used on all three diagrams to display multiple measurements, which could reflect different time points, mutations, or treatments.

Examples of the three Omics Viewers are shown in Figs. 6.1, 6.2, and 6.3.

A new tool for analysis of ChIP-chip datasets is shown in Fig. 6.4. This tool, which we call graph tracks, is an extension of the genome-browser tracks capability. A ChIP-chip dataset is loaded into the Eco-Cyc genome browser (like the Omics Viewers, a data file can be uploaded via the EcoCyc Web site, or loaded into the desktop version of EcoCyc and Pathway Tools; the latter is recommended for frequent users because it runs faster and provides more capabilities). The dataset must be in GFF format (see http://www.sanger.ac.uk/Software/formats/GFF/). Data is plotted against the genome with intensity values depicted both as the Y coordinate and as color. Multiple graph tracks and normal horizontal tracks can be displayed simultaneously to compare multiple datasets.

6.2.2 Leveraging EcoCyc to the Study of Other Microbes

EcoCyc sits at the core of the BioCyc collection of Pathway/Genome Databases (PGDBs) for 379 organisms (Karp et al. 2005). For each of those organisms, BioCyc

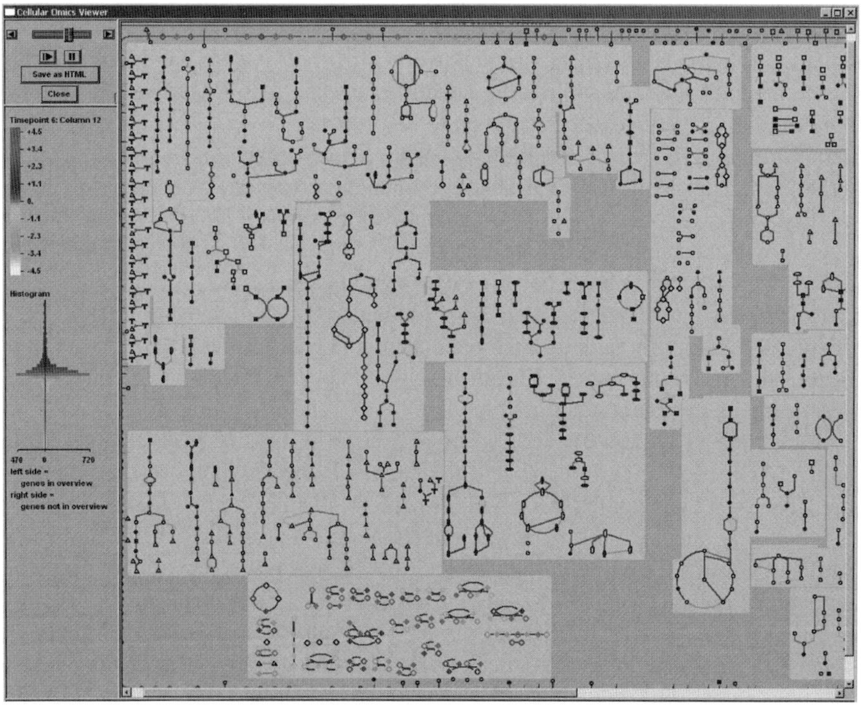

Fig. 6.1 The Cellular Omics Viewer. This image shows an *E. coli* gene expression dataset painted onto the *E. coli* metabolic network. The color assigned to each line (reaction) corresponds to the expression level of the gene coding for the enzyme that catalyzes that reaction. The controls in the upper left allow the user to stop and start animated displays within this diagram

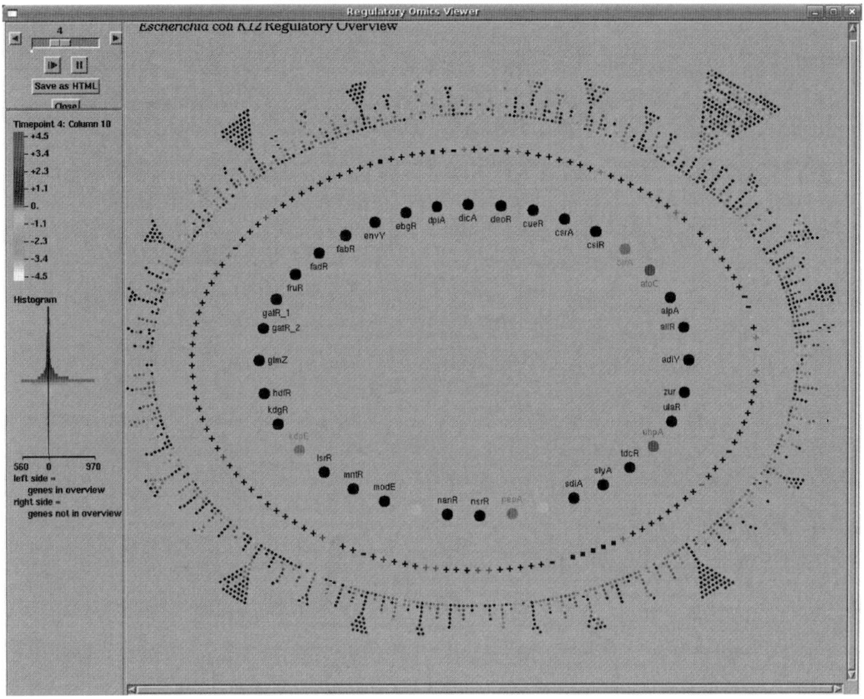

Fig. 6.2 The Regulatory Omics Viewer. This image shows an *E. coli* gene expression dataset painted onto the *E. coli* transcriptional regulatory network. The color assigned to each circle or square (genes) corresponds to the expression level of that gene. The innermost ring contains regulator genes (transcription factors and sigma factors) that have no regulatory inputs defined within EcoCyc. The middle ring contains regulator genes that do have defined regulatory inputs. The outer ring contains non-regulator genes. Genes in the outer ring are grouped into clusters such that two genes are assigned to the same cluster if those genes share the exact same set of regulators

contains their genomes, predicted metabolic pathways, and predicted pathway hole fillers (that is, genes that are predicted to code for enzymes missing from the metabolic pathways). For the bacteria, BioCyc also contains predicted operons. All the bioinformatics tools available for EcoCyc are also available for other organisms in BioCyc, including the genome browser and Omics Viewers.

The BioCyc.org site contains a powerful array of comparative genomics functionality that allows scientists who study other microbes to further leverage EcoCyc (46% of our users use EcoCyc to study other organisms besides *E. coli*), and also allows scientists who study *E. coli* to learn from its similarities to other organisms.

One comparative tool is the comparative genome browser. From a gene page in BioCyc (meaning, from a gene page for any organism in the BioCyc collection including EcoCyc), mid-way down the page is a button Align in Multi-Genome Browser. Clicking on the button will produce a list of all BioCyc organisms. Select the organisms of interest and click Submit. The resulting display will show

Fig. 6.3 The Genome Omics Viewer. This image shows the same *E. coli* gene expression dataset as shown in Fig. 6.1, painted onto the complete *E. coli* genome. Each "shark fin" represents a single gene. The color assigned to each gene corresponds to the expression level of the gene. Upward pointing genes code for proteins, downward pointing genes code for RNAs. The *left–right* directionality of each gene indicates its direction of transcription

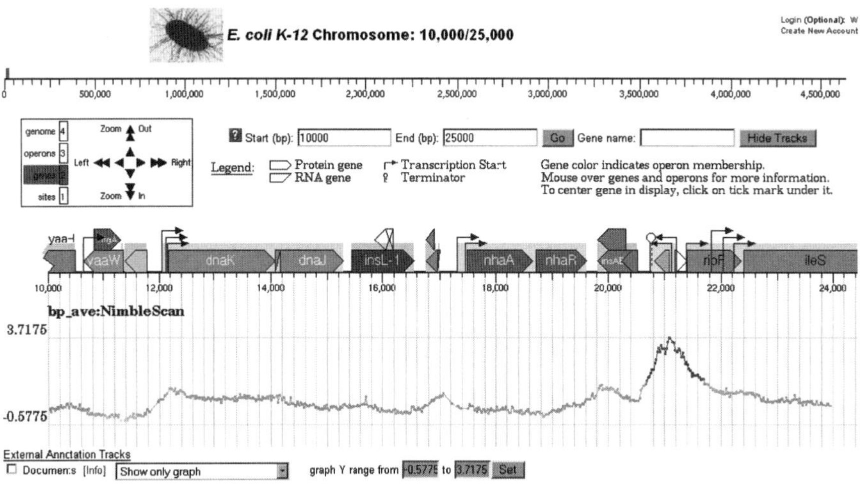

Fig. 6.4 The EcoCyc genome browser with a graph track displayed. The graph track shows an X–Y plot of the intensity of RNA polymerase binding along the *E. coli* genome near the bottom of the figure. This ChIP-chip dataset was kindly provided by Dr. R. Landick of the University of Wisconsin

chromosomal regions for each organism, aligned at the orthologs of the starting genes. Genes drawn in the same color in this display are orthologs of one another (orthologs are defined as best bi-directional BLAST hits). The genome browser navigation controls can be used to zoom or translate the genome map display.

A second set of comparative tools is available from the Comparative Analysis link on the main BioCyc query page (http://biocyc.org/server.html). These tools will generate comparative reports across many dimensions of a PGDB. Several report types are available. One report compares the metabolic pathway complements of the selected organisms. Another report compares the metabolic reaction complements. Another compares transporter complements. Reports are also available to compare proteins, metabolites, and transcription units.

Each report contains several sections. For example, Figs. 6.5 and 6.6 show sections of the pathway report. The reports contain tables that contain summary statistics. To drill down to the data from which those summary statistics were integrated, click on a cell within the table. For example, clicking on a row name will produce

Pathway Class	E. coli CFT073	E. coli K-12
Biosynthesis	144	127
- Amines and Polyamines Biosynthesis	7	7
- Amino acids Biosynthesis	38	27
- Aminoacyl-tRNA Charging	1	1
- Aromatic Compounds Biosynthesis	0	0
- Carbohydrates Biosynthesis	13	8
- Cell structures Biosynthesis	6	8
- Cofactors, Prosthetic Groups, Electron Carriers Biosynthesis	26	32
- Fatty Acids and Lipids Biosynthesis	11	14
- Hormones Biosynthesis	0	0
- Metabolic Regulators Biosynthesis	1	1
- Nucleosides and Nucleotides Biosynthesis	9	8
- Other Biosynthesis	2	3
- Secondary Metabolism	0	0
- Secondary Metabolites Biosynthesis	2	2
- Siderophore Biosynthesis	1	1
Degradation/Utilization/Assimilation	109	105
- Alcohols Degradation	2	4
- Aldehyde Degradation	1	5
- Amines and Polyamines Degradation	1	7
- Amino Acids Degradation	24	18
- Aromatic Compounds Degradation	4	3
- C1 Compounds Utilization and Assimilation	3	1
- Carbohydrates Degradation	25	20
- Carboxylates Degradation	5	9
- Chlorinated Compounds Degradation	0	0
- Cofactors, Prosthetic Groups, Electron Carriers Degradation	0	0
- Degradation/Utilization/Assimilation - Other	3	1
- Fatty Acid and Lipids Degradation	4	2
- Hormones Degradation	0	0
- Inorganic Nutrients Metabolism	6	3
- Nucleosides and Nucleotides Degradation and Recycling	2	4
- Secondary Metabolites Degradation	15	14
Generation of precursor metabolites and energy	26	15
Signal transduction pathways	0	21
Total	232	230

Fig. 6.5 This table presents statistics on the number of pathways present in each pathway class for the two *E. coli* strains under comparison. The two largest top-level classes, Biosynthesis and Degradation/Utilization/Assimilation, are broken down further to show the distribution of pathways among their next-level subclasses. The vast majority of pathways are assigned to only a single class. However, a small number may be assigned to more than one class

Pathway Holes	E. col CFT073	E. coli K-12
Number of Pathway Holes	328	22
Pathway Holes as a percentage of total reactions in pathways	43%	3%
Pathways with No Holes	68	190
Pathways with 1 Hole	64	9
Pathways with 2 Holes	31	1
Pathways with 3 Holes	23	2
Pathways with 4 Holes	13	0
Pathways with 5 Holes	14	0
Pathways with > 5 Holes	10	1
Total Pathways with Holes	'55	13

Fig. 6.6 A pathway hole is a reaction in a pathway for which no corresponding enzyme has been identified in the genome. Pathway holes may exist for a number of reasons: They may represent true enzymatic functions in the organism for which the gene has not yet been found, or they could represent false positive pathway predictions or cases in which the pathway in this organism differs slightly from the reference pathway in MetaCyc. This table counts all the pathway holes in each organism, and classifies pathways based on their number of pathway holes

a new table showing a list of all data elements within the columns of that row. For example, clicking on the row heading "Total Pathways with Holes" in Fig. 6.6 will produce Fig. 6.7, which shows every pathway containing at least one pathway hole (a reaction that has no assigned enzyme) in these *E. coli* strains.

Pathway Holes: Total Pathways with Holes	E. coli CFT073	E. coli K-12
(deoxy)ribose phosphate degradation	1	
acetoacetate degradation I (to acetyl CoA)	1	0
acetyl CoA fermentation to butyrate	4	
acrylonitrile degradation	2	
adenosylcobalamin salvage from ccbinamide and cobalamin	5	0
aerobic respiration -- electron donors reaction list	3	0
alanine biosynthesis I	1	0
alanine biosynthesis II	1	0
alanine degradation I	1	0
alanine degradation II (to D-lactate)	2	
aldoxime cegradation	2	
aminopropanol biosynthesis	1	0
aminopropylcadaverine biosynthesis	1	0
APS pathway of sulfate reduction	5	
arginine biosynthesis IV	1	
arginine degradation VII	2	
ascorbate biosynthesis I	5	
asparagine biosynthesis III	1	0
benzoyl-CoA degradation I (aerobic)	9	
chorismate biosynthesis	2	0
citrate fermentation to diacetyl	4	
CMP-KDO biosynthesis I	2	0
coenzyme A biosynthesis	2	0
colanic acid building blocks biosynthesis	2	0
D-allose degradation	2	0
D-arabinose degradation I	1	0
D-arabinose degradation II	1	
D-arabitol degradation	1	

Fig. 6.7 A listing of pathways in two *E. coli* PGDBs that contain pathway holes. The listing is truncated

6.3 Significance for Computational Biology

By computational biology we mean analysis of biological systems using computational methods; 51% of our survey responders said they use EcoCyc for computational biology, such as in the following areas.

6.3.1 Significance for Microbial Genome Analysis

A flood of nucleotide sequence data from microbial genomes is upon us. The genomes of more than 500 microorganisms—cultured and uncultured—have been completely sequenced, and many more will be completed in the next 5 years. Accurate, extensive analysis of these data is essential to permit them to be fully exploited in applications in medicine and biotechnology.

EcoCyc allows microbial-genome projects to produce more accurate annotations of sequenced genes, and to predict the metabolic pathways of their organisms. When gene function predictions are performed using sequence-similarity programs such as BLAST and FASTA, newly sequenced microbial genes often show similarity to *E. coli* genes. Researchers turn to EcoCyc as a source of information about *E. coli* gene function because EcoCyc is updated so frequently with literature-based information. Because *E. coli* is the genome with the highest fraction of its gene functions established experimentally, annotators for other microbial genomes are well advised to prefer sequence-similarity matches to *E. coli* genes over matches with similar scores from other organisms, to minimize the transitive annotation problem. Transitive annotation can decrease the accuracy of sequence annotation by transferring gene functions from one gene to another through long chains of similarity matches, each of which increases the likelihood of an incorrect functional prediction. Although EcoCyc curation in the 1990s focused on those genes whose products encode enzymes in metabolic pathways, it now contains rich annotations of all characterized *E. coli* genes.

In addition to predicting gene function, many scientists are using EcoCyc pathway data to predict the metabolic pathways of genomes they sequence. That prediction occurs by combining the PathoLogic module of Pathway Tools in combination with the larger MetaCyc pathway database (Caspi et al. 2008). Twice per year, SRI propagates updates to EcoCyc metabolic pathways and enzymes to MetaCyc. MetaCyc version 12.0 describes 1,036 experimentally elucidated pathways from 1,108 organisms. PathoLogic predicts the pathways of an organism by matching enzymes in the organism's annotated genome against enzymes in MetaCyc pathways, to predict which pathways from MetaCyc are present in the organism. To date, 1,300 groups have licensed Pathway Tools and MetaCyc from SRI, and tell us they are applying the software to at least 200 genomes.

As antibiotic-resistant bacteria become more prevalent, pharmaceutical companies are seeking novel microbial drug targets. Some companies are targeting enzymes within metabolic pathways (Karp 1997, 2003). Because EcoCyc improves our ability to predict the metabolic pathways of a microbe from its genomic sequence, it facilitates development of new pharmaceuticals (Karp 1997, 2003, Karp et al. 1999), such as its use by Bristol-Myers Squibb to find drug targets in *Streptococcus pneumoniae* (Thanassi et al. 2002).

6.3.2 Significance for Global Biological Studies

Because the EcoCyc data are structured within a sophisticated ontology that is amenable to computational analyses, EcoCyc allows scientists to ask questions spanning the entire genome of *E. coli*, the known metabolic network of *E. coli*, the known transport complement of *E. coli*, the known genetic regulatory network of *E. coli*, and combinations thereof. A surprisingly diverse array of systems biology studies is being fueled by EcoCyc: 40% of our survey responders said they use EcoCyc for large systematic biological studies. As we add new types of data to EcoCyc, we facilitate new types of global studies. For example, addition of new types of regulatory mechanisms will accelerate global studies of these mechanisms.

EcoCyc was used to develop methods for computing shortest path lengths within metabolic networks. These methods were used to study the topological organization of the *E. coli* metabolic network (Ravasz et al. 2002), and to investigate correlations between path lengths and factors such as genome distance between enzymes (Arita 2004, Simeonidis et al. 2003).

EcoCyc was used in several studies relating protein structure to the metabolic network. One study compared the small-molecule metabolism enzymes of yeast and *E. coli* to see which were conserved (Jardine et al. 2002). Two related studies surveyed the structural anatomy of EcoCyc pathways (Teichmann et al. 2001a,b). Two studies considered the organization of *E. coli* metabolic enzymes into protein families using EcoCyc (Rison and Thornton 2002, Tsoka and Ouzounis 2001). EcoCyc was used as a source of information on metabolic enzymes in a study that correlated sequence and functional relatedness in enzymes (Pellegrini et al. 1999).

EcoCyc was used as a source of transcriptional regulatory network information for analysis of genome-wide transcriptional regulatory networks (Ma et al. 2004), and was used to understand patterns in transcriptional control (Shen-Orr et al. 2002). EcoCyc pathways were used as a source of functionally related proteins for a study of the correlation between protein levels—evaluated based on codon bias—and functional relationship (Lithwick and Margalit 2005).

Van Dien et al. drew on EcoCyc to interpret label-tracing experiments in *Methylobacterium extorquens* to estimate flux rates through its metabolic network (Van Dien et al. 2003). Cases et al. used EcoCyc to investigate the fraction of the genome devoted to transcription-related proteins, small-molecule metabolism enzymes, and transport, for 60 bacterial genomes classified by lifestyle (Cases et al. 2003). Peregrin-Alvarez et al. used EcoCyc to study the phylogenetic extent of metabolic enzymes and pathways throughout all taxonomic domains (Peregrin-Alvarez et al. 2003).

6.4 Significance for Bioinformatics Research

The development of many new bioinformatics methods requires high-quality gold-standard datasets for training and validation of those methods; 21% of our survey responders said they use EcoCyc as a gold-standard dataset for developing bioinformatics algorithms, and 58% said they use EcoCyc for bioinformatics. As we add new types of data to EcoCyc, we facilitate development of new bioinformatics methods, for example, addition of new types of regulatory mechanisms will enable development of new predictors for those types of regulation.

Genome context methods for predicting gene function, such as phylogenetic profiles, conserved chromosomal adjacency, and the Rosetta Stone method, have been one of the major developments in bioinformatics in the last 5 years. EcoCyc played a key role in their development (Bowers et al. 2004, Enault et al. 2003, von Mering et al. 2003). EcoCyc was used to determine whether proteins that appear to share regulatory sequences might be functionally related (Studholme et al. 2004).

EcoCyc data were used to develop computational methods for predicting other key biological relationships, such as protein-protein interactions (Bowers et al. 2004, Tsoka and Ouzounis 2000), and to compute correspondences among atoms in reactants and products in biochemical reactions (Arita 2003).

EcoCyc was used as a gold standard for developing analytic and predictive computer programs. It has been used in operon prediction (Price et al. 2005, Romero and Karp 2004, Steinhauser et al. 2004) as well as for predicting promoters and transcription start sites (Burden et al. 2005, Gordon et al. 2003). EcoCyc was used as the source of metabolic pathways for genome-wide prediction of protein functions and interactions (Marcotte et al. 1999). The EcoCyc class hierarchy was used to categorize proteins for generating phylogenetic profiles (Pellegrini et al. 1999).

EcoCyc was consulted for compound-related information in a C-14-glucose radio-labeling study that followed the time dependence of various metabolite pools (Tweeddale et al. 1999). EcoCyc proved useful for investigating details of various proteins in a project to construct a whole-cell simulation (Tomita et al. 1999).

6.5 Significance for Model-Organism Database Development

In addition to *E. coli* serving as a model organism for microbial research, EcoCyc has become a model for development of bioinformatics database development for other organisms. The Pathway Tools software underlying EcoCyc is now being used in the development of many other organism-specific databases. Web links to these databases can be found at http://BioCyc.org.

Databases include

- Microbes: *Saccharomyces cerevisiae, Candida albicans, Streptomyces coelicolor, Pseudomonas aeruginosa, Rhizobium etli, Brucella suis, Coxiella burnetii, Rickettsia typhi*
- Plants: *Arabidopsis thaliana, Medicago truncatula,* multiple *Solanaceae species*
- Mammals: *Mus musculus*

6.6 Significance for Metabolic Engineering

Metabolic engineers alter microbes to produce biofuels, to produce flavor enhancers in food, to increase efficiency of production of bioproducts such as amino acids and vitamins, to produce pharmaceuticals, and to degrade toxic pollutants (Bailey 1991, Stephanopoulos and Vallino 1991). The Department of Energy GTL Project seeks to engineer microbes to solve problems of global carbon sequestration and environmental remediation (Frazier et al. 2003). The late Jay Bailey described many metabolic-engineering case studies in which heterologous proteins are introduced into cells to alter their metabolism (Bailey 1991). He wrote "No universal principles have emerged from metabolic engineering research to guide the choice of the next useful genetic alteration... there is no substitute for knowledge of the pathways involved, their regulation, and their kinetics" (Bailey 1991). Metabolic engineers consult EcoCyc and MetaCyc to select the optimal enzyme for an engineering problem, to predict undesirable side effects of a metabolic alteration, and to predict the metabolic network of their workhorse organism using Pathway Tools; 25% of our survey responders said they use EcoCyc for metabolic engineering.

The Palsson group has drawn heavily from EcoCyc to prepare quantitative flux balance models of the *E. coli* metabolic network (Edwards and Palsson 2000, Reed and Palsson 2003, Reed et al. 2003). We have recently collaborated with the Palsson group to further develop new versions of our respective models of the network (Feist et al. 2007). The Palsson group also used EcoCyc to validate results from *in silico* modeling of genome-scale *E. coli* metabolism (Reed and Palsson 2004). Other metabolic engineering studies making use of EcoCyc include (Chassagnole et al. 2002, Jardine et al. 2002, Weber et al. 2002)

6.7 Significance for Education

Of our survey responders, 20% said they use EcoCyc in graduate or undergraduate classes that they teach. The classes include Metabolic Network Analysis; Microbial Physiology; Introduction to Bioinformatics; Molecular Genetics; Genomics, Proteomics and Systems Biology; and Microbial Biotechnology. Dr. R. Gunsalus of the University of California Los Angeles is developing a Web portal to EcoCyc for use in undergraduate microbiology education.

Acknowledgments EcoCyc has greatly benefited from the involvement and guidance of M. Riley, J. Ingraham, R. Gunsalus, I. Paulsen, and J. Collado-Vides. The graph-tracks capability was suggested by R. Landick and implemented by J. Myers. EcoCyc is supported by grant GM077678 from the National Institutes of Health. The contents of this publication are solely the responsibility of the authors and do not necessarily represent the official views of the National Institutes of Health.

References

Arita M (2003) *In silico* atomic tracing by substrate-product relationships in *Escherichia coli* intermediary metabolism. Genome Res 13(11):2455–66

Arita M (2004) The metabolic world of *Escherichia coli* is not small. Proc Natl Acad Sci USA 101(6):1543–7

Bailey JE (1991) Toward a science of metabolic engineering. Science 252(5013):1668–75
Bowers PM, Pellegrini M, Thompson MJ et al. (2004) Prolinks: a database of protein functional linkages derived from coevolution. Genome Biol 5(5):R35
Burden S, Lin YX, Zhang R (2005) Improving promoter prediction for the NNPP2.2 algorithm: a case study using *Escherichia coli* DNA sequences. Bioinformatics 21(5):601–7
Cases I, de Lorenzo V, Ouzounis CA (2003) Transcription regulation and environmental adaptation in bacteria. Trends Microbiol 11(6):248–53
Caspi R, Foerster H, Fulcher CA et al. (2008) The MetaCyc Database of metabolic pathways and enzymes and the BioCyc collection of Pathway/Genome Databases. Nucleic Acids Res 36(Database issue):D623–31
Chassagnole C, Letisse F, Diano A et al. (2002) Carbon flux analysis in a pantothenate overproducing *Corynebacterium glutamicum* strain. Mol Biol Rep 29(1–2):129–34
Edwards JS, Palsson BO (2000) The *Escherichia coli* MG1655 *in silico* metabolic genotype: its definition, characteristics, and capabilities. Proc Natl Acad Sci USA 97(10):5528–33
Enault F, Suhre K, Poirot O et al. (2003) Phydbac (phylogenomic display of bacterial genes): An interactive resource for the annotation of bacterial genomes. Nucleic Acids Res 31(13): 3720–2
Feist AM, Henry CS, Reed JL et al. (2007) A genome-scale metabolic reconstruction for *Escherichia coli* K-12 MG1655 that accounts for 1260 ORFs and thermodynamic information. Mol Syst Biol 3:121
Frazier ME, Johnson GM, Thomassen DG et al. (2003) Realizing the potential of the genome revolution: the genomes to life program. Science 300(5617):290–3
Gama-Castro S, Jimenez-Jacinto V, Peralta-Gil M et al. (2008) RegulonDB (version 6.0): gene regulation model of *Escherichia coli* K-12 beyond transcription, active (experimental) annotated promoters and Textpresso navigation. Nucleic Acids Res 36(Database issue):D120–4
Gordon L, Chervonenkis AY, Gammerman AJ et al. (2003) Sequence alignment kernel for recognition of promoter regions. Bioinformatics 19(15):1964–71
Jardine O, Gough J, Chothia C et al. (2002) Comparison of the small molecule metabolic enzymes of *Escherichia coli* and *Saccharomyces cerevisiae*. Genome Res 12(6):916–29
Karp PD (1997) Use of metabolic databases to guide target selection for anti-microbial drug design. Blackwell Science Ltd., Oxford, UK
Karp PD (2003) The Pathway Tools software and its role in anti-microbial drug discovery. Marcel Dekker, Inc., New York
Karp PD, Keseler IM, Shearer A et al. (2007) Multidimensional annotation of the *Escherichia coli* K-12 genome. Nucleic Acids Res 35(22):7577–90
Karp PD, Krummenacker M, Paley S et al. (1999) Integrated pathway-genome databases and their role in drug discovery. Trends Biotechnol 17(7):275–81
Karp PD, Ouzounis CA, Moore-Kochlacs C et al. (2005) Expansion of the BioCyc collection of pathway/genome databases to 160 genomes. Nucleic Acids Res 33(19):6083–9
Lithwick G, Margalit H (2005) Relative predicted protein levels of functionally associated proteins are conserved across organisms. Nucleic Acids Res 33(3):1051–7
Ma HW, Kumar B, Ditges U et al. (2004) An extended transcriptional regulatory network of *Escherichia coli* and analysis of its hierarchical structure and network motifs. Nucleic Acids Res 32(22):6643–9
Marcotte EM, Pellegrini M, Thompson MJ et al. (1999) A combined algorithm for genome-wide prediction of protein function. Nature 402(6757):83–6
Pellegrini M, Marcotte EM, Thompson MJ et al. (1999) Assigning protein functions by comparative genome analysis: protein phylogenetic profiles. Proc Natl Acad Sci USA 96(8): 4285–8
Peregrin-Alvarez JM, Tsoka S, Ouzounis CA (2003) The phylogenetic extent of metabolic enzymes and pathways. Genome Res 13(3):422–7
Price MN, Huang KH, Alm EJ et al. (2005) A novel method for accurate operon predictions in all sequenced prokaryotes. Nucleic Acids Res 33(3):880–92
Ravasz E, Somera AL, Mongru DA et al. (2002) Hierarchical organization of modularity in metabolic networks. Science 297(5586):1551–5

Reed JL, Palsson BO (2003) Thirteen years of building constraint-based *in silico* models of *Escherichia coli*. J Bacteriol 185(9):2692–9
Reed JL, Palsson BO (2004) Genome-scale *in silico* models of *E. coli* have multiple equivalent phenotypic states: assessment of correlated reaction subsets that comprise network states. Genome Res 14(9):1797–805
Reed JL, Vo TD, Schilling CH et al. (2003) An expanded genome-scale model of *Escherichia coli* K-12 (iJR904 GSM/GPR). Genome Biol 4(9):R54
Rison SC, Thornton JM (2002) Pathway evolution, structurally speaking. Curr Opin Struct Biol 12(3):374–82
Romero PR, Karp PD (2004) Using functional and organizational information to improve genome-wide computational prediction of transcription units on pathway-genome databases. Bioinformatics 20(5):709–17
Salgado H, Gama-Castro S, Martinez-Antonio A et al. (2004) RegulonDB (version 4.0): transcriptional regulation, operon organization and growth conditions in *Escherichia coli* K-12. Nucleic Acids Res 32(Database issue):D303–6
Shen-Orr SS, Milo R, Mangan S et al. (2002) Network motifs in the transcriptional regulation network of *Escherichia coli*. Nat Genet 31(1):64–8
Simeonidis E, Rison SC, Thornton JM et al. (2003) Analysis of metabolic networks using a pathway distance metric through linear programming. Metab Eng 5(3):211–9
Steinhauser D, Junker BH, Luedemann A et al. (2004) Hypothesis-driven approach to predict transcriptional units from gene expression data. Bioinformatics 20(12):1928–39
Stephanopoulos G, Vallino JJ (1991) Network rigidity and metabolic engineering in metabolite overproduction. Science 252(5013):1675–81
Studholme DJ, Bentley SD, Kormanec J (2004) Bioinformatic identification of novel regulatory DNA sequence motifs in *Streptomyces coelicolor*. BMC Microbiol 4(14):14
Teichmann SA, Rison SC, Thornton JM et al. (2001a) The evolution and structural anatomy of the small molecule metabolic pathways in *Escherichia coli*. J Mol Biol 311(4):693–708
Teichmann SA, Rison SC, Thornton JM et al. (2001b) Small-molecule metabolism: an enzyme mosaic. Trends Biotechnol 19(12):482–6
Thanassi JA, Hartman-Neumann SL, Dougherty TJ et al. (2002) Identification of 113 conserved essential genes using a high-throughput gene disruption system in *Streptococcus pneumoniae*. Nucleic Acids Res 30(14):3152–62
Tomita M, Hashimoto K, Takahashi K et al. (1999) E-CELL: software environment for whole-cell simulation. Bioinformatics 15(1):72–84
Tsoka S, Ouzounis CA (2000) Prediction of protein interactions: metabolic enzymes are frequently involved in gene fusion. Nat Genet 26(2):141–2
Tsoka S, Ouzounis CA (2001) Functional versatility and molecular diversity of the metabolic map of *Escherichia coli*. Genome Res 11(9):1503–10
Tweeddale H, Notley-McRobb L, Ferenci T (1999) Assessing the effect of reactive oxygen species on *Escherichia coli* using a metabolome approach. Redox Rep 4(5):237–41
Van Dien SJ, Strovas T, Lidstrom ME (2003) Quantification of central metabolic fluxes in the facultative methylotroph *Methylobacterium extorquens* AM1 using 13C-label tracing and mass spectrometry. Biotechnol Bioeng 84(1):45–55
von Mering C, Zdobnov EM, Tsoka S et al. (2003) Genome evolution reveals biochemical networks and functional modules. Proc Natl Acad Sci USA 100(26):15428–33
Weber J, Hoffmann F, Rinas U (2002) Metabolic adaptation of *Escherichia coli* during temperature-induced recombinant protein production: 2. Redirection of metabolic fluxes. Biotechnol Bioeng 80(3):320–30

Chapter 7
Analysis of *E. coli* Network

Hawoong Jeong

Contents

Abstract Diverse complex systems such as cells, Internet and society can be mapped into networks by simplifying each constituent as a node and their interaction as a link. Traditionally it has been considered that these networks are random, but recent series of studies show that they are far from being random and have common inhomogeneous topology through generic self-organizing process. In this chapter, we briefly introduce the network analysis methods which were re-developed in statistical physics community recently. First, we introduce basic complex network models such as Erdős-Rényi model, small-world model, scale-free model which were developed to describe complex systems. And then, we applied these methods to biological system, such as metabolic network and protein-protein interaction network of *E. coli*. We measure the global and local characteristics of the network structure. Finally we briefly review recent works on biological networks especially on dynamic aspect.

H. Jeong (✉)
Department of Physics, Institute for the BioCentury, KAIST, Daejeon, 305-701, Korea
e-mail: hjeong@kaist.edu

S.Y. Lee (ed.), *Systems Biology and Biotechnology of Escherichia coli*,
DOI 10.1007/978-1-4020-9394-4_7,

7.1 Introduction: Complex Bio-Networks

During the latter half of the 20th century, biology has been dominated by reductionist approaches that have provided a wealth of knowledge about individual cellular components and their functions. Typically, these approaches have entailed careful examination of a limited number of individual components in a biological system, hypothesis building based on the empirical observations, and further experiments to test these hypotheses. Reflecting the value of following this approach, biomedical researchers from a range of disciplines have deliberately restricted their analyses to well-defined systems with relatively few components, implicitly attempting to reduce biological phenomena to the behavior of individual molecules.

Despite the enormous success of the reductionist approach, a discrete biological function can only rarely be attributed to an individual molecule. Indeed, most biological functions arise from complex interactions among its various components (individual proteins, nucleic acids, small molecules, etc.). The need for more comprehensive approaches that address the full complexity of a biological system has now surfaced, largely with the emergence of genomics, in which the entire DNA sequences for a number of organisms now allows the definition of their gene portfolios. Extrapolation between genomes has accelerated the definition of what amounts to a "parts catalog" of cellular components in a large number of organisms. Also, large-scale efforts for studying the effects of systematic gene disruptions and for measuring expression levels of all genes under different conditions by microarray and proteomics approaches for entire genomes are well underway.

In turn, these advances have created an unprecedented opportunity towards developing a comprehensive understanding of biological systems, in part through the identification of the fundamental logic and derivative constraints that limit cell behavior. While the datasets available to us are far from being complete, they do offer a critical mass and coherency for such analyses, and for the subsequent capacity for model development and prediction through simulation of the ensuing model. Therefore, it has been studied to identify such underlying constraints and to model in quantitative terms the structure and functional (including regulatory) properties of the complex biological networks that maintain proper functioning various organisms. This analysis is aided by the coincidence of two recent scientific developments: the emergence of databases containing integrated data on the topology of various networks of biological significance, and the recent advances in understanding and quantifying the topology of complex (non-biological) networks which we are going to review in the next sections.

This chapter has been organized as follows. In Section 7.2, we introduce several basic network models which were developed to describe the ubiquitous complex networks found in real world. In Section 7.3, we analyzed metabolic network and protein-protein interaction network of *E. coli* in details. Section 7.4 includes recent advance in *network-biology* especially about the dynamic aspect of bio-network analysis. Most of this chapter was taken from recent papers written by the author (Eom et al. 2006, Jeong 2003, Kim et al. 2007).

7.2 Simple Models of Complex Networks

Modeling complex networks has a long history, and has been particularly active as a branch of combinatorial graph theory. However, the study of random networks in association with the real-world networks such as information systems, economic systems, and biological systems has begun recently. In this section, we briefly review a few important theoretical network models, and discuss recent empirical results on the network topology, which indicate the need for new approaches in understanding network development and describing their topology.

7.2.1 Erdős-Rényi Random Network Model

The most investigated random network model has been introduced by two Hungarian mathematicians, Erdős and Rényi (ER) (Bollobas 1985, Erdős and Rényi 1960) (see Fig. 7.1a), who were the first to study the statistical aspect of random

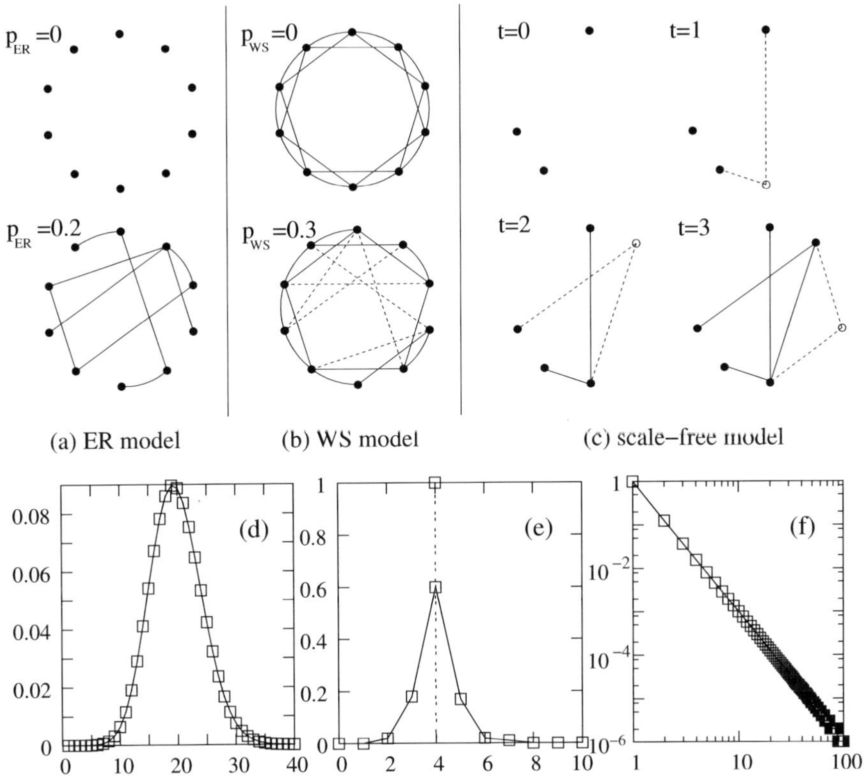

Fig. 7.1 Examples of model networks (**a**) Erdős-Rényi network, (**b**) Watts-Strogatz Small-world network, (**c**) Barabasi-Albert scale-free network. Typical degree distribution of (**d**) ER (**e**) SW (**f**) SF networks

graphs using the probabilistic method. The popularity of the ER model lies in its simplicity: It assumes that all vertices are equivalent, and any pair of vertices is connected with the same probability p_{ER}. ER discovered that many properties of random graphs, such as the emergence of trees or cycles, appear quite suddenly at a threshold value $p_{ER}(N)$. Within the physical literature, the ER model is known as infinite-dimensional percolation, belonging to the universality class of the mean field percolation (Stauffer and Aharony 1992). To compare the ER model with other network models, we need to focus on the connectivity distribution. As ER have shown, the probability that a vertex has k edges follows a Poisson distribution $P(k) = e^{-\lambda}k^{\lambda}/k!$, where the expectation value of degree $\langle k \rangle = \lambda$ is $(N-1)p_{ER}$, therefore ER network exhibits random and homogeneous structure (See Fig. 7.1d). However, it was found that degree distribution of most real world networks is far from being random which leads us to develop new network model.

7.2.2 Small-World Network Model

In 1998 Watts and Strogatz (WS) reported that many systems display both a high degree of local clustering reminiscent of finite-dimensional lattices (for example, a square lattice), and small-world phenomena characterizing random networks. Local clustering describes the tendency of groups of nodes to be all connected to each other, while small-world phenomena describes the property that any two nodes in the system can be connected by relatively short paths. To account for the transition from the local order to the small world behavior, they introduced the small-world network model (see Fig. 7.1b) (Watts and Strogatz 1998), which has been investigated rather intensely lately (Barthelemy and Amaral 1999a, Suki et al. 1998). In this model, starting from a regular lattice, each link between nodes is rewired with probability p_{WS}, such that long range link can be formed to ensure small-world characteristics. The connectivity distribution of the WS model depends on the parameter p_{WS}: for $p_{WS} = 0$, $P(k)$ is narrowly peaked at the average connectivity of the regular lattice, while for finite p_{WS}, $P(k)$ gets broader, converging to the Poisson connectivity distribution of the ER random graph (See Fig. 7.1e), which again turns out to be not appropriate to describe the inhomogeneous topology of the real world networks.

7.2.3 Barabasi-Albert Scale-Free Network Model

All existing network models we have considered so far fail to incorporate two generic aspects of real networks. First, they assume that networks have a fixed number of nodes. In contrast, most networks form and grow by the continuous addition of new nodes, that link to the nodes already present in the system. For example, the Internet expands by the attachment of new communication devices and routers to the system, and the World-Wide Web (WWW) grows by the addition of new web pages and domains. Second, the models assume that the probability that two vertices are connected is random and uniform. In contrast, most real networks exhibit preferential connectivity. For example, new Internet domains are preferentially linked to major highly connected routers (nodes) to obtain broader bandwidth, or a newly created webpage will more likely link to well known, popular webpages with already high connectivity. Consequently, the probability with which a new node is connected to the existing nodes is not uniform, but there is a higher probability to be connected to a node that already has a large number of links (Fig. 7.1c). Barabasi et al. demonstrated that these two ingredients are sufficient to explain the inhomogeneous power-law distribution observed in real networks (Barabasi and Albert 1999). The network generated by this model evolves into a scale-invariant state, the probability that a node has k edges following $P(k) \sim k^{-3}$, i.e., a power-law with an exponent $\gamma = 3$ (See Fig. 7.1f). Furthermore, the Barabasi's group showed that excluding any of the two ingredients will eliminate the power-law connectivity (Albert and Barabasi 2000) and they developed a continuum theory (Barabasi et al. 1999) that allowed them to calculate the exponent γ, and predict the dynamics of the scale-free network. And they have also shown that the power-law distribution is robust against various local actions on the network structure, such as establishing links between existing nodes, or rerouting existing links from one node to another (Albert and Barabasi 2000). While these events can modify the scaling exponent γ, they do not eliminate the inhomogeneous nature of the network connectivity. The user's main goal is to maximize the benefits of the online environment, which can be best achieved by connecting to nodes where the best service is available, a flocking attitude that eventually leads to a few highly connected nodes and power laws. Consequently, complex communication networks inevitably evolve to develop scale-free network connectivity, and thus display topological inhomogeneities. (Albert et al. 1999b, Huberman and Adamic 1999)

7.3 Topology of Biological Networks

It is increasingly appreciated that the robustness of various cellular processes is rooted in the dynamic interactions among its many constituents (Barkai and Leibler 1997, Bhalla and Iyengar 1999, Yi et al. 2000), such as proteins, DNA, RNA, and small molecules. The existence of complex interactions among various components of a cell or simple microorganisms has long been appreciated, but in the absence of large-scale databases and a sufficiently developed theoretical framework, no meaningful analysis of these interactions was deemed possible. However, recent large-scale sequencing projects coupled with systematic two-hybrid analyses have provided complete sequence information for a number of genomes, and also allowed the development of protein interaction-(Rain et al. 2001a, Uetz et al. 2000) and integrated pathway-genome databases (Kanehisa and Goto 2000, Karp et al. 1999, Overbeek et al. 2000) that provide organism-specific connectivity maps of metabolic- and, to a lesser extent, various other cellular networks. Yet, due to the large number and the diversity of the constituents and reactions forming such networks, these maps are extremely complex, offering only limited insight into the organizational principles of these systems. Our ability to address in quantitative terms the structure of these cellular networks, however, has benefited from recent advances in understanding the generic properties of complex networks (Albert et al. 2000, Watts and Strogatz 1998), which will be described in this section.

7.3.1 Network Analysis Methods

Until recently, complex networks have been modeled using the classical random network theory (Bollobas 1985, Erdős and Rényi 1960) which assumes that each pair of nodes (i.e., constituents) in the network is connected randomly with probability p. This process leads to a statistically homogeneous network, in which most nodes have approximately the same number of links, $\langle k \rangle$ (Fig. 7.1a). On the other hand, recent empirical studies on the structure of the World-Wide Web (Albert et al. 1999a), Internet (Faloutsos et al. 1999), and social networks (Barabasi and Albert 1999) have demonstrated that these systems are described by scale-free networks (Barabasi and Albert 1999) (Fig. 7.1c), for which degree distribution *P(k)* follows a power-law, i.e. $P(k) \sim k^{-\gamma}$. Unlike exponential networks, scale-free networks are extremely heterogeneous, their topology being dominated by a few highly connected nodes (hubs) which link the rest of the less connected nodes to the system (Fig. 7.1c). This degree distribution *P(k)* is a good measure for analyzing connectivity of the complex network and also has been applied to several biological networks as well.

Another basic measure for network analysis is a clustering coefficient. The clustering coefficient C_i of node i is the ratio of the total number y of the links connecting its nearest neighbors to the total number of all possible links between all these nearest neighbors, $C_i = {}^{2y}/_{k_i(k_i-1)}$ where k_i is the degree of node i. The clustering coefficient of a network, C, is the average of this value over all the nodes. Most real networks have much larger value of clustering coefficient than model networks such as ER or BA network due to, e.g., the community or modular structure (Dorogovtsev and Mendes 2002). Finally, the assortativity r, which measures the correlation between degrees of node linked to each other, is defined as the Pearson correlation coefficient of degrees between pairs of nodes (Newman 2002). Positive values of r stand for the positive degree-degree correlation which means that nodes with large degrees tend to be connected to one another. Most social networks have this positive degree correlation $r > 0$ (assortative mixing), like the co-authorship network of arxiv.org network (Newman 2001). On the other hand, most biological and technological networks show negative degree correlation $r < 0$ (disassortative mixing), including protein interaction network (PIN) and Internet AS network. If there is no degree correlation among nodes (neutral), as in the case of BA model, the value of r is in the vicinity of 0. There is another convenient way to check the degree correlation, which is measuring the quantity $\langle k_{nn} \rangle = \Sigma_{k'} k' p(k'/k)$, i.e. the average degree of nearest neighbors of nodes with degree k (Pastor-Satorras et al. 2001). Assortative mixing is represented by a positive slope of the $\langle k_{nn}(k) \rangle$ graph, while the others by horizontal (neutral) or a negative slope (disassortative).

While these quantities are measures for global properties of the network, local properties of the network have been analyzed via *motif* analysis. Subgraph patterns and network motifs have been applied recently to understand the local structure of complex networks (Milo et al. 2004, 2002, Vazquez et al. 2004). Subgraph patterns consist of more than three nodes and the links connecting only these nodes, which represent the minimum subnetworks of complex networks. Examples of triad subgraph patterns are shown in Fig. 7.4a. Network motifs are the subgraph patterns that occur in a complex network at numbers that are significantly higher than those in a random network (Milo et al. 2002). These are believed to represent the simplest building blocks of complex networks and the topologically characteristic interaction patterns within complex networks. Recently, it was also shown that certain motifs have been enhanced through the evolution of a network, which supports the functional importance of the motifs (Vazquez et al. 2004). For example, in transcription networks, a biochemical network responsible for regulating the expression of genes in cells, the network motifs are thought to be circuit elements that perform key information processing functions (Mangan and Alon 2003, Milo et al. 2002, Shen-Orr et al. 2002). The feed-forward loop, one motif of transcription networks, can act as a circuit that reduces noise and responds only to a persistent signal.

The following algorithm is used to obtain the network motifs (Milo et al. 2002). We scanned for all possible three-node subgraphs in the network and recorded the number of occurrences of each subgraph. To identify a statistically significant subgraph pattern, we compared the network to an ensemble of suitably randomized networks. Each node in the randomized networks contained the same number of incoming and outgoing links as the corresponding node in the original network. In addition, the randomized networks that were used to estimate the significance of n-node subgraphs were generated to preserve the same number of appearances of all $(n - 1)$ node subgraphs as in the original network. For each subgraph i, the statistical significance of the subgraph is described by the Z score $Z_i = (N_i^{real} - \langle N_i^{rand} \rangle)/std(N_i^{rand})$. N_i^{rand} is the number of appearances of the subgraph i in the network, and $\langle N_i^{rand} \rangle$ and $std(N_i^{rand})$ are the average and standard deviation of its appearances in the ensemble of randomized networks, respectively. The subgraph pattern exhibiting a high Z score is the statistically significant pattern. In this analysis, the network motifs were selected when those subgraph patterns have a Z score greater than 2.

With this well-developed theoretical framework in hand and with the availability of detailed databases, we are now in position to initiate the analyses of complex bio-networks. Some of the first questions we asked included the following: What is the topological structure of metabolic and other cellular networks in global and local perspective? (See Fig. 7.2) What are the biologically and topologically relevant quantities that characterize them? Are there generic and common structural characteristics that apply to all cells, including both prokaryotes and eukaryotes? How are the specificity and the differential properties of various organisms reflected in the structure of these networks? In the following section we will summarize our results obtained on the large-scale structure of biochemical reaction pathways and protein interaction networks, especially for the case of *E. coli*, main topic of this book.

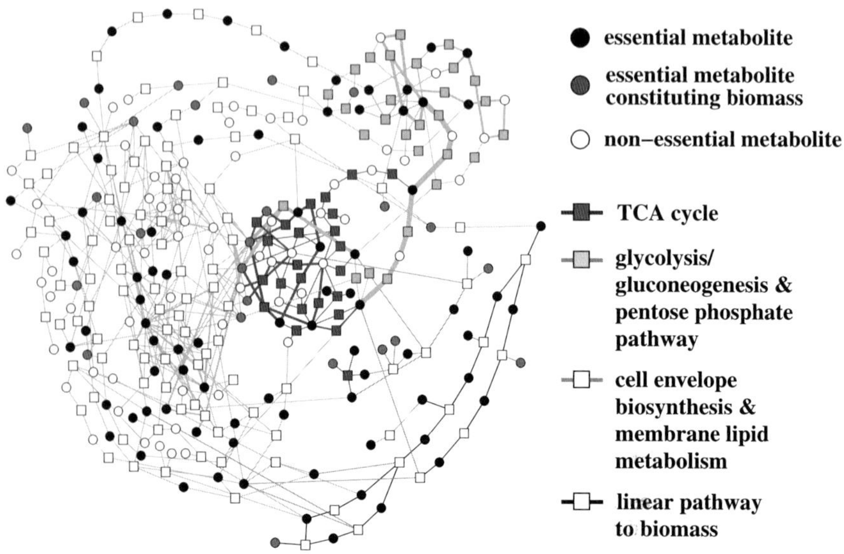

Fig. 7.2 Metabolic network including the central pathways and the membrane formation pathways. Circles denote essential and non-essential metabolites distinguished by the colors (*black*: essential metabolite; *gray*: essential metabolite constituting biomass; and *white*: non-essential metabolite). Cofactors are not drawn here because the number of the associated reactions is too large for visual examination. Each box represents the metabolic reaction for different functional classes specified by different colors and line styles

7.3.2 Metabolic Network of E. coli

To address the large-scale structural organization of metabolic networks, we have examined the topologic properties of the core metabolic network of 43 different organisms based on data deposited in the WIT (now ERGO) database (Jeong et al. 2000, Overbeek et al. 2000). In the metabolic network, nodes are substrates which are connected to each other through the actual metabolic reactions (Fig. 7.4B). As illustrated in Fig. 7.3a, results convincingly indicate that in *E. coli* the probability that a given substrate participates in k reactions follows a power-law distribution, i.e., the *E. coli* metabolic network belong to the class of scale-free networks. Furthermore, it is found that scale-free networks describe the metabolic networks in all organisms in all three domains of life, including 6 Archaea, 32 Bacteria, and 5 Eukaryotes, indicating the generic nature of this structural organization. Also, essentially identical results were obtained when we examined the topologic properties of the information transfer pathways of the 43 different organisms based on 'Information transfer' portions of data deposited in the WIT/ERGO database (Overbeek et al. 2000). Another general feature of many complex networks is their small-world character (Strogatz 2001, Watts and Strogatz 1998), i.e., any two nodes in the system can be connected by relatively short paths along existing links. In metabolic networks these paths correspond to the biochemical pathway connecting two substrates. The degree of interconnectivity of a metabolic network can be

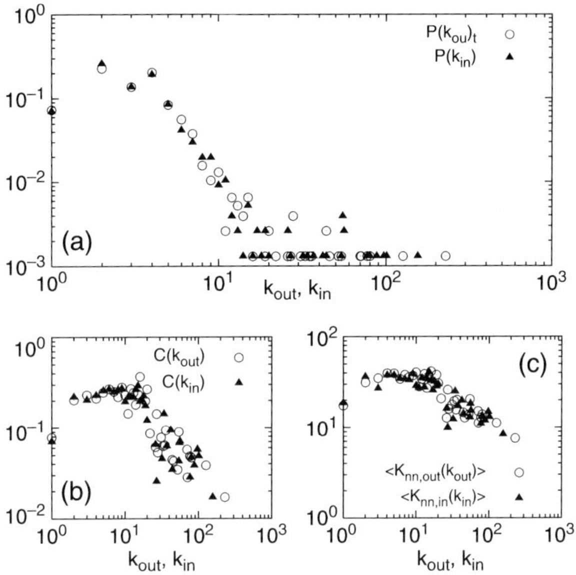

Fig. 7.3 Topological properties of *E. coli* metabolic network. **(a)** Degree distribution P(k), showing inhomogeneous structure for both in and out degrees, **(b)** Clustering coefficient C(k), showing typical decreasing behavior as a function of degree k like many other biological networks, **(c)** Assortativity, average degree of neighbor node $\langle K_{nn}(k)\rangle$, showing dissortative mixing again like many other biological networks

characterized by the network diameter, defined as the shortest biochemical pathway averaged over all pairs of substrates. For all non-biological networks they examined to date the average connectivity of a node is fixed, which implies that the diameter of the network increases logarithmically with the addition of new nodes (Barabasi and Albert 1999, Barthelemy and Amaral 1999b, Watts and Strogatz 1998). In contrast, we find that the diameter of the metabolic network is the same for all 43 organisms, irrespective of the number of substrates found in the given species (Jeong et al. 2000). This is surprising and unprecedented, and is possible only if with increasing organism complexity individual substrates are increasingly connected in order to maintain a relatively constant metabolic network diameter. Within the special characteristics of living systems this attribute may increase an organism's fitness to efficiently respond to external changes or internal errors. For example, the transition time between two metabolic steady states is apparently largely governed by time constants involved in changing the enzyme concentrations (Cascante et al. 1995), an attribute which could be best achieved when only a few alternative biochemical reactions need to be activated. In Fig. 7.3b, clustering coefficient of *E. coli* metabolic network shows $C(k) \sim k^{-\alpha}$ which represents the hierarchical and modular structures embedded in the biological networks (Ravasz et al. 2002). L ike other biological network, metabolic network of *E. coli* shows dissortative mixing (Fig. 7.3c), such that substrates with larger degrees (hubs) tend to interact with substrates with smaller degrees.

We also examined the triad subgraph patterns of metabolic networks of 43 organisms and identified their network motifs including *E. coli*. In this analysis, the direction of each link implies direction from an input substrate (educt) to an output substrate (product) (Fig. 7.4b) (Eom et al. 2006). We found that all metabolic networks have their own network motifs. To provide a more quantitative analysis,

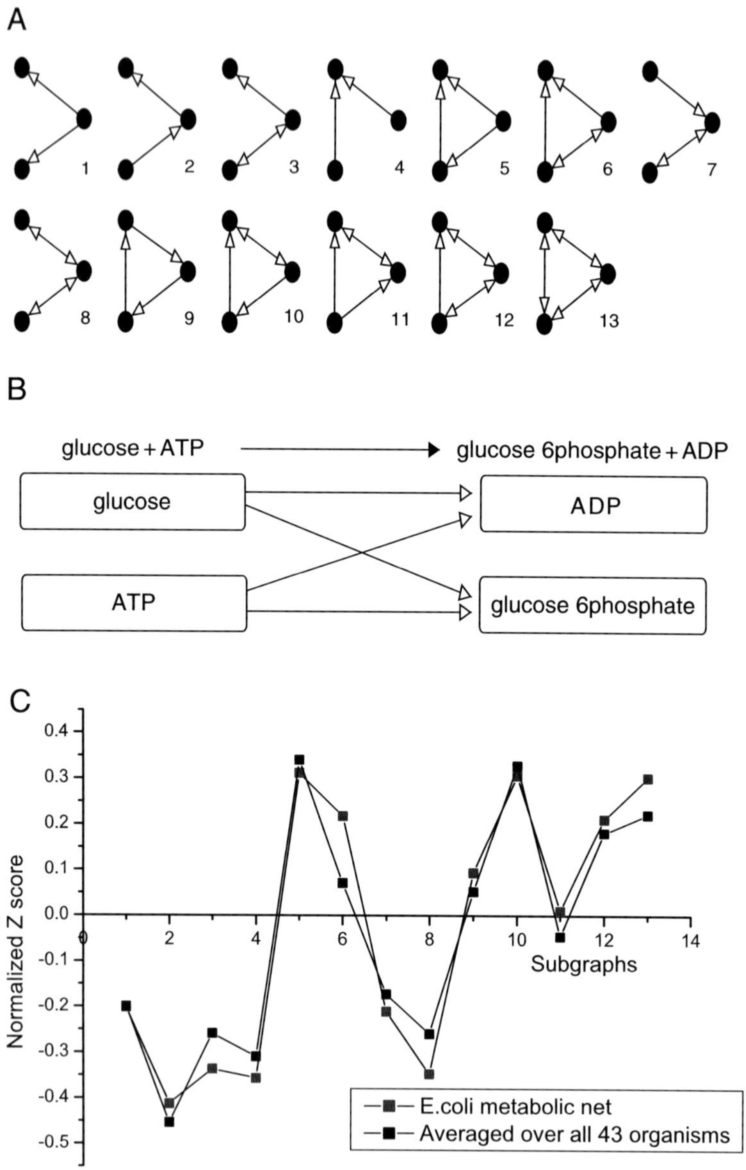

Fig. 7.4 Local properties of *E. coli* metabolic network. (**A**) Motif profile, all possible 13 types of three node connected subgraphs. (**B**) Graphical reorientation of a chemical reaction. (**C**) The triad significance profiles (TSPs) of metabolic networks. TSPs for *E. coli* and other organisms found in WIT database were plotted

we investigated the local structure of metabolic networks of each organism in detail and identified the significance profile (SP) of each metabolic network (Milo et al. 2004). The SP is the vector of Z scores normalized to a unit length, of which the i-th component is given by $SP_i = Z_i/(\Sigma_j Z_j^2)^{1/2}$. The SP of a given network represents the relative significance of the subgraphs in that network. It is important to compare networks of different sizes because network motifs in large networks tend to have higher Z scores than network motifs in small networks (Milo et al. 2004). The triad significance profile (TSP) for each metabolic network is presented in Fig. 7.4c. The TSPs of these networks are found to be almost insensitive to a removal of 20% of edges or to an addition of 20% new edges randomly, representing that the results are robust to possible missing or false-positive data errors. All metabolic networks showed similar TSPs and three network motifs of triads 5, 10, and 13 were found frequently. These motifs, especially 5 and 10, are well-known feed-forward loop and its variation of function is a prevalence of short detours in metabolic network (Gleiss et al. 2001, Heinrich and Schuster 1996). In contrast, triads 2, 4, and 8 were anti-motifs that were significantly underrepresented. The correlation coefficient between the TSPs of metabolic networks in 43 organisms was about 0.78 showing that metabolic networks have the same topological structure in both large-scale organization (inhomogeneous power-law degree distribution) and in local organization (sharing common topological substructures).

7.3.3 Protein Interaction Network of E. coli

Next example of biological network is protein interaction network (PIN). Proteins are traditionally defined by their individual actions as catalysts, signaling molecules, or building blocks of cells and microorganisms. However, recent integrative approaches view their role as an element in a network of protein–protein interactions with a ‘contextual’ or ‘cellular’ function within functional modules (Eisenberg et al. 2000, Hartwell et al. 1999). To uncover this role, it is important to assess the position of a protein within the protein–protein interaction network. We first have assessed the topologic characteristics of system-wide protein–protein interaction network found in the yeast, *S. cerevisiae*, and the bacterium, *H. pylori*, obtained mostly by systematic two-hybrid analyses (Ito et al. 2001, Rain et al. 2001b, Xenarios et al. 2000). Due to its size, a complete map of the yeast and *H. pylori* networks, while informative, in themselves offers little insight into their large-scale characteristics (See Fig. 7.5). Like other bio-networks, the probability that a given yeast protein interacts with k other yeast proteins follows a power-law (Jeong et al. 2001) with an exponential cutoff (Barthelemy and Amaral 1999a). This exponential cutoff is due to the physical limitation of the binding sites in the protein structure. A similar result was obtained for *H. pylori* as well. This indicates that the network of protein interactions in both a bacterium and an eukaryotic cell forms a highly inhomogeneous scale-free network. An important known consequence of the inhomogeneous structure is the network's simultaneous tolerance against random errors coupled with fragility

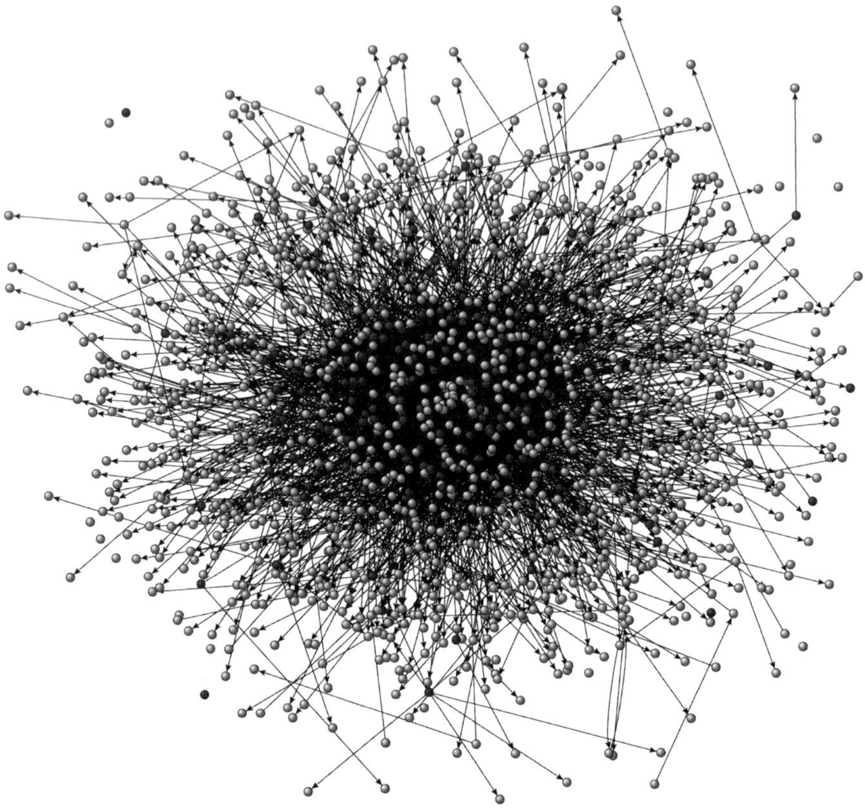

Fig. 7.5 Protein-Protein interaction network, essential (*gray*) and non-essential (*white*) proteins were connected through physical bindings

against the removal of the most connected nodes (Barabasi and Albert 1999). Yet, if there is indeed a biologically relevant functional link between topology and error tolerance, on average less connected proteins should prove less essential than highly connected ones. We calculated this correlation and showed that the likelihood that removal of a protein will prove to be lethal clearly correlates with the number of interactions the protein has. For example, while proteins with five or less links constitute 93% of the total number of proteins they find that only 21% of them are essential. In contrast, only 0.7% of the yeast proteins with known phenotypic profile have more than 15 links but single deletion of 62% of these proves lethal. This implies that highly connected proteins with a central role in the network's architecture are three times more likely to prove essential than proteins with low number of links to other proteins (Jeong et al. 2001).

We also analyzed PPI network of *E. coli* using protein complex data by G. Butland et al. (Butland et al. 2005). We found that again degree distribution of *E. coli* PPI network shows inhomogeneous scale-free degree distribution (Fig. 7.6a) and proteins with larger degrees are more essential than proteins with

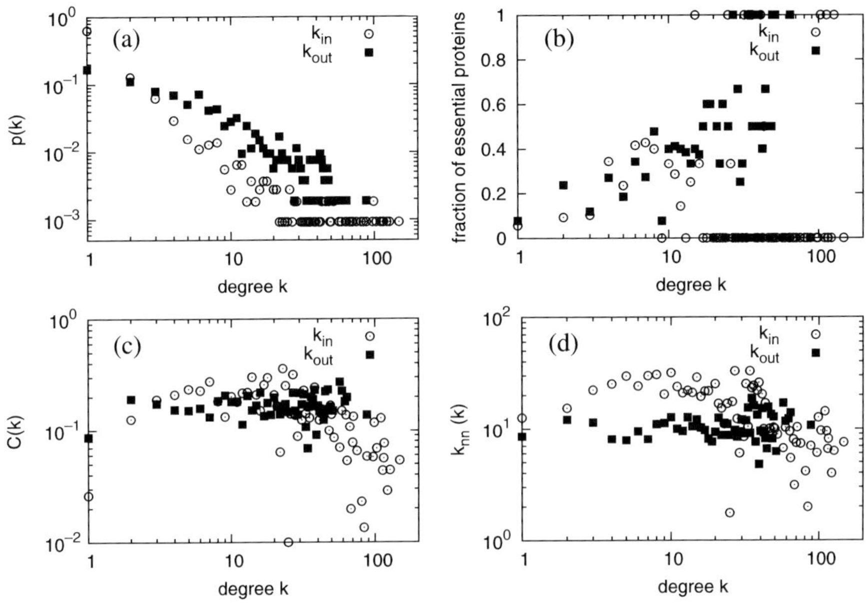

Fig. 7.6 Topological properties of protein-protein interaction network. (**a**) Degree distribution, (**b**) degree vs essentiality showing that as degree increases, lethality of the protein also increases. (**c**) Clustering coefficient C(k), (**d**) Assortativity, average degree of neighbor nodes

smaller degrees (Fig. 7.6b). Interestingly in Fig. 7.6c, it is observed that clustering coefficient *C(k)* shows relatively neutral behavior implying *E. coli* protein interaction network doesn't have hierarchical characteristics. Also the assortativity of *E. coli* protein interaction network seems to be neutral for outgoing link while it is dissortative for incoming link like many other biological networks (see Fig. 7.6d). Since PPI network by Han et al. is directed, we applied motif analysis algorithm to find relevant subgraph pattern hidden in *E. coli* protein interaction network. As seen in Fig. 7.7, quite different from the metabolic network, motif 11 is found more frequently and motif 10 is suppressed. However, motifs 5, 6, 12, 13 are shared with the metabolic network of *E. coli*.

7.4 Beyond Static Graph Analysis

So far, we have only considered spatial (geometrical) inhomogeneity of the complex networks, however it is also important to deal with temporal heterogeneity of the complex network. Links between nodes in the network can vary over time, for example, not every reaction in the metabolic network is active all the time. And the activity of each link in the metabolic network or regulatory network can be different in time such that some of them are highly active under most conditions while others are activated for certain specific conditions. Therefore, to fully understand the

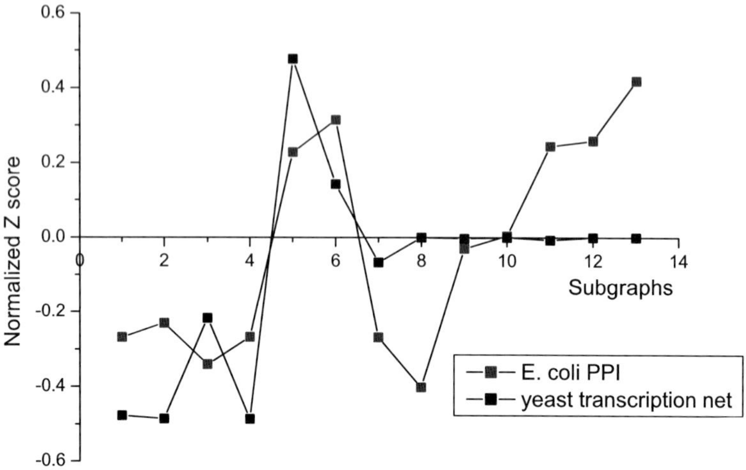

Fig. 7.7 Local properties of protein-protein interaction network of *E. coli*. TSPs for *E. coli* and yeast transcription network were plotted together

biological networks we have to consider the weight and direction and the temporal change of the network components. In this respect, we will introduce recent studies on dynamic aspect of metabolic networks using flux balance analysis (FBA) and protein interaction networks of *E. coli* in this section.

7.4.1 Understanding the Robustness of Metabolic Network

As complex biological systems are very robust to genetic and/or environmental changes on all levels of organization, their inherent robustness has been of great interest in biology as well as in engineering theory (Wagner 2005). The biological function of *E. coli* metabolism can be sustained against single-gene or even multiple-gene mutation possibly by utilizing the redundant pathways (Papp et al. 2004, Reed and Palsson 2004). While the investigations on the topological and functional/phenotypic properties of metabolic networks have been increasingly populated as shown in previous sections, (Almaas et al. 2004, Covert et al. 2004, Guimera and Nunes Amaral 2005, Papp et al. 2004) they still provide a limited understanding of the metabolic robustness despite its biological significance. In this section, we focus on the interplay between such robustness and the underlying metabolism, and how the robustness can be accomplished at the level of the metabolites which are the fundamental entities (Raymond and Segre 2006, Schmidt et al. 2003) integrated/dissipated by the metabolic processes. To this end, we constructed the computational models at a system level, and simulated them with a constraints-based flux analysis (Price et al. 2004).

To explore the robustness of *E. coli* metabolism from the metabolite perspective, we should identify the metabolites which are substantial in cellular functions. In this regard, all intracellular metabolites are classified into two categories, essential and non-essential metabolites according to the phenotypic effects on cell survival when the consumption rate of the given metabolite is suppressed to zero. The resultant list of essential metabolites is identified under many different environments which are specified by combinations of several C, P, N, and S sources, and aerobic/anaerobic conditions (Kim et al. 2007). By disrupting multiple genes around essential/non-essential metabolites *in vivo*, we could validate the predicted effects of the metabolite essentiality on cell survival. For example, the associated genes of an essential metabolite, tetrahydrofolate, were selected for the multiple-gene disruption. Each single and double gene deletion mutant (ΔpurN, ΔlpdA, ΔglyA, and ΔpurN ΔlpdA) could still survive albeit with some growth rate changes, but simultaneous deletions of the triple genes (ΔpurN ΔlpdA ΔglyA) did not allow the cell to grow at all, reflecting that the combinatory suppression of the tetrahydrofolate pool is indeed fatal to the cell. On the contrary, 1-deoxy-D-xylulose 5-phosphate had been identified as a non-essential metabolite *in silico*, and experimental removals of all the reactions producing the metabolite by constructing Δdxs ΔxylB caused the only slight change and even increase of growth rate compared with wild type. Throughout these experiments, the measured growth rates of the gene deletion mutants relative to that of the wild type were found to be consistent with the *in silico* predictions. These results indicate that deletion strains for essential metabolites can suffer from the deleterious impact on cellular functions, while those for non-essential metabolites show the negligible influence on the actual growth. We also investigated the inherent network property of essential metabolites to elucidate the correlation between the structural property and functional behavior from the metabolite perspective. We found that essential metabolites are likely to be connected with more reactions than non-essential ones. Furthermore, the metabolic networks of 227 organisms with fully sequenced genomes disclose that the metabolites essential for various growth conditions are commonly distributed across the organisms, showing the high degree of phylogenetic conservation.

To better understand the robustness of the cellular metabolism from the metabolite perspective, it is necessary to quantify the usage of all relevant fluxes to a single metabolite. In this sense, we introduce the flux-sum (Φ) of the metabolite, which is defined as the summation of all incoming or outgoing fluxes for given metabolite i as follows:

$$\Phi_i = \sum_{j \in P_i} S_{ij} v_j = -\sum_{j \in C_i} S_{ij} v_j = \frac{1}{2} \sum_j |S_{ij} v_j|$$

where S_{ij} is the stoichiometric coefficient of metabolite i in reaction j, and j is the flux of reaction j. P_i denotes the set of reactions producing metabolite i, C_i the set of reactions consuming metabolite i. Under the stationary assumption, Φ_i is the mass flow contributed by all fluxes producing (consuming) metabolite i. Based on this measure pertaining to the behavioral characteristic of metabolites, we can analyze the robustness of *E. coli* metabolism to maintain the cellular functions against the genetic mutations. The sensitivity to genetic perturbation for a given metabolite can be quantified by evaluating the relative fluctuation of Φ_i in response to each deletion of non-lethal reactions: $\sqrt{\langle\Phi_i^2\rangle - \langle\Phi_i\rangle^2} / \langle\Phi_i\rangle$ where $< \ldots >$ denotes the average over the reaction deletions. It turns out that the essential metabolites are more likely to have small relative fluctuations. This indicates that flux-sums of essential metabolites are relatively insensitive to genetic perturbation compared with those of non-essential ones. Indeed, 94.3% of total metabolites found in the fluctuation range of less than 0.0875 are essential, while there are only non-essential metabolites in the twenty highest ranks in relative fluctuations. Thus, essential metabolites are resistant to the internal variation compared with non-essential ones by maintaining the basal mass flow of the corresponding metabolite, thereby leading to the robustness of the cellular metabolism.

To clarify such resistance of essential metabolites against the internal disturbance, the severe perturbation was conducted by deleting the most contributing reaction to the flux-sum for a given essential metabolite. Remarkably, for many essential metabolites, the resultant flux loss is mostly recovered by the fluxes of other remaining reactions, thereby leading to very small change of the flux-sum, in spite of removing the dominant reaction5 with the largest flux value. For instance, the flux-sum of an essential metabolite, carbamoyl phosphate, is reproducible by other fluxes even when the largest flux from carbamate kinase is eliminated; other reaction, carbamoyl-phosphate synthase can compensate such flux loss fully, thus resulting in the recovery of 98.9% of the basal flux-sum. For many essential metabolites, the flux-sum is only changed much less than the reduced flux corresponding to the deleted reaction. Accordingly, even though the reaction with relatively high flux is eliminated, the flux-sum can be compensated by other fluxes around the essential metabolite, recovering such flux loss. Moreover, using the stoichio-similarity, we could develop the method to predict the most probable reaction which would recover the flux-sum after disruption. Hence, we believe that cellular robustness can be elucidated by such functional property of metabolic network manifesting the resilience of essential metabolites against the disturbed flux configuration.

Essential metabolites play a pivotal role in the cell survival, steadily maintaining the mass flow to produce or consume the metabolites against any internal disturbance within the cell. In other sense, this metabolite perspective on the robustness of *E. coli* provides us the cellular-level fragility: the failure of maintaining the flux-sum of a single essential metabolite can suppress the whole cellular growth drastically. Especially, for most essential metabolites (85%), reducing the flux-sum by half below the basal level intentionally leads to the growth rate down to half or even less, while only 28.9% of active non-essential metabolites have the same effect on the cell growth for such flux-sum perturbation.

The functional robustness of metabolic networks reflects the resistance towards internal defects and environmental fluctuations as an end product of a long evolutionary process. Such fault-tolerance or robustness may be a key to cell survival against environmental or genetic change. In this regard, a metabolite-based perspective could provide us a new guideline to interpret the cellular robustness. Essential metabolites substantial to the cell survival are capable of rerouting metabolic fluxes while sustaining their usage level. This capability of the essential metabolites leads to the quite dramatic tolerance to a wide range of internal disturbances. From a therapeutic point of view, disrupting (knock-out) the multiple non-lethal genes around an essential metabolite can lead to fatal cell damage; even attenuating (knock-down) the relevant genes may cause the same effect. Thus, synthetic lethal mutations (Tucker and Fields 2003, Wong et al. 2004) can be systematically identified in conjunction with experimental screening techniques available (Ooi et al. 2003, Tong et al. 2001), thereby facilitating the discovery of drug targets for the genetic therapy.

7.4.2 Beyond the Static Graph Analysis: Spatio-Temporal Dynamics

The topological data and approach discussed in previous sections represent a partial snapshot of the metabolism. Indeed, the topology of the metabolic network provides only the genome-encoded potential metabolic activity of an organism. The actual function of its metabolic network, however, is realized through the genetic regulatory network that functionally activates and inactivates various enzymes or groups of enzymes that catalyze biochemical reactions embedded in the metabolic network topology. Thus, for an in-depth characterization of metabolism we need to develop a better understanding of the regulatory network and its dynamics, as well. An important limitation of any modeling effort is the lack of availability of enzyme kinetic data, making impossible the full dynamic characterization of these pathways. However, already available microarray data does give us important qualitative information on the correlation between the enzymatic activities of different pathways. In this sense, there are several studies to analyze the available microarray data to infer information about correlations between the various components of the *E. coli* metabolism. These studies will offer valuable information on the dynamical features of its metabolism that has never been included in previous modeling efforts. One of simple but interesting works on temporal aspect of complex network was found in protein interaction network. For the case of protein interaction network, it was verified that considering dynamic aspect is crucial to understand the lethality of the node properly. That is, although it was shown that highly connected proteins (hubs) are more essential (lethal) than less connected proteins, recent study shows that all hub proteins are not equivalent. Han et al. showed (Han et al. 2004) that there are two different categories for the hub proteins, first one is 'party' hubs which interact with their partner proteins simultaneously, the other is 'date' hubs which in contrast, interact with different proteins at different locations and times using a filtered yeast interactome (FYI), compiled from different sets of yeast mRNA expression data to find the difference. (See Fig. 7.8) They found that date hub is more important than party hub such that when party hubs are removed from the system, general connectivity of the network remains still unaffected while the removal of date hubs breaks network into pieces so that proteins cannot interact with each other. Therefore, it is very important to consider spatial and temporal information when we analyze the

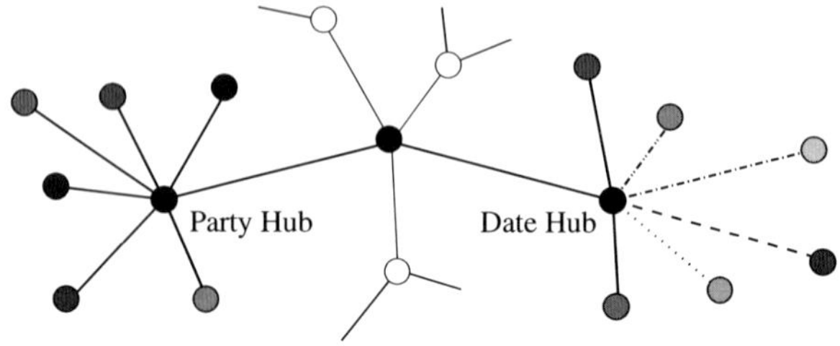

Fig. 7.8 Two different types of the hub. Party hub interacts with many proteins at the same time and location while date hub interacts with many proteins at different time and locations

bio-network. In this sense, spatio-temporal dynamic analysis should be applied to biological system along with static graph analysis.

Despite of the significant advance in network science during last decade, we are still far from understanding the biological system even for simple organism like *E. coli*. However, network biology which is still in its infancy, will give us an insight to find a way to understand the biological system along with large scale data sets generated and integrated into the database extensively.

Acknowledgments I'd like to thank many colleagues for their assistance in preparing this chapter, including Dr. P.-J. Kim, Y.-H. Eom, S. H. Lee for their help on data analyses and new illustrations, and Dr. R. Albert, Dr. B. Tomber, Dr. S. P. Mason, Dr. A.-L. Barabasi, Dr. Z. N. Oltvai, Dr. S. Lee, Dr. D.-Y. Lee, T. Y. Kim, K. H. Lee, Dr. S. Y. Lee, Dr. S. Park for their kind supports.

References

Albert R, Barabasi AL (2000) Topology of evolving networks: local events and universality. Phys Rev Lett 85(24):5234–7

Albert R, Jeong H, Barabasi A-L (1999a) The diameter of the World Wide Web. Nature 401:130–1

Albert R, Jeong H, Barabasi A-L (1999b) Emergence of Scaling in Random Networks Nature 400:130

Albert R, Jeong H, Barabasi AL (2000) Error and attack tolerance of complex networks. Nature 406(6794):378–82

Almaas E, Kovacs B, Vicsek T et al. (2004) Global organization of metabolic fluxes in the bacterium *Escherichia coli*. Nature 427(6977):839–43

Barabasi A-L, Albert R (1999) Emergence of Scaling in Random Networks Science 286:509

Barabasi A-L, Albert R, Jeong H (1999) Emergence of Scaling in Random Networks Physica A 272:173

Barkai N, Leibler S (1997) Robustness in simple biochemical networks. Nature 387(6636):913–7

Barthelemy M, Amaral LAN (1999a) Small-World Networks: Evidence for a Crossover Picture. Phys Rev Lett 82:3180–2

Barthelemy M, Amaral LAN (1999b) Small-World Networks: Evidence for a Crossover Picture. Phys Rev Lett 82:3180–3183

Bhalla US, Iyengar R (1999) Emergent properties of networks of biological signaling pathways. Science 283(5400):381–7

Bollobas B (1985) Random Graphs. Academic Press, London

Butland G, Peregrin-Alvarez JM, Li J et al. (2005) Interaction network containing conserved and essential protein complexes in *Escherichia coli*. Nature 433(7025):531–7

Cascante M, Melendez-Hevia E, Kholodenko B et al. (1995) Control analysis of transit time for free and enzyme-bound metabolites: physiological and evolutionary significance of metabolic response times. Biochem J 308(Pt 3):895–9

Covert MW, Knight EM, Reed JL et al. (2004) Integrating high-throughput and computational data elucidates bacterial networks. Nature 429(6987):92–6

Dorogovtsev SN, Mendes JFF (2002) Evolution of networks. Adv Phys 51:1079–1187

Eisenberg D, Marcotte EM, Xenarios I et al. (2000) Protein function in the post-genomic era. Nature 405(6788):823–6

Eom YH, Lee S, Jeong H (2006) Exploring local structural organization of metabolic networks using subgraph patterns. J Theor Biol 241(4):823–9

Erdős P, Rényi A (1960) On the evolution of random graphs. Publ Math Inst Hung Acad Sci 5A:17–61

Faloutsos M, Faloutsos P, Faloutsos C (1999) On power-law relationships of the internet topology. ACM SIGCOMM'99, Boston, MA

Gleiss PM, Stadler PF, Wagner A et al. (2001) Relevant Cycles in Chemical ReactionNetworks. Adv Complex Syst 1:1–000

Guimera R, Nunes Amaral LA (2005) Functional cartography of complex metabolic networks. Nature 433(7028):895–900

Han JD, Bertin N, Hao T et al. (2004) Evidence for dynamically organized modularity in the yeast protein-protein interaction network. Nature 430(6995):88–93

Hartwell LH, Hopfield JJ, Leibler S et al. (1999) From molecular to modular cell biology. Nature 402(6761 Suppl):C47–52

Heinrich R, Schuster S (1996) The Regulation of Cellular Systems. Chapman & Hall, New York

Huberman BA, Adamic LA (1999) Growth dynamics of the World-Wide Web. Nature 400:131

Ito T, Chiba T, Ozawa R et al. (2001) A comprehensive two-hybrid analysis to explore the yeast protein interactome. Proc Natl Acad Sci USA 98(8):4569–74

Jeong H (2003) Complex scale-free networks Physica A 321:226–37

Jeong H, Mason SP, Barabasi AL et al. (2001) Lethality and centrality in protein networks. Nature 411(6833):41–2

Jeong H, Tombor B, Albert R et al. (2000) The large-scale organization of metabolic networks. Nature 407(6804):651–4

Kanehisa M, Goto S (2000) KEGG: kyoto encyclopedia of genes and genomes. Nucleic Acids Res 28(1):27–30

Karp PD, Krummenacker M, Paley S et al. (1999) Integrated pathway-genome databases and their role in drug discovery. Trends Biotechnol 17(7):275–81

Kim PJ, Lee DY, Kim TY et al. (2007) Metabolite essentiality elucidates robustness of *Escherichia coli* metabolism. Proc Natl Acad Sci USA 104(34):13638–42

Mangan S, Alon U (2003) Structure and function of the feed-forward loop network motif. Proc Natl Acad Sci USA 100(21):11980–5

Milo R, Itzkovitz S, Kashtan N et al. (2004) Superfamilies of evolved and designed networks. Science 303(5663):1538–42

Milo R, Shen-Orr S, Itzkovitz S et al. (2002) Network motifs: simple building blocks of complex networks. Science 298(5594):824–7

Newman MEJ (2001) Scientific collaboration networks. Phys Rev E64:016131

Newman MEJ (2002) Assortative Mixing in Networks. Phys Rev Lett 89:208701

Ooi SL, Shoemaker DD, Boeke JD (2003) DNA helicase gene interaction network defined using synthetic lethality analyzed by microarray. Nat Genet 35(3):277–86

Overbeek R, Larsen N, Pusch GD et al. (2000) WIT: integrated system for high-throughput genome sequence analysis and metabolic reconstruction. Nucleic Acids Res 28(1):123–5

Papp B, Pal C, Hurst LD (2004) Metabolic network analysis of the causes and evolution of enzyme dispensability in yeast. Nature 429(6992):661–4

Pastor-Satorras R, Vázquez A, Vespignani A (2001) Dynamical and Correlation Properties of the Internet. Phys Rev Lett 87:258701

Price ND, Reed JL, Palsson BO (2004) Genome-scale models of microbial cells: evaluating the consequences of constraints. Nat Rev Microbiol 2(11):886–97

Rain J-C, Selig L, De Reuse H et al. (2001a) The protein-protein interaction map of *Helicobacter pylori*. Nature 409:211

Rain JC, Selig L, De Reuse H et al. (2001b) The protein-protein interaction map of *Helicobacter pylori*. Nature 409(6817):211–5
Ravasz E, Somera AL, Mongru DA et al. (2002) Hierarchical organization of modularity in metabolic networks. Science 297(5586):1551–5
Raymond J, Segre D (2006) The effect of oxygen on biochemical networks and the evolution of complex life. Science 311(5768):1764–7
Reed JL, Palsson BO (2004) Genome-scale *in silico* models of *E. coli* have multiple equivalent phenotypic states: assessment of correlated reaction subsets that comprise network states. Genome Res 14(9):1797–805
Schmidt S, Sunyaev S, Bork P et al. (2003) Metabolites: a helping hand for pathway evolution? Trends Biochem Sci 28(6):336–41
Shen-Orr SS, Milo R, Mangan S et al. (2002) Network motifs in the transcriptional regulation network of *Escherichia coli*. Nat Genet 31(1):64–8
Stauffer D, Aharony A (1992) Percolation Theory. Taylor & Francis, London
Strogatz SH (2001) Exploring complex networks. Nature 410(6825):268–76
Suki B, Alencar AM, Sujeer MK et al. (1998) Life-support system benefits from noise. Nature 393(6681):127–8
Tong AH, Evangelista M, Parsons AB et al. (2001) Systematic genetic analysis with ordered arrays of yeast deletion mutants. Science 294(5550):2364–8
Tucker CL, Fields S (2003) Lethal combinations. Nat Genet 35(3):204–5
Uetz P, Giot L, Cagney G et al. (2000) A comprehensive analysis of protein-protein interactions in *Saccharomyces cerevisiae*. Nature 403(6770):623–7
Vazquez A, Dobrin R, Sergi D et al. (2004) The topological relationship between the large-scale attributes and local interaction patterns of complex networks. Proc Natl Acad Sci 101:17940
Wagner A (2005) Robustness and Evolvability in Living Systems. Princeton University Press, Princeton
Watts DJ, Strogatz SH (1998) Collective dynamics of 'small-world' networks. Nature 393(6684):440–2
Wong SL, Zhang LV, Tong AH et al. (2004) Combining biological networks to predict genetic interactions. Proc Natl Acad Sci USA 101(44):15682–7
Xenarios I, Rice DW, Salwinski L et al. (2000) DIP: the database of interacting proteins. Nucleic Acids Res 28(1):289–91
Yi TM, Huang Y, Simon MI et al. (2000) Robust Perfect Adaptation in Bacterial Chemotaxis through Integral Feedback Control. Proc Natl Acad Sci USA 97:4649–53

Chapter 8
Sensory Transduction Network of *E. coli*

Michael Y. Galperin

Contents

Abstract The genome of *Escherichia coli* K12 encodes at least 6 classes of sensor proteins: 30 histidine protein kinases, 5 methyl-accepting chemotaxis proteins, 23 membrane components of the sugar:phosphotransferase system (PTS), 29 proteins with diguanylate cyclase and/or c-di-GMP-specific phosphodiesterase activity and two predicted serine/threonine protein kinases. The full signal transduction network additionally includes 32 response regulators, numerous chemotaxis proteins, PTS components, adenylate cyclase, CRP, and uncharacterized c-di-GMP-responsive components. Bacterial response to environmental signals can occur on several levels: the level of individual genes and proteins (changes in gene expression, post-translational regulation), the whole-cell level (chemotaxis), and the multicellular level (biofilm formation). All signal transduction systems are energy-dependent but their energy expenditure is miniscule compared to that of the processes they

M.Y. Galperin (✉)
National Center for Biotechnology Information, National Library of Medicine, National Institutes of Health, Bethesda, MD, USA
e-mail: galperin@ncbi.nlm.nih.gov

S.Y. Lee (ed.), *Systems Biology and Biotechnology of Escherichia coli*,
DOI 10.1007/978-1-4020-9394-4_8, © Springer Science+Business Media B.V. 2009

regulate. A better understanding of the signal transduction mechanisms and integration of these mechanisms into the metabolic pathway model of the *E. coli* cell will remain major challenges for systems biology.

8.1 Introduction

For many years, *Escherichia coli* K12 served as a favorite model organism for studying principles and mechanisms of bacterial signal transduction. As a result, the current understanding of the signal transduction machinery in *E. coli*, albeit obviously incomplete, is probably as good as that for any organism in the prokaryotic or eukaryotic world. The availability of complete genome sequences of three strains of *E. coli* K12 (Blattner et al. 1997, Hayashi et al. 2006, Durfee et al. 2008) and their pathogenic counterparts (Hayashi et al. 2001, Perna et al. 2001, Welch et al. 2002, Johnson et al. 2007) made it possible to enumerate all (known) components of the signal transduction machinery encoded in each *E. coli* genome. This, in turn, allowed identification, at least in terms of sequence, of those signal transduction proteins whose biological functions are still unknown and remain to be experimentally characterized. In many respects, *E. coli* K12 proved to be a very convenient model: its signal transduction machinery is far more complex than that of its relatives who are obligate pathogens, such as *Haemophilus influenzae* or *Legionella pneumophila*. On the other hand, *E. coli* encodes far fewer signal transduction proteins than its free-living relatives (and opportunistic pathogens), such as *Pseudomonas aeruginosa*, *Shewanella oneidensis*, or *Vibrio cholerae*, not to mention the enormous expansion of signaling systems in the genomes of such model organisms as *Anabaena* PCC7120, *Myxococcus xanthus, or Streptomyces coelicolor* (Galperin 2005). Thus, signal transduction in *E. coli* is an experimentally tractable system that is responsible for much of the progress in understanding the principles and mechanisms of prokaryotic signal transduction.

The difficult task of a systematic description of the bacterial signal transduction machinery has been greatly simplified by the availability of specialized public databases, such as the *Mi*crobial *S*ignal *T*ransduction database (MiST, http:// genomics.ornl.gov/mist) at the Oak Ridge National Laboratory in Tennessee (Ulrich and Zhulin 2007) and the Kyoto Encyclopedia of Genes and Genomes (KEGG, http://www.genome.ad.jp/kegg/) at the Kyoto University in Japan (Kanehisa et al. 2008). The web pages of these databases dedicated to *E. coli* K12 (http://genomics.ornl.gov/mist/view_organism.php?organism_id=99, and http://www.genome.ad.jp/dbget-bin/get_pathway?org_name=eco&mapno=02020, respectively) provide a bird's eye view of the composition and properties of signaling proteins encoded in the *E. coli* genome. In addition, the author maintains tables of Signal Transduction Census and Response Regulator Census at the web sites http://www.ncbi.nlm.nih.gov/Complete_Genomes/SignalCensus.html and http://www.ncbi.nlm.nih.gov/Complete_Genomes/RRcensus.html, respectively. These web sites provide an easy way to access up-to-date information on signal transduction mechanisms in *E. coli* and related bacteria.

8.2 Diversity of Bacterial Signal Transduction Pathways

The two best-studied classes of membrane-bound receptor proteins are sensory histidine kinases and methyl-accepting chemotaxis proteins (MCPs), discovered in *E. coli* in the mid 1980s (Grebe and Stock 1999, Stock et al. 2000, Inouye and Dutta 2003). In the past several years, analyses of microbial genomes, as well as experimental studies, revealed several additional classes of bacterial receptors, which include Ser/Thr protein kinases and protein phosphatases, adenylate cyclases, diguanylate cyclases and c-di-GMP-specific phosphodiesterases (Table 8.1).

The signaling pathways utilized by various receptors are shown on Fig. 8.1 Signaling by histidine kinases and MCPs is usually referred to as two-component signal transduction, as it includes phosphoryl transfer between two different proteins, a histidine kinase and a response regulator. Two-component signal transduction pathways are extremely diverse but always include the following three steps:

Table 8.1 Principal Classes of Sensory Proteins in *Escherichia coli* K12

Sensor type	No.	Function	Signaling mechanism
Histidine kinase	30	Transcriptional regulation, control of other processes	Phosphorylation of the REC domain of various response regulators
Methyl-accepting chemotaxis protein	5	Chemotaxis	Interaction with histidine kinase CheA, chemotaxis response regulator CheY
Ser/Thr protein kinase	1 + 1[a]	Transcriptional regulation, posttranslational regulation	Phosphorylation of Ser or Thr residues in target proteins
Ser/Thr protein phosphatase	2	Same as above	Dephosphorylation of Ser/Thr protein kinases or other target proteins
PTS membrane component	23	Sugar transport, chemotactic signaling, regulation of adenylate cyclase activity	Direct effect on chemotaxis, most likely through direct interaction of PTS enzyme I with the histidine kinase CheA
Adenylate cyclase	1	Global regulation of transcription	Synthesis of cAMP
Diguanylate cyclase	12+7[b]	Regulation of protein and polysaccharide secretion	Synthesis of c-di-GMP
c-di-GMP-specific phosphodiesterase	10+7[b]	Same as above	Hydrolysis of c-di-GMP

[a] While YegI is believed to function as a Ser/Thr kinase, it remains unclear whether UbiB is an enzyme of ubiquinone biosynthesis or a Ser/Thr kinase that regulates this pathway (see the text for details).
[b] Seven *E. coli* K12 proteins contain both GGDEF and EAL domains and could potentially catalyze both synthesis and hydrolysis of c-di-GMP (see the text for details).

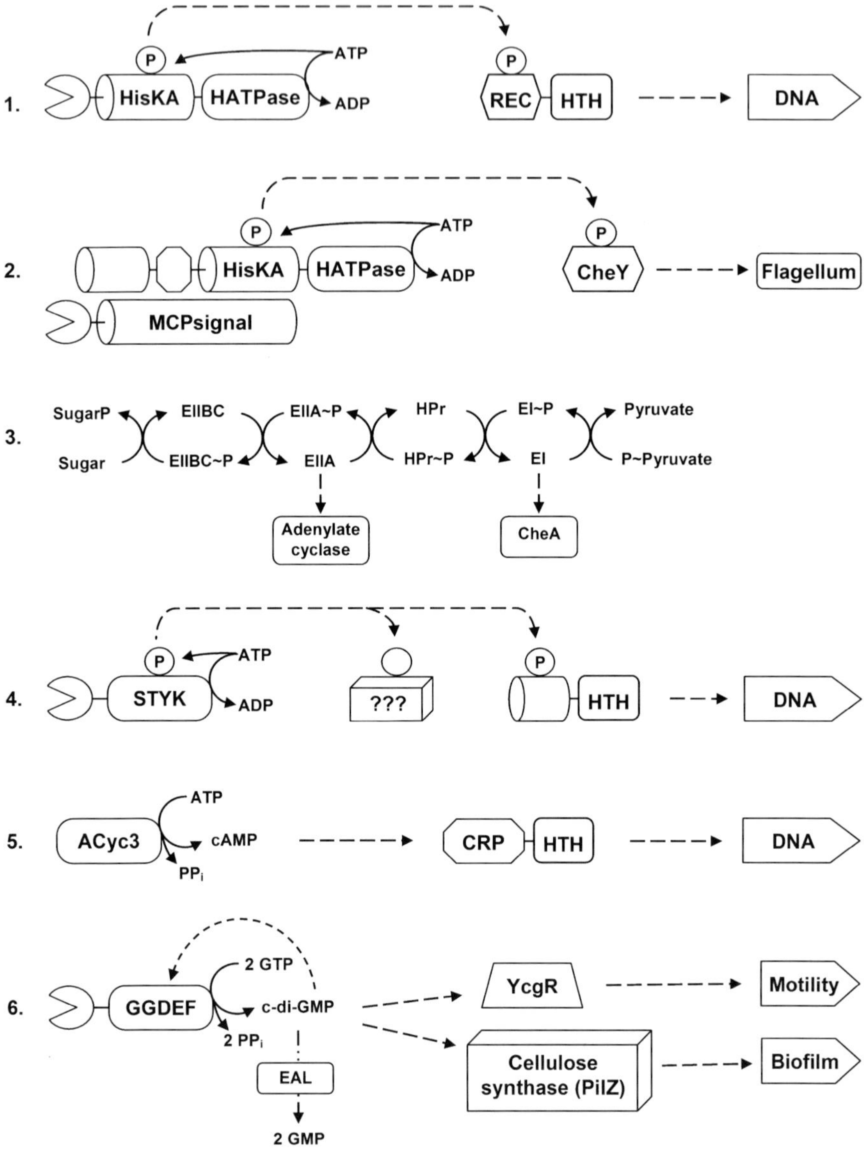

Fig. 8.1 Signal transduction pathways of the principal classes of bacterial receptors. Signal transduction from a two-component signal transduction system (1), methyl-accepting chemotaxis sensor protein (2), phospho*enol*pyruvate-dependent sugar:phosphotransferase system (3), Ser/Thr protein kinase (4), adenylate cyclase (5), and sensor diguanylate cyclase (6)

(i) phosphorylation of a His residue in the kinase molecule; (ii) phosphoryl transfer to an Asp residue in the molecule of the cognate response regulator; (iii) conformational change of the response regulator that alters its interaction with its target on the chromosomal DNA or bacterial flagellum or, in some cases, the enzymatic activity of its output domain (see below).

Chemotaxis signaling, which starts from MCPs, is a special kind of two-component signal transduction that involves a specialized histidine kinase CheA, which directly interacts with MCPs, and a specialized response regulator CheY that consists of stand-alone receiver domain without any output domains. Regulation of flagellar motility is based on the interaction of the phosphorylated form of CheY with the FliM protein at the base of the flagellum, which affects the direction of flagellar rotation and thus regulates the chemotaxis response (Aizawa et al. 2002, Szurmant and Ordal 2004).

Components of the PEP-dependent sugar:phosphotransferase system (PTS) participate in phosphorylative sugar uptake and traditionally have not been considered part of the signal transduction machinery. Nevertheless, two members of the PTS phosphorelay play key roles in signal transduction. The phosphorylation level of the PTS enzyme I (EI) directly affects the chemotaxis machinery, whereas the phosphorylation level of the glucose-specific enzyme IIA ($EIIA^{Glc}$) modulates the activity of the adenylate cyclase, at least in *E. coli* and its closest relatives (Postma et al. 1993, Deutscher et al. 2006).

Ser/Thr protein kinases phosphorylate Ser and Thr residues in various cellular proteins. Only a small fraction of their targets have been identified so far. Ser/Thr protein phosphatases reverse the effect of Ser/Thr protein kinases by dephosphorylating their target proteins or, in some cases, the Ser/Thr protein kinases themselves (Shi et al. 1998, Deutscher and Saier 2005).

The adenylate cyclase modulates the cellular level of cyclic adenosine monophosphate (cAMP), a key cellular second messenger that regulates transcription from a variety of relatively weak promoters. The mechanism of this regulation includes binding of cAMP to a specialized adaptor protein, CAP (also referred to as cAMP receptor protein, CRP), triggering a conformational change in CRP that increases its affinity to DNA and allows it to activate transcription of otherwise poorly expressed genes (operons).

Signaling through diguanylate cyclases includes modulation of the cellular level of another cellular second messenger, cyclic dimeric bis-(3′–5′)-guanosine monophosphate (c-di-GMP), which regulates a variety of function related to the cell surface elements, including motility, secretion of proteins and exopolysaccharides, biofilm formation, and production of certain virulence factors (Römling et al. 2005, Jenal and Malone 2006). Some of the c-di-GMP functions are mediated by its binding to the recently described PilZ domain, while others might involve other binding proteins, including diguanylate cyclases themselves. Cyclic-di-GMP-specific phosphodiesterases, which catalyze c-di-GMP hydrolysis, could also function as c-di-GMP-binding proteins.

8.3 Signal Transduction Machinery of *E. coli*

8.3.1 Two-component Sensors: Histidine Kinases

Histidine kinases are most numerous and most diverse membrane receptors encoded in bacterial genomes. Accordingly, they control the greatest variety of cellular responses. Most of the diversity of histidine kinases comes from the sensory (signal input) domains, which can be periplasmic, membrane-embedded or cytoplasmic. A single histidine kinase can contain several sensory domains, for example a periplasmic sensory domain and one or more ligand-binding PAS domains in the cytoplasm. In contrast, cytoplasmic signal transduction modules of histidine kinases are rather uniform and consist of two structural domains, dimerization/phosphorylation HisKA domain that consists of long alpha-helices and a C-terminal globular ATPase domain. Signal transmission by histidine kinases involved formation of dimers, so that an ATPase domain of one molecule binds ATP and transfers its γ-phosphate onto a conserved histidine residue in the HisKA domain of the other molecule in the dimer. This phosphoryl residue is subsequently transferred to an aspartyl residue in the receiver domain of the cognate response regulator. Analysis of sequence similarities between different histidine kinases by Parkinson and Kofoid (1992) revealed five conserved sequence motifs, referred to as H, N, G1, F and G2 boxes. The first of these boxes corresponded to the sequence motif around the conserved phosphoryl-accepting histidine residue.

A cell of *E. coli* K12 encodes 30 histidine kinases; functions of six of them (AtoS, RstB, YehU, YpdA, YfhK, and YedV) still remain unknown (Hagiwara et al. 2004, Yamamoto et al. 2005), see Table 8.2 Among the remaining 24, by far the most (six, namely, BaeS, BasS, CpxA, EvgS, RcsC, and RscD), are involved in response to the envelope stress. Two more, EnvG and KdpD, are responsible for osmotic stress and adjustment of the magnitude of K^+ gradient. Other perceived signals include phosphate and/or its Ca^{2+} or Mg^{2+} salts (PhoQ, PhoR); nitrate and nitrite (NarQ, NarX); oxygen and/or hydrogen peroxide (ArcB, BarA); heavy metals, such as Cu^+/Ag^+, (CusS) or Zn^{2+} and Pb^{2+} (ZraS); di- and tricarboxylates (CitA, DcuS); glucose-6-phosphate (UhpB), glutamine (GlnL), and trimethylamine N-oxide (TorS). One more histidine kinase sensor, QseC, is responsible for quorum sensing.

It is remarkable how many histidine kinases are sensing either envelope and osmotic stress or the redox state of the cell and the availability of terminal electron acceptors. The fact that these histidine kinases coexist in the same cell suggests a certain degree of sophistication in their interactions, seen, for example, in the complex division of functions between NarQ and NarX (Stewart 2003). In most cases, however, the hierarchy between different sensors, if any, remains unknown.

8.3.2 Two-component Transmitters: Response Regulators

Two-component response regulators are diverse proteins that share the common phosphoacceptor REC domain, often referred to as the CheY-like domain, after its best-known representative (Galperin 2006, Gao et al. 2007). This domain catalyzes phosphoryl transfer from the His residues of the histidine kinase HisKA domains to its own aspartate residues, as well as its own dephosphorylation (Thomas et al. 2008). The combination of these two activities in the REC domains of each particular response regulator determines the half-life of the phosphorylated form of

Table 8.2 Two-component signal transduction in *E. coli*

Histidine kinase	Response regulator	Signal	Regulated system or process (genes)
ArcB[a]	ArcA[b]	Redox state of the respiratory chain component(s)	Aerobic/anaerobic respiration
AtoS	AtoC[d]	Unknown (expression induced by acetoacetate)	Short-chain fatty acid metabolism (*atoDAEB*)
BaeS	BaeR[b]	Envelope stress	Multidrug efflux (*mdtABCD, acrD*)
BarA[a]	UvrY[c]	O_2, H_2O_2, oxidative stress	Carbon storage (*csrB*), catalase (*katE*)
BasS	BasR[b]	Envelope stress (high Fe^{2+})	Multidrug efflux
CheA	CheY, CheB	MCPs, PTS sugars	Chemotaxis
CitA (DpiB)	CitB[c] (DpiA)	Citrate	Citrate metabolism (*citCDEFG, citT*)
CpxA	CpxR[b]	Envelope stress, misfolded proteins	Protein degradation (*htrA*)
CreC (PhoM)	CreB[b]	Unknown (induced by growth in minimal media)	Central metabolism
CusS	CusR[b]	Cu^+, Ag^+	Efflux transporters
DcuS	DcuR[c]	Fumarate, C4-dicarboxylates	Fumarate respira-tion (*dcuB*)
EnvZ	OmpR[b]	Envelope stress	Outer membrane (*ompC, ompF*)
EvgS[a]	EvgA[c]	Envelope stress	Multidrug efflux
GlnL (NtrB)	GlnG[d] (NtrC)	Nitrogen starvation	Glutamine metabolism
KdpD	KdpE[b]	Osmotic stress	K^+ transport (*kdpABC*)
NarQ	NarP[c]	Nitrite/nitrate	Nitrate reductase (*narGHIJ*), formate dehydrogenase
NarX	NarL[c]	Nitrite/nitrate	Nitrate reductase (*narGHIJ*), formate dehydrogenase
PhoQ	PhoP[b]	Low Mg^{2+}	Various genes
PhoR	PhoB[b], PhoP	Low phosphate	Phosphate assimilation (*phoA, phoB*)
QseC	QseB[b]	Cell density (autoinducer-2), epinephrine, norepinephrine	Flagellar biosynthesis
RcsC[a]	RscB[c]	Unknown	Colanic acid biosynthesis
RscD	RscB[c]	Unknown	Colanic acid biosynthesis
RstB	RstA[b]	Unknown	Acid resistance, flagellar and capsular biosynthesis
TorS[a]	TorR[b]	Trimethylamine-N-oxide	TMAO reductase (*torCAD*)
UhpB	UhpA[c]	UhpC, glucose-6-phosphate	Hexose phosphate uptake (*uhpT*)
ZraS (HydH)	ZraR[d] (HydG)	Heavy metals (Zn^{2+}/Pb^{2+})	Efflux transporter
YedV		Unknown	Unknown
YehU	YehT[e]	Unknown	Unknown
YfhK	YfhA[d]	Unknown	Unknown
YpdA	YpdB[e]	Unknown	Unknown
	FimZ[c] (YbcA)	Unknown	Fimbriae biosynthesis
	RssB (Hnr)	Unknown	Proteolysis of RpoB by ClpXP

[a] A hybrid histidine kinase that contains a receiver domain at its C-terminus.
[b] DNA-binding transcriptional regulator, OmpR/PhoB (winged helix) family.
[c] DNA-binding transcriptional regulator, NarL/FixJ (helix-turn-helix) family.
[d] DNA-binding transcriptional regulator, NtrC (enhancer-binding) family.
[e] DNA-binding transcriptional regulator, LytR/AgrA family.

the domain (CheY∼P or, more generally, REC∼P) and hence, the fraction of the response regulator molecules that are in the active (phosphorylated) conformation at any given time. A great majority of response regulators combine the REC domain with some kind of a signal output domain. However, some response regulators, such as the chemotaxis response regulator CheY, consist of a stand-alone REC domain. Chemotactic signal transduction through CheY relies solely on protein-protein interactions. Phosphorylation of CheY by phosphoryl transfer from the chemotaxis histidine kinase CheA shifts the CheY molecule into the active conformation that has an increased affinity to its target molecule FliM in the flagellar basal body. Non-phosphorylated CheY is also capable of interacting with FliM, albeit not as strongly. Thus, phosphorylation of CheY merely shifts the equilibrium of its two principal forms (there appear to be intermediate forms as well (Dyer and Dahlquist 2006)), leading to a change in the rotation pattern of the flagellum, which is reflected in an altered motility pattern of the whole cell.

With the exception of members of the CheY protein family, all other response regulators are two-domain (or three-domain) proteins that combine the REC domain with a signal output domain, which is usually located at the C-terminus of the polypeptide chain. Most of these proteins (in *E. coli*, 29 out of 32) are transcriptional regulators that activate or repress transcription of specific target genes. Accordingly, the most common output domains bind DNA, although some response regulators have enzymatic or ligand-binding output domains. The most common DNA-binding response regulators belong to the OmpR/PhoB family and have a winged helix-turn-helix DNA-binding domain. In *E. coli*, this family includes 14 proteins of the total of 32 response regulators (Table 8.2). The second in abundance with 9 representatives in *E. coli* is the NarL/FixJ family of response regulators with a typical helix-turn-helix DNA-binding output domain. Less common DNA-binding response regulators contain DNA-binding output domains of the Fis type (NtrC family) and LytTR type (LytR/AgrA family) with 4 and 2 representatives, respectively, encoded in the *E. coli* genome. Despite the differences in the structures of the DNA-binding response regulators, they all appear to follow a general mechanism of activation in response to the environmental signals. In each case, phosphorylation of the REC domain favors its transition into an active conformation and/or its dimerization (Toro-Roman et al. 2005, Gao et al. 2007). Dimerization of response regulators is a key mechanism of the transcriptional regulation by two-component systems, as response regulator dimers have a higher affinity to the tandem (or palindromic) transcriptional regulator binding sites on the chromosome. Within each family of response regulators, the signaling specificity is determined by the tight interaction of the REC domains with their cognate histidine kinases and of the HTH domains with the target sites on the DNA. As a result, transcriptional regulators with similar sequences (e.g., OmpR and PhoB) may have dramatically different biological functions. Some response regulators consist of more than two domains. In transcriptional regulators of the NtrC family (4 members in *E. coli*), the N-terminal REC domain and the C-terminal DNA-binding Fis-like domain are separated by the central AAA-type ATP-binding domain, whose ATPase activity is required for the DNA binding. In summary, bacterial response regulators contain a wide variety of output domains that put the histidine kinases at the top of signaling hierarchy, allowing the cell to control its metabolism and behavior in response to various environmental challenges.

In some response regulators, the output domains are enzymatic. In *E. coli*, there is only one such response regulator, CheB, whose output domain is a methyl esterase of MCP proteins that takes part in chemotactic adaptation. Finally, *E. coli* and several closely related bacteria encode an unusual response regulator RssB (or Hnr), which regulates proteolysis of the stress sigma factor RpoS (Muffler et al. 1996, Zhou et al. 2001, Hengge-Aronis 2002). Its C-terminal domain is a degraded version of the Ser/Thr protein phosphatase domain which has apparently lost its catalytic activity and participates solely in protein-protein interactions (Galperin 2006).

8.3.3 Methyl-accepting Chemotaxis Proteins

Escherichia coli K12 encodes 5 methyl-accepting chemotaxis proteins (MCPs). The signals sensed by each of them have been experimentally characterized as follows: Tsr – serine; Tar – aspartate, maltose; Trg – ribose, galactose; Tap – dipeptides; and Aer – redox state of the respiratory chain (Szurmant and Ordal 2004). The last of these MCPs, Aer, is obviously important for sensing the presence of usable terminal electron acceptors, reflecting the choice between a respiratory and fermentative metabolism (Repik et al. 2000, Zhulin 2001). All these MCPs appear to interact with the chemotaxis histidine kinase CheA and transmit the respective signals thorough the two-component phosphorelay to the chemotactic response regulator CheY.

8.3.4 Phosphotransferase System Components

An MCP-independent mechanism of regulating chemotaxis is provided by the phospho*enol*pyruvate-dependent sugar:phosphotransferase system (PTS), which catalyzes uptake of certain sugars, coupling membrane transport of its substrates with their phosphorylation (Postma et al. 1993, Deutscher et al. 2006). Transport of sugar substrates by the PTS is coupled to signaling, both to the chemotaxis machinery and to the adenylate cyclase. Like histidine kinases, PTS proteins are phosphorylated on the histidine residue. However, in contrast to the ATP-His-Asp or ATP-His-Asp-His-Asp phosphorelay, typical for the two-component signaling, the PTS phosphorelay starts from phospho*enol*pyruvate (PEP) and includes only His residues, (at least, in EI, HPr and EIIA components). The high free energy of PEP hydrolysis ensures that in the absence of carbohydrate substrates all PTS components stay in the phosphorylated form. The limiting step in the whole phosphorelay appears to be PEP-dependent autophosphorylation of the first component, EI. Therefore, in the presence of carbohydrate substrates, phosphoryl flow through the PTS components occurs at a higher rate than re-phosphorylation of EI by PEP. As a result, EI, HPr and EIIA components become partly dephosphorylated, which serves as a signal both for the chemotaxis machinery and for the *E. coli* adenylate cyclase.

Although any direct interaction between PTS components and MCP or CheA remains to be demonstrated, the available data suggest that unphosphorylated EI can interact with CheA, modulating its activity and, hence, the cellular level of CheY∼P. The second mechanism of signal transduction from the PTS involves $EIIA^{Glc}$. This protein has been shown to interact with the adenylate cyclase and other targets, including the lactose permease. The *E. coli* cell encodes 23 membrane components of the PTS, five of which (FruA, FrvB, FrwC, HrsA, and YpdG) are apparently specific to fructose. The other ones, according to the existing experimental data and sequence-based predictions, are specific to the following sugars: glucose (PtsG), mannose (ManX/Y), mannitol (MtlA, CmtA), N-acetylglucosamine (NagE), cellobiose (AscF, CelB), galactitol (GatC, SgcC), N-acetylgalactosamine (AgaC/D, AgaW), sorbitol (SrlA), maltose (MalX), trehalose (TreB), α-glucosides (GlvC), β-glucosides (BglF), ascorbate (SgaB), and N-acetylmuramic acid (YfeV).

Thus, *E. coli* carries in its genome genes encoding chemotaxis receptors for almost any commonly found monosaccharide and several disaccharides. Whether these genes are constitutively expressed at sufficient levels to contribute to the cell behavior remains an open question. It appears that at least for some of the PTS receptor genes need to be induced by the corresponding sugar.

8.3.5 Ser/Thr Protein Kinases and Protein Phosphatases

Reversible protein phosphorylation on serine, threonine, or tyrosine residues is a key regulatory mechanism in eukaryotic cells. In the past several years, Ser/Thr protein kinases have been recognized in a variety of prokaryotic cells but are still often referred to as "eukaryotic-type" protein kinases. In certain groups of bacteria (e.g., actinobacteria) and archaea, Ser/Thr protein kinases appear to be the principal, if not the only (known) type of receptor proteins (Galperin 2005).

Most enterobacteria, including *E. coli*, encode just one or two Ser/Thr protein kinases and phosphatases, which remain poorly characterized. One of the predicted Ser/Thr protein kinases, UbiB, has been shown to be required for a hydroxylation step in ubiquinone biosynthesis and was initially thought to function as 2-octaprenylphenol hydroxylase (Poon et al. 2000). However, this enzymatic activity has not been experimentally demonstrated. In contrast, it has been identified as a member of the Ser/Thr protein kinase superfamily and has all the key active site residues intact. Thus, it remains unknown at this time whether UbiB is an enzyme of ubiquinone biosynthesis or a Ser/Thr protein kinase that regulates this process. The functions of the second predicted Ser/Thr protein kinase, YegI, also remain unknown.

8.3.6 Adenylate Cyclases

Bacteria encode several different variants (referred to as classes) of adenylate (adenylyl) cyclase, the enzyme that produces cAMP from ATP. The enzyme from *E. coli* is considered class I adenylate cyclase. It is a soluble enzyme that does not appear to sense any environmental signals by itself. However, its activity is modulated by the $EIIA^{Glc}$ component of the glucose-specific phosphotransferase system. The phosphorylated form of $EIIA^{Glc}$ appears to activate adenylate cyclase, whereas the dephosphorylated form, accumulating in the presence of extracellular glucose, does not bind to the adenylate cyclase or even inhibits it (Krin et al. 2002, Park et al. 2006). Thus, in the presence of glucose or other PTS sugars, adenylate cyclase activity decreases, leading to a drop in the cellular level of cAMP. This is one of the mechanisms contributing to the phenomenon of catabolite repression.

8.3.7 Diguanylate Cyclases and C-di-GMP Phosphodiesterases

A recently identified group of bacterial receptors includes proteins with so-called GGDEF and EAL domains that, respectively, synthesize and hydrolyze the second messenger c-di-GMP. Recent studies implicated c-di-GMP in regulating biofilm formation, development of flagellar apparatus, and a variety of other processes. The GGDEF domain has been shown to function as a diguanylate cyclase that produces a c-di-GMP molecule from two molecules of GTP (Paul et al. 2004, Ryjenkov et al. 2005). The EAL domain functions as c-di-GMP-specific phosphodiesterase, hydrolyzing c-di-GMP to a linear pGpG, and, eventually, to two molecules of GMP (Christen et al. 2005, Schmidt et al. 2005). *Escherichia coli* encodes 12 proteins with the GGDEF domain, 10 proteins with the EAL domain and 7 proteins that contain both of them and could potentially catalyze both reactions (Galperin et al. 2001, Galperin 2005). It appears, however, that in most of such fusion proteins, at least one of the domains is enzymatically inactive and serves to regulate the catalytic activity of the other one. In some cases, however, both domains appear to be active.

Our current knowledge of the functions of *E. coli* diguanylate cyclases and c-di-GMP-specific phosphodiesterases is very limited. The sensed ligand, oxygen (and/or CO and NO), has been established only for one of them, YddU, which was accordingly renamed 'direct oxygen sensor', or Dos (Delgado-Nixon et al. 2000). Several other GGDEF and EAL domain proteins, such as YaiC (AdrA), YdaM, YciR, and YhdA, have been shown to regulate, respectively, cellulose biosynthesis (Zogaj et al. 2001), production of curli fimbriae, and carbon storage, although the signal they respond to remains unknown. For other GGDEF and/or EAL domain proteins (Rtn, YcdT, YddV, YdeH, YeaI, YeaJ, YeaP, YedQ, YegE, YfeA, YfgF, YfiN, YhjK, YliF, YneF, YahA, YcgF, YcgG, YdiV, YhjH, YjcC, YlaB, YliE, YoaD), neither the sensed signal nor the regulated process are known at this time.

8.4 A System-level Look at the *E. coli* Signal Transduction

8.4.1 Multiple Responses to Multiple Signals

The above discussion shows that signal transduction machinery of *E. coli* is a complex network of interconnected pathways that underlie the ability of the cell to respond to environmental challenges. These responses are elicited by a variety of environmental parameters and occur on several different levels, including the level of individual genes and operons (changes in gene expression), at the level of the whole cell (chemotaxis), and at the level of multicellular communication (quorum sensing, biofilm formation). The regulation of gene expression, in turn, is multi-faceted and can occur at the transcriptional level (changes in expression of certain genes, operons, or even global regulons), and at the levels of post-transcriptional (e.g. modulation of the mRNA decay rate) and post-translational regulation (e.g. modulation of enzyme activity, protein stability, or protein-protein interactions).

However, a closer look at the mechanisms actually utilized by *E. coli* shows that most of the cell responses occur either at the level of transcriptional regulation or at the level of whole-cell behavior (chemotaxis). The two-component signal transduction in *E. coli* is primarily targeted towards transcriptional regulation (29 of 32 response regulators are DNA-binding). Chemotaxis involves just two response regulators, CheY and CheB, and the single remaining response regulator (RssB or Hnr) acts post-translationally, at the level of proteolysis of RpoS (Hengge-Aronis 2002), and ultimately affecting transcription of RpoS-dependent genes. Another way transcription can be regulated by environmental signals is through the cAMP-CRP system. As mentioned above, sugar uptake by the PTS affects the adenylate cyclase activity and, hence, transcription from a variety of catabolite repression-sensitive promoters. Predicted Ser/Thr protein kinase YegI contains a C-terminal helix-hairpin-helix DNA-binding domain and is probably also involved in transcriptional regulation. There is a distinct possibility that transcription can also be regulated by signaling pathways leading from the cellular diguanylate cyclases. However, there is currently no experimental data to support that possibility.

The whole-cell behavioral changes include (i) chemotaxis in response to a variety of sugars, several amino acids, and/or changes in the redox state of the cell and (ii) production of exopolysaccharide and curli fimbriae, eventually leading to biofilm formation. This dichotomy might reflect the critical choice between "stay" and "run" survival modes, which appears to be governed by the c-di-GMP-mediated signaling.

The same cellular responses can be classified in terms of the environmental parameters that cause them. Although we still know very little about the signals sensed by several histidine kinases, predicted Ser/Thr protein kinase, diguanylate cyclases, and c-di-GMP-specific phosphodiesterases, the listing of the environmental parameters sensed by experimentally characterized histidine kinases, MCPs, and membrane components of the PTS shows two interesting trends. On one hand, the *E. coli* cell monitors (or, rather, is capable of monitoring) a variety of environmental stress conditions and extracellular concentrations of a variety of nutrients. The first group includes, among others, envelope stress, osmotic stress, presence of heavy metals, and presence of membrane-penetrating acids, such as acetate or benzoate. The second group includes a variety of hexoses, most disaccharides, di- and tricarboxylates, but apparently only one pentose (ribose) and only a minimal selection of amino acids (glutamine, serine, aspartate). While all these compounds are obviously important for *E. coli* metabolism, it is hard to rationalize why *E. coli* senses primarily hexoses and not pentoses or these particular amino acids and not glutamate or asparagine. Some of these traits probably reflect simplification of the effector panel during adaptation of *E. coli* and other enterobacteria to the high-nutrient intestinal environment. Others might provide clues to the functional specialization of enteric bacteria within that specific ecological niche.

8.4.2 Energy Expenditure Considerations

It is important to note that environmental sensing is almost never energy-neutral: transmission of environmental signals requires significant energy expenditures, although minor in comparison to the energy requirements for motility, transcription of new genes (operons) or polysaccharide secretion,

As shown on Fig. 8.1, transmission of a signal through the two-component system takes an ATP molecule to phosphorylate a single molecule of a response regulator. Transcriptional regulation usually requires dimerization of response regulators, so two ATP molecules are being spent to convert an inactive response regulator into the active phosphorylated dimeric form. Autocatalytic spontaneous dephosphorylation of the receiver domains of response regulators weakens their protein-protein interaction, leading to the dissociation of dimers and a significant decrease in the DNA-binding ability. Therefore, the energy is spent here to achieve a rapid but relatively short-term activation (or, in some cases, repression) of transcription of certain genes (operons). Obviously, these energy expenditures are minor in comparison to the energy requirements of the transcription process, not to mention protein translation.

Transmission of the chemotactic signal includes a histidine kinase-response regulator pair and follows the same general principle as above. However, in case of CheB, as well as in the methylation-demethylation cycle, additional energy is being spent to regulate the adaptation time, i.e. to achieve a more precise timing of the signal. Again, these energy expenditures are minor in comparison to the energy spent on flagellar rotation that is required for motility of *E. coli*.

The lack of knowledge of the targets for Ser/Thr protein phosphorylation does not allow us to calculate the energy costs of this type of regulation. Nevertheless, they appear to be comparable to that of two-component signal transduction.

cAMP-mediated signaling requires a molecule of ATP to produce cAMP, which is then hydrolyzed to AMP by various phosphodiesterases. Converting the resulting AMP back to ATP requires two more ATP equivalents. Thus, transcriptional regulation of catabolite-sensitive operons requires at least 6 molecules of ATP per cAMP-CRP dimer.

In contrast, signal transmission through the PTS is remarkably energy efficient. As far as we know, chemotactic signaling by dephosphorylated EI and inhibition of the adenylate cyclase by dephosphorylated $EIIA^{Glc}$ do not require additional energy expenditure. However, the energy price here is paid in synthesizing all the components of the PTS and keeping them phosphorylated in the absence of the sugar substrate.

Finally, formation of a single c-di-GMP molecule consumes two molecules of GTP and four more ATP equivalents are required to restore these two molecules of GTP from pGpG. Further, the available crystal structures of c-di-GMP bound to proteins suggest that the active conformation of c-di-GMP is its dimer. The mechanisms of c-di-GMP-mediated regulation are still not fully understood, but both activation of cellulose biosynthesis through binding of c-di-GMP to the PilZ domain of the cellulose synthase (Amikam and Galperin 2005) and inhibition of flagellar formation through binding of the YcgR protein to the flagellar basal body (Ryjenkov et al. 2006) seem to occur solely by conformational changes, without any further energy-consuming reactions. Again, in these cases, energy expenditure seems to be minimal compared to that of the regulated process, that is, cellulose biosynthesis and export of flagellin.

In conclusion, despite the recent progress, there remain major puzzles in signal transduction pathways of even such well-studied organism as *Escherichia coli* K12. Determination of the range of signals sensed by this organism and the range of cellular responses elicited by these signals is an important goal of the ongoing experimental studies. A complete understanding of the signal transduction mechanisms and full integration of these mechanisms into the metabolic pathway model of the *E. coli* cell will probably remain a challenge for the nearest future.

Acknowledgments M.Y.G. is supported by the Intramural Research Program of the NIH, National Library of Medicine. The author's opinions do not necessarily reflect the views of NCBI, NLM, or the National Institutes of Health.

References

Aizawa S-I, Zhulin IB, Marquez-Magana L et al. (2002) Chemotaxis and motility. In: Sonenshein AL, Hoch JA, Losick R (eds) *Bacillus subtilis* and its closest relatives: from genes to cells. ASM Press, Washington, D.C., pp. 437–452

Amikam D, Galperin MY (2005) PilZ domain is part of the bacterial c-di-GMP binding protein. Bioinformatics 22:3–6

Blattner FR, Plunkett G, 3rd, Bloch CA et al. (1997) The complete genome sequence of *Escherichia coli* K-12. Science 277:1453–1474

Christen M, Christen B, Folcher M et al. (2005) Identification and characterization of a cyclic di-GMP-specific phosphodiesterase and its allosteric control by GTP. J Biol Chem 280: 30829–30837

Delgado-Nixon VM, Gonzalez G, Gilles-Gonzalez MA (2000) Dos, a heme-binding PAS protein from *Escherichia coli*, is a direct oxygen sensor. Biochemistry 39:2685–2691

Deutscher J, Saier MH, Jr. (2005) Ser/Thr/Tyr protein phosphorylation in bacteria – for long time neglected, now well established. J Mol Microbiol Biotechnol 9:125–131

Deutscher J, Francke C, Postma PW (2006) How phosphotransferase system-related protein phosphorylation regulates carbohydrate metabolism in bacteria. Microbiol Mol Biol Rev 70:939–1031

Durfee T, Nelson R, Baldwin S et al. (2008) The complete genome sequence of *Escherichia coli* DH10B: Insights into the biology of a laboratory workhorse. J Bacteriol 190:2597–2606

Dyer CM, Dahlquist FW (2006) Switched or not?: the structure of unphosphorylated CheY bound to the N terminus of FliM. J Bacteriol 188:7354–7363

Galperin MY, Nikolskaya AN, Koonin EV (2001) Novel domains of the prokaryotic two-component signal transduction systems. FEMS Microbiol Lett 203:11–21

Galperin MY (2005) A census of membrane-bound and intracellular signal transduction proteins in bacteria: bacterial IQ, extroverts and introverts. BMC Microbiol 5:35

Galperin MY (2006) Structural classification of bacterial response regulators: diversity of output domains and domain combinations. J Bacteriol 188:4169–4182

Gao R, Mack TR, Stock AM (2007) Bacterial response regulators: versatile regulatory strategies from common domains. Trends Biochem Sci 32:225–234

Grebe TW, Stock JB (1999) The histidine protein kinase superfamily. Adv Microb Physiol 41: 139–227

Hagiwara D, Yamashino T, Mizuno T (2004) A genome-wide view of the *Escherichia coli* BasS-BasR two-component system implicated in iron-responses. Biosci Biotechnol Biochem 68:1758–1767

Hayashi K, Morooka N, Yamamoto Y et al. (2006) Highly accurate genome sequences of *Escherichia coli* K-12 strains MG1655 and W3110. Mol Syst Biol 2:2006 0007

Hayashi T, Makino K, Ohnishi M et al. (2001) Complete genome sequence of enterohemorrhagic *Escherichia coli* O157:H7 and genomic comparison with a laboratory strain K-12. DNA Res 8:11–22

Hengge-Aronis R (2002) Signal transduction and regulatory mechanisms involved in control of the s^S (RpoS) subunit of RNA polymerase. Microbiol Mol Biol Rev 66:373–395

Inouye M, Dutta R (eds) (2003) Histidine kinases in signal transduction. Academic Press, San Diego – London

Jenal U, Malone J (2006) Mechanisms of cyclic-di-GMP signaling in bacteria. Annu Rev Genet 40:385–407

Johnson TJ, Kariyawasam S, Wannemuehler Y et al. (2007) The genome sequence of avian pathogenic *Escherichia coli* strain O1:K1:H7 shares strong similarities with human extraintestinal pathogenic *E. coli* genomes. J Bacteriol 189:3228–3236

Kanehisa M, Araki M, Goto S et al. (2008) KEGG for linking genomes to life and the environment. Nucleic Acids Res 36:D480–484

Krin E, Sismeiro O, Danchin A et al. (2002) The regulation of Enzyme IIA^{Glc} expression controls adenylate cyclase activity in *Escherichia coli*. Microbiology 148:1553–1559

Muffler A, Fischer D, Altuvia S et al. (1996) The response regulator RssB controls stability of the s^S subunit of RNA polymerase in *Escherichia coli*. EMBO J 15:1333–1339

Park YH, Lee BR, Seok YJ et al. (2006) In vitro reconstitution of catabolite repression in *Escherichia coli*. J Biol Chem 281:6448–6454

Parkinson JS, Kofoid EC (1992) Communication modules in bacterial signaling proteins. Annu Rev Genet 26:71–112

Paul R, Weiser S, Amiot NC et al. (2004) Cell cycle-dependent dynamic localization of a bacterial response regulator with a novel di-guanylate cyclase output domain. Genes Dev 18:715–727

Perna NT, Plunkett G, 3rd, Burland V et al. (2001) Genome sequence of enterohaemorrhagic *Escherichia coli* O157:H7. Nature 409:529–533

Poon WW, Davis DE, Ha HT et al. (2000) Identification of *Escherichia coli ubiB*, a gene required for the first monooxygenase step in ubiquinone biosynthesis. J Bacteriol 182:5139–5146

Postma PW, Lengeler JW, Jacobson GR (1993) Phosphoenolpyruvate:carbohydrate phosphotransferase systems of bacteria. Microbiol Rev 57:543–594

Repik A, Rebbapragada A, Johnson MS et al. (2000) PAS domain residues involved in signal transduction by the *aer* redox sensor of *Escherichia coli*. Mol Microbiol 36:806–816

Römling U, Gomelsky M, Galperin MY (2005) C-di-GMP: The dawning of a novel bacterial signalling system. Mol Microbiol 57:629–639

Ryjenkov DA, Tarutina M, Moskvin OM et al. (2005) Cyclic diguanylate is a ubiquitous signaling molecule in *Bacteria*: Insights into biochemistry of the GGDEF protein domain. J Bacteriol 187:1792–1798

Ryjenkov DA, Simm R, Römling U et al. (2006) The PilZ domain is a receptor for the second messenger c-di-GMP: the PilZ domain protein YcgR controls motility in enterobacteria. J Biol Chem 281:30310–30314

Schmidt AJ, Ryjenkov DA, Gomelsky M (2005) Ubiquitous protein domain EAL encodes cyclic diguanylate-specific phosphodiesterase: Enzymatically active and inactive EAL domains. J Bacteriol 187:4774–4781

Shi L, Potts M, Kennelly PJ (1998) The serine, threonine, and/or tyrosine-specific protein kinases and protein phosphatases of prokaryotic organisms: a family portrait. FEMS Microbiol Rev 22:229–253

Stewart V (2003) Nitrate- and nitrite-responsive sensors NarX and NarQ of proteobacteria. Biochem Soc Trans 31:1–10

Stock AM, Robinson VL, Goudreau PN (2000) Two-component signal transduction. Annu Rev Biochem 69:183–215

Szurmant H, Ordal GW (2004) Diversity in chemotaxis mechanisms among the bacteria and archaea. Microbiol Mol Biol Rev 68:301–319

Thomas SA, Brewster JA, Bourret RB (2008) Two variable active site residues modulate response regulator phosphoryl group stability. Mol Microbiol 69:453–465

Toro-Roman A, Wu T, Stock AM (2005) A common dimerization interface in bacterial response regulators KdpE and TorR. Protein Sci 14:3077–3088

Ulrich LE, Zhulin IB (2007) MiST: a microbial signal transduction database. Nucleic Acids Res 35:D386–D390.

Welch RA, Burland V, Plunkett G, 3rd et al. (2002) Extensive mosaic structure revealed by the complete genome sequence of uropathogenic *Escherichia coli*. Proc Natl Acad Sci USA 99:17020–17024

Yamamoto K, Hirao K, Oshima T et al. (2005) Functional characterization *in vitro* of all two-component signal transduction systems from *Escherichia coli*. J Biol Chem 280:1448–1456

Zhou Y, Gottesman S, Hoskins JR et al. (2001) The RssB response regulator directly targets σ^S for degradation by ClpXP. Genes Dev 15:627–637

Zhulin IB (2001) The superfamily of chemotaxis transducers: from physiology to genomics and back. Adv Microb Physiol 45:157–198

Zogaj X, Nimtz M, Rohde M et al. (2001) The multicellular morphotypes of *Salmonella typhimurium* and *Escherichia coli* produce cellulose as the second component of the extracellular matrix. Mol Microbiol 39:1452–1463.

Chapter 9
Genome-Scale Reconstruction, Modeling, and Simulation of *E. coli's* Metabolic Network

Adam M. Feist, Ines Thiele, and Bernhard Ø. Palsson

Contents

Abstract Since the release of the first genome-scale metabolic reconstruction of the *E. coli* metabolic network in 2000, there has been a growing number of researchers around the world adapting it for a broad range of studies (Feist and Palsson 2008). The uses range from practical applications to obtaining basic biological understanding of cellular behavior. This range of uses is further expected to expand as the reconstruction broadens in scope and as new *in silico* methods are developed, implemented, and put to use.

In this chapter, we will describe foundational concepts central to the reconstruction process and model formulation, the history of reconstruction of the *E. coli* metabolic network, the development of reconstruction technology, genome-scale constraint based modeling with key exemplary case studies of uses of the *E. coli* metabolic reconstruction, and insights into the future of the field. As such,

B.Ø. Palsson (✉)
Department of Bioengineering, University of California San Diego, 9500 Gilman Drive, La Jolla, CA 92093-0412, USA
e-mail: palsson@ucsd.edu

S.Y. Lee (ed.), *Systems Biology and Biotechnology of Escherichia coli*,
DOI 10.1007/978-1-4020-9394-4_9, © Springer Science+Business Media B.V. 2009

this chapter should serve as a guide to those interested in either expanding the application of the *E. coli* reconstruction or adapting established applications to other organisms.

9.1 Foundational Concepts

The reconstruction of the *E. coli* metabolic network has led to the development of *'bottom-up'* reconstruction technology, genome-scale modeling methods, and basic and practical uses. A number of foundational concepts have also been developed during the period that we introduce here and provide background and a conceptual framework for the reader (see Palsson 2006, Price et al. 2004a).

Forming a BiGG knowledge base: A network reconstruction is based on a highly curated set of primary biological information for a particular organism; a biochemically, genetically and genomically structured (BiGG) knowledge base (Reed et al. 2006a). Such a knowledge base represents a large body of experimental data that is meticulously assembled and curated through the systems biology and reconstruction approaches detailed herein.

Genome-scale network reconstruction (GENRE): An organism-specific BiGG knowledge base is the basis for a GENRE. A GENRE is specific to a particular organism, for example, GENRE of *Escherichia coli* (below we will see four of these, specifically called *i*JE660, *i*JR904, *i*MBEL979, and *i*AF1260). A GENRE contains a list of all the known (and some predicted) chemical transformations that are believed to take place in the particular network (e.g. metabolic, transcriptional regulatory network, etc.).

The central role of network reconstruction in systems biology: Systems biology research generally can be conceptualized as a four-step process (Fig. 9.1). Foundational to the field is the generation of global, or genome-scale, data. The growing number of available 'omics' data types has created the need for formal and structured multi-'omic' data integration (Joyce and Palsson 2006). Omics data, along with legacy information (i.e., the 'bibliome') and detailed small-scale experiments, can be used to define the interactions among biological components that are used to reconstruct networks in particular organisms (Reed et al. 2006a). Network reconstruction is also an iterative, on-going process that continually integrates data in a formal fashion as it becomes available (Reed and Palsson 2003). These characteristics render the network reconstruction as a common denominator for those studying systems biology. The reconstruction effectively represents a 2-D annotation of a genome detailing not only the parts for an organism, but the interactions between specific components (Palsson 2004). Genome-scale reconstruction technologies for metabolic (Reed et al. 2006a), transcriptional regulation (Covert et al. 2004, Gianchandani et al. 2006, Herrgard et al. 2004) and signaling networks (Papin et al. 2005) have been established, and transcriptional/translational network reconstruction methods are currently under development (Thiele et al. 2009). An in depth review on the bottom-up reconstruction process (Palsson 2006) as well as a current review of biological network reconstruction (Feist et al. 2009) have been generated.

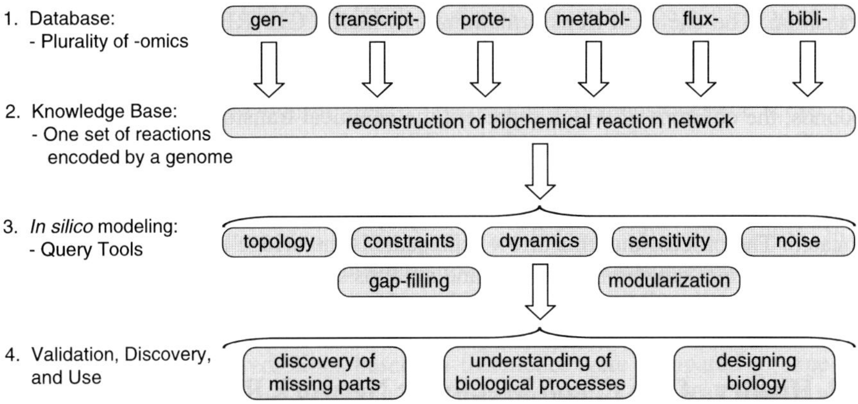

Fig. 9.1 Systems Biology as a 4-step Process. Step 1, the process is based on a variety of high-throughput data sets (i.e., 'omics' data) and a comprehensive assessment of the literature (i.e., bibliomic data). Step 2, all of the data types are used to reconstruct the list of biochemical transformations that make up a network as well as their genetic basis (Reed et al. 2006a). In principal, the network is unique. Step 3, the data contained in the reconstruction can be formally represented (i.e., in the form of matrices and logical statements) that can be mathematically characterized by a variety of methods. Step 4, the computational model enables a broad spectrum of applications, as reviewed in this chapter. Figure adapted from (Feist and Palsson 2008, Palsson 2006)

Constraint-based reconstruction and analysis (COBRA): COBRA is the overall philosophy and approach of applying constraints to limit the range of achievable functional (phenotypic) states of GENREs (outlined below). A GENRE operates under defined constraints. These constraints fall into at least four categories (Palsson 2006): physico-chemical, topological, regulatory, and environmental. Such constraints can be mathematically represented and imposed on the functional states that a GENRE can take on. Functional states can be assessed using a variety of computational methods (Palsson 2006, Price et al. 2004a) and have been disseminated in the form of a COBRA Toolbox (Becker et al. 2007) that is a MATLAB (The MathWorks Inc., Natick, MA) based software package.

Converting network reconstructions into a Genome-scale Model (GEM): A GENRE can be converted into a mathematical form (i.e., an *in silico* model) and used to computationally assess phenotypic properties (reviewed in (Price et al. 2004a)). The COBRA approach is used to analyze the properties of GENREs by assessing allowable functional states. Genome-scale reconstructions are thus a key step in quantifying the genotype-phenotype relationship and can be used to 'bring genomes to life' (Frazier et al. 2003). The availability of reconstructed metabolic networks for microorganisms has increased rapidly in recent years and a growing number of research groups are synthesizing GENREs for target organisms of interest (see Fig. 9.4) (Feist et al. 2009, Reed et al. 2006a).

The conversion of a reconstruction (GENRE) to an *in silico* model (GEM), represented by the arrow from step 2 to step 3 in Fig. 9.1, involves a subtle, but critical,

An updated annotation of the *E. coli* K-12 MG1655 genome (Serres et al. 2001) and continual functional characterization of *E. coli* metabolic content enabled an expansion of the reconstruction in 2003, which consisted of 931 reactions catalyzed by 904 gene products (Reed et al. 2003). This reconstruction, titled *i*JR904, was an improvement over previous efforts in that contained both charge and elemental balancing of all reactions, expanded the various carbon source utilization pathways, contained a larger number of characterized transport systems and their encoding genes, better accounted for quinone usage in the electron transport chain, and better detailed the relationship between given genes, proteins, and reactions contained in the reconstruction (the GPR associations).

This reconstruction has been utilized for a broad number of applications reviewed later in this chapter. Utilizing the *i*JR904 (Reed et al. 2003) reconstruction, an expanded reconstruction of *E. coli* was generated (containing 979 reactions and titled *i*MBEL979) for the purpose of designing overproducing strains in the software framework MetaFluxNet (Lee et al. 2005).

The most recent metabolic reconstruction for *E. coli*, titled *i*AF1260, incorporates data from the most recent *E. coli* K-12 MG1655 genome annotation (Riley et al. 2006) and consists of 2,077 reactions and 1,260 genes (Feist et al. 2007). The advancements represented by *i*AF1260 over *i*JR904 lie in five main areas: an increased scope with the inclusion of 357 additional ORFs; compartmentalization into three distinct compartments (cytoplasmic, periplasmic and extra-cellular); the detailing of all grouped, or lumped, reactions (most often associated with lipid and lipopolysaccharide biosynthesis); the incorporation of reaction thermodynamics, calculated Gibbs free energy (Δ G°) values for 950 metabolites and 1935 reactions; and alignment with the EcoCyc database (Keseler et al. 2005) which provided expanded coverage for the network and content mappings for further computational analyses.

This 18-year history of reconstruction of the *E. coli* metabolic network has culminated in a network containing a total number of 1,260 metabolic genes covering 28% of the 4,453 identified ORFs on the *E. coli* genome. More importantly, the 1260 ORFs represent 48% of the functionally annotated ORFs that have been confirmed by experimental data (Table 9.1). Thus, 92% of the 1,260 gene products included in *i*AF1260 have been experimentally verified (Riley et al. 2006) with the balance of 8% having a computationally predicted function which necessitate confirmation with focused experimentation. Model-aided gap-filling and discovery will aid in this process (see Section 9.5.2). In addition, protein structures (computed or experimental) are available for a large fraction of the proteins in *i*AF1260 (Berman et al. 2000). Integration of protein structural data with the functional content of the reconstruction will lead to a better understanding of structural motifs and their properties.

Reconstruction of the *E. coli* metabolic network is thus approaching exhaustion of known metabolic gene functions and is now being used in a prospective fashion to discover new metabolic capabilities in *E. coli* (see below). As a result of this endeavour, the reconstruction of the *E. coli* metabolic network represents the best-developed genome-scale network to date.

Table 9.1 Properties of the most current *E. coli* metabolic reconstruction

	*i*AF1260 this study
Included genes	1260 (28%)[d]
Experimentally–based function	1161 (92%)
Computationally predicted function	99 (8%)
Unique functional proteins	1148
Multigene complexes	167
Genes involved in complexes	415
Instances of isozymes[a]	346
Reactions	2077
Metabolic Reactions	1387
Unique metabolic reactions[b]	1339
Cytoplasmic	1187
Periplasmic	192
Extracellular	8
Transport Reactions	690
Cytoplasm to periplasm	390
Periplasm to extracellular	298
Cytoplasm to extracellular	2
Gene - protein - reaction associations	
Gene associated (met./trans.)	1294/625
Spontaneous/diffusion reactions[c]	16/9
Total gene associated and no association needed (met./trans.)	1310/634 (94%)
No gene association (metabolic/transport)	77/56 (6%)
Exchange reactions	304
Metabolites	
Unique Metabolites[b]	1039
Cytoplasmic	951
Periplasm	418
Extracellular	299

[a] tabulated on a reaction basis, not counting outer membrane non-specific porin transport.
[b] reactions can occur in or between multiple compartments and metabolites can be present in more than one compartment.
[c] diffusion reactions do not include facilitated diffusion reactions and are not included in this total if they can also be catalyzed by a gene product at a higher rate.
[d] overall genome coverage based on 4453 total ORFs in *E. coli*; *i*AF1260 contains 48% of the ORFs in *E. coli* that have been characterized experimentally (2403 ORFs).

9.3 Continuing Development of Reconstruction Technology

Development of the reconstruction process for metabolic networks: As illustrated in the previous section, the reconstruction process for metabolic networks is an iterative procedure that requires different types of experimental data and techniques at each phase of reconstruction. The experience with *E. coli* has led to

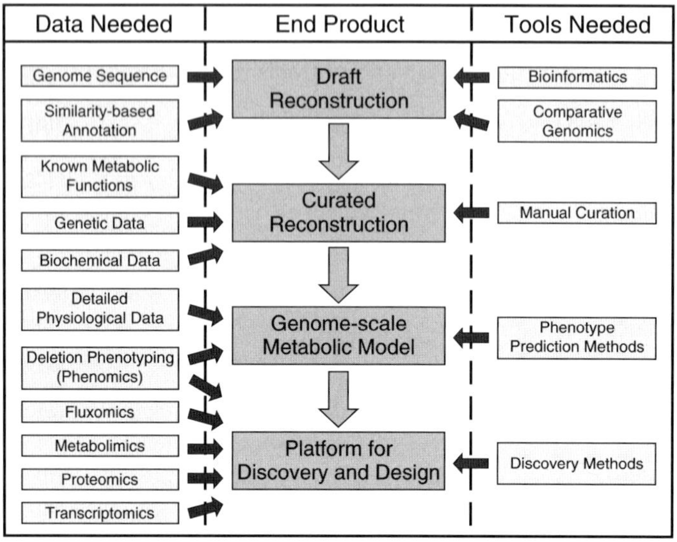

Fig. 9.3 The phases and tools necessary to generate a metabolic reconstruction. The genome-scale metabolic reconstruction process can be broken down into four major phases (center column), with each of the latter phases building off the previous. This process is iterative and driven by experimental data (primarily in the three latter phases). For each phase, specific data types are necessary and these range from high-throughput data types (e.g., phenomics, metabolomics, etc.), to detailed studies characterizing individual components (e.g., biochemical data for a particular reaction). For example, the genome annotation can provide a parts list of a cell, whereas genetic data can provide information about the contribution of each gene product towards a phenotype (e.g., when removed or mutated). The product generated from each reconstruction phase can be utilized and applied to examine a growing number of questions with the final product having the broadest applications

the formulation of the workflows that underlie metabolic reconstruction. The four phases of the reconstruction process are depicted in Fig. 9.3 and the product at each phase can be used for different applications, with the number of applications increasing with network development. This procedure represents the current status of network reconstruction, and the most recent *E. coli* reconstruction, *i*AF1260, was built accordingly (Feist et al. 2007) with the advantage of starting from an already well-established reconstruction, *i*JR904. The end product of this reconstruction effort is a platform for design and discovery, and key examples of use are given later in this chapter. More extensive descriptions exist, which outline the conceptual basis (Reed et al. 2006a) and the detailed process to generate genome-scale biological networks, (Feist et al. 2009) and these will not be repeated here.

Development of the reconstruction process beyond metabolism: The development and use of genome-scale reconstruction was rapid and many computational models were developed to address a growing spectrum of basic research and applied problems. Still, further development of reconstruction technology is necessary. The scope of reconstructions is bound to grow, representing more and more BiGG knowledge in the structured format of GEMs (Breitling et al. 2008). Growth in scope is likely to proceed in phases (Feist and Palsson 2008). Growth in scope in the near-term will involve the transcriptional and translational machinery (Allen and Palsson 2003, Mehra and Hatzimanikatis 2006, Thiele et al. 2009, Thomas et al. 2007). Such an extension will enable a range of studies including the direct inclusion of proteomic data, fine graining of growth requirements, and the explicit consideration of secreted protein products.

Another expansion in scope in the near-term is the reconstruction of the genome-scale transcriptional regulatory network (TRN). Such reconstruction at the genome-scale is now enabled by new experimental technologies, such as ChIP-chip (Lee et al. 2002). Experimental interrogation of the currently available TRN suggests that we know about one-fourth to one-third of its content (Covert et al. 2004), indicating that there is much to be discovered. This expectation is being confirmed with high-resolution ChIP-Chip data for *E. coli* (Cho et al. 2008). Once reconstructed, the TRN will allow computational predictions of the context-specific uses of the *E. coli* genome and the responses of two-component signaling systems.

Mid-term expansions in scope are likely to include the growth cycle, shock responses (e.g. heat and acid shock), and additional cellular functions (e.g. DNA replication and flagellar biosynthesis). Such a reconstruction should eventually be a comprehensive representation of the chemical reactions and transformations enabled by *E. coli*'s gene products.

Longer-term reconstruction may begin to address the 3-dimensional organization of the bacterial cell. In particular, high-resolution ChIP-chip data on the DNA binding protein could enable the estimation of the topological arrangement of the genome, and potentially elucidate the structure of the cell wall and other cellular structures that will allow a full 3-dimensional reconstruction of *E. coli*.

The two near-term expansions in content will encompass the activity of approximately 2000 ORFs in the *E. coli* genome. Clearly, quality-controlled reconstructions will help in guiding us to comprehensive genome-scale representation of all major cellular processes in bacteria at the BiGG data level of resolution that, in turn, enables GEMs of growing coverage and resolution. The scope of this effort has been described as being; "... 10 times more ambitious and 100 times more important for mankind [compared with Human Genome Project]..." Hans Westerhoff (Holden 2002).

Influence of the *E. coli* reconstruction on the *in silico* analysis of other micro-organisms: The metabolic network reconstruction of *E. coli* has been influential in the generation of other organism-specific metabolic networks. The *E. coli* metabolic reconstruction has served: As a content database where stoichiometrically and charge balanced reactions, and even pathways, have been incorporated into new reconstructions, As a database for defined metabolites, and as a source for a biomass objective function to query network content and functionality.

This influence has sparked an increase in the number of genome-scale network reconstructions that have been generated to formulate GEMs for a number of organisms. A detailed list of GEMs that have been developed, curated, and used for computation is given in Table 9.2. This table is a current snapshot of the

Table 9.2 Available predictive genome-scale metabolic network reconstructions

		Organism properties	Reconstruction properties				
Name	Strain	Genes	Genes	Metabolites	Reactions	Compartments	References
BACTERIA							
Bacillus subtilis		4, 225	844	988	1020	2 (c,e)	(Oh et al. 2007)
Clostridium acetobutylicum	ATCC 824	3, 848	474	422	552	2 (c,e)	(Senger and Papoutsakis 2008)
Clostridium acetobutylicum	ATCC 824	3, 848	432	479	502	2 (c,e)	(Lee et al. 2008)
Escherichia coli	K12 MG1655	4, 405	660	438	627	2 (c,e)	(Edwards and Palsson 2000)
Escherichia coli	K12 MG1655	4, 405	904	625	931	2 (c,e)	(Reed et al. 2003)
Escherichia coli	K12 MG1655	4, 405	1260	1039	2077	3 (c,e,p)	(Feist et al. 2007)
Geobacter sulfurreducens		3, 530	588	541	523	2 (c,e)	(Mahadevan et al. 2006)
Haemophilus influenzae	Rd	1, 775	296	343	488	2 (c,e)	(Edwards and Palsson 1999)
Haemophilus influenzae	Rd	1, 775	400	451	461	2 (c,e)	(Schilling and Palsson 2000)
Helicobacter pylori	26695	1, 632	341	485	476	2 (c,e)	(Thiele et al. 2005b)
Helicobacter pylori	26695	1, 632	291	340	388	2 (c,e)	(Schilling et al. 2002)
Lactobacillus plantarum	WCFS1	3, 009	721	531	643	2 (c,e)	(Teusink et al. 2006)
Lactococcus lactis	ssp. lactis IL1403	2, 310	358	422	621	2 (c,e)	(Oliveira et al. 2005)
Mannheimia succiniciproducens	MBEL55E	2, 384	425	519	686	2 (c,e)	(Kim et al. 2007)
Mycobacterium tuberculosis	H37Rv	4, 402	726	739	849	2 (c,e)	(Beste et al. 2007)
Mycobacterium tuberculosis	H37Rv	4, 402	661	828	939	2 (c,e)	(Jamshidi and Palsson 2007)
Mycoplasma genitalium	G-37	521	189	276	264	2 (c,e)	Personal Comm.: Patrick F. Suthers
Neisseria meningitidis	serogroup B	2, 226	555	471	496	2 (c,e)	(Baart et al. 2007)
Pseudomonas aeruginosa	PA01	5, 640	1056		883	2 (c,e)	(Oberhardt et al. 2008)
Pseudomonas putida	KT2440	5, 350	746	911	950	3 (c,e,p)	(Nogales 2008)
Rhizobium etli	CFN42	3, 168	363	371	387	2 (c,e)	(Resendis-Antonio et al. 2007)

Table 9.2 (continued)

		Organism properties	Reconstruction properties				
Name	Strain	Genes	Genes	Metabolites	Reactions	Compartments	References
Staphylococcus aureus	N315	2, 588	619	571	641	2 (c,e)	(Becker and Palsson 2005)
Staphylococcus aureus	N315	2, 588	551	604	712	2 (c,e)	(Heinemann et al. 2005)
Streptomyces coelicolor	A3(2)	8, 042	700	500	700	2 (c,e)	(Borodina et al. 2005)
ARCHAEA							
Methanosarcina barkeri	Fusaro	5, 072	692	558	619	2 (c,e)	(Feist et al. 2006)
Halobacterium salinarum	R-1	2, 867	490	557	711	2 (c,e)	(Gonzalez et al. 2008)
EUKARYOTES							
Aspergillus nidulans		9, 451	666	732	794	4	(David et al. 2008)
Homo sapiens		28, 783	1, 496	2, 766	3, 311	8	(Duarte et al. 2007)
Leishmania major	Friedlin	8, 370	560	1, 101	1, 112	8	(Chavali et al. 2008)
Mus musculus		28, 287	473	872	1, 220	3 (c,e,m)	(Sheikh et al. 2005)
Saccharomyces cerevisiae	Sc288	6, 183	708	584	1, 175	3 (c,e,m)	(Forster et al. 2003)
Saccharomyces cerevisiae	Sc288	6, 183	750	646	1, 149	8	(Duarte et al. 2004)
Saccharomyces cerevisiae	Sc288	6, 183	672	636	1, 038	3 (c,e,m)	(Kuepfer et al. 2005)

This list includes genome-scale metabolic network reconstructions that have been converted into predictive genome-scale models and whose predictive power has been validated against experimental data. Compartments: c – cytosol, e – extraorganism, p – periplasm, m – mitochondrion.

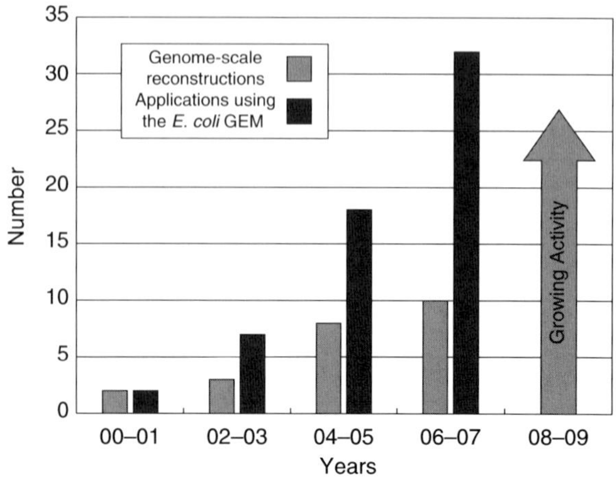

Fig. 9.4 Appearance of organism-specific genome-scale reconstructions and applications of the E. coli metabolism reconstruction. The genome-scale reconstructions for metabolic networks that have appeared every two years since the release of the first GEMs in 2000 (see Table 9.2) and the number of published studies the have appeared utilizing the *E. coli* GEM (Feist and Palsson 2008). Since the release of the first GEMs for *E. coli* (Edwards and Palsson 2000) and that of *Haemophilus influenzae* (Edwards and Palsson 1999), there has been a significant increase in both the number of genome-scale reconstructions and studies focused on the *E. coli* GEM for every time period

available reconstructions and a continually updated version can be found online (http://systemsbiology.ucsd.edu/*In_Silico*_Organisms/Other_Organisms). Additionally, Fig. 9.4 shows the number of genome-scale reconstructions that have been developed over two year periods (for the reconstructions listed in Table 9.2). The number of reconstructions generated for each period has increased since the release of the first genome-scale reconstructions for *Haemophilus influenzae* in 1999 (Edwards and Palsson 1999) and *E. coli* in 2000 (Edwards and Palsson 2000). Furthermore, the number of published studies utilizing the *E. coli* GEM has also increased significantly over time resulting in the applications outlined in the sections below (Feist and Palsson 2008).

Modeling strategy and philosophy: Models are a formal way of accounting for our knowledge about the phenomena being described. When describing biochemical reaction networks formally, we need to deal with the 'links' (i.e., the reactions) between 'nodes' (i.e., the compounds). Our knowledge about links between biological molecules varies; from the abstract to the specific (Fig. 9.5). Statistical models are built on correlations and a black box approach that is not mechanism based. Specific mechanism-based models are based on knowledge of chemistry, kinetics, and thermodynamics. Given the fact that kinetic and thermodynamic information is hard to obtain on a large-scale, stoichiometric models stop one step short of full specification (in the spectrum conveyed in Fig. 9.5). The result is that we have chemistry (and its genetic basis) and network structure used as the foundation for building a mathematical description of network functions. Such models do not have a unique solution

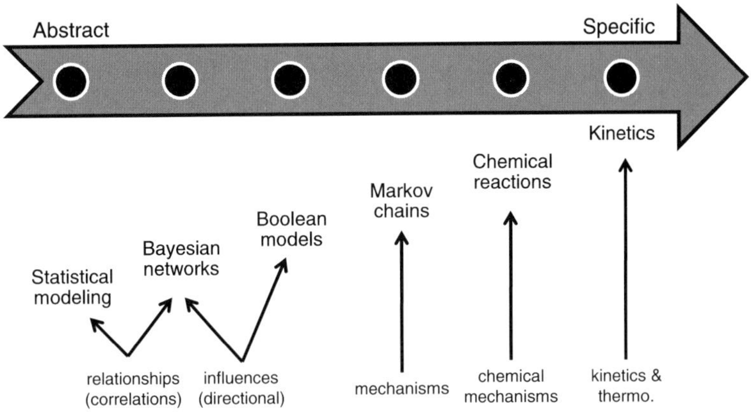

Fig. 9.5 The different levels of knowledge used to generate biological models. Our knowledge about links in biochemical networks varies. At one extreme, the information is abstract and often takes the form of *black-box* correlations. At the other, we have detailed chemical mechanisms with kinetic and thermodynamic information. Stoichiometric models would be second from the *right*, accounting for mechanisms, but not incorporating kinetic and thermodynamic information

(e.g., see (Palsson 2006) and below). The lack of kinetic information can be dealt with by: (1) examining the properties of the entire set of solutions (i.e., the solutions space) or (2) by using constraint-based optimization to find specific solutions in the space (Price et al. 2004a). The latter can be successful if we know the prevailing selection pressure on an organism. The combination of a network reconstruction that is based on a knowledge-base at the genome-scale and the inherent optimality properties of the selection process underlie the success of COBRA for a number of applications.

Constraint-based modeling methods: Over the past quarter century, there has been a growing number of computational tools developed to interrogate biological networks and models (Breitling et al. 2008, Palsson 2006, Price et al. 2004a). Owing to its early development, the *E. coli* reconstruction and model has been a popular target for initial screening and development of a number of these methods. In this section, we introduce basic concepts common to most of these methods and describe in more detail those methods that were used in the studies presented in this chapter. The interested reader is encouraged to refer to recently published reviews presenting the constraint-based modeling methods in more detail (Breitling et al. 2008, Palsson 2006, Price et al. 2004a).

Mathematical description of the reconstruction: The metabolic reconstruction consists of a list of biochemical transformations known to take place in the target organism. This reaction list can be readily converted into a mathematical, computable format by using any available parser (e.g. in COBRA toolbox (Becker et al. 2007)). Using a parser, the stoichiometric coefficients are extracted for the individual reactions and entered in the cell of the stoichiometric matrix, also called the S matrix (Fig. 9.6). In this S matrix, every row corresponds to a metabolite and every column corresponds to a network reaction. Note that a typical S matrix

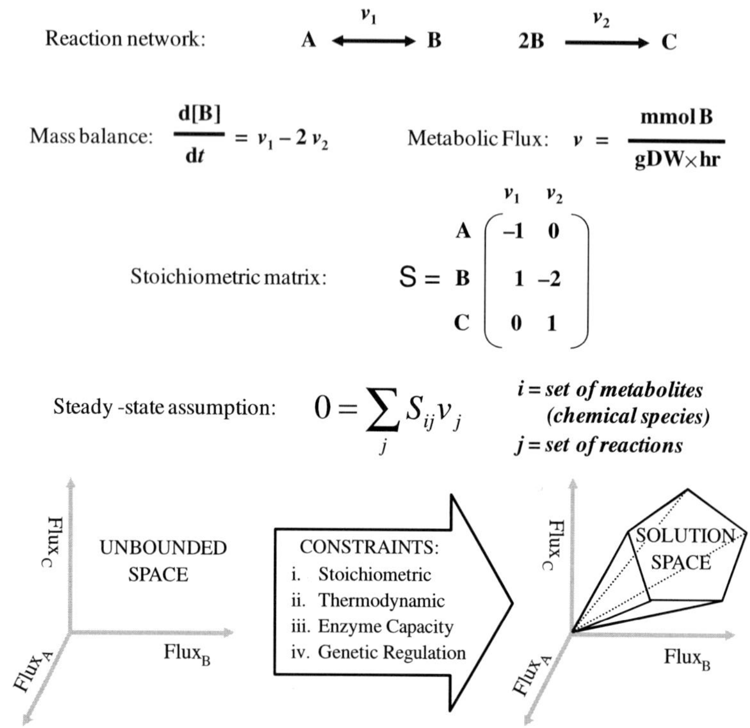

Fig. 9.6 The structure and application of constraints to networks. Shown are the components (Reaction network) and the engineering approaches and equations used to model a reconstructed network. The stoichiometric matrix is a mathematical representation of a reconstructed network and the steady-state assumption is used in a number of COBRA approaches, including flux balance analysis. The *bottom* of the diagram depicts how an unbound space can be confided to a solution space in which a network must behave by imposing the governing physiochemical constraints on a system (e.g., thermodynamic constraints)

is very sparse (< 1% non-zero entries) as many biochemical transformations are bi-linear, and the majority of metabolites appear only in few metabolic reactions. Only a few metabolites, such as protons, water, and ATP, are highly connected in a metabolic network, and participate in many metabolic reactions. Many studies have concentrated on studying the topological features of metabolic networks and the S matrix (see Section 9.5.4 or (Feist and Palsson 2008)).

The multiplication of this S matrix with a flux vector v, containing flux values for all reactions v_j in S, results in a vector listing the changes in concentrations of all metabolites x_i over time:

$$S \bullet v = \frac{dx}{dt} \tag{9.1}$$

The constraint–based modeling approaches are based on the steady state assumption (Fig. 9.6), which assumes that the change of metabolite concentration over time is zero:

$$S \bullet v = \frac{dx}{dt} = 0 \tag{9.2}$$

This assumption is valid for the metabolic reactions as the time scale of the reaction rates is much smaller (milliseconds range) than the doubling time of a cell, which is on the order of hours. Due to this time-scale separation, the metabolic network is essentially in a steady state during cell replication, and as a consequence, intracellular metabolites are not allowed to accumulate. This restriction, imposed by Equation (9.2), is known as the mass-balance constraint (Fig. 9.6).

Further constraints may be added to the reconstruction, leading to the conversion of the reconstruction to a condition-specific model. Such constraints can include thermodynamic (i.e., reaction reversibility), regulatory (e.g., expression of an enzyme), topological (i.e., composition and connectivity of network), and environmental (e.g., presence/absence of a specific carbon source).

Interrogation of the steady state solution space: In most cases, the set of equations encoded in the S matrix are underdetermined, meaning that there are more variables (fluxes v_j for $j = 1 \ldots n$) than there are equations (mass-balances for each metabolite x_i for $i = 1 \ldots m$). As a consequence, there is no single solution or flux vector v satisfying all the equations, but rather there are many possible flux vectors. This set of possible flux vectors is called the steady-state solution space. Each flux vector v, satisfying the given model constraints, is called a functional state of the network. This term functional state can be seen as analogous to the traffic pattern of the road mesh in a large city. The road mesh would correspond to the metabolic network and the traffic pattern, which shows high traffic and low traffic on the highways, corresponds to the functional state of the road system. Clearly this traffic pattern will be very different in the afternoon during rush hour versus the traffic pattern found late into the night. This example highlights the idea that one network can have many distinct functional states.

Functional states of a network can be determined using different mathematical approaches. In the COBRA approach, there is a distinction between biased and unbiased methods. Biased methods require the statement of an objective function, such as a biomass formation reaction or a byproduct secretion reaction by the metabolic network. This objective function is then maximized (or minimized) to obtain a functional state leading to the maximal (or minimal) flux value of the objective function. In contrast, unbiased methods explore the entire steady state solution space by determining a representative subset of possible functional states that can be analyzed in a statistical manner. Examples of unbiased methods are uniform sampling (Almaas et al. 2004, Price et al. 2004b, Thiele et al. 2005a, Wiback et al. 2004) and extreme pathway analysis (Papin et al. 2002, Price et al. 2003).

In many COBRA applications, it is assumed that the aim of a living cell is to grow as fast as possible to outgrow competitors and thus to use available nutrients mainly for biomass production. Hence, many COBRA applications are used in conjunction with the maximization of the biomass production rate. For example, gene essentiality can be determined *in silico* where the essentiality of every gene is tested to see whether the metabolic network is still able to produce biomass despite the *in silico* disruption of a gene (see Fig. 9.9). Other examples in this chapter discuss metabolic engineering applications, where the metabolic network is modified in such way that it produces a desired byproduct while maintaining a certain biomass production capability. Many industrially-interesting byproducts are produced by cells when they cannot produce biomass (e.g., due to nitrogen or phosphate limitations). Thus, the byproduct and biomass production are competitive, or 'orthogonal' to each other. COBRA has been successfully used to couple the byproduct production with the biomass production by deleting certain metabolic genes, thereby redirecting carbon fluxes in the metabolic networks (see below). The byproduct coupling to biomass production forces the organism to produce the desired byproduct in order to obtain the cellular objective of biomass production.

9.4 Applications and Uses of the *E. coli* Metabolic Reconstruction

Ask not what you can do for a reconstruction, but what a reconstruction can do for you: The *E. coli* reconstruction and GEM has been adapted for a broad number of uses by research groups around the world. Studies utilizing the reconstructed *E. coli* network range from pragmatic to theoretical applications and address a wide range of questions. These uses can be further categorized into five areas which include: (1) metabolic engineering, (2) biological discovery, (3) assessment of phenotypic behavior, (4) biological network analysis, and (5) studies of bacterial evolution (Fig. 9.7). A more extensive review of these uses has recently appeared (Feist and Palsson 2008), as well as an additional review on metabolic engineering efforts with *E. coli* and other organisms (Kim et al. 2008). Here, key examples of uses of the *E. coli* reconstruction in each of these fields will be presented to demonstrate the utility of the reconstruction and modeling process.

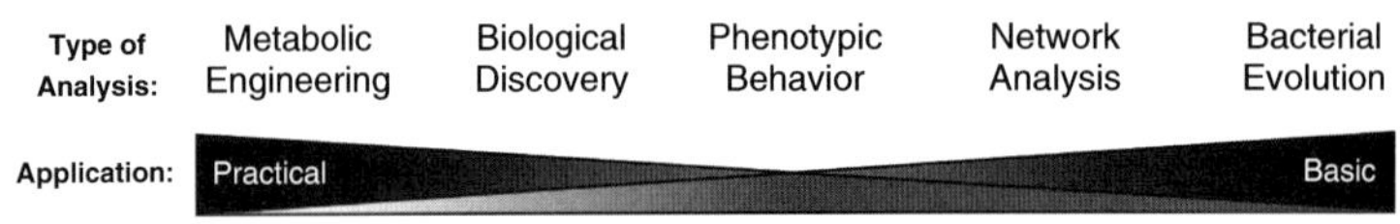

Fig. 9.7 Spectrum of uses of the of the genome-scale E. coli metabolic network reconstruction. Uses of the *E. coli* metabolic reconstruction can be categorized into 5 different areas. Furthermore, these categories can be arranged in order of addressing more practical (e.g., generating a production strain) or more basic (e.g., understanding horizontal gene transfer) questions

9.4.1 Metabolic Engineering

Metabolic engineering efforts utilizing the GEM of *E. coli* have focused on exploring overproduction for a number of products. Three examples in which computation and experimental construction were used to achieve overproduction will be discussed here. The first two examples utilized the *E. coli* GEM to explore the production of the amino acids L-valine (Park et al. 2007) and L-threonine (Lee et al. 2007) in *E. coli*, and each has demonstrated the broad usage of GEM-aided computation for strain design.

Production of L-threonine: In the first study, GEM-aided modeling was employed in three different areas to increase the production of L-threonine to industrial titers (Fig. 9.8) (Lee et al. 2007). In one instance, *in silico* modeling was used to identify the optimal activity of a key enzymatic reaction towards maximum L-threonine production using a parametric sensitivity analysis that compared reaction activity to L-threonine production rate. The optimal activity prediction was subsequently used to tune the over-expression of the gene that encodes for this enzymatic reaction through comparison to base line activity, and the result was a production increase. This method proved to be vital to the success of this strain, as a previous transcription profiling guided attempt at over-expression resulted in an undesirable surplus of activity that was detrimental to L-threonine production.

For the same strain, a GEM-aided flux analysis in conjunction with mRNA expression data levels guided the elimination of negative regulation on a gene, which encoded for a reaction that channeled flux towards the final product. The third use of the GEM for the design of this strain occurred when an unwanted byproduct was observed in the culture medium and computation was utilized to divert the flux from this byproduct to L-threonine (Lee et al. 2007) through over-expression of another key gene encoded activity.

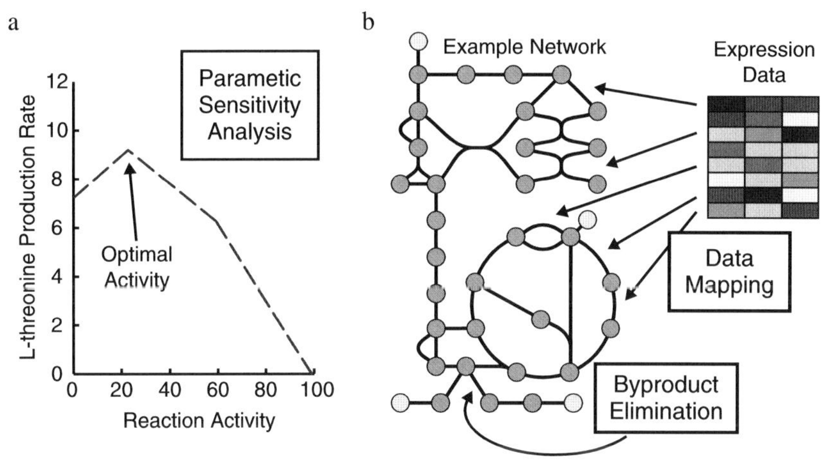

Fig. 9.8 Three different areas where modeling was incorporated to increase strain production. Areas of model-driven strain improvement utilized to overproduce L-threonine in *E. coli* (Lee et al. 2007). (**a**) Shown is a graph that provides the computed relationship between L-threonine production and the activity of particular reaction. This *in silico* parametric sensitivity analysis guided the level of expression necessary for increased production of the amino acid in the strain. (**b**) Given is a map of central metabolism representing the metabolic reconstruction of *E. coli*. In the analysis, expression data was mapped onto the network to guide the elimination of negative regulation and the network was used to overexpress a reaction that diverted flux away from a byproduct (byproduct elimination) towards the desired product

Production of lycopene: Lycopene is an important intermediate in the biosynthesis of many carotenoids, and it is used for food coloring as it possesses a strong color (bright red) and is non-toxic. To increase the production of an already high-producing strain, a systematic computational search was developed (Alper et al. 2005b) to explore the *E. coli* metabolic network and report gene deletions that diverted metabolic flux towards the desired product. This process resulted in a knock-out strain that, when constructed, showed a two fold increase in the production of lycopene over the parental strain. In this analysis, the minimization of metabolic adjustment (MOMA) computational algorithm (Segre et al. 2002) and the IJE660 (Edwards and Palsson 2000) *E. coli* GEM were utilized to sequentially examine additive genetic deletions that would improve lycopene production while maintaining cell viability. It was found that this computational approach yielded a twofold increase in production rate over a previously engineered overproducing strain and an eightfold increase over wild-type production harboring only a lycopene biosynthesis plasmid (Alper et al. 2005b). In addition, the strain designs identified computationally were compared to mixed combinatorial transposon mutagenesis, and it was found that the maximum production observed could be designed solely using the systematic GEM-aided computational method (Alper et al. 2005a,b). Furthermore, a deleterious effect was observed when targets identified in individual computational designs were combined in an attempt to achieve an overall more desirable phenotype. Thus, the overall systematic effects from individual designs were not additive and needed to be interpreted in the context of the entire network.

Production of L-valine: This model-driven example of metabolic engineering demonstrates the use of applying a systematic computational search algorithm (Alper et al. 2005b) to the updated *E. coli* GEM MBEL979 (Lee et al. 2005) (similar to the *i*JR904 GEM (Reed et al. 2003)) to improve L-valine production. In this analysis, the *in silico* computation of beneficial knock-outs to divert flux towards the desired product once again resulted in a significant increase (greater than twofold) in the production of the desired metabolite over an existing overproducing strain (Park et al. 2007). A number of additional metabolic engineering approaches to increase overproduction were performed by, (i) relieving feedback inhibition and regulation through attenuation, (ii) removing competing pathways, (iii) up-regulation of primary biosynthetic pathways, and (iv) over-expression of export machinery. When compared to each of the other individual strain modifications, the *in silico* GEM aided interventions resulted in the greatest increase in L-valine production (Park et al. 2007). Taken together, this and the previous study demonstrate the broad applications for which GEMs can be utilized to design strains not only in a *de novo* fashion, but to make further improvements on strains through integrating and interpreting experimental data.

9.4.2 Biological Discovery

The GEM of *E. coli* can be used as a guide to discovery. There is still a significant amount of information missing relating to gene functions in *E. coli* (Riley et al.

2006), and the content contained in the *E. coli* reconstruction can be queried and analyzed to first, determine the current gaps in our knowledge of the organism and second, design experiments to specifically fill uncovered gaps in the knowledge landscape. Two examples of model-driven discovery are presented, and these studies should form the basis for further analysis. To uncover the genetic basis for experimentally observed functions in *E. coli*, the studies combined GEM-aided computation with guided experimentation.

Systems approach to refining genome annotation: The first study utilized an iterative process (Reed et al. 2006b) in which, (i) differences in modeling predictions and high-throughput growth phenotype data were identified, (ii) potential missing reactions that remedy these disagreements were algorithmically determined, (iii) bioinformatics was utilized to identify likely encoding ORFs, and (iv) resulting targeted ORFs were cloned and experimentally characterized. Application of this process led to the functional characterization of eight ORFs that are involved in transport, regulatory and metabolic functions in *E. coli* (Reed et al. 2006b). The discovery process was aided by a high-throughput growth phenotyping analysis and the genome-wide single-gene mutant collection (Baba et al. 2006), along with other characterization analyses such as targeted expression profiling. This work was the first such example of model-driven discovery of genome content aided by a metabolic network reconstruction.

Genetic basis of orphan reactions: The second GEM-based analysis that resulted in ORF discovery utilized network topology to examine orphan reactions in the *E. coli* network (i.e., reactions known to exist in *E. coli* that have not been linked to an encoding gene) identified by network topology-based gap-filling algorithms (Chen and Vitkup 2006, Kharchenko et al. 2006, 2004). The basic premise behind these algorithms is the utilization of an orphan reaction's network neighbors as constraints to assign metabolic function. With the resulting tentative ORF assignments, biochemical characterization studies utilizing genetic mutants (Baba et al. 2006), analysis of growth under different substrate conditions, and expression data were all utilized to characterize and assign function to an orphan ORF that is responsible for a metabolic conversion that has been known for 25 years (Fuhrer et al. 2007). These two studies are early examples of how GEM computation can lead to the discovery of new genetic and biochemical content in an organism.

9.4.3 Assessment of Phenotypic Behavior

Researchers have utilized the *E. coli* GEM to better understand the coordinated functions of the cell and observed physiological outcomes. Computations seeking to predict cellular phenotypes have been performed under a range of genetic and environmental conditions, and phenotypic assessment has received the most attention in terms of publication and tool development. Here, we outline computational tools developed to analyze the *E. coli* GEM in each of the two major areas of phenotypic assessment, studies of (i) network perturbation/essentiality, and (ii) the incorporation of thermodynamic information.

a

Gene	Glucose	Glycerol	Succinate	Acetate
aceA	+ / +		+ / +	− / −
aceB				− / −
aceEF	− / +			
ackA				+ / +
acn	− / −			− / −
acs				+ / +
cyd	+ / +			
cyo	+ / +			
eno	− / +	− / +	− / −	− / −
fba	− / +			
fbp	+ / +	− / −	− / −	− / −
frd	+ / +		+ / +	+ / +
gap	− / −	− / −	− / −	− / −
glk	+ / +			
gltA	− / −			− / −
gnd	+ / +			
idh	− / −			− / −
mdh	+ / +	+ / +	+ / +	
ndh	+ / +	+ / +		
nuo	+ / +	+ / +		
pfk	− / +			
pgi	+ / +	+ / −	+ / −	
pgk	− / −	− / −	− / −	− / −
pgl	+ / +			
pntAB	+ / +	+ / +	+ / +	
ppc	± / +	− / +	+ / +	
pta				+ / +
pts	+ / +			
pyk	+ / +			
rpi	− / −	− / −	− / −	− / −
sdhABCD	+ / +		− / −	− / −
sucAB	+ / +		− / +	− / +
tktAB	− / −			
tpi	− / +	− / −	− / −	− / −
unc	+ / +		± / +	− / −
zwf	+ / +	+ / +	+ / +	

b

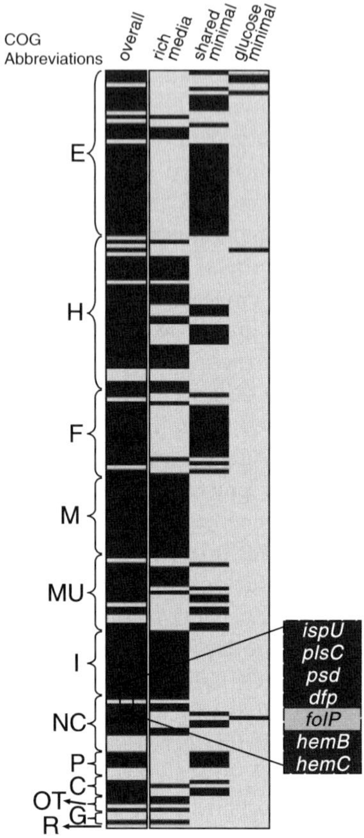

Fig. 9.9 Gene-deletion analyses utilizing the E. coli GEM. Analyses of gene essentiality in the *E. coli* metabolic network. (**a**) A table of results from an analysis performed using the *i*JE660 GEM of *E. coli* where experimental phenotypes were collected from bibliomic data. Results are scored as + or − meaning growth or no growth determined from *in vivo/in silico data*. The ± indicates that suppressor mutations have been observed that allow the mutant strain to grow. In 68 of 79 cases the *in silico* behavior is the same as the experimentally observed behavior. Each column represents a different carbon source. (**b**) This heat map characterizes the agreement between ORFs predicted to be essential using the *i*AF1260 GEM of *E. coli* (Feist et al. 2007) and those experimentally determined (Baba et al. 2006, Joyce et al. 2006). The enlarged region details how each row corresponds to a computationally predicted essential ORF (188 total). A dark row indicates the condition under which each ORF was found to be essential. For example, *folP* was predicted to be an essential ORF for the biosynthesis of folate in *i*AF1260 under these conditions, but was not identified as essential by Baba et al. (2006). The different columns show at which level each gene in the overall column was found to be essential on. With the advancement of both experimental data and model coverage, analyses of this type have reached the genomic scale

***In silico* perturbations**: A set of distinct computational methods using GEMs has been developed to determine the physiological state of *E. coli* (and other cells for which a GEM exists) after genetic perturbations (Segre et al. 2002, Shlomi et al. 2005, Wunderlich and Mirny 2006). These methods were analyzed to examine the effectiveness of predictions when compared to experimental data (Fig. 9.9). Whereas comparisons to flux data from wild-type and *E. coli* mutants reveals that one of the computational algorithms, MOMA (Segre et al. 2002), provided better predictions for transient growth rates (early post perturbation state), another algorithm, ROOM (Shlomi et al. 2005) (and basic FBA), was found to be more successful in predicting final steady-state growth rates and overall lethality (Shlomi et al. 2005). These two algorithms have been utilized, in addition to basic FBA, for genome-wide essentiality screens. Aiding the effort is the recent availability of a comprehensive single-gene knock-out library for *E. coli* (Baba et al. 2006) which has been utilized for comparison with GEM computation (Feist et al. 2007, Joyce et al. 2006). Touching on the predictive capability of GEM computations, it was found that the *E. coli* GEM was able to predict the outcomes of adaptively evolved strains to a high degree (78%) when knock-out *E. coli* strains were grown in a number of different substrate environments by examining growth rates at the beginning and end of adaptive evolution (Fong and Palsson 2004). Genetic perturbations have played a key role in the study of the genotype-phenotype relationship in biology, and GEMs can be used to mechanistically interpret the results and predict the outcomes of such perturbations.

Adding thermodynamic information: The incorporation of thermodynamic information with GEMs is an effort that is progressing rapidly and should increase the predictive capabilities of genome-scale modeling through the addition of further governing physico-chemical constraints. Furthermore, the addition of thermodynamics enables the analysis of metabolomic data in the context of a reconstruction. A study utilizing high-throughput metabolomic data and GEMs resulted in the proposition of likely regulatory interactions by deciphering the metabolite concentrations in the context of overall network functionality (Kümmel et al. 2006). Not only did the metabolomic data benefit computations by constraining the system using physiological measurements, but the computational predictions were also able to validate quantitative metabolomic data sets for consistency through providing a functional context to relate metabolite concentrations. This application is one example of how metabolomic data will directly influence modeling. Metabolite concentration data is likely to greatly influence future metabolic modeling due to its intimate connection with GEM content.

9.4.4 Biological Network Analysis

Although there is still much to learn about the metabolism of *E. coli* and how a model-driven approach can be used to uncover these unknowns, the wealth of knowledge collected and represented in the current *E. coli* reconstruction makes it an ideal platform for network analyses. Researchers have been taking advantage of this fact and have centered network analyses on probing and uncovering the properties of biological networks in general. In this section, we discuss a key analysis based on the *E. coli* GEM and the implications drawn from such analyses.

One noteworthy study utilizing the *E. coli* network examined thousands of different potential growth conditions and resulted in the observation of a 'high-flux backbone' in *E. coli* that both, (i) carried high levels of flux across the different environmental conditions, and (ii) was composed of a relatively small set of enzymatic reactions (Almaas et al. 2004). This result can be of practical importance for synthetic biology efforts aimed towards manipulating flux within biological systems. Furthermore, this finding was hypothesized to be a universal feature of metabolic activity in all cells and was consistent with flux measurements from 13C labeling experiments (Almaas et al. 2004).

Overall, studies of network analyses have a common systems biology theme: the development and subsequent demonstration of methods that identify sets of reactions or metabolites with correlated or coordinated functions and systematic relationships. The systems biology that these methods enable and demonstrate has the potential to influence the more practical applications already outlined. The role that the *E. coli* GEM has taken is a comprehensive and curated set of up-to-date metabolic knowledge that provides a scaffold for large-scale computations.

9.4.5 Studies of Bacterial Evolution

The GEMs of *E. coli* have been used to examine the process of bacterial evolution (Pal et al. 2005a,b, 2006). Specifically, the network reconstructions have been used to interpret adaptive evolution events (Pal et al. 2005a), horizontal gene transfer (Pal et al. 2005a,b) and evolution to minimal metabolic networks (Pal et al. 2006). These studies, which utilize the *E. coli* reconstruction as an organism-specific genetic and metabolic content database and the corresponding GEM have been able to provide insight into evolutionary events through combining known physiological data (e.g., in various environmental conditions) with hypotheses and *in silico* computation. Examination of the evolution of minimal metabolic networks through simulation demonstrated that it was possible to predict the gene content of close relatives of *E. coli* by examining the necessity of genes and reactions in the overall context of the system functionality for a specific lifestyle (Pal et al. 2006). Similarly, by re-examining network functionality in a number of different environments, and through the utilization of comparative genomics, it was shown that recent evolutionary events (i.e., horizontal gene transfer) likely resulted from a response to a change in environment (Pal et al. 2005a). Furthermore, computational analysis led to the additional conclusion that these horizontal gene transfer events are more likely when the host organism contains an enzyme that catalyzes a coupled metabolic flux related to the transferred enzyme's function (Pal et al. 2005a,b). Taken together, these studies demonstrate the importance of having high-quality curated reconstructions to enable studies on an organism's response to environmental changes and on the fundamental forces driving bacterial evolution.

9.5 Need for New *In Silico* Methods and Applications

We now know how to represent BiGG data in either a stoichiometric format or in the form of causal relationships (Gianchandani et al. 2006) and how to use this data to perform several lines of computational inquiries. Computational query tools of GEMs will continue to be developed. New advances in these query tools will likely include, (i) modularization methods, (ii) use of fluxomic data, and (iii) eventually, kinetic information.

Modularization: As the scope and content of the reconstruction grows, the need to modularize its content becomes more pressing. Fine or coarse-grained views of cellular processes are needed for different applications.

Fluxomics: Currently, computational limitations force the reduction in network size for the analysis of isotopomer data. Given the systemic nature of fluxomic data and its phenotypic relevance, there is a pressing need to increase the size of the networks that can be utilized for experimental measurement and estimation of flux states. A network reconstruction will both guide the content that is needed for analyzing fluxomic data and offer a starting point for a rational reduction to generate relevant models in the meantime.

Kinetics/thermodynamics: Although detailed kinetic models of microbial functions may currently be mostly of academic interest, they will most likely be able to be constructed in the mid-term based on advances with metabolomic and fluxomic data, in addition to the developments that are occurring with the incorporation of thermodynamic information. Such large-scale kinetic models are likely to differ from those resulting from traditional approaches for construction of kinetic models as they come with different challenges.

9.6 Closing

The process underlying the *E. coli* metabolic reconstruction has pioneered many approaches, methods, and studies in the systems biology of microbial metabolism. This effort has effectively put a mechanistic basis into the genotype-phenotype relationship. In fact, this relationship is now broken down into four steps:

(1) Components (a large knowledge base, BiGG), leading to networks (the reconstruction process resulting GENRE), leading to *In silico* Models (GEMs), leading to Phenotypic States (estimated by COBRA methods).
(2) GEMs will allow for gap-filling and systematic biological discovery (Breitling et al. 2008) and for understanding of complex biological processes (see Chapter 15).

Predictive models also allow for experimental strain design. In fact, in engineering, there is '*nothing more practical than a good theory*.' As this chapter demonstrated, genomics and high-throughput technologies have enabled the construction of predictive computational models. The scope of such predictions is limited at the moment, but with the growing scope and coverage of genome-scale reconstructions and advancements in the development of computational tools, this scope will broaden. Not only will GEMs influence design in synthetic biology, but also their help with discovering cellular content will provide a more complete picture of the intra-cellular environment in which future synthetically engineered constructs and circuits will be placed. The impact of GEMs on synthetic biology is thus likely to be notable, ranging from the provision of the cellular context of a small-scale gene circuit design to engineering of the entire genome-scale network towards fundamentally new and useful (i.e., production) phenotypes.

Finally, we can speculate about the deep scientific impact that comprehensive predictive GEMs will have on our understanding of the living process. A comprehensive view of cellular functions will allow us to study the fundamental properties of both the underlying energy and information flows in living organisms. Such a view is likely to deeply affect our understanding of both distal and proximal causation in biology.

References

Allen TE, Palsson BO (2003) Sequenced-Based Analysis of Metabolic Demands for Protein Synthesis in Prokaryotes. J Theor Biol 220(1):1–18

Almaas E, Kovacs B, Vicsek T et al. (2004) Global organization of metabolic fluxes in the bacterium *Escherichia coli*. Nature 427(6977):839–43

Alper H, Jin YS, Moxley JF et al. (2005a) Identifying gene targets for the metabolic engineering of lycopene biosynthesis in *Escherichia coli*. Metab Eng 7(3):155–64

Alper H, Miyaoku K, Stephanopoulos G (2005b) Construction of lycopene-overproducing *E. coli* strains by combining systematic and combinatorial gene knockout targets. Nat Biotechnol 23(5):612–6

Baart GJ, Zomer B, de Haan A et al. (2007) Modeling Neisseria meningitidis metabolism: from genome to metabolic fluxes. Genome Biol 8(7):R136

Baba T, Ara T, Hasegawa M et al. (2006) Construction of *Escherichia coli* K-12 in-frame, single-gene knockout mutants: the Keio collection. Mol Syst Biol 2:2006.0008

Becker SA, Feist AM, Mo ML et al. (2007) Quantitative prediction of cellular metabolism with constraint-based models: The COBRA Toolbox. Nat Protocols 2(3):727–38

Becker SA, Palsson BO (2005) Genome-scale reconstruction of the metabolic network in Staphylococcus aureus N315: an initial draft to the two-dimensional annotation. BMC Microbiol 5(1):8

Berman HM, Westbrook J, Feng Z et al. (2000) The Protein Data Bank. Nucleic Acids Res 28(1):235–42

Beste DJ, Hooper T, Stewart G et al. (2007) GSMN-TB: a web-based genome-scale network model of Mycobacterium tuberculosis metabolism. Genome Biol 8(5):R89

Blattner FR, Plunkett G, 3rd, Bloch CA et al. (1997) The complete genome sequence of *Escherichia coli* K-12. Science 277(5331):1453–74

Borodina I, Krabben P, Nielsen J (2005) Genome-scale analysis of Streptomyces coelicolor A3(2) metabolism. Genome Res 15(6):820–9

Breitling R, Vitkup D, Barrett MP (2008) New surveyor tools for charting microbial metabolic maps. Nat Rev Microbiol 6(2):156–61

Chavali AK, Whittemore JD, Eddy JA et al. (2008) Systems analysis of metabolism in the pathogenic trypanosomatid Leishmania major. Mol Syst Biol 4:177

Chen L, Vitkup D (2006) Predicting genes for orphan metabolic activities using phylogenetic profiles. Genome Biol 7(2):R17

Cho BK, Knight EM, Barrett CL et al. (2008) Genome-wide Analysis of Fis Binding in *Escherichia coli* Indicates a Causative Role for A-/AT-tracts. Genome Res 18(6):900–10

Covert MW, Knight EM, Reed JL et al. (2004) Integrating high-throughput and computational data elucidates bacterial networks. Nature 429(6987):92–6

David H, Ozcelik IS, Hofmann G et al. (2008) Analysis of Aspergillus nidulans metabolism at the genome-scale. BMC Genomics 9:163

Duarte NC, Becker SA, Jamshidi N et al. (2007) Global reconstruction of the human metabolic network based on genomic and bibliomic data. Proc Natl Acad Sci USA 104(6):1777–82

Duarte NC, Herrgard MJ, Palsson B (2004) Reconstruction and Validation of *Saccharomyces cerevisiae* iND750, a Fully Compartmentalized Genome-Scale Metabolic Model. Genome Res 14(7):1298–309

Edwards JS, and Palsson, B.O. (2000a) Metabolic flux balance analysis and the in silico analysis of *Escherichia coli* K-12 gene deletions. BMC Bioinformatics 1(1)

Edwards JS, Ibarra RU, Palsson BO (2001) *In silico* predictions of *Escherichia coli* metabolic capabilities are consistent with experimental data. Nat Biotechnol 19(2):125–30

Edwards JS, Palsson BO (1999) Systems properties of the Haemophilus influenzae Rd metabolic genotype. J Biol Chem 274(25):17410–6

Edwards JS, Palsson BO (2000b) The *Escherichia coli* MG1655 *in silico* metabolic genotype: Its definition, characteristics, and capabilities. Proc Natl Acad Sci USA 97(10):5528–33

Feist AM, Henry CS, Reed JL et al. (2007) A genome-scale metabolic reconstruction for *Escherichia coli* K-12 MG1655 that accounts for 1260 ORFs and thermodynamic information. Mol Syst Biol 3(121)

Feist AM, Herrgard MJ, Thiele I et al. (2009) Reconstruction of biochemical networks in microbial organisms. Nat Rev Microbiol 7(2)

Feist AM, Palsson BO (2008) The growing scope of applications of genome-scale metabolic reconstructions using *Escherichia coli*. Nat Biotech 26(6):659–67

Feist AM, Scholten JCM, Palsson BO et al. (2006) Modeling methanogenesis with a genome-scale metabolic reconstruction of *Methanosarcina barkeri*. Mol Syst Biol 2(2006.0004):1–14

Fong SS, Palsson BO (2004) Metabolic gene-deletion strains of *Escherichia coli* evolve to computationally predicted growth phenotypes. Nat Genet 36(10):1056–58

Forster J, Famili I, Fu PC et al. (2003) Genome-Scale Reconstruction of the *Saccharomyces cerevisiae* Metabolic Network. Genome Res 13(2):244–53

Frazier ME, Johnson GM, Thomassen DG et al. (2003) Realizing the potential of the Genome Revolution: The Genomes to life Program. Science 300(5617):290–3

Fuhrer T, Chen L, Sauer U et al. (2007) Computational prediction and experimental verification of the gene encoding the NAD+/NADP+-dependent succinate semialdehyde dehydrogenase in *Escherichia coli*. J Bacteriol 189(22):8073–8

Gianchandani EP, Papin JA, Price ND et al. (2006) Matrix Formalism to Describe Functional States of Transcriptional Regulatory Systems. PLoS Comput Biol 2(8):e101

Gonzalez O, Gronau S, Falb M et al. (2008) Reconstruction, modeling & analysis of *Halobacterium salinarum R-1* metabolism. Mol Biosyst 4(2):148–59

Heinemann M, Kummel A, Ruinatscha R et al. (2005) In silico genome-scale reconstruction and validation of the *Staphylococcus aureus* metabolic network. Biotechnol Bioeng 92(7):850–64

Herrgard MJ, Covert MW, Palsson BO (2004) Reconstruction of Microbial Transcriptional Regulatory Networks. Curr Opin Biotechnol 15(1):70–7

Holden C (2002) Alliance launched to model *E. coli*. Science 297(5586):1459–60

Ibarra RU, Edwards JS, Palsson BO (2002) *Escherichia coli* K-12 undergoes adaptive evolution to achieve *in silico* predicted optimal growth. Nature 420(6912):186–9

Jamshidi N, Palsson BO (2007) Investigating the metabolic capabilities of *Mycobacterium tuberculosis H37Rv* using the in silico strain iNJ661 and proposing alternative drug targets. BMC Syst Biol 1:26

Joyce AR, Palsson BO (2006) The model organism as a system: integrating 'omics' data sets. Nat Rev Mol Cell Biol 7(3):198–210

Joyce AR, Reed JL, White A et al. (2006) Experimental and Computational Assessment of Conditionally Essential Genes in *Escherichia coli*. J Bacteriol 188(23):8259–71

Kümmel A, Panke S, Heinemann M (2006) Putative regulatory sites unraveled by network-embedded thermodynamic analysis of metabolome data. Mol Syst Biol 2:2006.0034

Keseler IM, Collado-Vides J, Gama-Castro S et al. (2005) EcoCyc: a comprehensive database resource for *Escherichia coli*. Nucleic Acids Res 33(Database Issue):D334–7

Kharchenko P, Chen L, Freund Y et al. (2006) Identifying metabolic enzymes with multiple types of association evidence. BMC Bioinformatics 7(177)

Kharchenko P, Vitkup D, Church GM (2004) Filling gaps in a metabolic network using expression information. Bioinformatics 20(Suppl 1):I178-I185

Kim HU, Kim TY, Lee SY (2008) Metabolic flux analysis and metabolic engineering of microorganisms. Mol BioSyst 4(2):113–20

Kim TY, Kim HU, Park JM et al. (2007) Genome-scale analysis of *Mannheimia succiniciproducens* metabolism. Biotechnol Bioeng 97(4):657–71

Kuepfer L, Sauer U, Blank LM (2005) Metabolic functions of duplicate genes in *Saccharomyces cerevisiae*. Genome Res 15(10):1421–30

Lee J, Yun H, Feist AM et al. (2008) Genome-scale reconstruction and in silico analysis of the *Clostridium acetobutylicum* ATCC 824 metabolic network. Appl Microbiol Biotechnol 80(5):849–52

Lee KH, Park JH, Kim TY et al. (2007) Systems metabolic engineering of *Escherichia coli* for L-threonine production. Mol Syst Biol 3:149

Lee SY, Woo HM, Lee D-Y et al. (2005) Systems-level analysis of genome-scale *in silico* metabolic models using MetaFluxNet. Biotechnol Bioproc Eng 10:425–31

Lee TI, Rinaldi NJ, Robert F et al. (2002) Transcriptional regulatory networks in *Saccharomyces cerevisiae*. Science 298(5594):799–804.

Mahadevan R, Bond DR, Butler JE et al. (2006) Characterization of Metabolism in the Fe(III)-Reducing Organism *Geobacter sulfurreducens* by Constraint-Based Modeling. Appl Environ Microbiol 72(2):1558–68

Majewski RA, Domach MM (1990) Simple constrained optimization view of acetate overflow in *E. coli*. Biotechnol Bioeng 35:732–8

Mehra A, Hatzimanikatis V (2006) An algorithmic framework for genome-wide modeling and analysis of translation networks. Biophys J 90(4):1136–46

Nogales J, Thiele, I.*, Palsson, B. Ø. (2008) A genome-scale metabolic reconstruction for *P. putida* KT2440: *i*JN746 as cell factory

Oberhardt MA, Puchalka J, Fryer KE et al. (2008) Genome-scale metabolic network analysis of the opportunistic pathogen *Pseudomonas aeruginosa PAO1*. J Bacteriol 190(8):2790–803

Oh YK, Palsson BO, Park SM et al. (2007) Genome-scale reconstruction of metabolic network in bacillus subtilis based on high-throughput phenotyping and gene essentiality data. J Biol Chem 282(89):28791–9

Oliveira AP, Nielsen J, Forster J (2005) Modeling Lactococcus lactis using a genome-scale flux model. BMC Microbiol 5:39

Pal C, Papp B, Lercher MJ (2005a) Adaptive evolution of bacterial metabolic networks by horizontal gene transfer. Nat Genet 37(12):1372–5

Pal C, Papp B, Lercher MJ (2005b) Horizontal gene transfer depends on gene content of the host. Bioinformatics 21 Suppl 2:ii222–3

Pal C, Papp B, Lercher MJ et al. (2006) Chance and necessity in the evolution of minimal metabolic networks. Nature 440(7084):667–70

Palsson BO (2004) Two-dimensional annotation of genomes. Nat Biotechnol 22(10):1218–9

Palsson BO (2006) Systems biology: properties of reconstructed networks. Cambridge University Press, New York

Papin JA, Hunter T, Palsson BO et al. (2005) Reconstruction of cellular signalling networks and analysis of their properties. Nat Rev Mol Cell Biol 6(2):99–111

Papin JA, Price ND, Palsson BO (2002) Extreme pathway lengths and reaction participation in genome-scale metabolic networks. Genome Res 12(12):1889–900

Park JH, Lee KH, Kim TY et al. (2007) Metabolic engineering of *Escherichia coli* for the production of L-valine based on transcriptome analysis and in silico gene knockout simulation. Proc Natl Acad Sci USA 104(19):7797–802

Pramanik J, Keasling JD (1997) Stoichiometric model of *Escherichia coli* metabolism: Incorporation of growth-rate dependent biomass composition and mechanistic energy requirements. Biotechnol Bioeng 56(4):398–421

Pramanik J, Keasling JD (1998) Effect of *Escherichia coli* biomass composition on central metabolic fluxes predicted by a stoichiometric model. Biotechnol Bioeng 60(2):230–8

Price ND, Reed JL, Palsson BO (2004a) Genome-scale models of microbial cells: evaluating the consequences of constraints. Nat Rev Microbiol 2(11):886–97

Price ND, Schellenberger J, Palsson BO (2004b) Uniform Sampling of Steady State Flux Spaces: Means to Design Experiments and to Interpret Enzymopathies. Biophys J 87(4):2172–86

Price ND, Reed JL, Papin JA et al. (2003) Analysis of metabolic capabilities using singular value decomposition of extreme pathway matrices. Biophys J 84(2):794–804

Reed JL, Famili I, Thiele I et al. (2006a) Towards multidimensional genome annotation. Nat Rev Genet 7(2):130–41

Reed JL, Palsson BO (2003) Thirteen Years of Building Constraint-Based In Silico Models of *Escherichia coli*. J Bacteriol 185(9):2692–9

Reed JL, Patel TR, Chen KH et al. (2006b) Systems approach to refining genome annotation. Proc Natl Acad Sci USA 103(46):17480–4

Reed JL, Vo TD, Schilling CH et al. (2003) An expanded genome-scale model of *Escherichia coli* K-12 (*i*JR904 GSM/GPR). Genome Biol 4(9):R54.1–R54.12

Resendis-Antonio O, Reed JL, Encarnacion S et al. (2007) Metabolic reconstruction and modeling of nitrogen fixation in Rhizobium etli. PLoS Comput Biol 3(10):1887–95

Riley M, Abe T, Arnaud MB et al. (2006) *Escherichia coli* K-12: a cooperatively developed annotation snapshot-2005. Nucleic Acids Res 34(1):1–9

Schilling CH, Covert MW, Famili I et al. (2002) Genome-scale metabolic model of *Helicobacter pylori* 26695. J Bacteriol 184(16):4582–93

Schilling CH, Palsson BO (2000) Assessment of the Metabolic Capabilities of *Haemophilus influenzae Rd* through a Genome-scale Pathway Analysis. J Theor Biol 203(3):249–83

Segre D, Vitkup D, Church GM (2002) Analysis of optimality in natural and perturbed metabolic networks. Proc Natl Acad Sci USA 99(23):15112–7

Senger RS, Papoutsakis ET (2008) Genome scale model for *Clostridium acetobutylicum*. Part 1: Metabolic network resolution and analysis. Biotechnol Bioeng 101(5):1036–52

Serres MH, Gopal S, Nahum LA et al. (2001) A functional update of the *Escherichia coli* K-12 genome. Genome Biol 2(9):RESEARCH0035

Sheikh K, Forster J, Nielsen LK (2005) Modeling hybridoma cell metabolism using a generic genome-scale metabolic model of *Mus musculus*. Biotechnol Prog 21(1):112–21

Shlomi T, Berkman O, Ruppin E (2005) Regulatory on/off minimization of metabolic flux changes after genetic perturbations. Proc Natl Acad Sci USA 102(21):7695–700

Suthers PF, Burgard AP, Dasika MS et al. (2007) Metabolic flux elucidation for large-scale models using 13C labeled isotopes. Metab Eng 9(5–6):387–405

Teusink B, Wiersma A, Molenaar D et al. (2006) Analysis of growth of *Lactobacillus plantarum* WCFS1 on a complex medium using a genome-scale metabolic model. J Biol Chem 281(52):40041–8

Thiele I, Jamshidi N, Fleming RMT et al. (2009) Genome-scale reconstruction of *E. coli*'s transcriptional and translational machinery: A knowledge-base its mathematical formulation, and its functional characterization. PLOS Comp Biol. In press

Thiele I, Price ND, Vo TD et al. (2005a) Candidate metabolic network states in human mitochondria: Impact of diabetes, ischemia, and diet. J Biol Chem 280(12):11683–95

Thiele I, Vo TD, Price ND et al. (2005b) An Expanded Metabolic Reconstruction of *Helicobacter pylori* (*i*IT341 GSM/GPR): An *in silico* genome-scale characterization of single and double deletion mutants. J Bacteriol 187(16):5818–30

Thomas R, Paredes CJ, Mehrotra S et al. (2007) A model-based optimization framework for the inference of regulatory interactions using time-course DNA microarray expression data. BMC Bioinformatics 8:228

Varma A, Boesch BW, Palsson BO (1993a) Biochemical production capabilities of *Escherichia coli*. Biotechnol Bioeng 42(1):59–73

Varma A, Boesch BW, Palsson BO (1993b) Stoichiometric interpretation of *Escherichia coli* glucose catabolism under various oxygenation rates. Appl Environ Microbiol 59(8):2465–73

Varma A, Palsson BO (1993) Metabolic capabilities of *Escherichia coli*: I. Synthesis of biosynthetic precursors and cofactors. J Theor Biol 165(4):477–502

Varma A, Palsson BO (1994) Stoichiometric flux balance models quantitatively predict growth and metabolic by-product secretion in wild-type *Escherichia coli* W3110. Appl Environ Microbiol 60(10):3724–31

Varma A, Palsson BO (1995) Parametric sensitivity of stoichiometric flux balance models applied to wild-type *Escherichia coli* metabolism. Biotechnol Bioeng 45(1):69–79

Wiback SJ, Famili I, Greenberg HJ et al. (2004) Monte Carlo Sampling Can Be Used to Determine the Size and Shape of the Steady State Flux Space. J Theor Biol 228(4):437–47

Wunderlich Z, Mirny LA (2006) Using the topology of metabolic networks to predict viability of mutant strains. Biophys J 91(6):2304–11

Chapter 10
Kinetic Modeling of *E. coli* Enzymes: Integration of *in vitro* Experimental Data

Ekaterina A. Mogilevskaya, Kirill V. Peskov, Eugeniy A. Metelkin, Galina V. Lebedeva, Tatiana Y. Plyusnina, Igor I. Goryanin, and Oleg V. Demin

Contents

Abstract The metabolic network of *E. coli* is one of the most well studied biochemical systems, with an abundance of *in vitro* and *in vivo* data available for quantitative estimation of its kinetic parameters. In this chapter, we present our approach to developing mathematical description of individual enzymatic reactions within bacterial metabolic networks. This description is based on the detailed consideration of enzyme catalytic mechanisms and includes several stages: reconstruction of the enzyme catalytic cycle, derivation of the reaction rate equation, and validation of its parameters on the basis of available *in vitro* experimental data. We illustrate our strategy with the models developed for three *E. coli* enzymes with rather complicated regulatory mechanisms: allosteric tetramer phosphofructokinase-1, citrate synthase with its regulation by ATP and pH, and β-galactosidase validated against time dependencies of its substrates. The modeling results clearly demonstrate that developing detailed enzyme kinetic models is essential to capture key regulatory properties of enzymes. The kinetic models allow to integrate large sets of *in vitro*

I.I. Goryanin (✉)
Centre for Systems Biology, University of Edinburgh, Edinburgh, EH9 3JU, 0131 6519063 Scotland, UK; Informatics Forum, University of Edinburgh, Edinburgh, EH8 9LE Scotland, UK
e-mail: goryanin@inf.ed.ac.uk

S.Y. Lee (ed.), *Systems Biology and Biotechnology of Escherichia coli*,
DOI 10.1007/978-1-4020-9394-4_10,

experimental data available for *E. coli* enzymes and to get insight into important regulatory features of their catalytic mechanism.

10.1 Introduction

The last several years have seen substantial progress in molecular biology and genetic research of *E. coli* (Han and Lee 2006, Ishii et al. 2007, Perna et al. 2001). Sequence information on the genomes of hundreds of different organisms has stimulated the emergence of functional genomics, a discipline that sets out to understand the meaning of sequenced data using high throughput small molecule, gene and protein expression data. Life scientists have transformed old-style protein chemistry into proteomics, and traditional biochemistry into metabolomics. These new fields provide essential clues to the underlying metabolic, gene regulatory and signaling networks that operate in cells, tissues and organisms under different conditions.

Cellular metabolism of *E. coli*, the integrated interconversion of more then two thousands of metabolic substrates through more then one thousand enzyme-catalyzed biochemical reactions, is the most investigated system of intracellular molecular interactions (Feist et al. 2007). When one has knowledge of most, or all, of the major biological entities and stoichiometry of their interactions an illusion could appear that this voluminous knowledge will enable us to predict whole cell behavior for the purposes of mechanistic understanding and bioengineering control. Indeed, in some cases, it is possible to make plausible predictions based on "static" non temporal information without relying upon kinetic data (Edwards et al. 2001, Kim et al. 2008, Price et al. 2004, Schütz et al. 2007). Recently, stoichiometric metabolic models (SMM) and metabolic engineering techniques have been successfully applied to improve production of succinic acid, lactic acid, L-threonine, and L-valine by *E. coli*. (Lee et al. 2007, Lee et al. 2005, Park et al. 2007). Unfortunately, SMM techniques cannot predict cellular bahaviour in non-steady state conditions. It results in high level of false positive predictions (Lee et al. 2006). Time series data cannot be easily integrated to SM models as well, so it is difficult to validate SM models, as steady state or chemostat cultures are reqired. It is difficult to incorporate to SMM real data from batch or fed batch experiments, and time series after systems perturbations (immediately or short after the intervention) (Ishii et al. 2007).

The discrepancy between SMM predictions and experimental data is usually explained by the statement that ultimately the real cell will mimic *in silico* behavior after process of evolution. But, the questions why and how the process of evolution itself is happening, which physico-chemical properties of proteins are subject of selective pressure are not addressed at all.

Indeed, overall cellular behavior is determined not only by available biological entities, but mainly by their dynamic interactions and individual properties. Activities of most if not all of the enzymes involved in cellular metabolism are often regulated by the end products and intermediates of corresponding pathways. This complex network of positive and negative feedbacks, as well as genetic regulation of expression levels, provides flexible adaptation of the metabolic network to fast and slow changes in environmental conditions. It is the overall dynamic nature of the whole cell that determines not only its present properties, but its future ones as well. The cellular regulatory system is responsible for maintenance of homeostasis and for transitions between different physiological states. That is why when modeling cellular metabolism, it is essential to consistently describe its key regulatory properties. The regulatory system of *E. coli* metabolism is known to have an hierarchical architecture, including regulatory effects at the levels of enzyme activity, gene transcription and translation. Consistent mathematical description of this complex multilevel regulatory system requires accurate consideration of the regulatory properties of individual components involved in the metabolic network – enzymes, which further contribute to the regulatory properties of the whole system. This task becomes extremely important with the recent expansion of a new discipline – synthetic biology, the ultimate goal of which is to design and build engineered biological systems with predefined properties (Barrett et al. 2006, Endy 2005).

In this paper we present kinetic modelling approach applied to modeling the individual enzymatic reactions within metabolic networks of *E. coli*, which allows capture of the key regulatory properties of these networks. Our approach is based on the detailed consideration of the enzyme catalytic cycle and on the utilization of all available experimental data characterizing the kinetics of the enzyme being studied. This modeling approach includes several stages. The first is the reconstruction of a catalytic cycle of the enzymes. This cycle represents both interaction of the enzyme with substrates and products and the effect of different inhibitors and activators. The second stage is the derivation of the reaction rate equation that defines quantitative dependence of the rate of the enzyme performance on concentrations of substrates, products and effectors. The third and last stage is the identification of the parameters for into the rate equation based on the available experimental data. To illustrate our approach and its stages, as well as to demonstrate how different types of experimental data can be incorporated into the kinetic model, we present our developed kinetic models for several *E. coli* enzymes which are known to have complicated patterns of regulation of their activity: phosphofructokinase 1, β-galactosidase and citrate synthase

10.2 Methods

10.2.1 Basic Principles of Kinetic Description of Enzymatic Reactions Using In Vitro *Experimental Data*

As a part of our strategy to make the models scalable and comparable with different kinds of experimental data, we develop both *detailed* and *reduced* descriptions for every appropriate biochemical process to make the models scalable and comparable with different kinds of experimental data. The *detailed* reaction description includes the exact molecular mechanism of the reaction, i.e. enzyme catalytic cycle. Usually, the detailed kinetic model of an enzyme reaction represents a set of ordinary differential equations describing the totality of elementary reactions within the enzyme catalytic cycle, such as substrate binding, catalytic transformation of substrates into products, product release, etc. It defines the dynamics of all possible enzyme intermediate states (free enzyme, enzyme-substrate, enzyme-product complexes), as well the time course of substrates and products consumption/production. The *reduced* description represents the reaction rate as an explicit analytic function of the concentration of substrates and products.

In our approach, for each active protein involved in the model of metabolic network, i.e. enzyme with catalytic function, we identify from the literature or hypothesize the catalytic cycle based on 3D structures and other relative biological information. Basing on the developed scheme of the catalytic mechanism we can construct a detailed kinetic model of the enzyme catalytic cycle. In most cases such a detailed model can further be replaced with a reduced description of the reaction rate. To derive the corresponding rate equations from the catalytic cycle, we use quasi-steady state and rapid equilibrium approaches (Demin and Goryanin 2008). The catalytic cycle of each enzyme is described by non-linear differential equations. Initially, concentrations of substrates, products and effectors (inhibitors and activators) are assumed to be buffered, i.e. do not change with time.

The quasi-steady state of the system is calculated as a function of substrates, products, inhibitors, activators, total protein concentrations and all kinetic constants of the processes. The rate law for every process is derived as a flux from the catalytic cycle for this quasi-steady state. Finally, the rate law depends on temporal changes of the total concentration of the protein, concentrations of the effectors (activators, inhibitors, agonists, and antagonists), substrates, products and the values of the kinetic parameters (K_m, K_i, K_d and elementary rate constants).

The level of detailed elaboration of the catalytic cycles of selected enzymes and subsequent derivation of rate equations are fully determined by the available experimental data on the structural and functional organization of the enzyme. Indeed, if the catalytic cycle of the enzyme is established and proved experimentally then we use it to derive the rate equation. If the mechanism underlying enzyme operation is unknown we infer a "minimal" catalytic cycle that

1. satisfies all structural and stoichiometric data available from literature
2. allows us to derive a rate equation describing all available kinetic experimental data
3. is the mathematically simplest catalytic cycle of all possible ones satisfying clauses 1 and 2.

Another challenge in developing a mathematical description of enzyme catalysis based on *in vitro* data is that the kinetic experimental data available from literature are usually obtained under different conditions (pH, temperature). This means that we should account for these parameters in our model, i.e.

4. construct such a catalytic cycle and derive such a rate equation that satisfies available experimental data describing the dependence of enzyme activity on pH, temperature and other experimental conditions

5. the mechanism describing the dependence of the reaction rate on pH and temperature should be taken into account in the catalytic cycle of the enzyme in the simplest of all possible ways

Parameter estimation is the third stage of model development. To estimate the kinetic parameter values we use the following sources:

1. literature data on the values of K_m, K_i, K_d, rate constants, pH optimum, etc;
2. electronic databases; only a few databases with specific kinetic content are available at the moment, in particular EMP (Selkov et al. 1996) and BRENDA (Shomburg et al. 2002)
3. Experimentally measured dependencies of the initial reaction rates on concentrations of substrates, products, inhibitors and activators
4. Time series data from enzyme kinetics

However, many processes, such as enzyme reactions, have not been studied kinetically. Moreover, many kinetic parameters cannot be estimated from the literature or databases due to a lack of available experimental data. One remedy is to express these unknown or "free" parameters via other available measured kinetic parameters. The result is the establishment of functional relationships between "free" parameters and measured kinetic parameters. Each parameter value, of course, is constrained by physico-chemical properties and any other information available from other organisms or related processes. The more constraints available, the more defined is the system.

To illustrate the basic principles of construction of catalytic cycles and derivation of rate equations described above we present the results of the modeling of three enzymes of *E. coli* metabolism: phosphofructokinase-1, β-galactosidase and citrate synthase. We demonstrate how kinetic data measured at different conditions (pH, temperature and others) can be taken all together to construct a quantitative description of enzyme catalytic activity and its regulation. The method developed in this section allows us to predict kinetic behaviour of the enzymes at any set of experimental or cellular conditions.

10.3 Results

10.3.1 Kinetic Modeling of Phosphofructokinase-1 (pfkA) from E. coli Cells

Phosphofructokinase-1 (PfkA) catalyzes the transfer of γ-phosphate from ATP to fructose-6-phosphate (F6P) resulting in ADP and fructose-1,6-biphosphate (F16bP) production (Babul 1978, Blangy et al. 1968, Kotlarz and Buc 1982):

$$ATP + F6P \xrightarrow[Mg^{2+}]{PfkA} ADP + F16bP$$

This reaction is of importance in regulation of glycolysis and gluconeogenesis (Ausat et al. 1997, Babul 1978, Berger and Evans 1991, Blangy et al. 1968, Deville-Bonne et al. 1991a, Deville-Bonne et al. 1991b, Kotlarz and Buc 1982, Rye et al. 1995, Saier and Ramseier 1996, Saier et al. 1996, Waygood and Sanwal 1974). *E. coli* cells contain two isozymes of this enzyme: PfkA and PfkB (Babul 1978, Kotlarz and Buc 1982). PfkA, studied in this work, is considered to be a key phosphofructokinase in *E. coli* metabolism (Kotlarz and Buc 1977, Torres and Babul 1991, Vinopal and Fraenkel 1974, 1975).

Unlike PfkB, PfkA has a rather complicated regulatory profile: purine nucleotide diphosphates, ADP (Babul 1978, Blangy et al. 1968, Kotlarz and Buc 1982) and GDP (Ausat et al. 1997), are acting as activators, phosphoenolpyruvate (PEP) is the inhibitor (Babul 1978, Blangy et al. 1968, Kotlarz and Buc 1982). In this case, all the regulators of this enzyme are allosteric, by virtue of the fact that binding sites found for effectors do not overlap with catalytic ones (Reeves and Sols 1973). Moreover, phosphorylation of fructoso-6-phosphate is carried out in the presence of Mg^{2+} ions (Babul 1978, Blangy et al. 1968, Kotlarz and Buc 1982).

The evidence in favor this protein being an allosteric enzyme is as follows: the quaternary structure of the enzyme (tetramer); complex regulatory profile due to the presence of complementary allosteric sites at the monomer; and sigmoid dependencies of initial rates of the reaction on F6P concentrations (Babul 1978, Blangy et al. 1968, Kurganov 1978).

In this part we constructed the kinetic model of phosphofructokinase-1, that describes the majority of the experimental data known for this enzyme. This model examins the properties of phosphofructokinase-1 such as substrate and product inhibition, cooperativity and competition at joint action of allosteric effectors.

10.3.1.1 Catalytic Cycle of Phosphofructokinase-1 Construction

An approach to modeling the kinetics of allosteric enzymes has been developed by Monod, Wyman, Changeux (MWS) (Monod et al. 1965) and it is traditionally used by a majority of modellers. At the same time this modeling approach is valid only for the enzymes which catalyze the reactions of irreversible isomerization, since enzyme-binding of only one substrate is taken into account and the catalytic stage is thought to be irreversible. Moreover, MWS-based modeling includes several strong assumptions that make it practically impossible to use it while modeling real enzymes (Kurganov 1978). There exist a number of generalizations for the MWS-based modeling techniques, which allow the inclusion of allosteric regulation in models of enzymes with more complicated mechanisms. We took a generalization offered in (Ivanitsky et al. 1978), which assumes that the enzyme can exist in two states: R (relaxed) and T (tense). Unlike standard MWS-based modeling, in this approach the functional difference between R and T states is not only in different affinities of substrates, products and effectors with respect to the enzyme, but also in the catalytic properties of these states. In other words, not only Michaelis constants and dissociation constants as in MWS-based modeling (Monod et al. 1965), but also catalytic constants (Deville-Bonne et al. 1991b) are different for R and T states. One another important difference is as follows: a catalytic cycle of separate subunits is taken into consideration in the generalization alluded. This gives us a possibility to take into account detail kinetic mechanism of an enzyme and estimate a contribution of product inhibition.

The rate of reaction may be modulated by variation of the relationship between the states of the enzyme. By this means allosteric regulation is introduced, since enzyme-binding of the effectors in a site, which is separate from catalytic one, will disturb the equilibrium. For instance, activators (that demonstrate the best affinity to R-form) shift the balance to the R-state; and vice versa, inhibitors (which bind more with the T-form) shift the balance to the T-state. In this case, the action of the effectors (activation or inhibition) will be determined only by the ratio of dissociation constants of different forms of the enzyme (Kurganov 1978). The existing experimental data on regulation of phosphofructokinase-1 hold that its regulation is carried out in just this manner, since binding sites both for PEP, and for ADP (GDP), have been found separately from the catalytic cycle (Ausat et al. 1997, Babul 1978, Blangy et al. 1968, Reeves and Sols 1973).

Catalytic Cycle of Separate Subunit

Since we have found no unambiguous opinions as to the mechanism of action of a single subunit in the literature on phosphofructokinase-1, we proposed that the monomer of phosphofructokinase-1 and its isozyme PfkB have a similar mechanism of action. In (Campos et al. 1984) the monomer of phosphofructokinase-2 was shown to operate by the mechanism of Ordered Bi Bi, in accordance with the classification offered by Cleland (Cleland 1963). First F6P and then ATP binds with the enzyme resulting in phosphorylation (Campos et al. 1984, Ewings and Doelle 1980, Guixe and Babul 1985). Furthermore, as $ATPMg^{2-}$ was a substrate of the reaction, we added to the catalytic cycle a competitive inhibition of the free form of ATP^{4-}. This inhibition can be registered when the complete ATP concentration increases in a system with a fixed Mg^{2+} concentration. The pH effect on the activity of the enzyme has been taken into account in a standard way suggested by Cornish-Bowden (Cornish-Bowden 2001), as was shown for the *E. coli* enzyme imidazole glycerol phosphate synthase (Demin et al. 2004). Having summarized the above material, it is possible to construct the profile of the catalytic cycle of a separate subunit of phosphofructokinase-1.

10.3.1.2 Derivation of the Rate Equation of Phosphofructokinase-1

The rate equation, a generalization of MWS modeling by Popova and Sel'kov for multisubstrate reactions is written as follows (Ivanitsky et al. 1978):

$$V = n \cdot f \left(1 + ({}^{f'}\!/_{f})Q\right)/(1+Q)$$
$$Q = L_0 \left(\frac{\left(1 + {}^{I}\!/_{K_i^t}\right)}{\left(1 + {}^{I}\!/_{K_i^r}\right)} \cdot \frac{E_r}{E_t} \right)^n \qquad (10.1)$$

where f is the rate equation derived on the basis of catalytic cycle of single subunit in r-state, f' – the rate equation derived on the basis of catalytic cycle of single subunit in t-state, E_r – concentration of free enzyme in r-state, E_t – concentration of free enzyme in t-state, L_0 – constant equilibrium for r/t-transition, I – allosteric effecter, n – number of enzyme's subunits.

Q is a function of a state, it determines the relation between R and T forms of the enzyme. In order to write the function of a state of the enzymes under study, we should, in accordance with a regulatory profile, take into account the action of all allosteric effectors (Ivanitsky et al. 1978, Kurganov 1978):

$$
\begin{aligned}
f &= \frac{V_{mr}^{forward} \cdot \left(ATPMg^{2-} \cdot F6P - {ADPMg^{-} \cdot F16bP}/{Keq} \right)}{Z_{SP}^{R} \cdot Z_{pH}};\\
Z_{pH} &= 1 + {H^{+}}/{K_{d\_H\_1}} + {K_{d\_H\_2}}/{H^{+}};\\
Z_{SP}^{R} &= K_{ir\_F6P} \cdot K_{mr\_ATPMg^{2-}} + K_{mr\_ATPMg^{2-}} \cdot F6P \cdot \left(1 + \frac{ATP^{4-}}{K_{ir\_ATP^{4-}}}\right) +\\
&+ K_{mr\_F6P} \cdot ATPMg^{2-} + ATPMg^{2-} \cdot F6P +\\
&+ K_{mr\_F6P} \cdot ATPMg^{2-} \cdot {FbP}/{K_{ir\_FbP}} +\\
&+ ATPMg^{2-} \cdot F6P \cdot {ADP}/{K_{ir\_ADP}} +\\
&+ K_{mr\_F6P} \cdot ATPMg^{2-} \cdot ADP \cdot {FbP}/{Wr \cdot K_{mr\_ADP} \cdot K_{ir\_FbP}} +\\
&+ {Wr}/{Keq} \cdot \left(\begin{array}{l} K_{mr\_FbP} \cdot F6P \cdot {ADP}/{K_{ir\_F6P}} + K_{mr\_ADP} \cdot FbP + \\ + K_{mr\_FbP} \cdot ADP + ADP \cdot FbP \end{array} \right);
\end{aligned}
\tag{10.2}
$$

f is the rate equation derived on the basis of catalytic cycle of a single subunit. The expression of reaction rate (the reaction runs by the mechanism of Ordered Bi Bi), borrowed from Cleland (Cleland 1963), was slightly modified, and competitive inhibition of ATP^{4-} and pH effects were added. Moreover, experimental findings suggest that the reaction, catalyzed by phosphofructokinase-1, is virtually irreversible (Babul 1978):

$$
Keq = Wr \frac{K_{ir\_F6P} \cdot K_{mr\_ATPMg^{2-}}}{K_{ir\_FbP} \cdot K_{mr\_ADP}} = Wt \frac{K_{it\_F6P} \cdot K_{mt\_ATPMg^{2-}}}{K_{it\_FbP} \cdot K_{mt\_ADP}} \tag{10.3}
$$

An expression for f' appears in much the same manner, with the only difference from f that the expression for f' contains the constants of binding and catalytic constants of the T-state.

Er and *Et* is the expression for a free form of the enzyme in R-and T-states (Ivanitsky et al. 1978), respectively:

$$Q = Lo\left(\frac{\left(1+\dfrac{ADP}{K_{eft\_ADP}}+\dfrac{GDP}{K_{eft\_GDP}}\right)\left(1+\dfrac{PEP}{K_{eft\_PEP}}\right)}{\left(1+\dfrac{ADP}{K_{efr\_ADP}}+\dfrac{GDP}{K_{efr\_GDP}}\right)\left(1+\dfrac{PEP}{K_{efr\_PEP}}\right)}\frac{E_r}{E_t}\right)^n$$
$$Er = \frac{K_{ir\_F6P}\cdot K_{mr\_ATPMg^{2-}}\cdot Eo}{Z_{SP}^{R}}$$
$$Et = \frac{K_{it\_F6P}\cdot K_{mt\_ATPMg^{2-}}\cdot Eo}{Z_{SP}^{T}} \tag{10.4}$$

Substitution of expressions (10.2, 10.3, and 10.4) in (10.1) gives a complete equation of the rate of a reaction, catalyzed by phosphofructokinase-1.

10.3.1.3 Estimation of the Parameters of the Rate Equation of Phosphofructokinase-1

As a result of the above, the model contains 20 parameters, two of which we could take from the literature data – $Kd\_ATPMg = 0.0588$ (Taquikhan and Martell 1962), $w\_pfk1 = 0,08$ (Babul 1978). In order to determine the remaining parameters appearing in the rate equation, we fitted the model with experimental data. In total, the 11 experimental curves published in (Ausat et al. 1997, Babul 1978, Deville-Bonne et al. 1991a) were used to determine 18 parameters. It should be noted that the curves obtained in our model correlate rather well with the experimental data (Figs. 10.1a–d, and 10.2a). In addition, the obtained parameter values and analysis of behavior of the model may lead to the following conclusions:

1. Phosphofructokinase-1 has a distinct allostericity associated with different affinities of substrates to states of the enzyme. At the same time, the difference between Michaelis constants of R- and T-states reaches several orders. The above has a great influence on the shape of the curves of initial reaction rate dependence on substrate concentrations (Figs. 10.1b, and 10.2a).
2. Figure 10.2a clearly shows that substrate inhibition appears in the experimental *in vitro* system at a total ATP concentration of more than 10 mM and at a fixed concentration of Mg^{2+} ions (10 mM); at a total ATP concentration of 20 mM, the phosphofructokinase reaction rate in the *in vitro* system is only 20% of the maximum. The mechanism of this inhibition is coupled with emergence in the system of a free form, ATP^{4-} (Fig. 10.2b), which, as mentioned above, acts as an inhibitor by competing with the substrate, magnesium form $ATPMg^{2-}$, for the catalytic site. The values of identified parameters also point to a possibility of substrate inhibition (Table 10.1). So, the ATP^{4-} molecule free from magnesium ions has better affinity to the enzyme than the magnesium form.

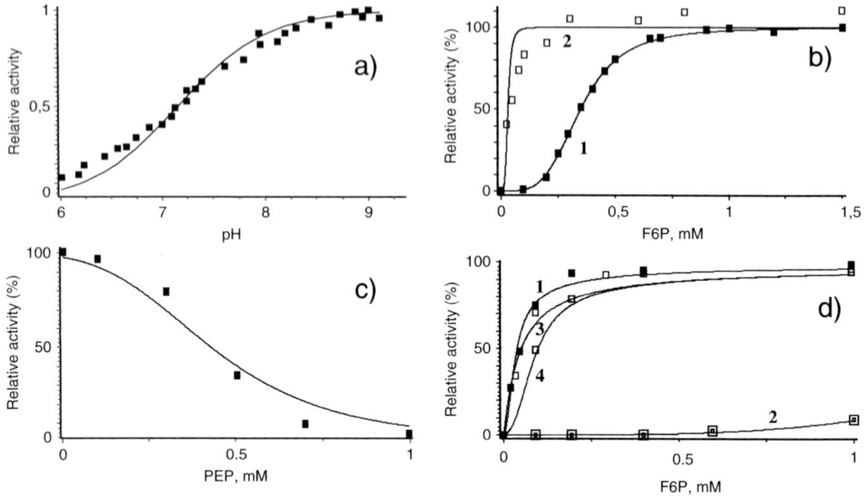

Fig. 10.1 The comparison of experimental data on PfkA-1 and model results. (**a**) PfkA relative maximal activity dependence on pH described by the model and experimental data (Park et al. 2007); (**b**) PfkA relative activity dependence on F6P concentration described by the model and experimental data (Barrett et al. 2006): curve 1 (■) – ATP = 1 mM, Mg^{2+} = 10 mM, GDP = 0 mM, pH = 8.2; curve 2 (□) – ATP = 1 mM, Mg^{2+} = 10 mM, GDP = 2 mM, pH = 8.2; (**c**) PfkA relative activity dependence on PEP concentration described by the model and experimental data (Lee et al. 2005): ATP = 1 mM, Mg^{2+} = 10 mM, F6P = 1 mM, pH = 8.2; (**d**) PfkA relative activity dependence on F6P concentration described by the model and experimental data (Lee et al. 2005): curve 1 (■) – ATP = 1 mM, Mg^{2+} = 10 mM, ADP = 0.5 mM, PEP = 0 mM, pH = 8.2; curve 2 (⊡) – ATP = 1 mM, Mg^{2+} = 10 mM, ADP = 0 mM, PEP = 1 mM, pH = 8.2; curve 3 (□) – ATP = 1 mM, Mg^{2+} = 10 mM, ADP = 1 mM, PEP = 0.1 mM, pH = 8.2; curve 4 (▣) – ATP = 1 mM, Mg^{2+} = 10 mM, ADP = 1 mM, PEP = 1 mM, pH = 8.2

3. Another property of phosphofructokinase-1, the significant effect of which on the kinetic curves we have shown, is that, besides allosteric ADP activation, there is an apparent product inhibition by this metabolite. In other words, ADP can compete with $ATPMg^{2+}$ for the catalytic site. Although product inhibition has no effect on reaction rate that would be noticeable under experimental conditions, it shows up in the combined action of allosteric effectors (curves 2, 3, 4 on Fig. 10.1d). Parameter values obtained from fitting also suggest the possibility of significant inhibition by ADP. So, $Kmr\_ADP = 0.69$ mM, evidencing fair affinity of ADP molecule to phosphofructokinase-1.
4. The analysis of fitted parameter values for the binding of effectors in allosteric sites shows that the activators ADP and GDP bind almost exclusively with the R-state of the enzyme, while the inhibitor PEP binds with the T-state. This proves that any significant synergy due to a combined action of allosteric effectors is impossible, because it is impossible for antagonistic regulators to bind simultaneously with a single subunit.
5. Activation of phosphofructokinase-1 by GDP (Fig. 10.2b) is probably also modulated by competitive inhibition, provided that GDP is bound in the catalytic

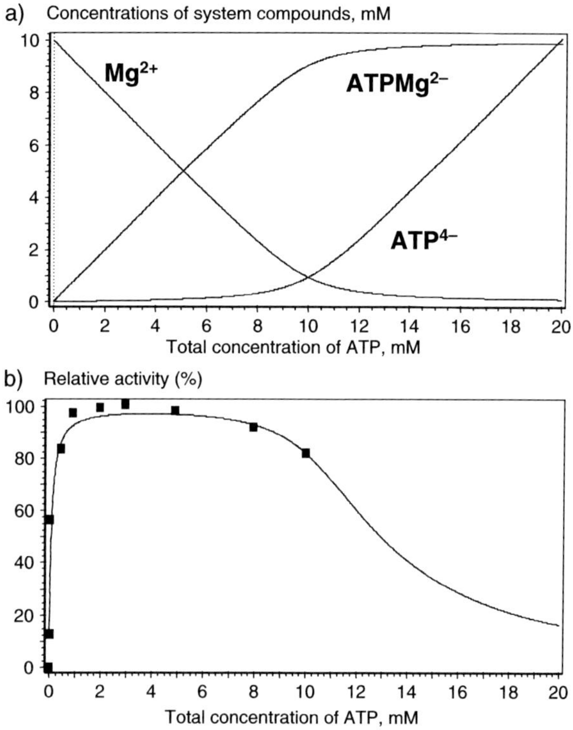

Fig. 10.2 Effect of PfkA inhibition by ATP^{4-}. (**a**) ATP^{4-}, Mg^{2+} and $ATPMg^{2-}$ concentration dependence on the total ATP concentration; (**b**) PfkA relative activity dependence on the total ATP concentration described by the model and experimental data (Lee et al. 2005): $F6P = 1\,mM$, $Mg^{2+} = 10\,mM$, $pH = 8.2$

site instead of $ATPMg^{2+}$. Such a situation is observed in some kinases that use different nucleotide phosphates in their action, e.g., pyruvatekinase-1 from *E. coli* (Waygood and Sanwal 1974). Most probably, GDP and ADP are bound in the same regulatory site (Ausat et al. 1997), therefore competition is possible in the presence of two effectors in the medium. The results of fitting (Table 10.1) show that ADP has nearly two times better affinity to the allosteric site than GDP: $Kir\_ADP = 0.0737\,mM$, whereas $Kir\_GDP = 0.122\,mM$. However, the absence of experimental data prevents any unambiguous conclusions.

10.3.2 The Kinetic Model of β-galactosidase from E. coli *Cells*

β-galactosidase (EC 3.2.1.23) is an important enzyme of *Escherichia coli* involved in sugar utilization. Together with lac-permease, this protein is encoded in the region of the lac-operon, which is the most popular model system for the study of transcription regulation in prokaryotes.

The enzyme has a complex catalytic cycle and catalyzes several reactions in a single catalytic site. As has been shown previously, the main catalytic activity of β-galactosidase at the addition of lactose to the medium reverts to hydrolysis of the latter with the formation of glucose and galactose monosaccharides (Huber

Table 10.1 PfkA model parameters values estimated from experimental data

Model parameter	Value (mM)	Reference
Kmr_ATPMg	8.13e-05	(Kim et al. 2008, Lee et al. 2006)
Kmr_F6P	2.05e-05	(Kim et al. 2008, Lee et al. 2006)
Kir_F6P	1.84	(Kim et al. 2008, Lee et al. 2006)
Kir_ATP	3.17e-05	(Kim et al. 2008, Lee et al. 2006)
Kefr_PEP	200	(Kim et al. 2008)
Kefr_ADP	7.37e-02	(Kim et al. 2008)
Kefr_GDP	0.197	(Lee et al. 2006)
Kir_FbP	2.58e-02	(Kim et al. 2008)
Kmr_FbP	5.5	(Kim et al. 2008)
Kir_ADP	1000	(Kim et al. 2008)
Kmr_ADP	0.69	(Kim et al. 2008)
Kmt_ATPMg	3.35	(Kim et al. 2008, Lee et al. 2006)
Kit_ATP		
Kmt_F6P	780	(Kim et al. 2008, Lee et al. 2006)
Kit_F6P		
Keft_PEP	33.1	(Kim et al. 2008, Lee et al. 2006)
Keft_ADP		
Keft_GDP	8.56e-03	(Kim et al. 2008, Lee et al. 2006)
Kit_FbP	2.6e-01	(Kim et al. 2008)
Kmt_FbP	1	(Kim et al. 2008)
Kmt_ADP	61	(Lee et al. 2006)
Kit_ADP	143	(Kim et al. 2008)
Lo	660	(Kim et al. 2008)
Kd_H_1	1e+03	(Kim et al. 2008)
Kd_H_2	1e+03	(Kim et al. 2008)
	14.4	(Kim et al. 2008, Lee et al. 2006)
	3.78e-12	(Park et al. 2007)
	6.97e-05	(Park et al. 2007)

et al. 1976), as well as reaction of isomerization with the formation of allolactose (Burstein et al. 1965, Jobe and Bourgeois 1972). This work (Huber et al. 1976) has also shown that a certain amount of tri- and tetrasaccharides appear in the medium.

In this section we have constructed a kinetic model of β-galactosidase, which describes both hydrolysis and transgalactosidase activity of the enzyme. On the basis of experimental data available currently, kinetic parameters of the model have been found. Using the constructed model, we analysed correlation between different enzyme activities under different conditions.

10.3.2.1 Construction of β-galactosidase Catalytic Cycle

The catalytic cycle of *E. coli* β-galactosidase (Fig. 10.3) was constructed based on the scheme proposed in (Huber et al. 1976), extended by the addition of trisaccharide formation stages. The scheme describes both hydrolase and transgalactosidase activities of β-galactosidase. Stages 1a and 2a show the binding of lactose (*lac*) and allolactose (*alac*), respectively, with the catalytic center of free enzyme (E) with the formation of enzyme-substrate complex (*E_lac* and *E_alac*, respectively). Stages 1b and 2b describe decomposition of disaccharides to glucose and galactose in the catalytic center of the enzyme, which results in the formation of a ternary complex

Fig. 10.3 The catalytic cycle of β-galactosidase

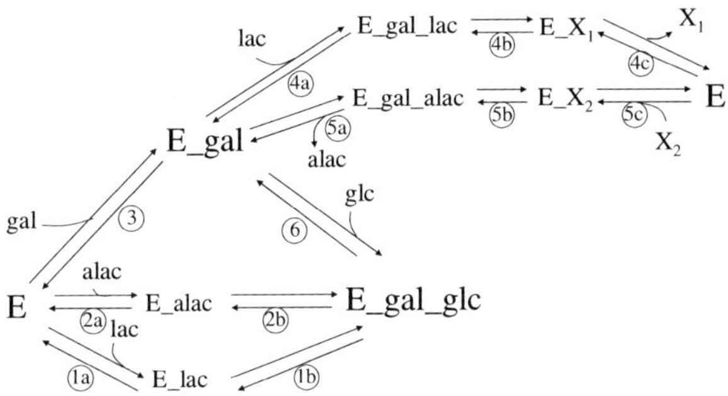

(*E_gal_glc*). According to the literature data (Huber et al. 1984, Huber et al. 1976), the process of glucose (*glc*) and galactose (*gal*) dissociation occurs in sequence (Stages 6, 3), via an intermediate enzyme form bound with galactose (*E_gal*). It follows from the scheme that the processes of lactose and allolactose hydrolysis are described by stages 1a-1b-6-3 and 2a-2b-6-3, and the transgalactosidase reaction proceeds through stages 1a-1b-2b-2a.

The reactions of galactosilation of lactose (4a-4b-4c) and allolactose (5a-5b-5c) are shown to be the main reactions of oligosaccharide formation in *E. coli* (Reeves and Sols 1973). We propose that, by analogy with glucose galactosilation (stages 6-1b-1a), these processes are realized through binding of lactose (stage 4a) or allolactose (stage 5a) to the enzyme form *E_gal* with the formation of trisaccharides *X1* and *X2* (specifying two types of trisaccharides: $gal + lac = X1,\ gal + alac = X2$).

10.3.2.2 Derivation of the Rate Equation of β-galactosidase

To simplify derivation and analysis of the equation for the rate of *beta-galactosidase* functioning, each process of trisaccharide formation (4a-4b-4c and 5a-5b-5c) was described by a single integral stage, neglecting intermediate enzyme forms (Fig. 10.4). It was an enforced approximation, because at present rather little is known about rate constants of these elementary stages, and the lack of kinetic data prevented us from assessing the contribution of each of them.

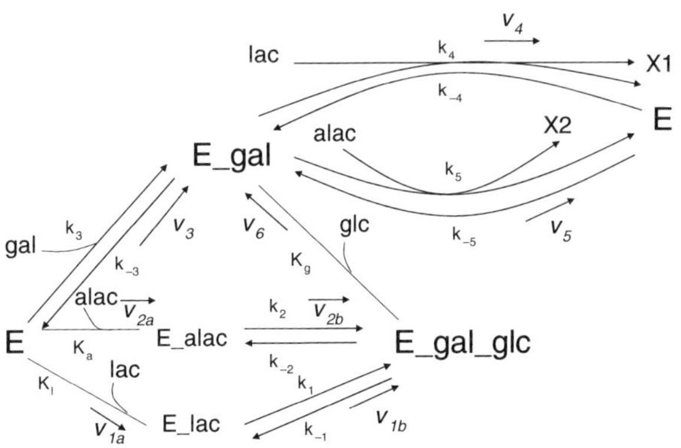

Fig. 10.4 The reduced catalytic cycle of β-galactosidase used for derivation of the rate equation

The simplified scheme of the catalytic cycle for the enzyme is given in Fig. 10.4. The corresponding rate as well as the rate or dissociation constant is indicated near each arrow.

With approximation of quasi stationary concentrations (Cornish-Bowden 2001), and entering a condition of the stationary for all the forms of the enzyme, the following system of equations was obtained

$$\begin{cases} (E\_lac)' = v_{1a} - v_{1b} = 0, \\ (E\_alac)' = v_{2a} - v_{2b} = 0, \\ (E\_gal\_glc)' = v_{1b} + v_{2b} - v_6 = 0, \\ (E)' = v_4 + v_5 - v_{1a} - v_{2a} - v_3 = 0, \\ (E\_gal)' = v_3 + v_6 - v_4 - v_5 = 0, \end{cases} \tag{10.5}$$

where dash is a complete time derivative. Solving this system with respect to the rate of reactions, we found that in a stationary (steady) state the rates of separate stages of catalytic cycle correlate with each other as follows:

$$\begin{cases} v_{1a} = v_{1b}, \\ v_{2a} = v_{2b}, \\ v_6 = v_{1b} + v_{2b}, \\ v_3 = -v_{1b} - v_{2b} + v_4 + v_5. \end{cases} \tag{10.6}$$

Thus, it turned out that all rates of elementary stages can be expressed via four velocities v_{1b}, v_{2b}, v_4 и v_5, hereinafter called «independent».

To derive the equation of the velocity of β-galactosidase, we formulated the velocities for the elementary stages of the catalytic cycle using mass action law assumption (Kotlarz and Buc 1977).

$$\begin{cases} v_{1b} = k_1 \cdot E\_lac - k_{-1} \cdot E\_gal\_glc, \\ v_{2b} = k_2 \cdot E\_alac - k_{-2} \cdot E\_gal\_glc, \\ v_3 = k_3 \cdot E \cdot gal - k_{-3} \cdot E\_gal, \\ v_4 = k_4 \cdot E\_gal \cdot lac - k_{-4} \cdot E \cdot X_1, \\ v_5 = k_5 \cdot E\_gal \cdot alac - k_{-5} \cdot E \cdot X_2. \end{cases} \tag{10.7}$$

The expressions of the velocities (10.7) were substituted into the fourth equation of the system (10.6) and the following equation was obtained:

$$\begin{aligned} & k_3 \cdot gal \cdot E + k_1 \cdot E\_lac + k_2 \cdot E\_alac + k_{-4} \cdot X_1 \cdot E + k_{-5} \cdot X_2 \cdot E - \\ & - k_{-3} \cdot E\_gal - k_{-1} \cdot E\_gal\_glc - k_{-2} \cdot E\_gal\_glc - \\ & - k_4 \cdot lac \cdot E\_gal - k_5 \cdot alac \cdot E\_gal = 0. \end{aligned} \tag{10.8}$$

We also accounted for the law of mass conservation for the total enzyme concentration:

$$E - E\_lac - E\_alac - E\_gal\_glc - E\_gal = E_0. \tag{10.9}$$

According to (Huber et al. 1984) the velocities of binding of lactose, allolactose and glucose are well above the rates of catalysis. So, we used approximation of fast equilibrium for stages 1a, 2a and 6. This allowed us, using the ratio of the constants of equilibrium for all quasi equilibrium stages, to express concentrations of different forms of the enzyme as concentrations of free form of the enzyme E and the complex of the enzyme and galactose E_gal:

$$\begin{aligned} E\_lac &= \frac{E \cdot lac}{K_l}, \\ E\_alac &= \frac{E \cdot alac}{K_a}, \\ E\_gal\_glc &= \frac{E\_gal \cdot glc}{K_g}. \end{aligned} \tag{10.10}$$

Substitution of the expression (10.10) in the formulae (10.8, 10.9) gives a system of two linear equation with respect to E и E_gal

$$\begin{cases} E\left(1 + \dfrac{lac}{K_l} + \dfrac{alac}{K_a}\right) + E\_gal\left(1 + \dfrac{glc}{K_g}\right) = E_0, \\ E\left(k_3 gal + k_1\dfrac{lac}{K_l} + k_2\dfrac{alac}{K_a} + k_{-4}X_1 + k_{-5}X_2\right) - \\ -E\_gal\left(k_{-3} + k_{-1}\dfrac{glc}{K_g} + k_{-2}\dfrac{glc}{K_g} + k_4 lac + k_5 alac\right) = 0. \end{cases}$$

Solving the system, the expressions for stationary concentrations of the enzyme states E and E_gal were obtained

$$\begin{aligned} E &= \frac{E_0}{\Delta}\left\{k_{-3} + \frac{glc}{K_g}(k_{-1} + k_{-2}) + k_4 lac + k_5 alac\right\}, \\ E\_gal &= \frac{E_0}{\Delta}\left\{k_3 gal + k_1\frac{lac}{K_l} + k_2\frac{alac}{K_a} + k_{-4}X_1 + k_{-5}X_2\right\}, \end{aligned} \tag{10.11}$$

where

$$\begin{aligned} \Delta = &\left\{k_3 + \frac{glc}{K_g}(k_{-1} + k_{-2}) + k_4 lac + k_5 alac\right\}\left(1 + \frac{lac}{K_l} + \frac{alac}{K_a}\right) + \\ &+ \left\{k_3 gal + k_1\frac{lac}{K_l} + k_2\frac{alac}{K_a} + k_{-4}X_1 + k_{-5}X_2\right\}\left(1 + \frac{glc}{K_g}\right). \end{aligned}$$

The concentrations of other stationary forms of the enzyme can be expressed using the formula (10.10). With knowledge of stationary concentrations of all the forms of the enzyme, it is possible to calculate the velocity of any elementary stage. Using (10.7, 10.10, and 10.11) we obtained the following

$$\begin{aligned} v_{1b} = \frac{E_0}{\Delta}\Bigg\{ & k_1\frac{lac}{K_l}\left(k_{-3} + (k_{-1}+k_{-2})\frac{glc}{K_g} + k_4 lac + k_5 alac\right) - \\ & - k_{-1}\frac{glc}{K_g}\left(k_3 gal + k_1\frac{lac}{K_l} + k_2\frac{alac}{K_a} + k_{-4}X_1 + k_{-5}X_2\right)\Bigg\} \end{aligned} \qquad (10.12)$$

in much the same manner for other key velocities

$$\begin{aligned} v_{2b} = \frac{E_0}{\Delta}\Bigg\{ & k_2\frac{alac}{K_a}\left(k_{-3} + (k_{-1}+k_{-2})\frac{glc}{K_g} + k_4 lac + k_5 alac\right) - \\ & - k_{-2}\frac{glc}{K_g}\left(k_3 gal + k_1\frac{lac}{K_l} + k_2\frac{alac}{K_a} + k_{-4}X_1 + k_{-5}X_2\right)\Bigg\} \end{aligned} \qquad (10.13)$$

$$\begin{aligned} v_4 = \frac{E_0}{\Delta}\Bigg\{ & k_4 lac\left(k_3 gal + k_1\frac{lac}{K_l} + k_2\frac{alac}{K_a} + k_{-4}X_1 + k_{-5}X_2\right) \\ & - k_{-4}X_1\left(k_{-3} + (k_{-1}+k_{-2})\frac{glc}{K_g} + k_4 lac + k_5 alac\right)\Bigg\} \end{aligned} \qquad (10.14)$$

$$\begin{aligned} v_5 = \frac{E_0}{\Delta}\Bigg\{ & k_5 alac\left(k_3 gal + k_1\frac{lac}{K_l} + k_2\frac{alac}{K_a} + k_{-4}X_1 + k_{-5}X_2\right) \\ & - k_{-5}X_2\left(k_{-3} + (k_{-1}+k_{-2})\frac{glc}{K_g} + k_4 lac + k_5 alac\right)\Bigg\} \end{aligned} \qquad (10.15)$$

The foregoing presents equations of the velocities for all the elementary stages. These expressions include the rate and dissociation constants. In order to describe the behavior of the real enzyme, the parameters of equations should be determined based on experimental data. In this case the experimental data were taken from (Burstein et al. 1965). The experiment was as follows. To a solution, containing 0.5 M of lactose, at zero point of time, β-galactosidase was added in such a way that its final concentration was equal to 130 μg per 1 ml. Concentration of lactose, allolactose, galactose, glucose and total concentration of oligosaccharides measured for 10 h have been measured parameters of the system.

Since the system described contains neither influxes nor effluxes of the matter, then in accordance with a kinetic profile depicted in Fig. 10.5 the change of

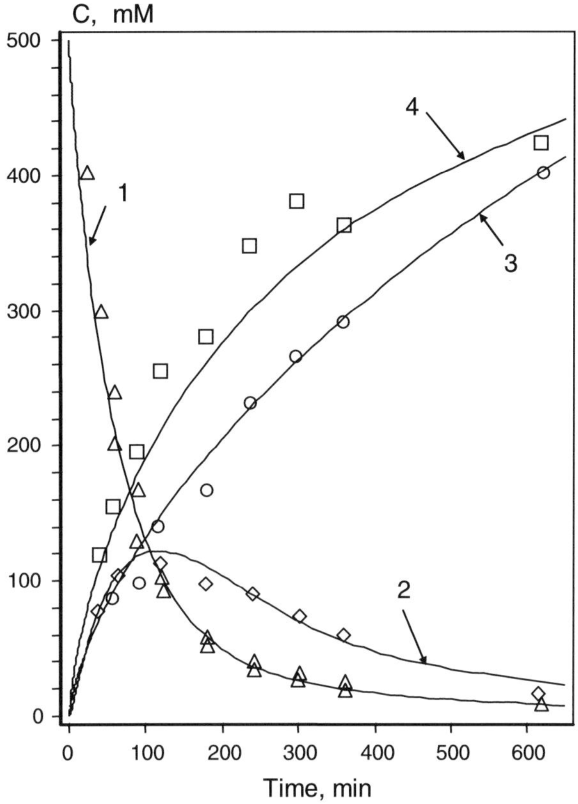

Fig. 10.5 β-galactosidase metabolites concentrations dependence on time described by the model and experimental data (Saier and Ramseier 1996): triangles (1-lactose concentration), rhombuses (2-allolactose concentration), circles (3-galactose concentration), squares (4-glucose concentration). Experimental conditions: 130 μg of *β-galactosidase* on 1 ml of the solution, T = 30 °C, $pH = 7.2$, $MgSO_4$ 6.7 mM, $NaCl$ 10 mM

concentrations of the metabolites in time is determined only by the activity of the enzyme and the following can be written

$$\begin{cases} (lac)' = -v_{1a} - v_4, \\ (alac)' = -v_{2a} - v_5, \\ (gal)' = -v_3, \\ (glc)' = -v_6, \\ (X_1)' = v_4, \\ (X_2)' = v_5. \end{cases} \tag{10.16}$$

In terms of the expression (10.6), which resulted from the approximation of quasistationary concentrations for all the forms of the enzyme, the system transforms into the following one

$$\begin{cases} (lac)' = -v_{1b} - v_4, \\ (alac)' = -v_{2b} - v_5, \\ (gal)' = -v_{1b} - v_{2b} - v_4 - v_5, \\ (glc)' = v_{1b} + v_{2b}, \\ (X_1)' = v_4, \\ (X_2)' = v_5. \end{cases} \tag{10.17}$$

The given system is a system of differential equations with concentrations of metabolites as variables. The system of Equations (10.18) represents two first linear integrals which correspond to the mass conservation laws for two monosaccharides-glucose and galactose:

$$\begin{aligned} glc + lac + alac + (X_1 + X_2) = const_1, \\ gal + lac + alac + 2(X_1 + X_2) = const_2. \end{aligned} \tag{10.18}$$

With the values of concentrations of metabolites in the system under study at zero point of time, it is possible to determine the values of the parameters $const_1$ и $const_2$. So, quantitative description of the above experiment reduces to a solution of Cauchy problem

$$\begin{cases} (lac)' = -v_1 - v_4, \\ (alac)' = -v_2 - v_5, \\ (X_1)' = v_4, \\ (X_2)' = v_5, \\ glc = 0,5M - lac - alac - (X_1 + X_2), \\ gal = 0,5M - lac - alac - 2(X_1 + X_2), \end{cases} \tag{10.19}$$

with initial terms

$$\begin{aligned} lac_0 = 0,5M, \\ alac_0 = glc_0 = gal_0 = X_{10} = X_{20} = 0M. \end{aligned}$$

10.3.2.3 Identification of the Parameters of β-galactosidase Rate Equation

In order for the model to describe the behavior of the real system it is necessary to identify the parameters included in the equation of the rate of an enzyme based on experimental data. In our work we used as a criterion of adequacy of the model constructed to the real enzyme a sum of quadratic deviations of theoretical values, the results of modelling from the experimental points from the work (Huber et al. 1976). In this connection a search of optimal values of the model parameters determined by the system of Equation (10.19) was in designating of such a set of constants, where the criterion should reach a minimum. Minimization of deviation was made according to the Hook-Jeeves technique within the wide range of possible values of the constants of rate and dissociation.

All the algorithms used for solving the system of differential equations and searching optimal values of the parameters were developed using the DBsolve software 7.01 (Gizzatkulov et al. 2004). The parameters identified for β-galactosidase are given in Table 10.2. Experimental estimates available for some constants are given in brackets. Figure 10.5 represents the results of fitting of experimental data with our developed model.

Table 10.2 β-galactosidase model parameters values estimated from experimental data published in (Huber et al. 1976)

Model parameter	Value	Model parameter	Value
k_1	$1,0 \cdot 10^4$ min^{-1}	k_{-1}	$0,8 \cdot 10^3$ min^{-1}
			$1,0 \cdot 10^4$ min^{-1}
k_2	$4 \cdot 10^4$ min^{-1}	k_{-2}	($2,3\ 10^4$ min^{-1} (Kurganov 1978))
k_3	$3 \cdot 10^1$ min^{-1} mM^{-1}	k_{-3}	$1,6 \cdot 10^4$ min^{-1}
k_4	$2 \cdot 10^1$ min^{-1} mM^{-1}	k_{-4}	$0,8 \cdot 10^3$ min^{-1} mM^{-1}
k_5	$2 \cdot 10^1$ min^{-1} mM^{-1}	k_{-5}	$0,8 \cdot 10^3$ min^{-1} mM^{-1}
K_I	0,7 mM (1,3 mM (Kotlarz and Buc 1977))	K_g	14 mM (17 mM (Kurganov 1978))
K_a	0,8 mM		

We used the model to study the ratio between different activities of an enzyme. β-galactosidase has several activities: formation of glucose and galactose, transformation of lactose into allolactose and synthesis of trisaccharides. Which of these activities are dominating and how does the contribution of each activity depend on the concentration of the substrates and products? The answers to these questions enable us to understand better how the functioning of an enzyme is controlled by the substrate. To solve this problem we used the equations of the rates at fixed zero values of the concentration of allolactose, galactose, glucose and trisaccharides and changing concentration of lactose within the range of 0–5 mM (Fig. 10.6). It is possible to interpret this study as an attempt to forecast theoretically the experimental

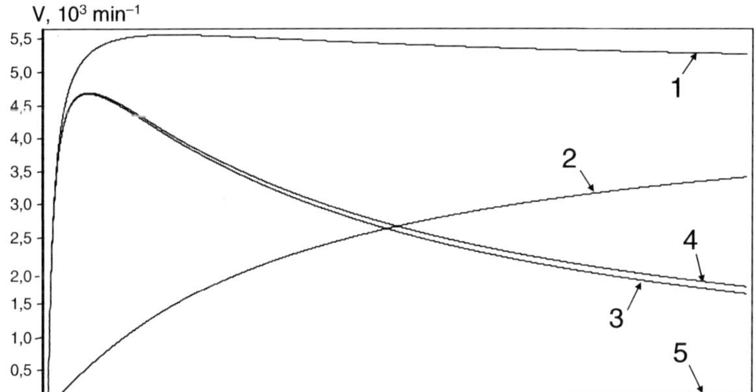

Fig. 10.6 The model fluxes distribution and their dependences on lactose concentration: 1- stationary rate of lactose consumption. 2,3,4,5-stationary synthesis rates of allolactose, galactose, glucose, and trisaccharides correspondingly. The model conditions: alac = oligo = 0 mM, glc = 0–1000 mM, gal = 0–1000 mM

data on measuring initial rates of lactose utilization and synthesis of various products depending on the concentration of the main substrate.

As seen from Fig. 10.6 the dependence of the rate of lactose consumption on the concentration of the latter in the conditions has clear deviation from the classical law of Michaelis-Menten, namely, at lactose concentrations higher than 10 mM a small decrease of the consumption rate is seen. The rates of the synthesis of glucose and galactose have a pronounced optima; at lactose concentration of 25 mM the yield is maximal. The rate of allolactose synthesis under the same conditions increases monotonically within the whole range studied. At a lactose concentration of 25 mM the rate of allolactose synthesis becomes equal to the rate of monosaccharides synthesis. It should be noted that the situation where the rate of allolactose synthesis can exceed the rate of monosaccharides formation *in vivo* is most probably impossible, since actual lactose concentration inside the bacterium is deliberately lower than the modeled one.

Within the whole range of lactose concentrations the outflow of the substance for the synthesis of byproducts (trisaccharides) turns out to be very low and does not exceed 3% of the rate of lactose consumption. We can conclude that under conditions close to the intracellular ones the substance outflow for the synthesis of trisaccharides can be ignored without any loss in exactness described.

Besides the data shown in Fig. 10.6 we studied how the permanent reaction rates depended on the concentrations of glucose and galactose (data are not available), since metabolites could be in higher concentrations in actual bacteria. It turned out that at concentrations up to 1 mM of each of the monosaccharides the values of the fixed rates remained unchanged. It is possible to explain this by the fact that equilibrium of lactose hydrolysis is significantly shifted towards higher values, hence decomposition is, in fact, irreversible in the range studied resulting in weak sensitivity of the enzyme to monosaccharides.

We constructed a kinetic model of *E. coli* β-galactosidase, and determined the parameters included in the rate expression. The model obtained enables us to describe not only experimental data used for identification of the parameters, but to predict the behavior of the enzyme for other conditions (for example *in vivo*). Besides we showed in the work that the model used can "simulate" another type of experiment, like measuring of initial rates at variable concentrations of one of the substrates.

10.3.3 Kinetic Model of the E. coli *Citrate Synthase*

Citrate Synthase (EC 2.3.3.1, *gltA*) is the key enzyme of the Krebs cycle – the central part of the cell's energy metabolism. It catalyzes the first step of carbon atoms entering to the cycle, i.e. acetyl coenzyme A (AcCoA) condensates with oxaloacetate (OAA) resulting in citrate (Cit) production and release of coenzyme A (CoA) (Neidhardt and Curtiss 1996): AcCoA + OAA = Cit + CoA. The enzyme has complex regulation by the key metabolites (ATP, NADH) which reflect energy state of the cell. That is why it is so imprtant to obtain realistic description of the enzyme as a part of the whole model of *E. coli* central metabolism. To date there is no generally accepted opinion in literature about the mechanism of the *E. coli* citrate synthase reaction. There exists disagreement in assumptions about the mechanism of citrate synthase reaction: whether substrates binding to the enzyme is arbitrary (Wright and Sanwal 1971a) or ordered (Pereira et al. 1994). The order of substrates binding is also disputed. In our model we have assumed AcCoA being the first substrate according to ATP inhibition studies (Jangaard et al. 1968a). Citrate synthase is known to have a complex regulatory pattern – the activity of the enzyme is pH-dependent and is modulated by inhibitors – ATP, NADH, 2-ketoglutarate. We have accounted for these regulators in our developed catalytic cycle for citrate synthase. We have derived the rate equation, and estimated the enzyme's kinetic parameters. Their values have been verified using an independent set of experimental data published in (Jangaard et al. 1968a). The enzyme's concentrations in *E. coli* cells grown under aerobic conditions on acetate and glucose have been estimated from *E. coli* cell extracts' specific activities.

When constructing the model we have used the following facts about *E. coli* citrate synthase functioning, known from literature:

1. Citrate synthase of gram-negative bacteria is a hexamer. Sigmoid dependence of initial rate on AcCoA was obtained under zero concentration of KCl in (Pereira et al. 1994). The effect was not observed under 0.1 M KCl addition (Pereira et al. 1994). As the concentration of 0.1 M KCl is physiological for *E. coli* we have not described the sigmoid dependence observed under KCl = 0.
2. The reaction is practically irreversible with an equilibrium constant of $2.24^{*}10^{6}$ (Guynn et al. 1973).
3. NADH and a-ketoglutarate are citrate synthase's inhibitors noncompetitive to oxaloacetate (Jangaard et al. 1968a).
4. ATP is the inhibitor competitive in respect to AcCoA and noncompetitive to oxaloacetate (Jangaard et al. 1968a).
5. The enzyme's maximal activity depends on pH without inhibitors and with ATP addition (Jangaard et al. 1968a). It was shown that ATP moves the maximum of the bell-shaped pH-dependence to the right and decreases the rate in its maximum (Jangaard et al. 1968a).
6. *E. coli* Citrate synthase kinetic parameters known from literature:

$$\begin{aligned} K_m^{AcCoA} &= 0.11 \text{ mM (Faloona and Srere 1969a);} \\ K_m^{OAA} &= 0.021 \text{ (pH 8.1, t } = 21^{\circ}\text{C in the presence of 0.1M KCl)} \\ &\quad \text{(Faloona and Srere 1969a);} \\ K_d^{AcCoA} &= 0.7 \text{ mM (pH 8.0) (Faloona and Srere 1969b);} \\ \mathrm{k_{cat}} &= 48601/\min \text{ (pH 8.0) (Donald et al. 1991);} \\ \mathrm{pH_{opt}} &= 7.3 \text{(Jangaard et al. 1968a).} \end{aligned} \tag{10.20}$$

10.3.3.1 Construction of Citrate Synthase Catalytic Cycle

As there is disagreement about citrate synthase's mechanism we have used experimental data on the enzyme inhibition by ATP. ATP was shown to be a competitive inhibitor with respect to AcCoA and noncompetitive to oxaloacetate (Jangaard et al. 1968a). This can be observed only when AcCoA is the first substrate. So we have assumed that citrate synthase functions according to Irreversible Ordered Bi Bi mechanism by Cleland classification (Cleland 1963), with AcCoA binding first. The scheme of this catalytic cycle is presented in Fig. 10.7. We also have taken into account inhibitors α-ketoglutarate and NADH which bind to two enzyme forms – enzyme bound with AcCoA, and enzyme bound with AcCoA and OAA (Fig. 10.7). This assumption allowed us to describe experimental data on enzyme inhibition. ATP binding to free enzyme form was also included into the scheme (Fig. 10.7). Moreover we have described dependence of the enzyme activity on pH (pH-dependence). The classic assumption (Cornish-Bowden 2001) was applied that enzyme can be protonated in its active site and the singly protonated form is active whereas non protonated and doubly protonated forms are inactive (Fig. 10.7). To describe ATP effects on pH-dependence (see clause 5) we have assumed that the active form is ATP bound to doubly protonated enzyme as only in this case the maximum could decrease and its shift to the higher values of proton concentration could be observed.

10.3.3.2 Derivation of Citrate Synthase Rate Equation

Derivation of Rate Equation in Terms of Catalytic Cycle Parameters

As the reaction is almost irreversible (Guynn et al. 1973) we have described the enzyme working only in a forward direction in respect of rate dependence on substrates and effectors concentrations. On the scheme (Fig. 10.7) the stages of substrates binding are shown as reversible and the stage of product formation is irreversible with rate constant k_3. The rate equation has been derived based on the assumption that stages of effector and proton binding are much faster than the catalytic stage

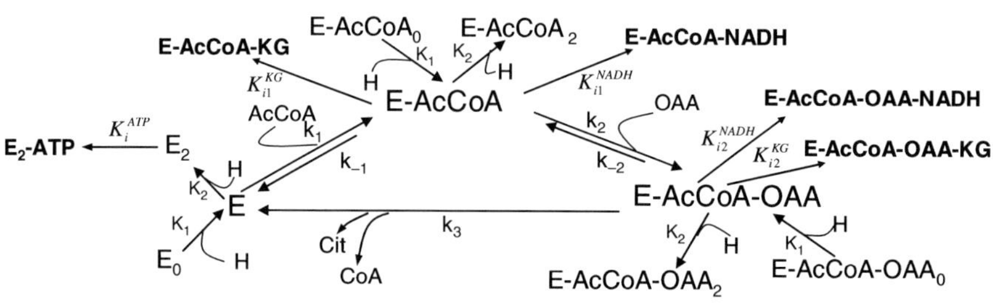

Fig. 10.7 The scheme of *E. coli* citrate synthase catalytic cycle. Designations: E – Citrate Synthase; E_0, E-AcCoA_0, E-AcCoA-OAA_0 – deprotonated enzyme forms; E, E-AcCoA, E-AcCoA-OAA – once protonated enzyme forms; E_2, E-AcCoA_2, E-AcCoA-OAA_2 – twice protonated enzyme forms

and the stages of substrates binding. The stages of protons binding were characterized by two parameters: dissociation constants for proton dissociation from doubly and singly protonated enzyme forms. With these assumptions we have derived the following rate equation for citrate synthase:

$$V = CS\frac{k_1k_2k_3 \cdot AcCoA \cdot OAA}{\begin{array}{l}(k_{-1}k_3 + k_{-1}k_{-2} + k_2k_3 \cdot OAA)\left(1 + \frac{K_{d1}^H}{H} + \frac{H}{K_{d2}^H}\left(1 + \frac{ATP}{K_i^{ATP}}\right)\right) + \\ + AcCoA(k_1k_3 + k_1k_{-2})\left(1 + \frac{K_{d1}^H}{H} + \frac{H}{K_{d2}^H} + \frac{KG}{K_{i1}^{KG}} + \frac{NADH}{K_{i1}^{NADH}}\right) + \\ + k_1k_2 \cdot AcCoA \cdot OAA\left(1 + \frac{K_{d1}^H}{H} + \frac{H}{K_{d2}^H} + \frac{KG}{K_{i2}^{KG}} + \frac{NADH}{K_{i2}^{NADH}}\right)\end{array}} \tag{10.21}$$

Here $k_{1,-1}$, $k_{2,-2}$, k_3 are rate constants for corresponding stages of the catalytic cycle, CS is the enzyme concentration, $K_{d1}^H, K_{d2}^H, K_i^{ATP}, K_{i1}^{KG}, K_{i2}^{KG}, K_{i1}^{NADH}$, K_{i2}^{NADH} are dissociation constants for corresponding inhibitors and protons.

Derivation of Rate Expression in Terms of Kinetic Parameters (Michaelis Constants, Inhibition Constants, Catalytic Constants etc.)

To express rate equation (10.21) in terms of experimentally measured kinetic parameters we have found their expressions from Equation (10.21):

$$V_{\max} = CS\frac{k_3}{\left(1 + \frac{K_{d1}^H}{H} + \frac{H}{K_{d2}^H}\right)}; K_m^{AcCoA} = \frac{k_3}{k_1}; K_m^{OAA} = \frac{k_3 + k_{-2}}{k_2} \tag{10.22}$$

Using these expressions (10.22) we could rewrite rate equation in the following form:

$$V = CS\frac{k_{cat0} \cdot AcCoA \cdot OAA}{\begin{array}{l}(K_d^{AcCoA} \cdot K_m^{OAA} + K_m^{AcCoA} \cdot OAA)\left(1 + \frac{K_{d1}^H}{H} + \frac{H}{K_{d2}^H}\left(1 + \frac{ATP}{K_i^{ATP}}\right)\right) + \\ + AcCoA \cdot K_m^{OAA} \cdot \left(1 + \frac{K_{d1}^H}{H} + \frac{H}{K_{d2}^H} + \frac{KG}{K_{i1}^{KG}} + \frac{NADH}{K_{i1}^{NADH}}\right) + \\ + AcCoA \cdot OAA \cdot \left(1 + \frac{K_{d1}^H}{H} + \frac{H}{K_{d2}^H} + \frac{KG}{K_{i2}^{KG}} + \frac{NADH}{K_{i2}^{NADH}}\right)\end{array}} \tag{10.23}$$

here,

$$k_{cat0} = k_3; K_d^{AcCoA} = \frac{k_{-1}}{k_1}$$

10.3.3.3 Estimation of the Citrate Synthase Kinetic Parameters

There were 12 parameters in rate equation (10.23): three of them were known from literature, 8 parameters have been estimated from *in vitro* experimental data whereas enzyme concentration in *E. coli* cells could not be found from *in vitro* data and was estimated from *E. coli* cell extract specific activity. As the catalytic constant of the enzyme was estimated in literature under fixed pH value conditions ($k_{cat} = 4860$ 1/ min, pH 8.0 (Donald et al. 1991)), we have used it for parameter k_{cat0} determination: $k_{cat0} = k_{cat}\left(1 + \frac{K_{d1}^H}{H} + \frac{H}{K_{d2}^H}\right)$. So we have reduced the number of unknown parameters to 7. Further, we have determined some of the parameters from experimental data obtained in the absence of inhibitors. We have used initial rate dependencies on substrate AcCoA from two sources (Faloona and Srere 1969b, Wright and Sanwal 1971b): five curves with different concentrations of oxaloacetate (Fig. 10.8a,b), two initial rate dependencies on oxaloacetate with fixed concentration of AcCoA (Faloona and Srere 1969b) (Fig. 10.8c), and pH-dependence (Jangaard et al. 1968b) (Fig. 10.8d). We have found such a set of parameters which allowed us to describe all these data with a rate equation (10.23). Fitting results are presented on Fig. 10.8 Here is the list of parameter values:

$$K_{d1}^H = 1e - 5\,mM;\ K_{d2}^H = 2e - 4\,mM;\ K_m^{AcCoA} = 0.18\,mM;$$
$$K_m^{OAA} = 0.04\,mM;\ K_d^{AcCoA} = 0.1\,mM;\ K_i^{ATP} = 0.58\,mM$$

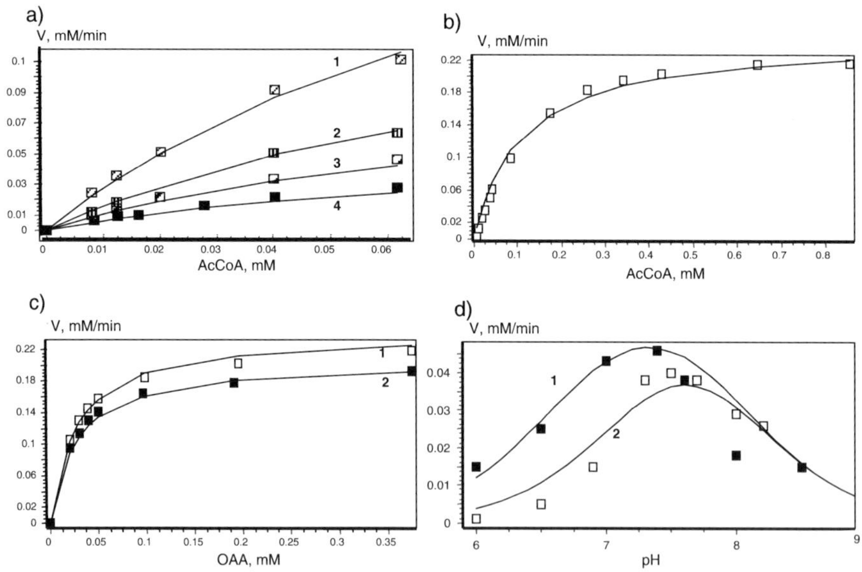

Fig. 10.8 Citrate synthase initial rate dependence on substrates concentrations and pH described by experimental points and rate equation (10.23): (**a**) OAA concentration: 1–0.1; 2–0.02; 3–0.005; 4–0.01 mM (Ivanitsky et al. 1978); (**b**) OAA concentration was 0.5 mM (Ivanitsky et al. 1978); (**c**) AcCoA concentration: 1–0.5; 2–0.25 mM (Ewings and Doelle 1980); (**d**) AcCoA = 0.05 mM; OAA = 0.1 mM; ATP concentration: 1–0; 2–2 mM (Cleland 1963)

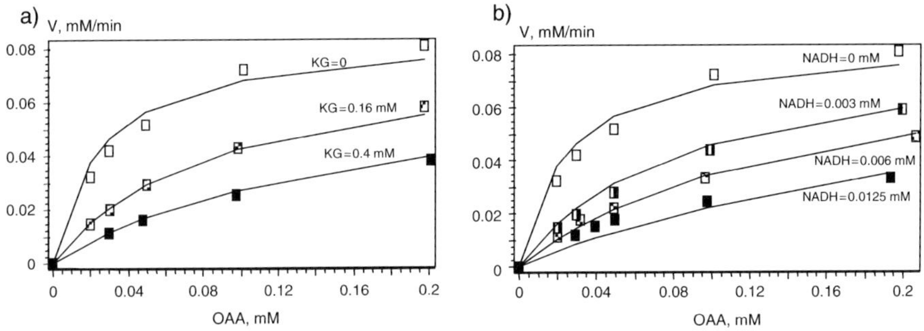

Fig. 10.9 Citrate synthase initial rate dependence on concentrations of substrates in the presence of inhibitors described by experimental points and rate equation (10.23)

To estimate the inhibition constants we have used experimental data on initial reaction rate dependence on substrates concentrations in the presence of different concentrations of the inhibitors α-ketoglutarate and NADH (Wright and Sanwal 1971b):

$$K_{i1}^{KG} = 0.015 \text{ mM}; K_{i2}^{KG} = 0.256 \text{ mM}; K_{i1}^{NADH} = 3.3e - 4 \text{ mM};$$
$$K_{i2}^{NADH} = 8.4e - 3 \text{ mM}.$$

The results of fitting are shown on Fig. 10.9.

On the next stage we have verified our model of citrate synthase functioning on experimental data which were not used for fitting. We have used the data on initial rate dependence on substrates concentration in the presence of ATP (Jangaard et al. 1968b). It was shown that the rate equation (10.23) and estimated parameters values allowed us to describe the independent set of experimental curves (Fig. 10.10).

Estimation of citrate synthase concentration depending on *E. coli* growth conditions

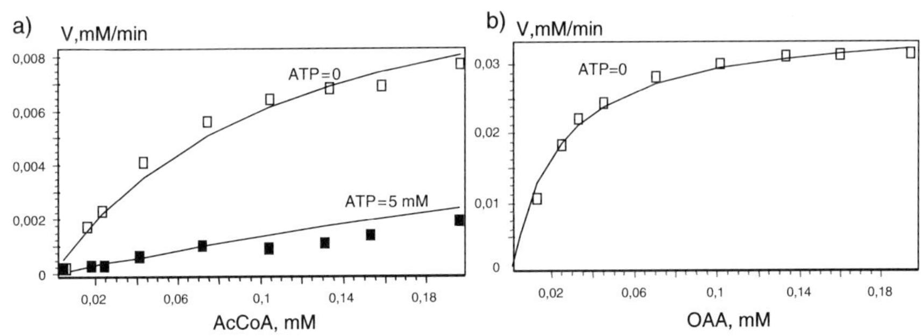

Fig. 10.10 Model verification. Citrate synthase initial rate dependence on concentrations of substrates described by experimental points (Cleland 1963) and rate equation (10.23)

Specific citrate synthase activity (SA) of *E. coli* cells extract grown on acetate has been measured in (Cornish-Bowden 2001):

$$\text{SA} = 0.25\ \text{micromoles/min}^{*}\ \text{mg}_{\text{extract protein}}$$

Assuming that 1 mg of *E. coli* cells protein corresponds to 5.5 microL of intracellular volume (Jangaard et al. 1968b) we have calculated the enzyme's maximal rate from the specific activity: $V_{max} = 45.5\,\text{mM/min}$. Further citrate synthase concentration can be calculated by dividing of maximal rate by catalytic constant of the enzyme. In this case, however, we should use the value of catalytic constant obtained at the same pH as the maximal velocity value, i.e. at physiological pH of 7.3 (Padan et al. 1981). We have calculated the required value of the catalytic constant in accordance with the obtained pH-dependence of the enzyme:

$$k_{cat}^{pH7.3} = \frac{k_{cat0}}{\left(1 + \frac{K_{d1}^{H}}{H} + \frac{H}{K_{d2}^{H}}\right)} = \frac{9941}{\left(1 + \frac{1e-5}{1e-4.3} + \frac{1e-4.3}{2.2e-4}\right)} = 6966\ (1/\text{min})$$

So we could calculate citrate synthase concentration in *E. coli* grown aerobically on acetate:

$$CS_{acetate} = V_{\max}^{acetate} / k_{cat}^{pH7.3} = 6.5\ (microM)$$

In the same way we have calculated the enzyme concentration which should be observed in *E. coli* cell grown aerobically on glucose. We have used citrate synthase specific activity measured on the extract of *E. coli* cells grown on glucose:

$$\text{SA} = 0{,}05\ \text{micromoles/min}^{*}\text{mg}_{\text{extract protein}}\ \text{(Peng and Shimizu 2003).}$$

Maximal velocity has been calculated as $V_{max} = 9\,\text{mM/min}$, and citrate synthase concentration in the cell in these conditions was estimated:

$$CS_{glucose} = V_{\max}^{glucose} / k_{cat}^{pH7.3} = 1.3\ (microM)$$

So we have constructed a kinetic model of *E. coli* citrate synthase functioning – the rate equation has been derived and kinetic parameters have been estimated. We have taken into account known inhibitory effects and pH dependence of the enzyme activity. This allowed us to describe a set of experimental data obtained under different pH values. Plausibility of the model was confirmed by its ability to describe an independent data set (Jangaard et al. 1968b) which had not been used for model parameters determination. Citrate synthase concentrations in *E. coli* cells grown aerobically on acetate and glucose have been obtained.

10.4 Conclusion

We have illustrated our kinetic modeling approach using three *E. coli* enzymes. We propose that only a detailed description of individual enzymes allows realistic kinetic models of the whole pathways in the cell to be obtained. The ultimate goal is to integrate all available *in vitro* experimental data in the description of the enzyme. First, we use relevant data to reconstruct the enzyme's catalytic cycle. Then we derive the rate equation of the enzyme. The next and most complicated step is to define such values of kinetic parameters from the rate equation that would allow us to describe all experimental dependencies measured *in vitro*. Here we have illustrated our strategy with three non-trivial and rather complicated enzyme models: allosteric tetramer phosphofructokinase-1, citrate synthase with its regulation by ATP and pH, and β-galactosidase validated against time dependencies of its substrates.

Analysis of the phosphofruktokinase-1 model allowed us to predict new operational properties of phosphofructokinase-1, such as cooperative action of allosteric effectors (PEP, ADP and GDP), competitive inhibition by free form of ATP and influence of magnesium ions on the enzyme rate.

We used the modelling to study the ratio between different activities of β-galactosidase. It turned out, that at lactose concentration of 25 mM the rate of allolactose synthesis becomes equal to the rate of monosaccharides synthesis. Within the whole range of lactose concentrations the outflow of the substance for the synthesis of trisaccharides does not exceed 3% of the lactose consumption. We also found that concentrations of glucose and galactose up to 1 mM did not change the consumption and production rates.

The kinetic model of *E. coli* citrate synthase allowed us to get insight into some important regulatory features of the enzyme catalytic mechanism. According to ATP inhibition studies (Jangaard et al. 1968b) we have proposed Ordered Bi Bi citrate synthase's mechanism with AcCoA binding first. Inhibition experimental data allowed us to accept the hypothesis that the inhibitors alpha-ketoglutarate and NADH binds to two enzyme forms (the enzyme bound with AcCoA and with both AcCoA and OAA). To describe ATP effects on pH-dependence (Jangaard et al. 1968b) we have assumed that the active enzyme form corresponds to the complex of twice protonated enzyme with ATP. With the use of our model we managed to estimate the concentration of citrate synthase in *E. coli* cells grown aerobically on acetate and glucose.

The models we presented in this paper prove that developing of detailed enzyme kinetic models can be essential to capture the enzyme regulatory properties. We illustrated how the detailed kinetic model of the enzyme can be further reduced to derive a reaction rate equation which inherits key regulatory effects included in the original detailed description, and allows to consistently approximate large sets of *in vitro* experimental data. Individual reaction rates derived in such a way can be further integrated into higher level kinetic models of *E. coli* metabolic pathways. Pathway models will in their turn allow investigating higher level regulatory effects in bacterial metabolic networks, observed in cellular extracts and *in vivo*. We hope that the kinetic modeling approach in general, and three kinetic models of *E. coli* enzymes in particular, will be useful for future whole cell models of *E. coli*, and practical applications in metabolic engineering and synthetic biology.

Acknowledgments For support of this work we acknowledge GlaxoSmithKline, Institute for Systems Biology Sankt Petersburg, EU FP6 EC-MOAN and Centre for Systems Biology at Edinburgh (CSBE). CSBE is a Centre for Integrative Systems Biology supported by BBSRC and EPSRC.

Abbreviations

PfkA	Phosphofructokinase-1
F6P	fructose-6-phosphate
F16bP	fructose-1,6-biphosphate
PEP	phosphoenolpyruvate
$ATPMg^{2-}$	magnesium form of ATP
lac	lactose
alac	allolactose
glc	glucose
gal	galactose
oligo	oligosaccharides
E_gal, E_lac, E_alac	β-galactosidase enzyme form bound with galactose, lactose, allolactose
E_gal_glc	ternary complex of β-galactosidase enzyme form bound with galactose and glucose
gltA	Citrate Synthase
CS	Citrate synthase concentration
AcCoA	acetyl coenzyme A
OAA	oxaloacetate
Cit	citrate
CoA	coenzyme A
KG	2-ketoglutarate
H	proton
SA	specific activity of the enzyme

References

Ausat I, Le Bras G, Garel J-R (1997) Allosteric Activation Increases the Maximum Velocity of *E. coli* Phosphofructokinase. J Mol Biol 267:476–80

Babul J (1978) Phosphofructokinases from *Escherichia coli*. Purification and characterization of the nonallosteric isozyme. J Biol Chem 253(12):4350–5

Barrett CL, Kim TY, Kim HU et al. (2006) Systems biology as a foundation for genome-scale synthetic biology. Curr Opin Biotechnol 17(5):488–92

Berger SA, Evans PR (1991) Steady-state fluorescence of *Escherichia coli* phosphofructokinase reveals a regulatory role for ATP. Biochemistry 30(34):8477–80

Blangy D, Buc H, Monod J (1968) Kinetics of the allosteric interactions of phosphofructokinase from *Escherichia coli*. J Mol Biol 31(1):13–35

Burstein C, Cohn M, Kepes A et al. (1965) [Role of Lactose and Its Metabolic Products in the Induction of the Lactose Operon in *Escherichia Coli.*]. Biochim Biophys Acta 95:634–9

Campos G, Guixe V, Babul J (1984) Kinetic mechanism of phosphofructokinase-2 from *Escherichia coli*. A mutant enzyme with a different mechanism. J Biol Chem 259(10):6147–52

Cleland WW (1963) The kinetics of enzyme-catalyzed reactions with two or more substrates or products. I. Nomenclature and rate equations. Biochim Biophys Acta 67:104–37

Cornish-Bowden A (2001) Fundamentals of Enzyme kinetics. London

Demin O, Goryanin I (2008) Kinetic modelling in systems biology. Chapman & Hall/CRC, Virginia Beach, VA

Demin OV, Goryanin, II, Dronov S et al. (2004) Kinetic model of imidazologlycerol-phosphate synthetase from *Escherichia coli*. Biochemistry (Mosc) 69(12):1324–35

Deville-Bonne D, Bourgain F, Garel JR (1991a) pH dependence of the kinetic properties of allosteric phosphofructokinase from *Escherichia coli*. Biochemistry 30(23):5750–4

Deville-Bonne D, Laine R, Garel JR (1991b) Substrate antagonism in the kinetic mechanism of *E. coli* phosphofructokinase-1. FEBS Lett 290(1–2):173–6

Donald LJ, Crane BR, Anderson DH et al. (1991) The role of cysteine 206 in allosteric inhibition of *Escherichia coli* citrate synthase. Studies by chemical modification, site-directed mutagenesis, and 19F NMR. J Biol Chem 266(31):20709–13

Edwards JS, Ibarra RU, Palsson BO (2001) *In silico* predictions of *Escherichia coli* metabolic capabilities are consistent with experimental data. Nat Biotechnol 19(2):125–30

Endy D (2005) Foundations for engineering biology. Nature 438(7067):449–53

Ewings KN, Doelle HW (1980) Further kinetic characterization of the non-allosteric phosphofructokinase from *Escherichia coli* K-12. Biochim Biophys Acta 615(1):103–12

Faloona GR, Srere PA (1969a) *Escherichbia coli* citrate synthase. Purification and the effect of potassium on some properties. Biochemistry 8:4497–503

Faloona GR, Srere PA (1969b) *Escherichia coli* citrate synthase. Purification and the effect of potassium on some properties. Biochemistry 8(11):4497–503

Feist AM, Henry CS, Reed JL et al. (2007) A genome-scale metabolic reconstruction for *Escherichia coli* K-12 MG1655 that accounts for 1260 ORFs and thermodynamic information. Mol Syst Biol 3:121

Gizzatkulov N, Klimov A, Lebedeva G, Demin O (2004) DBSolve7: New update version to develop and analyse models of complex biological systems. 12th International Conference on Intelligent Systems for Molecular Biology: 210

Guixe V, Babul J (1985) Effect of ATP on phosphofructokinase-2 from *Escherichia coli*. A mutant enzyme altered in the allosteric site for MgATP. J Biol Chem 260(20):11001–5

Guynn RW, Gelberg HJ, Veech RL (1973) Equilibrium Constants of the Malate Dehydrogenase, Citrate Synthase, Citrate Lyase, and Acetyl Coenzyme A Hydrolysis Reactions under Physiological Conditions. J Biol Chem 248:6957–65

Han MJ, Lee SY (2006) The *Escherichia coli* proteome: past, present, and future prospects. Microbiol Mol Biol Rev 70(2):362–439

Huber RE, Gaunt MT, Hurlburt KL (1984) Binding and reactivity at the "glucose" site of galactosyl-beta-galactosidase (*Escherichia coli*). Arch Biochem Biophysics 234:151–60

Huber RE, Kurz G, Wallenfels K (1976) A quantitation of the factors which affect the hydrolase and transgalactosylase activities of beta-galactosidase (*E. coli*) on lactose. Biochemistry 15:1994–2001

Ishii N, Nakahigashi K, Baba T et al. (2007) Multiple high-throughput analyses monitor the response of *E. coli* to perturbations. Science 316:593–7

Ishii N, Nakahigashi K, Baba T et al. (2007) Multiple high-throughput analyses monitor the response of *E. coli* to perturbations. Science 316(5824):593–7

Ivanitsky GR, Krinsky V, Selkov EE (1978) Mathematical biophysics of the cell. Nauka, Moscow

Jangaard NO, Unkeless J, Atkinson DE (1968a) The inhibition of Citrate Synthase by adenosine triphosphate. Biochim Biophys Acta 151:225–35

Jangaard NO, Unkeless J, Atkinson DE (1968b) The inhibition of citrate synthase by adenosine triphosphate. Biochim Biophys Acta 151(1):225–35

Jobe A, Bourgeois S (1972) lac Repressor-operator interaction. VI. The natural inducer of the *lac* operon. J Mol Biol 69(3):397–408
Kim HU, Kim TY, Lee SY (2008) Metabolic flux analysis and metabolic engineering of microorganisms. Mol Biosyst 4(2):113–20
Kotlarz D, Buc H (1977) Two *Escherichia coli* fructose-6-phosphate kinases. Preparative purification, oligomeric structure and immunological studies. Biochim Biophys Acta 484(1): 35–48
Kotlarz D, Buc H (1982) Phosphofructokinases from *Escherichia coli*. Methods Enzymol 90 Pt E:60–70
Kurganov BI (1978) Allosteric enzymes. Moscow
Lee JM, Gianchandani EP, Papin JA (2006) Flux balance analysis in the era of metabolomics. Brief Bioinform 7(2):140–50
Lee KH, Park JH, Kim TY et al. (2007) Systems metabolic engineering of *Escherichia coli* for L-threonine production. Mol Syst Biol 3:149
Lee SJ, Lee DY, Kim TY et al. (2005) Metabolic engineering of *Escherichia coli* for enhanced production of succinic acid, based on genome comparison and *in silico* gene knockout simulation. Appl Environ Microbiol 71(12):7880–7
Monod J, Wyman J, Changeux JP (1965) On the Nature of Allosteric Transitions: a Plausible Model. J Mol Biol 12:88–118
Neidhardt FC, Curtiss R (1996) *Escherichia coli* and *Salmonella*: cellular and molecular biology. ASM Press, Washington, DC
Padan E, Zilberstein D, Schuldiner S (1981) pH homeostasis in bacteria. Biochim Biophys Acta 650(2–3):151–66
Park JH, Lee KH, Kim TY et al. (2007) Metabolic engineering of *Escherichia coli* for the production of L-valine based on transcriptome analysis and *in silico* gene knockout simulation. Proc Natl Acad Sci USA 104(19):7797–802
Peng L, Shimizu K (2003) Global metabolic regulation analysis for *Escherichia coli* K12 based on protein expression by 2-dimensional electrophoresis and enzyme activity measurement. Appl Microbiol Biotechnol 61(2):163–78
Pereira DS, Donald LJ, Hosfield DJ et al. (1994) Active site mutants of *Escherichia coli* citrate synthase. Effects of mutations on catalytic and allosteric properties. J Biol Chem 269:412–7
Perna NT, Plunkett G 3rd, Burland V et al. (2001) Genome sequence of enterohaemorrhagic *Escherichia coli* O157:H7. Nature 409:529–33
Price N, Reed J, Palsson B (2004) Genome-scale models of microbial cells: evaluating the consequences of constraints. Nat Rev Microbiol 2:886–97
Reeves RE, Sols A (1973) Regulation of *Escherichia coli* phosphofructokinase in situ. Biochem Biophys Res Commun 50(2):459–66
Rye S, Ramseier TM, Michotey V et al. (1995) Effect of the FruR regulator on transcription of *pts* operon in *Escherichia coli*. J Biol Chem 270:2489–96
Saier MH, Jr., Ramseier TM (1996) The catabolite repressor/activator (*Cra*) protein of enteric bacteria. J Bacteriol 178(12):3411–7
Saier MH Jr, Crasnier M (1996) Inducer exclusion and the regulation of sugar transport. Res Microbiol 147:482–9
Schütz R, Küpfer L, Sauer U (2007) Systematic evaluation of objective functions for predicting intracellular fluxes in *E. coli*. Mol Sys Biol 3:119
Selkov E, Basmanova S, Gaasterland T et al. (1996) The metabolic pathway collection from EMP: the enzymes and metabolic pathways database. Nucleic Acids Res 24(1):26–8
Shomburg I, Chang A, Shomburg D (2002) BRENDA, enzyme data and metabolic information. Nucleic Acids Res 30:47–9
Taquikhan MM, Martell AE (1962) Metal Chelates of Adenosine Triphosphate. J Phys Chem 66:10–5
Torres JC, Babul J (1991) An *in vitro* model showing different rates of substrate cycle for phosphofructokinases of *Escherichia coli* with different kinetic properties. Eur J Biochem 200(2):471–6

Vinopal RT, Fraenkel DG (1974) Phenotypic suppression of phosphofructokinase mutations in *Escherichia coli* by constitutive expression of the glyoxylate shunt. J Bacteriol 118(3): 1090–100

Vinopal RT, Fraenkel DG (1975) *PfkB* and *pfkC* loci of *Escherichia coli.* J Bacteriol 122(3): 1153–61

Waygood EB, Sanwal BD (1974) The control of pyruvate kinases of *Escherichia coli.* I. Physicochemical and regulatory properties of the enzyme activated by fructose 1,6-diphosphate. J Biol Chem 249(1):265–74

Wright JA, Sanwal BD (1971a) Regulatory Mechanisms involving nicotinamide adenine nucleotides as allosteric effectors. J Biol Chem 246:1689–99

Wright JA, Sanwal BD (1971b) Regulatory mechanisms involving nicotinamide adenine nucleotides as allosteric effectors. IV. Physicochemical study and binding of ligands to citrate synthase. J Biol Chem 246(6):1689–99

Chapter 11
Dynamic Modeling of the Central Metabolism of *E. coli* – Linking Metabolite and Regulatory Networks

Timo Hardiman, Karin Lemuth, Martin Siemann-Herzberg, and Matthias Reuss

Contents

Abstract Coupling complex regulatory and metabolic networks for the purpose of dynamic modeling requires knowledge of the quantitative kinetics of the participating reactions as well as the variation of parameters in the context of the physiological state of the system. This chapter aims at demonstrating the integration of the different networks for *E. coli* exposed to an increasing carbon limitation of a fed-batch process with constant feeding of the carbon and energy source glucose. Starting from a global observation of the response of the bacteria in terms of flux distribution

---

M. Reuss (✉)
Institute of Biochemical Engineering and Centre Systems Biology, University of Stuttgart, Allmandring 31, 70569, Stuttgart, Germany
e-mail: reuss@ibvt.uni-stuttgart.de

S.Y. Lee (ed.), *Systems Biology and Biotechnology of Escherichia coli*,
DOI 10.1007/978-1-4020-9394-4_11, © Springer Science+Business Media B.V. 2009

and gene expression in the central metabolism, emphasis is given to the dynamic modeling of regulation phenomena in the catabolism. The *cra* regulon which is linked to the dynamic response of the metabolite fructose 1,6- bis(phosphate) serves as an example to introduce a new concept, in which the binding constants are estimated from DNA-binding site sequences of the regulatory proteins. By comparison of the nucleotide frequencies within the DNA-binding sites for the individual target genes of the regulon, it is possible to perform a reasonable estimation of the kinetic parameters. Results of these estimations are compared with experimentally observed transcript concentrations measured with the aid of quantitative PCR. In addition it is shown how these outputs of the regulatory networks can be linked to the maximal rates of the enzymes for the metabolic system of interest. The discussion of this issue is embedded within a critical assessment of different conceptual frameworks for modeling the metabolic network, which covers the spectrum of dynamic modeling at different levels of complexity, such as genome scale, modular approaches and reduced models.

11.1 Introduction

Systems biology as an emerging field of research in bio-, engineering and systems sciences aims at a systems-level understanding of biological processes – and ultimately whole cells and organisms. The grand, and currently unrealistic, hope to even continue these efforts into a whole cell *in silico* model time and again shapes the conceptual framework of this endeavour. There are several reasons that the present state of affairs still falls short of this euphoric expectation. The first is concerned with the fundamental question of a comprehensive definition of a "whole cell model" and closely related to this uncertainty the query about the purpose of such a model. Referring to the fundamental ideas and discussion of Casti (1992a,b) about a model and its intended application, Bailey (1998) reminded us that "mathematical modeling does not make sense without defining, before making the model, what its use is and what problem it is intended to help to solve". The second reason originates from a critical assessment of part of the experimental work in the field of holistic measurements and related top down approaches in inverse engineering for network inference. In spite of spectacular developments in high-throughput technologies such as genome sequencing, transcriptomics, metabolomics, fluxomics etc. – platforms which have monopolized systems biology research in recent years – there is a tendency to fragment the whole into various sub-omes and a great deal of arguments exists about what ome is more important. However, due to multiple border crossings these omes are inseparable parts of a single process – the complex and interwoven dynamics of the living organisms.

Another issue to be addressed in the context of fragmentation is the often observed focus on specific networks and treatment in separated and isolated territories, such as metabolism, regulation and signal transduction. In the course of this partition and kind of downward analysis, levels are reached where the whole meaning of the system is destroyed because of neglected interactions and missing integration. In order to underline the systemic thinking, the exchanges of material and information between the heuristically isolated modules of a system to be investigated may also be termed "intra-actions".

In this chapter we will highlight with a few examples the importance of integration of regulatory and metabolic networks in *Escherichia coli* and discuss the framework of how this process of integration can be portrayed dynamically in the structure of the mathematical model.

Taking up the aforementioned attenuation of the importance of defining first the purpose of the mathematical model, the environmental changes triggering the regulation of the metabolism have to be introduced. The example deals with the regulation of the central metabolism of *E. coli* during a fed-batch process with constant feeding rate of the carbon and energy source glucose (Fig. 11.1). This process operation is important for technical processes for production of heterologous proteins as well as bacterial metabolites. For large-scale applications, fed-batch, high cell density cultivation strategies have proven suitable for considerably increasing the volumetric productivity of these processes (Lee 1996, Yee and Blanch 1992). Irrespective of more sophisticated closed-loop strategies, fed-batch cultivations are usually carried out with open loop control via exponential or constant feeding. Exponential feeding maintains the specific growth rate at a constant level. The maximum biomass concentration that can be achieved with this strategy depends on sufficient

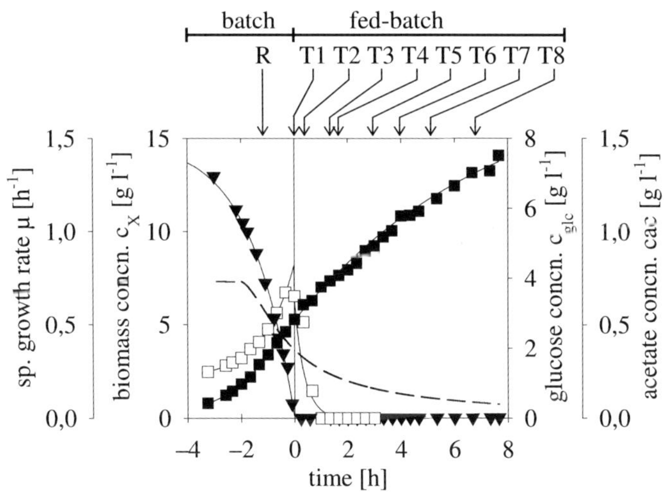

Fig. 11.1 Glucose limited fed-batch cultivation of *E. coli* K-12 W3110 with constant feed rate. The vertical solid line at $t = 0$ indicates glucose limitation. The concentrations of biomass (filled squares), glucose (triangles) and acetate (open squares) are given as well as the time course of the specific growth rate (μ) (broken line). Arrows above the graph indicate the time when the samples were removed for microarray analysis (R, reference; T1 to T8, time series samples)

oxygen supply and heat transfer capacities. At a constant feed rate, the specific growth rate gradually decreases due to declining carbon and energy source levels (Dunn and Mor 1975). The proceeding carbon limitation also leads to a range of serious starvation phenomena with manifold regulatory responses of the cells. These processes macroscopically manifest themselves in a loss of viability, such as was illustrated by Hewitt et al. (2000, 1999, Hewitt and Nebe-Von-Caron 2001).

Bacteria control metabolism and growth rate through global genetic regulatory systems, i.e. regulons and modulons (Lengeler et al. 1999, Neidhardt and Savageau 1996). Prominent examples in *E. coli* are the catabolite repression (*crp* modulon) and the stringent response (*relA/spoT* modulon), two processes that are active under carbon-limiting conditions. During stringent response (reviewed in Braeken et al. (2006), Cashel et al. (1996) and Lengeler et al. (1999)), the limitation of nutrients leads to the intracellular accumulation of ppGpp (guanosine 3′, 5′-bis(diphosphate)), which is supposed to bind to the RNA polymerase (Artsimovitch et al. 2004).

The transcription of genes involved in the translation process – in particular of ribosomal RNA and ribosomal proteins – is negatively regulated by ppGpp. As a result, the protein biosynthesis rate declines, which in turn also leads to a reduction in growth rate (Cashel et al. 1996, Lengeler et al. 1999). During amino acid limitation, the synthesis of ppGpp or guanosine pentaphosphate (pppGpp), collectively referred to as (p)ppGpp, is mediated by RelA (GDP pyrophosphokinase/GTP pyrophosphokinase). Under amino acid-limiting conditions, the ribosome-bound RelA protein is stimulated by uncharged tRNAs at the A site of ribosomes (Wendrich et al. 2002). However, the accumulation of (p)ppGpp depends also on the dual activity of the SpoT protein as (p)ppGpp-hydrolase or (p)ppGpp-synthetase. Although it is known from a homologous protein of *Streptococcus dysgalactiae subsp. equisimilis* that the opposing activities of SpoT are reciprocally regulated (Hogg et al. 2004, Mechold et al. 2002), the regulation of the SpoT protein in *E. coli* is still hypothetical. The most important issue for understanding growth control is the signalling mechanism, which leads to accumulation of ppGpp under carbon-limiting conditions, an aspect that is still not entirely clarified.

Besides various effects on growth-related functions (Cashel et al. 1996), the alarmone ppGpp is known to be involved in the regulation of the sigma S factor concentration (σ^S; *rpoS* gene) on the transcriptional and posttranscriptional level (Hengge-Aronis 2002). As an alternative subunit of RNA polymerase, σ^S is involved in the regulation of transcription in the general stress response in *E. coli* (also designated as 'stationary phase response'). It is assumed that elevated levels of σ^S negatively regulate σ^D-dependent housekeeping genes, such as the TCA cycle genes (Patten et al. 2004). Moreover, ppGpp influences the competition between different stress-related sigma factors in the binding of the RNA polymerase core enzyme at the expense of the sigma factor σ^D (Jishage et al. 2002) and the RNA polymerase availability (Barker et al. 2001a,b, Cashel et al. 1996, Jensen and Pedersen 1990, Traxler et al. 2006).

The *crp* modulon belongs to a group of global genetic regulatory systems, which can be subsumed under the term catabolite control. One basic feature of these systems is that the presence or absence of an extracellular carbon source is indicated by an intracellular metabolite (catabolite) that serves as a signal for derepression (catabolite activation) or deactivation (catabolite repression) of catabolic genes (Saier et al. 1996). The *crp* modulon includes catabolic operons for the utilization of various carbon sources and is regulated by the Crp-cAMP complex. The synthesis of the alarmone cAMP (cyclic $3'$, $5'$-AMP) by the enzyme adenylate cyclase (CyaA) is stimulated by the phophorylated $EIIA^{Glc}$ protein, a component of the *E. coli* phosphoenolpyruvate:carbohydrate phosphotransferase system (PTS) (reviewed in Lengeler et al. (1999) and Postma et al. (1993)). It is assumed that a low glucose uptake rate by the PTS and a high ratio of phosphoenolpyruvate and pyruvate concentrations (c_{pep}/c_{pyr}) lead to the phosphorylation of the $EIIA^{Glc}$ protein (Hogema et al. 1998). Consequently, limited glucose availability leads to the synthesis of cAMP and the transcriptional regulator complex Crp-cAMP is formed. Catabolite control is also exerted by the catabolite repressor/activator protein Cra (formerly designated FruR), which regulates numerous genes involved in the carbon and energy metabolism (the *cra* modulon) (reviewed in Ramseier 1996, Saier and Ramseier 1996, Saier et al. 1996). The regulator protein Cra is inactivated by the catabolites fructose 1-phosphate and fructose 1,6-bis(phosphate) (Saier and Ramseier 1996).

Most of the aforementioned investigations have been performed during the shift from exponential to stationary growth phase in batch cultivations. The dynamic perturbation during these experiments is characterized by a rapid drop of glucose concentration to zero in a short time period. As distinguished from this very fast perturbation the fed-batch cultivation with constant feeding rate prolongs the period of declining glucose concentrations towards a time span of several hours. This prolongation of the proceeding carbon limitation initiates a process of transient adaptation during which the organisms dynamically changes activities of enzymes in the catabolism and regulates the anabolism to adjust the synthesis of macromolecules and reduce growth rate. The result of this concerted action of global regulation differs from the short term regulation during the transient period from batch to stationary phase and the subsequent starvation as well as the behaviour of the organisms during steady state conditions at varying dilution rate during continuous operation.

With the goal to obtain a more in-depth understanding of these complex regulation phenomena and their impact on the flux distribution in the central metabolism experimental work has been initiated which covers three main areas, namely microarray analysis, flux analysis and selected quantitative measurements of metabolites and mRNA via PCR analysis. Results of this work have been already summarized in the papers of Hardiman et al. (2007a) and Lemuth et al. (2008). Part of these results will be presented once more within this chapter to support the tight link between experimental work and computational approach.

11.2 Reconstruction of the Global Regulatory Structure of Carbon Limitation

Current systems biology research in dynamic modeling of the central carbon metabolism of *Escherichia coli* aims at the comprehensive understanding of its global regulation in response to carbon limitation. The long-term goal is of course the support of rational producer strain optimization based on mathematical modeling. Much knowledge about regulatory processes during carbon limitation has accumulated and is available from literature and databases. However, it is not clarified which regulators are dominant under these conditions and thus, which regulators must be considered in a mathematical model. For the assembly of the global regulatory network underlying and for explaining the transient metabolic response to carbon limitation it is necessary to link this *a priori* knowledge with experimental observations in order to identify the relevant components of the network. As mentioned above, observing a single 'ome' alone is not adequate when such complex dynamic processes are being investigated.

The works of Hardiman et al. (2007a) and Lemuth et al. (2008) demonstrate the simultaneous experimental observation of concentrations of signaling molecules (cAMP and ppGpp) and a time series of metabolic flux and transcriptome analyses of *Escherichia coli* K-12 W3110 in a fed-batch cultivation applying a constant feed rate (Fig. 11.1). These omic approaches were employed for the reconstruction of the model structure, focussing on the most relevant parts that must be considered when dynamic modeling the regulatory and metabolic behaviour.

The constant feeding strategy applied, provided an appropriate approach for separating the time-dependent events during the transition from exponential to carbon-limited growth (Fig. 11.1). Both intracellular alarmones ppGpp and cAMP accumulated in large quantities after the onset of nutrient limitation, subsequently declining to basal levels (Hardiman et al. 2007a). The limited supply of the carbon and energy source glucose led to significantly decreasing fluxes in glycolysis, pentose phosphate pathway and biosynthesis, whereas TCA cycle fluxes remained constant (Fig. 11.2a,b). The flux redistribution resulted in an enhanced energy generation in the TCA cycle and consequently, in a 20 % lower biomass yield (Hardiman et al. 2007a). From the correlations of gene expression levels with the metabolic fluxes that were observed (Fig. 11.2), this behaviour can be interpreted as follows and transformed into a model structure (Hardiman et al. 2007a, Lemuth et al. 2008).

The flux through the upper part of glycolysis is favoured whereas the flux through the pentose phosphate pathway is minimized, which is most likely due to the reduced synthesis of *gnd* mRNA. The flux entering the pentose phosphate pathway is used for biosynthesis at the expense of the reflux into the glycolysis pathway, which might be regulated by the RpiA/Rpe split ratio. The reaction rates in the lower glycolysis decrease due to decreasing mRNA levels, thereby providing a sufficient, though minimal, efflux into the pentose phosphate pathway. The regulation of *pfkA*, *fbaA*, *pgk*, *pykF*, *gapA* and *eno* transcription by the Cra regulator protein (*cra* modulon) is suggested to lead to this behaviour (Fig. 11.3). Signalling occurs through

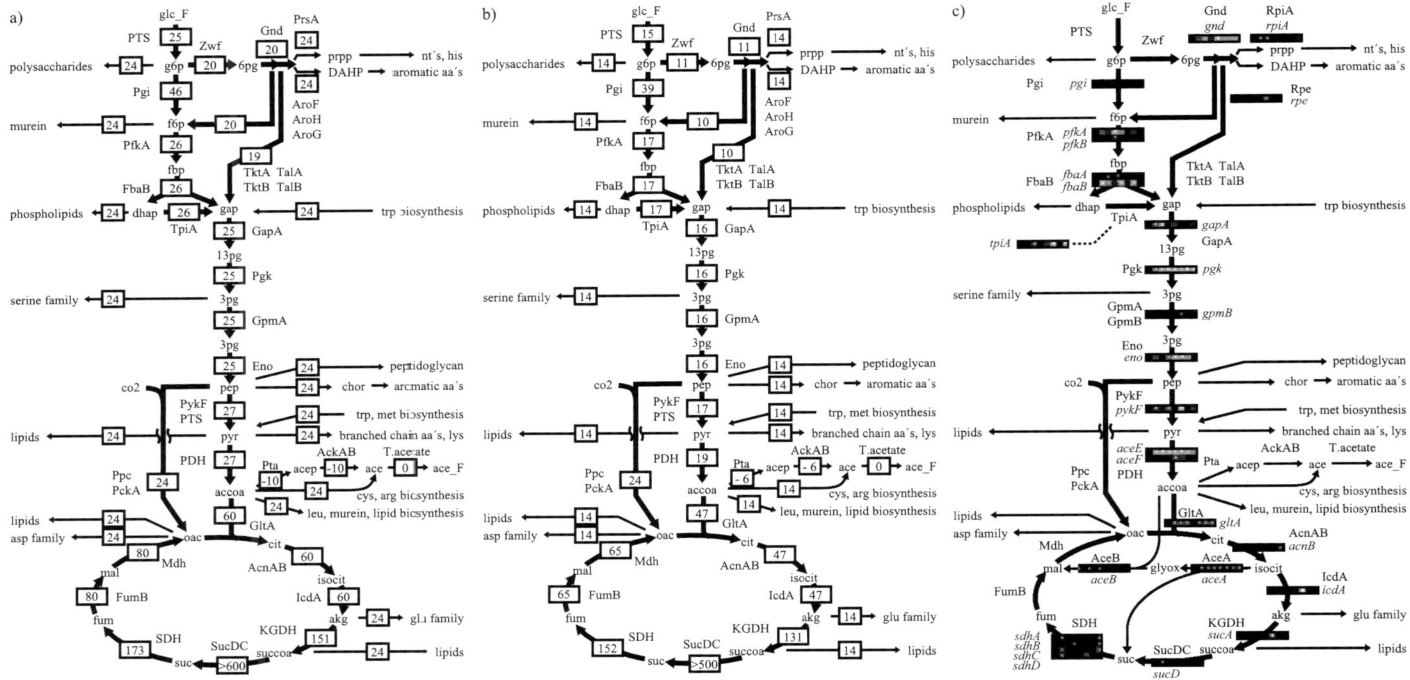

Fig. 11.2 Time series of metabolic flux and DNA microarray analyses of the central carbon metabolism in *E. coli* K-12 W3110. (**a**) and (**b**), metabolic fluxes at $\mu = 0.13\,h^{-1}$ and $\mu = 0.08\,h^{-1}$, respectively, during glucose-limited fed-batch growth applying a constant feed rate. Fluxes are mean values of five independent cultivations determined from stoichiometric metabolite balancing and given as molar percentages of the corresponding fluxes in the reference state during unlimited growth in the batch phase (0.3 h before start of the feed; $\mu = 0.4\,h^{-1}$; see Fig. 11.1). The model used includes also the respiration, biosynthetic pathways and polymerisation reactions. Only those reactions involving metabolites from central carbon metabolism are shown. (**c**) The time courses of the transcript levels are given for the samples T1 to T8 relative to the reference sample in the batch phase (R, see Fig. 11.1). Green: mRNA level is lower than that at the reference state. Red: higher mRNA level. Statistical significance ($p \leq 0.05$) is indicated by white asterisks

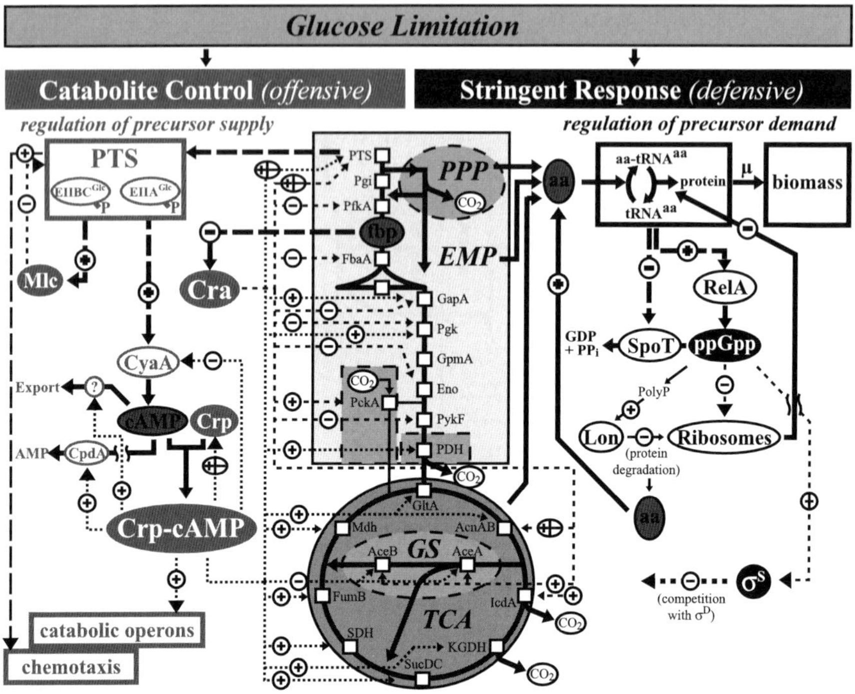

Fig. 11.3 Reconstruction of the global regulatory and metabolic network of carbon limitation. *Left*: Catabolite repression can be seen as an offensive strategy since various catabolic operons are induced, which encode transporters and metabolic pathways for the consumption of sugars other than glucose. Moreover, many genes of the TCA cycle, glyoxylate shunt (GS), PTS system and glycolysis are regulated by the Crp-cAMP regulator complex (*crp* modulon). Additionally, the Cra protein represses genes of glycolysis and activates transcription of GS genes (*cra* modulon). Fbp inactivates the Cra protein. The fbp concentration reflects the availability of extracellular glucose. *Right*: Stringent response is an defensive strategy since it regulates many components of the translational and transcriptional machinery, most prominently, the reduction of rRNA transcription by ppGpp (*relA/spoT* modulon). The dedicated reader is referred to Hardiman et al. (2007a) for a detailed analysis of the major mechanisms that lead to the accumulation of the alarmones cAMP and ppGpp and to the reduction of the fbp concentration during carbon limitation. The major negative feedback regulation mechanisms leading to a resetting of the signals are also discussed therein

the metabolite fructose 1,6-bis(phosphate) (fbp; Fig. 11.3), whiches concentration is proposed to reflect the availability of glucose. A reduction in the enzyme levels of the lower glycolysis concomitantly with the observed decreasing flux levels might be a hint for the control of metabolite concentrations (homeostasis). The carbon flux entering the TCA cycle (influx is enhanced via *gltA* expression) is split into the glyoxylate shunt (GS), the phosphoenolpyruvate(pep)-GS and the full TCA cycle. GS and pep-GS provide a better pep, pyr and oac precursor supply. It is proposed that the global regulation via the *crp* and *cra* modulons is the most relevant in this respect – i.e. the Crp-cAMP regulator complex mainly induces the transcription of the TCA cycle genes, whereas the glyoxylate shunt (GS) genes are regulated by the Cra regulator protein (positive) and the Crp-cAMP complex (negative) (Fig. 11.3).

In summary, the omic approaches reported by Hardiman et al. (2007a) and Lemuth et al. (2008) demonstrate that the substrate is extensively oxidized in the TCA cycle to enhance energy generation. However, the general rate of oxidative decarboxylation within the pentose phosphate pathway and the TCA cycle is restricted to a minimum. Fine regulation of the carbon flux through these pathways, i.e. the EMP/PPP, RpiA/Rpe and TCA/GS/pep-GS split ratios, supplies sufficient precursors for biosyntheses. The network topology regulating the central carbon metabolism provided in (Hardiman et al. 2007a) is novel inasmuch as it comprehensively explains the obtained systems-level data of the metabolic transition from exponential to carbon-limited growth typical of fed-batch processes – considering not only signal transduction, transcriptional regulation and metabolic behaviour but also the resetting of the signals (the two intracellular alarmones cAMP and ppGpp) and the effect of the respective feedback mechanisms (ascribed to catabolite repression and stringent response) on the dynamics in the central carbon metabolism (Fig. 11.3).

Besides the reported correlating transcript levels and metabolic fluxes in the central carbon metabolism, a picture of interesting interconnections between enhancement and attenuation of further cellular functions is drawn in (Lemuth et al. 2008), highlighting the importance of this adaptive behaviour for mathematical modeling and optimizing biotechnical production processes. Most of the physiological rearrangements, if not all of them, can clearly be linked to the regulation of the intracellular availability of precursors and energy, i.e., not only the supply and demand rates, but also the (resulting) concentrations of precursors are discussed to be tightly controlled. This physiologically highly important task is exemplified by the tempting proposal that the global regulation of diverse functions such as chemotaxis, transport and flagellar systems as well as glycolysis, TCA cycle and glyoxylate shunt are interconnected in controlling the availability of the precursor phosphenolpyruvate (pep). This and further major findings of Lemuth et al. (2008) are condensed in the following: (i) A cluster of high-affinity transporters is synthesized, while the activity of medium-affinity transporters is maintained. This is mainly due to their regulation by the Crp-cAMP complex. The glucose flux entering the cell is directed via transporters that do not use pep for phosphorylation. This preserves the pool of this metabolite (homeostasis) and affects the $EIIA^{Glc}{\sim}P$-dependent activation of cAMP synthesis through the enzyme adenylate cyclase (CyaA). (ii) These transport systems in particular depend on a membrane proton gradient for proper function. The expression of the proton gradient-dependent chemotaxis system is reduced, thereby enabling the transport system effectively utilise the energy available. (iii) Cellular growth is regulated predominantly by the stringent response (alarmone ppGpp, *relA/spoT* modulon), however, no extensive induction of the general *rpoS*-dependent response could be observed. This is attributed to the opposing regulation via the *crp* and *relA/spoT* modulons (see also Lapin et al. 2006). It is expected that slow substrate concentration changes do not trigger a strong starvation response Teich et al. (1999). However, other stress responses were detected.

Thus, a model topology has been reconstructed of the global regulation of the *E. coli* central carbon metabolism through the *crp*, *cra* and *relA/spoT* modulons that can be used for mathematical modeling metabolism and regulation (Fig. 11.3). In a second step, physiological functions that are important for precursor and energy availability (transport, chemotaxis, stringent and stress response) are suggested to be implemented as further modules of the mathematical model.

11.3 Basic Principles of Deterministic Modeling the Dynamics of Gene Expression

The development of deterministic models describing the regulation of gene expression (transcription, mRNA degradation and protein biosynthesis) has a long tradition. Already in a 1968 review, Rosen (1968) summarized important methods and approximations essential for modeling and simulation of gene regulatory networks. The majority of the models are similar in mathematical nature and more or less rest upon the concept suggested by Yagil and Yagil (1971) and Yagil (1975). Based on the operon model of Jacob and Monod (1961) these authors illustrated how to derive the probability of transcription initiation if a gene is regulated by a repressor or activation protein.

In the case of negative regulation it is defined as the ratio of the concentration of operators free to be transcribed, c_O, to the total concentration of operators, $c_{O,t}$:

$$\phi_{neg} = \frac{c_O}{c_{O,t}}. \tag{11.1}$$

Accordingly, the ratio of the concentration of activator proteins bound to DNA-binding sites, $c_{A.DNSbs}$, to the total concentration of DNA-binding sites, $c_{DNAbs,t}$, gives the probability:

$$\phi_{pos} = \frac{c_{A.DNSbs}}{c_{DNAbs,t}}. \tag{11.2}$$

The maximal rate of transcription can be achieved for $\phi \rightarrow 1$. In both cases the probability is derived from the equilibrium assumption for the biochemical binding reactions of the regulator protein and its DNA-binding site. This is reasonable because the initiation and the subsequent transcript and peptide elongation occur on different time scales (McClure 1985, Stephanopoulos et al. 1998, Uptain et al. 1997). In case of effectors inhibiting or enhancing the binding activity of regulator proteins (inducers or co-repressors), additional equilibrium reactions can be formulated. Equation (11.3, 11.4) exemplify the inactivation of the repressor protein R by binding the inducer molecule E and binding of the active repressor to the operator DNA sequence O. Equation (11.5) depicts the equilibrium (binding) constants and the derived probability of induction for negative regulation.

$$R + n \cdot E \underset{k_{-1}}{\overset{k_{+1}}{\rightleftharpoons}} R.E_n \tag{11.3}$$

$$R + O \underset{k_{-2}}{\overset{k_{+2}}{\rightleftharpoons}} R.O \tag{11.4}$$

$$\phi_{neg} = \frac{c_O}{c_{O,t}} = \frac{1 + K_1 c_E^n}{1 + K_1 c_E^n + K_2 c_{R,t}} \tag{11.5}$$

with

$$K_1 = \frac{k_{+1}}{k_{-1}} = \frac{c_{R.E_n}}{c_R \cdot c_E^n} \text{ and } K_2 = \frac{k_{+2}}{k_{-2}} = \frac{c_{R.O}}{c_R \cdot c_O}$$

The transcription rate is then obtained from

$$r_{tc,mRNA_i} = r_{tc,\max} \prod_j \phi_j f(\mu) - k_{Degradation}\, c_{mRNA_i} - \mu c_{mRNA_i} \tag{11.6}$$

and the translation rate of the protein of interest is calculated from

$$r_{TL,\mathrm{Protein}_i} = r_{\max,TL} c_{mRNA_i} - \mu c_{\mathrm{Protein}_i}. \tag{11.7}$$

The term $f(\mu)$ considers the impact of the specific growth rate on the transcription rate. Roels (1983) suggested the following form:

$$f(\mu) = \frac{a + b\mu}{a + b\mu_{max}}, \tag{11.8}$$

which reflects the linear dependency between mRNA biosynthesis and the specific growth rate.

The illustrated approach enables modeling of superimposed regulation mechanisms by several regulators and can be extended by the binding of RNA polymerase to the promoter DNA sequence. It is therefore suitable for implementation of gene expression kinetics in large metabolic models.

With increasing amount of knowledge available about the details of catabolite repression (reviewed by Deutscher et al. 2006) more sophisticated models have been developed. Many of these modifications are based on the approach of Lee and Bailey (1984a,b) in which a transcription efficiency is defined as:

$$\eta = \psi_P (1 - \psi_R)(1 + \alpha\psi_A) \tag{11.9}$$

with the fraction of occupied promoters ψ_P, the influence of a repressor $(1 - \psi_R)$ and an activator $(1 + \alpha\psi_A)$. In addition to the comprehensive models suggested by Kremling et al. (2007, 2001, 2000) (Kremling and Saez-Rodriguez 2007, Kremling and Gilles 2001) and Bettenbrock et al. (2006) this approach has been applied by Wong et al. (1997) as well as Van Dien and Keasling (1998) to mention a few.

In a different line of approaches Boolean networks are used for modeling regulatory phenomena. These models have been already introduced in the 1960's by Stuart Kauffman (1969). The conceptual framework of Boolean networks is based on the assumption that binary on/off switches functioning in discrete time steps can describe important aspects of gene regulation (Albert 2004, McAdams and Arkin 1998). In the context with the intended coupling of regulatory and metabolic networks, such a Boolean approach for description of the regulatory network would eventually lead to a hybrid model in which the concentrations of metabolites are expressed as continuous values and connected via enzyme kinetics to describe the dynamics of the metabolic networks described by a system of ODEs.

An alternative option to avoid the computational effort with the hybrid models is to approximate the switch like behavior of the expression with the aid of Hill kinetics. In case of a repression the rate of transcription can be represented by

$$r = r_{\max,transcription} \frac{1}{1 + \left(\frac{c_R}{K_R}\right)^{n_R}}, \tag{11.10}$$

whereas for the event of an activation

$$r = r_{\max,transcription} \frac{1}{1 + \left(\frac{K_A}{c_A}\right)^{n_A}} \tag{11.11}$$

could be an appropriate approximation. A more generic formulation based on the "general" Hill equation suggested by Cornish-Bowden (1995) and Hofmeyr and Cornish-Bowden (1997) for reversible reactions in case of metabolic reactions leads to a very useful rate expression for the concerted action of multiple activators and repressors (Likhoshvai and Ratushny 2007):

$$\frac{dc_{mRNA}^{Targetgene(s)}}{dt} = r_{maca,TC} \frac{k + \sum_{si_1}^{c_{As,1}} \left(\frac{R_{si_1}}{K_{si_1}}\right)^{h_{si_1}} + \sum_{si_1,2}^{c_{As,2}} \frac{R_{si_1}^{h_{s1_1}} R_{si_2}^{h_{si_2}}}{K_{si_{1,2}}^{h_{si_1}+h_{si_2}}} + \cdots + \sum_{si_1,\ldots,si_M}^{c_{As,M}} \frac{\prod_{k=1}^{M} R_{si_k}^{h_{si_k}}}{K_{si_{1\ldots M}}^{\sum_{k=1}^{M} h_{i_k}}}}{1 + \sum_{sj_1}^{c_{Is,As,1}} \left(\frac{R_{sj_1}}{K_{sj_1}}\right)^{h_{sj_1}} + \sum_{sj_1,sj_2}^{c_{Is,As,2}} \frac{R_{sj_1}^{h_{sj_1}} R_{sj_2}^{h_{sj_2}}}{K_{sj_{1,2}}^{h_{sj_1}+h_{sj_2}}} + \cdots + \sum_{sj_1,\ldots,sj_N}^{N} \frac{\prod_{w=1}^{N} R_{si_w}^{h_{si_w}}}{K_{sj_{1\ldots N}}^{\sum_{w=1}^{N} h_{sj_w}}}}. \tag{11.12}$$

Here the binding of regulatory proteins R includes inhibition (binding sites Is) and activation (bindig sites As). Figure 11.4 depicts the application of this equation for an example of joint regulation of two genes through two repressors – and one activator molecule. Starting from the framework of statistical mechanics Bintu et al.

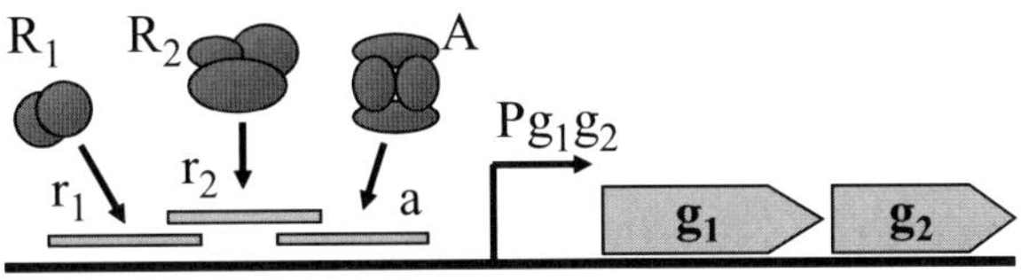

$$\frac{dG_1(t+\tau_{G_1})}{dt} = \frac{dG_2(t+\tau_{G_2})}{dt} = \frac{k_{g12}\cdot\left(\delta_0+\left(\frac{A}{K_a}\right)^{h_a}\right)}{1+\left(\frac{A}{K_a}\right)^{h_a}+\left(\frac{R_1}{K_{r_1}}\right)^{h_{r_1}}+\left(\frac{R_2}{K_{r_2}}\right)^{h_{r_2}}+\frac{(A)^{h_a}\cdot(R_1)^{h_{r_2}}}{K_{ar_1}}}$$

Fig. 11.4 Dynamic modeling of gene expression regulated by two repressors and one activator based on general hill kinetics (Ilya Peshkov, Novosibirsk, Russia: personal communication)

(2005) derived various "regulatory factors" for several different regulatory motifs very similar to the generic structure of Equation (11.12).

For portraying the sigmoid character of the dynamic response alternative approaches are based on generic sigmoidal functions (Weaver et al. 1999), such as

$$f(x) = \frac{1}{1+e^{-x}}. \tag{11.13}$$

With the aid of additional terms representing system and measurement noise, Haixin et al. (2007) have used this approach in connection with Kalman Filtering for the problem of genetic regulatory network inference from time series microarray data.

Another powerful method in the context of sigmoidal functions is built on the conceptual framework of neural networks (Vohradsky 2001a,b). The model has the form

$$\frac{dz_i}{dt} = r_{\max}\frac{1}{1+\exp\left[-\left(\sum_j w_{ij}y_j+b_i\right)\right]} - k_{\deg radation}z_i \tag{11.14}$$

with connection weights w_{ij}, delay parameter b_i and rate constant for degradation k. z_i is the target gene regulated by the genes y_j connected to the target (predictor genes).

The focus of application of most of the aforementioned approaches for dynamic modeling of gene regulatory networks is on network inference based on time series "profiles" of microarray data. A crucial point in the evaluation of the majority of these applications is the missing distinction and the rigorous mathematical description of the two processes of transcription and translation. Using nonlinear stability analysis Hatzimanikatis and Lee (1999) have shown that a combination of gene expression information at the mRNA level and at the protein level is required to describe even simple models of gene networks. This issue is all the more important for coupling gene regulatory networks with metabolic networks because at least the output of the regulatory network is linked at the protein level to change enzyme concentrations in the metabolic rate expressions. If balance equations for the translation process are neglected, the overall dynamics are corrupted by a mixture of characteristic time constants for transcription and translation.

11.4 Dynamic Model for the Intra-actions Between the Regulatory and Central Metabolic Networks of *Escherichia coli*: Translation of Sequence Information into Kinetic Parameters

The focus of this chapter is on the dynamic modeling of the intra-actions between the regulatory and metabolic networks depicted in Figs. 11.2 and 11.3. The ultimate goal of this approach is to quantitatively describe the dynamical changes of traffic patterns and variations in flux distributions in response to the environment changes caused by the diminishing supply of carbon and energy source glucose. The kinetics to describe the dynamics of the regulation phenomena is modeled in terms of probabilities of transcription as described in Section 11.3. The approach is based on a translation of gene sequence information into parameters of binding constants for the individual regulator protein-DNA-binding site interaction of interest. The methodology will be exemplified for the Cra-modulon, illustrated in Fig. 11.3.

The usage of Equations (11.1, 11.2, 11.3, 11.4, and 11.5) for modeling gene expression in large metabolic networks as illustrated in Fig. 11.3 requires the availability of the parameters K_1 and K_2 from literature, data bases or their identification from experimental observations. For estimation of the binding constant K_1 for the reaction between the regulatory protein and its effector E (Equation 11.3) this is of course feasible. However, $K_{2,i}$ has to be determined for each individual gene i coding for the enzymes or regulatory proteins being components of the network. For large networks or large regulons/modulons such an approach is not practicable because of the experimental effort. This is one of the reasons that verification of such models is most often dominated by identification methods for the estimation of large sets of parameters. To circumvent this kind of problems we therefore choose an approach in which the individual binding constants are estimated from the gene sequence information of the DNA-binding side (Hardiman et al. 2007b).

11.4.1 Decomposition of the Binding Reaction

For the purpose of derivation of $K_{2,i}$ from the DNA-binding site sequence, the regulator protein R is first assumed to bind to the mononucleotides, $b \in \{A, C, G, T\}$, of the binding site sequence and that these interactions are independent and additive according to Stormo (1988, 1990):

$$R + b \rightleftharpoons R.b \tag{11.15}$$

Assuming again an equilibrium reaction, the binding constant is proportional to the ratio of the bound pool to the unbound pool of bases:

$$K_b = \frac{c_{R.b}}{c_R \cdot c_b} \propto \frac{c_{R.b}}{c_b} = \frac{f_b}{p_b} \tag{11.16}$$

Equation (11.16) also illustrates that this ratio is equal to the ratio of the frequency at which the base b occurs at the considered position in the DNA-binding site sequence, f_b, to the frequency of this base in the genome of the considered organism, p_b, which was proposed by Stormo (1988, 1990). Considering that the binding to each nucleotide of the sequence is assumed to be independent, the binding constant for the total DNA-binding site, K_2, can be formulated as

$$K_2 = \prod_n K_{b,n} \propto \prod_n \frac{f_{b,n}}{p_b} \tag{11.17}$$

where n corresponds to the position of the nucleotide b in the sequence. Various scientific groups have investigated this relationship and found reasonable correlations between calculated and experimentally determined binding affinities or the equivalent free energy of binding (Equation 11.18). For instance, Berg and von Hippel ((1987)) developed a statistical-mechanical theory based on the assumption that specific DNA sequences have been selected according to their protein binding affinity and that all sequences that show equal affinities are equally likely to occur in the genome. The theory Berg and von Hippel (1987) was able to predict the correlation between the activities ($k_2 K_B$ values) of *E. coli* promoter sequences assuming that nucleotides at different positions in the promoter sites contribute independently to their activities. Many more contributions to the field demonstrated that there is a strong linear relation between base frequency and binding strength (Berg and von Hippel 1988, Fields et al. 1997, Stormo and Fields 1998, Takeda et al. 1989). For an overview the dedicated reader is referred to (Stormo 1990, 2000).

$$\Delta G_b = -RT \ln K_b \propto -\ln\left(\frac{f_b}{p_b}\right) \tag{11.18}$$

The findings of these authors are not surprising, because Equations (11.17, 11.18) simply express that highly conserved DNA sequences are bound stronger than less conserved ones by the respective regulator protein. Therefore, Equations (11.17, 11.18) provide a simple and valuable tool for the quantitative evaluation of any DNA-binding site sequence with respect to a reference sequence.

11.4.2 Application to the cra *Regulon of Escherichia coli*

The regulator protein Cra is a major component of the global regulation of the metabolic fluxes in glycolysis (EMP), the TCA cycle and the glyoxylate shunt (GS) in glucose-limited fed-batch processes of *E. coli* (see Section 11.2 and Fig. 11.3).

Binding of the Cra protein to the DNA-binding site of the transcription units i ($DNAbs_i$; Equation 11.21) of the *cra* modulon is inhibited by high concentrations of fructose 1,6-bis(phosphate) (fbp; Equation 11.19).

$$Cra + fbp \overset{K_1}{\rightleftharpoons} Cra\,.\,fbp \tag{11.19}$$

$$4\,Cra + DNAbs_i \overset{K_{2,i}}{\rightleftharpoons} Cra_4\,.\,DNAbs_i \tag{11.20}$$

$$\phi^{neg}_{Cra.DNAbs,i} = \frac{c_{DNAbs,i}}{\left(c_{DNAbs,i}\right)_{total}} = \frac{1}{1 + K_{2,i}\left(\frac{(c_{Cra})_{total}}{1 + K_1\,c_{fbp}}\right)^4} \tag{11.21}$$

$$\phi^{pos}_{Cra.DNAbs,i} = \frac{c_{Cra_4.DNAbs,i}}{\left(c_{DNAbs,i}\right)_{total}} = 1 - \phi^{neg}_{Cra.DNAbs,i} \tag{11.22}$$

$$r_{tc,mRNA_i} = r_{tc,\max}\prod_j \phi_j - k_{Degradation}c_{mRNA_i} - \mu c_{mRNA_i} \tag{11.23}$$

The probability of transcription initiation, ϕ, is determined by the fraction of unbound (Equation 11.21) or bound (Equation 11.22) DNA-binding sites when transcription is repressed or activated, respectively.

11.4.3 Comparison Between Model Prediction and Experimental Observations

Figure 11.5 illustrates the mRNA concentrations of central carbon metabolism genes measured using qPCR analysis during glucose-limited fed-batch cultivation of *E. coli* (see Fig. 11.1) as well as concentrations predicted by the model described by Equations (11.19, 11.20, 11.21, 11.22, and 11.23). The genes *eno* (encoding enolase), *pfkA* (6-phosphofructokinase I) and *pykF* (pyruvate kinase I) are known to be regulated by the Cra regulator protein (see Section 11.2). The repression of their transcription (Fig. 11.5) results in a strong decrease of the respective mRNA concentrations.

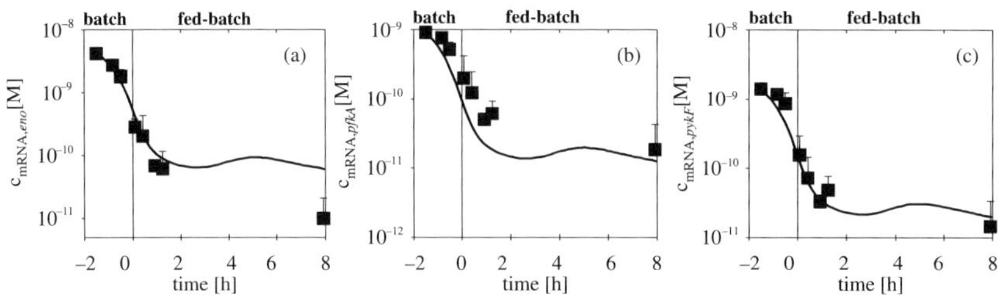

Fig. 11.5 mRNA concentrations during glucose limited fed-batch cultivation of *E. coli* K-12 W3110. The concentrations of mRNA (■) were determined by qPCR analysis (standard deviation, 3 independent samples). Simulation data are indicated by solid lines. (**a**) *eno* mRNA (encoding enolase), (**b**) *pfkA* mRNA (6-phosphofructokinase I) and (**c**) *pykF* mRNA (pyruvate kinase I)

Obviously, the DNA-binding activity of the Cra protein is high due to the low concentration of fructose 1,6-bis(phosphate) (fbp) during the fed-batch process (Fig. 11.6). The strong decrease in fbp concentration (Fig. 11.6a,b) can be attributed to the limited carbon supply (Section 11.2). However, according to Fig. 11.6a the concentration of fbp apparently increases after two hours of fed-batch cultivation, when the experimental data is related to the biomass concentration. Only when

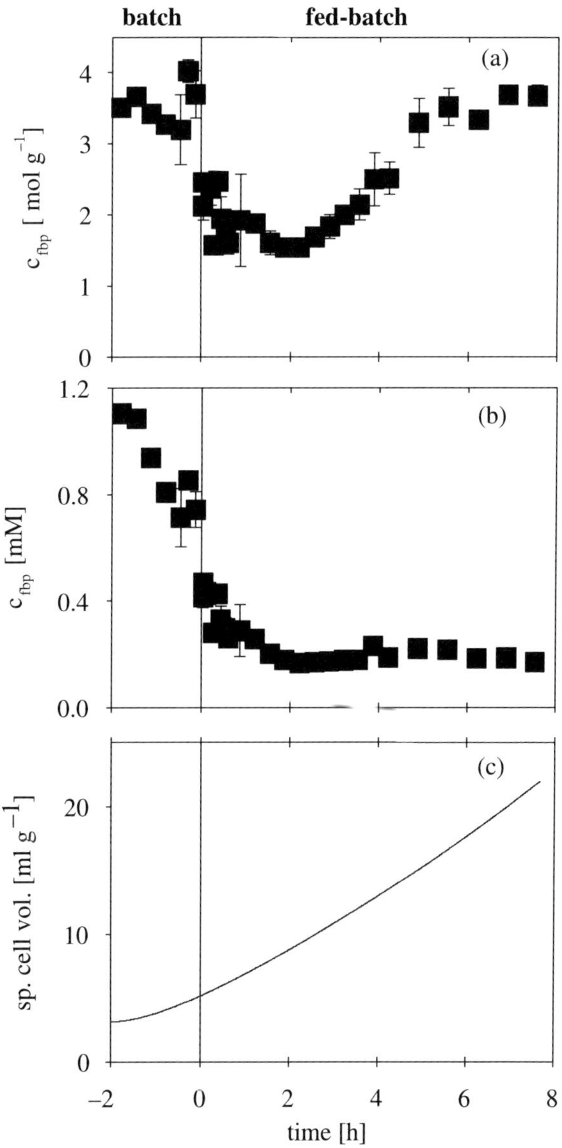

Fig. 11.6 Fructose 1,6-bis(phosphate) concentration during glucose-limited fed-batch cultivation of *E. coli* K-12 W3110. Concentrations were determined after quenching and extraction using perchloric acid as published in Hardiman et al. (2007a). (**a**) Fbp concentration related to biomass [μmol (g dry weight)$^{-1}$]. (**b**) Fbp concentration related to the cell volume [mmol (l cytosol)$^{-1}$] that is obtained by deviding the concentration given in (**a**) by (**c**) the specific cell volume v_x [l cytosol (g dry weight)$^{-1}$], and which is in turn approximated using the growth rate-dependent function $\hat{v}_X = \dfrac{0.4860 \cdot 2^{(1.144\hat{\mu})}}{-0.636 + 0.635 \cdot 2^{(0.718\hat{\mu})}}$ (Hardiman et al. 2007a)

the growth rate-dependent variation of the cell volume is considered a meaningful result may be obtained from the data (Fig. 11.6b). The time profile of the molar intracellular concentration given in [mmol (l cytosol)$^{-1}$] enables to explain the transcriptome and metabolic flux data as described in Section 11.2. That is, the persisting low concentration of fbp leads to the repression of glycolysis genes by the Cra regulator protein and activation of transcription of glyoxylate shunt genes.

The model predicts the mRNA concentration satisfactorily during the batch and the beginning of the fed-batch process and also at a later process phase where the growth rate is very low (Figs. 11.1 and 11.6). Note, that the model used for the simulations differs from the one introduced in Section 11.3. Equation (11.23) does not take into account the growth rate dependence of transcription initiation, whereas Equation (11.6) considers the impact of the specific growth rate on the transcription rate. Although the Equations (11.19, 11.20, 11.21, 11.22, and 11.23) are sufficient for a rough simulation of the mRNA concentrations (Fig. 11.5), the extension of the model by growth rate dependent variables and further regulons/modulons is needed. This is expected to make an important contribution to the understanding the global regulation of the central carbon metabolism during carbon limitation.

11.5 Conceptual Framework for Dynamic Models of Metabolic Networks of *E. coli* Suitable for Links to Regulatory Networks

A multitude of approaches is available for dynamic modeling of the metabolism of *E. coli*. Here, we shall limit our discussion on continuous and deterministic models, which are derived by considering the balance equations of the individual metabolites and can be represented in the compact form:

$$\frac{d\mathbf{x}}{dt} = \mathbf{Nr}\left(\mathbf{x}\left(t\right), \mathbf{P}\right) - \mu\mathbf{x}. \tag{11.24}$$

N is the $m\ x\ n$ stoichiometric matrix an **r** is the n-dimensional rate vector.

Based on dynamic measurements of intra- and extracellular metabolites in response to a perturbation of a continuous culture with a pulse of glucose Chassagnole et al. (2002) derived a rigorous dynamic model of the central metabolism of *E. coli* (Fig. 11.7). The model is based on kinetic rate expressions for the individual enzymes, the original structures of which have been derived from investigations with isolated enzymes at *in vitro* conditions. The key to afterwards generate the dynamic *in vivo* model is, to extract the kinetic parameters of the biochemical reactions from the *in vivo* metabolite measurements and, as such, considering the reactions in their "systemic" context (Reuss et al. 2007).

To describe the dynamic systems behaviour, deterministic kinetic rate equations of the form

$$r_i = r_{max,i}\, f\left(\mathbf{c}, \mathbf{p}\right) \tag{11.25}$$

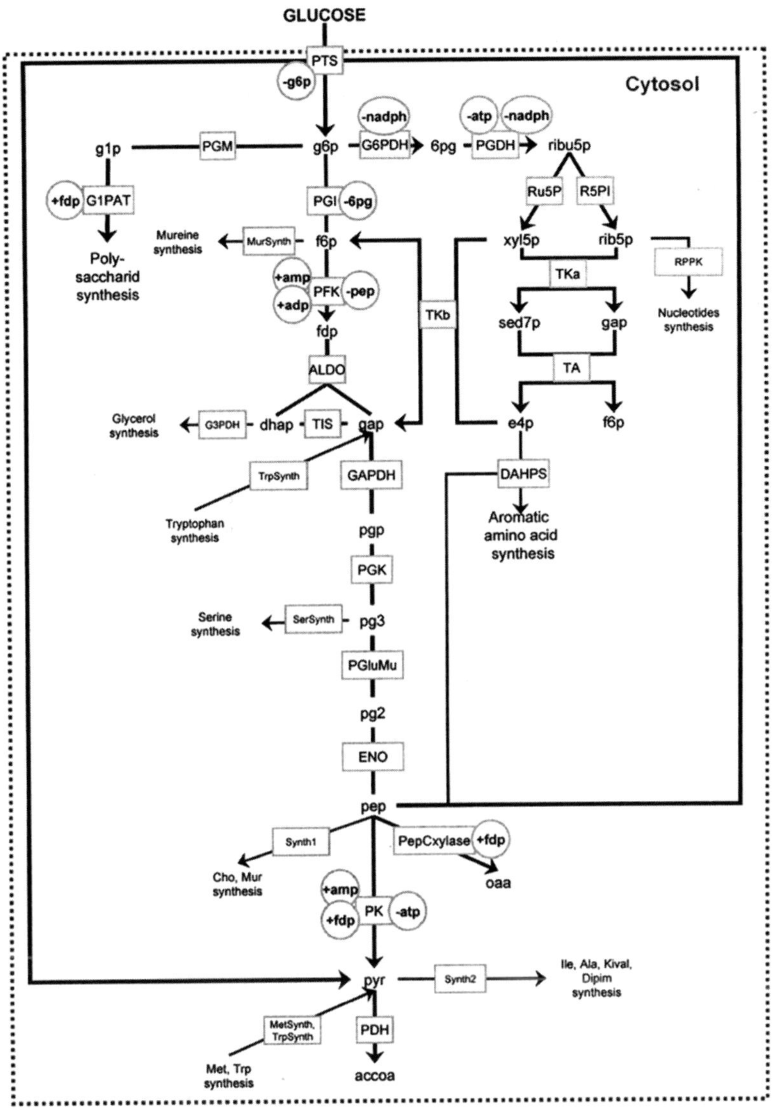

Fig. 11.7 Structure of the metabolic model of glycolysis and pentose phosphate pathway in *Escherichia coli* (Chassagnole et al. 2002)

are formulated, where the capacity of the reaction is characterized by its maximal rate and the kinetic function f represents the kinetic properties of the reaction. Substrates, products and other metabolic effectors influencing the rate of the reaction are represented by the state vector of metabolite concentrations **c**. The parameters of the reaction are summarized in the vector **p**.

If the maximal rate of reaction can be assumed to be proportional to the concentration of the enzyme, Equation (11.25) provides a simple way to integrate the output of the regulatory network with respect to the concentration of the individual enzymes.

The first step to embed the behaviour of the subsystem into the metabolic network as a whole is provided by the estimation of the maximal rates of the individual reactions. Applying the rate Equation (11.25) to the steady state leads to

$$\tilde{r}_{\max,i} = \frac{r_{i,steady\ state}}{f\left(\mathbf{c}_{steady\ state}, \mathbf{p}\right)}. \tag{11.26}$$

Let us assume that reaction rate $\tilde{r}_i$ at steady has been estimated from metabolic flux analysis. Let us further assume that a first estimate of the structure of the kinetics as well as the parameter vector **p** is available from *in vitro* measurements. If the components of the concentration vector **c** influencing the rate of the reaction have been measured at steady state, the unknown maximal rates are given as depicted in Equation (11.26).

If the stoichiometric model used for metabolic flux analysis has a genome scale or a metabolic submodel in case of ^{13}C analysis is linked to such a model (Schaub et al. 2008), the maximal rates estimated from Equation (11.26) are invariant to the scale of the submodule used for the dynamic model. As such, these rates are intrinsic properties of the system as a whole and in a meaningful way only depend on the physiological state of the system. Further details of the strategy to identify the *in vivo* kinetics from the measured stimulus-response date are discussed in the original papers (Chassagnole et al. 2002, Rizzi et al. 1997) and summarized in a review (Reuss et al. 2007).

The model structure depicted in Fig. 11.7 accounts for the enzymatic rate expressions for the glycolytic enzymes and therefore allows for connection of the most important output signals of the Cra and Crp modulon (Fig. 11.3). Apart of the necessary model extension for incorporation of TCA and glyoxylate shunt reactions, however, interactions between the regulatory and metabolic networks exceed the central metabolism by far. Particularly the precursor demand via e. g. amino acid synthesis and the subsequent polymerisation reactions are regulated through the alarmone ppGpp (Fig. 11.3) and demand further extension of the model structure.

Aside from the possibility to assign the large number of additional reactions with mechanistic enzyme kinetics, which is an excessively laborious and time consuming approach, conceptual frameworks based on canonical formulations of rate expressions leading to less detailed large- scale models may prove to be useful. Such an approach has been introduced by Reuss et al. (2007) and successfully applied for a large-scale dynamic model for *E. coli*. The dynamic model follows from the reaction network model of *Escherichia coli* introduced by Chassagnole et al. (2002). The network comprises both catabolic and anabolic routes with protein, DNA, RNA, polysaccharides, murein, and lipids building up biomass. Sequential reaction steps and parallel routes are lumped. With 129 reactions, 133 balanced metabolites, and seven conserved moieties, the degree of freedom of the null-space of the network is fixed to $129–133 + 7 = 3$. Additional informations regarding inhibition and activation (metabolic regulation) have been gathered from the MetaCyc data base (www.metacyc.org, (Caspi et al. 2006)). The kinetic behaviour of the individual reactions is assigned according to the universal linlog approach (Visser and Heijnen 2003, Visser et al. 2004, 2000):

$$r = J\frac{c_E}{c_E^0}\left(1+\underbrace{\sum_i \varepsilon_{S,i}\ln\frac{c_{S,i}}{c_{S,i}^0}}_{\text{substrates}}+\underbrace{\sum_j \varepsilon_{P,j}\ln\frac{c_{P,j}}{c_{P,j}^0}}_{\text{products}}+\underbrace{\sum_k \varepsilon_{A,k}\ln\frac{c_{A,k}}{c_{A,k}^0}}_{\text{activators}}+\underbrace{\sum_l \varepsilon_{I,l}\ln\frac{c_{I,l}}{c_{I,l}^0}}_{\text{inhibitors}}\right). \tag{11.27}$$

The variables are defined to the relative reference steady state, with concentration levels state c^0, fluxes J^0, and enzyme level c_E^0. The parameters are the elasticity coefficients

$$\varepsilon_M = \frac{c_M}{r}\left(\frac{\partial r}{\partial c_M}\right). \tag{11.28}$$

In total the network holds 921 kinetic parameters (elasticities). The dynamic simulation of the non-linear and stiff system of differential equations was performed with the aid of the extrapolation solver LIMEX from the Konrad-Zuse-Centre for Information Technology in Berlin (Ehrig et al. 1999). For estimation of the parameters the evolutionary algorithm developed by the Computer Science Department of the University of Tuebingen (Streichert and Ulmer 2005) has been applied. Results of the comparisons between model simulations and experimental observation from stimulus response experiments in which a pulse of glucose is added to the steady state of a continuous culture have been presented by Reuss et al. (2007).

One key to understanding how these large scale models do compare with dynamic models based on mechanistic rate expression is to carefully examine the differences between the simulation results of the two approaches. Visser et al. (2000) compared the outcome of the linlog approach with the dynamic model of Chassagnole et al. (2002). These authors noted a reasonable agreement for not to large dynamic perturbations with respect to the external glucose concentration. An important observation from this comparison and associated identification of the elasticity coefficients in Equation (11.27) concerns the expected behaviour of the reversible near-equilibrium reactions in the glycolysis. First, the individual elasticity coefficients of such reversible near-equilibrium reactions are not independent. Furthermore, it can be easily shown that the values of the elacticities must be very high and, in consequence, the flux control coefficient tends to zero. In essence then, these reactions are suited candidates for model reduction.

The issue of this model reduction should be always addressed in the context of the purpose of the model as emphasized in the beginning of this chapter. A first, well-proven concept for model reduction in metabolic engineering is based on the time hierarchy of the metabolism. The kernel of this method is a model analysis, which considers the eigenvalues and eigenvectors of the Jacobian associated to the dynamic model (Heinrich and Schuster 1996). The application of this time-scale separation for the Cassagnole model (Chassagnole et al. 2002) results in assumptions of quasi-steady state conditions for 11 eigenvectors possessing the highest values.

The result of this reduction, which shows reasonable agreement between the dynamic response of the original and reduced model, yields, however, a differential-algebraic system. Because the algebraic equations do not allow an explicit analytical solution it is necessary to resort to advanced and efficient solver for differential-algebraic systems.

As a promising alternative to the modal analysis we employed a sensitivity analysis based on the flux control coefficients (Lapin et al. 2006). These coefficients relate the fractional change of the steady state fluxes to the infinitesimal changes in the total enzyme concentrations (Heinrich and Schuster 1996). From the hierarchy of these flux control coefficients predicted from the original model reactions with the highest values in relation to the flux control coefficient of the glucose uptake were selected. The resulting network is depicted in Fig. 11.8 Because of low flux control

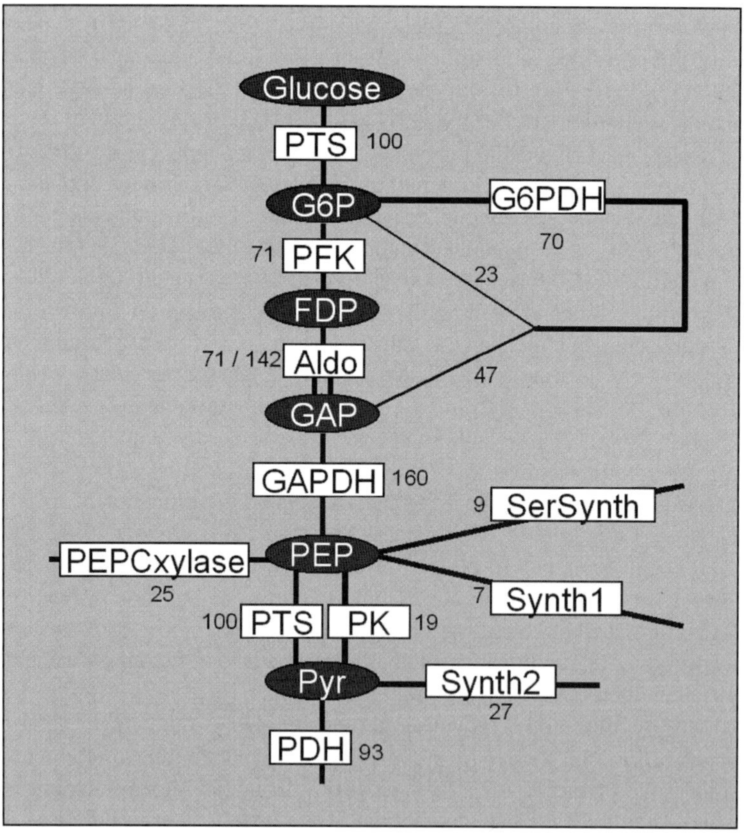

Fig. 11.8 Reduced metabolic network model for the sugar uptake system, glycolysis and pentose phosphate pathway. Reduction of the original model (Chassagnole et al. 2002) is based on the hierarchy of flux control coefficients. The numbers alongside the enzymes depict the metabolic fluxes related to glucose uptake rate 100

coefficients the reactions for the phosphoglucoisomerase, the triose phosphate isomerase, the phosphoglycerate kinase, the phosphoglyceromutase and the enolase could be neglected. The low flux control coefficients result from the reversibility of aforementioned reactions leading to very high values of the elasticity coefficient. For the purpose of model reduction a rapid equilibrium is assumed for these reactions and the dynamics of the metabolites are linked via equilibrium constants.

To summarize the efforts for designing the dynamic model for the central metabolism it is important to emphasize that systems biology modeling of these networks should not be restricted to the task of aggregating and integrating quantitative information on individual enzyme kinetics to "whole-cell models". An equally important challenge is to reduce the complexity and to tailor the model structure for the intended application. Thus, depending on its specific objectives, a model may involve details at different levels.

11.6 Conclusions

The framework for integration of regulatory and metabolic networks provides significant insights on the dynamic response of microorganisms to perturbations of the environmental condition with characteristic times relevant for variations in gene expression. This issue is of particular importance for process operations with dynamic variations in the supply of the carbon and energy source with high relevance for high cell density fermentations. The importance of these regulation phenomena in response to increasing carbon limitation is not restricted to the catabolism of the cell. The strong impact on anabolic reactions (Fig. 11.3) leads to serious variation of the protein expression dynamics with consequences on specific productivities in case of production of recombinant proteins. Future work in our group aims at the extension of integration of regulatory and metabolic networks for these important anabolic phenomena based on dynamic models for protein and ribosome synthesis linked to precursor supply from the central metabolism (Arnold et al. 2005, Elf and Ehrenberg 2005, Elf et al. 2005, Götz and Reuss 1997).

As far as the integration of regulatory networks with modules of the central carbon metabolism is concerned the main contribution of this chapter arises from the fact that a plausible conceptual framework has been developed which enable us to link existing dynamic models for the metabolism with simple models for regulation of transcription and translation of important target enzymes. The approach contains a concise method for the formulation of gene expression. It is demonstrated how the necessary model parameters regarding the gene regulation, i.e. the binding constants of regulator proteins to the DNA-binding site of the individual genes of the regulon, can be derived from the DNA sequence of the sites and minimal literature information.

The overall approach may also serve as an example of how to successfully bridge the top down and bottom up approach for the purpose of modeling and simulation in systems biology. After application of top down analysis for identification of the target genes in the central metabolism, the modeling cycle of the bottom up approach is initiated. This includes quantitative measurements of concentrations of key compounds such as single mRNA molecules, metabolites and even incorporation of "reductionistic" sequence information. This quantitative information at the compound level is afterwards used for the verification of the dynamic model. The ultimate goal of such a hybrid approach is that the characterization of the behavior of the parts of the system should be consistent with the expected and/or observed behavior of the system as a whole.

References

Albert R (2004) Boolean modeling of genetic regulatory networks. In: (ed) Complex Networks, Springer, Berlin, Heidelberg

Arnold S, Siemann-Herzberg M, Schmid J et al. (2005) Model-based inference of gene expression dynamics from sequence information. Adv Biochem Eng Biotechnol 100:89–179

Artsimovitch I, Patlan V, Sekine S et al. (2004) Structural basis for transcription regulation by alarmone ppGpp. Cell 117(3):299–310

Bailey JE (1998) Mathematical modeling and analysis in biochemical engineering: past accomplishments and future opportunities. Biotechnol Prog 14(1):8–20

Barker MM, Gaal T, Gourse RL (2001a) Mechanism of regulation of transcription initiation by ppGpp. II. Models for positive control based on properties of RNAP mutants and competition for RNAP. J Mol Biol 305(4):689–702

Barker MM, Gaal T, Josaitis CA et al. (2001b) Mechanism of regulation of transcription initiation by ppGpp. I. Effects of ppGpp on transcription initiation *in vivo* and *in vitro*. J Mol Biol 305(4):673–88

Berg OG, von Hippel PH (1987) Selection of DNA binding sites by regulatory proteins. Statistical-mechanical theory and application to operators and promoters. J Mol Biol 193(4):723–50

Berg OG, von Hippel PH (1988) Selection of DNA binding sites by regulatory proteins. II. The binding specificity of cyclic AMP receptor protein to recognition sites. J Mol Biol 200(4): 709–23

Bettenbrock K, Fischer S, Kremling A et al. (2006) A quantitative approach to catabolite repression in *Escherichia coli*. J Biol Chem 281(5):2578–84

Bintu L, Buchler NE, Garcia HG et al. (2005) Transcriptional regulation by the numbers: models. Curr Opin Genet Dev 15(2):116–24

Braeken K, Moris M, Daniels R et al. (2006) New horizons for (p)ppGpp in bacterial and plant physiology. Trends Microbiol 14(1):45–54

Cashel M, Gentry DR, Hernandez VJ et al. (1996) The stringent response. In: Neidhardt FC, et al. (ed) *Escherichia coli* and *Salmonella*: cellular and molecular biology, American Society for Microbiology Press, Washington DC

Caspi R, Foerster H, Fulcher CA et al. (2006) MetaCyc: a multiorganism database of metabolic pathways and enzymes. Nucleic Acids Res 34(Database issue):D511–6

Casti JL (1992a) Reality rules. I Picturing the world in mathematics. The fundamentals. John Wiley & Sons, New York, Chichester, Brisbane, Toronto, Singapore

Casti JL (1992b) Reality rules. II Picturing the world in mathematics. The frontier. John Wiley & Sons, New York, Chichester, Brisbane, Toronto, Singapore

Chassagnole C, Noisommit-Rizzi N, Schmid JW et al. (2002) Dynamic modeling of the central carbon metabolism of *Escherichia coli*. Biotechnol Bioeng 79(1):53–73

Cornish-Bowden A (1995) Fundamentals of enzyme kinetics. Portland Press Limited, London

Deutscher J, Francke C, Postma PW (2006) How phosphotransferase system-related protein phosphorylation regulates carbohydrate metabolism in bacteria. Microbiol Mol Biol Rev 70(4):939–1031

Dunn IJ, Mor JR (1975) Variable-volume continuous cultivation. Biotechnol Bioeng 17(12): 1805–22

Ehrig R, Nowak U, Oeverdieck L et al. (1999) Advanced extrapolation method for large scale differential algebraic problems. High performance scientific and engineering computing. In: H.-J. Bungartz et al. (eds) Computational Science and Engineering, Springer, Berlin

Elf J, Ehrenberg M (2005) Near-critical behavior of aminoacyl-tRNA pools in *E. coli* at rate-limiting supply of amino acids. Biophys J 88(1):132–46

Elf J, Paulsson J, Berg O et al. (2005) Mesoscopic kinetics and its applications in protein synthesis. In: Alberghina FA, Westerhoff HV (eds) Systems Biology, Springer, Berlin, Heidelberg

Fields DS, He Y, Al-Uzri AY et al. (1997) Quantitative specificity of the Mnt repressor. J Mol Biol 271(2):178–94

Götz P, Reuss M (1997) Dynamics of microbial growth: modeling time delays by introducing a polymerization reaction. J Biotechnol 58(2):101–114

Haixin W, Lijun Q, Dougherty E (2007) Modeling genetic regulatory networks by sigmoidal functions: a joint genetic algorithm and Kalman filtering approach. In Third International Conference on Natural Computation (ICNC), pp 324–8

Hardiman T, Siemann-Herzberg M, Reuss M (2007b) Derivation of kinetic parameters for coupled regulatory and metabolic network modeling from DNA-binding site sequences. In Foundations in Systems Biology in Engineering (FOSBE), Conference Proceedings, Stuttgart, Germany, pp 255–9

Hardiman T, Lemuth K, Keller MA et al. (2007a) Topology of the global regulatory network of carbon limitation in *Escherichia coli*. J Biotechnol 132(4):359–74

Hatzimanikatis V, Lee KH (1999) Dynamical analysis of gene networks requires both mRNA and protein expression information. Metab Eng 1(4):275–81

Heinrich R, Schuster S (1996) The regulation of cellular systems. Chapman & Hall, New York

Hengge-Aronis R (2002) Signal transduction and regulatory mechanisms involved in control of the σ^S (RpoS) subunit of RNA polymerase. Microbiol Mol Biol Rev 66(3):373–95

Hewitt CJ, Nebe-Von-Caron G (2001) An industrial application of multiparameter flow cytometry: assessment of cell physiological state and its application to the study of microbial fermentations. Cytometry 44(3):179–87

Hewitt CJ, Nebe-Von Caron G, Axelsson B et al. (2000) Studies related to the scale-up of high-cell-density *E. coli* fed-batch fermentations using multiparameter flow cytometry: effect of a changing microenvironment with respect to glucose and dissolved oxygen concentration. Biotechnol Bioeng 70(4):381–90

Hewitt CJ, Nebe-Von Caron G, Nienow AW et al. (1999) Use of multi-staining flow cytometry to characterise the physiological state of *Escherichia coli* W3110 in high cell density fed-batch cultures. Biotechnol Bioeng 63(6):705–11

Hofmeyr JH, Cornish-Bowden A (1997) The reversible Hill equation: how to incorporate cooperative enzymes into metabolic models. Comput Appl Biosci 13(4):377–85

Hogema BM, Arents JC, Bader R et al. (1998) Inducer exclusion in *Escherichia coli* by non-PTS substrates: the role of the PEP to pyruvate ratio in determining the phosphorylation state of enzyme IIA^{Glc}. Mol Microbiol 30(3):487–98

Hogg T, Mechold U, Malke H et al. (2004) Conformational Antagonism between Opposing Active Sites in a Bifunctional RelA/SpoT Homolog Modulates (p)ppGpp Metabolism during the Stringent Response. Cell 117(1):57–68

Jacob F, Monod J (1961) Genetic regulatory mechanisms in the synthesis of proteins. J Mol Biol 3:318–56

Jensen KF, Pedersen S (1990) Metabolic growth rate control in *Escherichia coli* may be a consequence of subsaturation of the macromolecular biosynthetic apparatus with substrates and catalytic components. Microbiol Rev 54(2):89–100

Jishage M, Kvint K, Shingler V et al. (2002) Regulation of σ factor competition by the alarmone ppGpp. Genes Dev 16(10):1260–70

Kauffman S (1969) Homeostasis and differentiation in random genetic control networks. Nature 224(5215):177–8

Kremling A, Bettenbrock K, Gilles ED (2007) Analysis of global control of *Escherichia coli* carbohydrate uptake. BMC Syst Biol 1(1):42

Kremling A, Bettenbrock K, Laube B et al. (2001) The organization of metabolic reaction networks. III. Application for diauxic growth on glucose and lactose. Metab Eng 3(4):362–79

Kremling A, Gilles ED (2001) The organization of metabolic reaction networks. II. Signal processing in hierarchical structured functional units. Metab Eng 3(2):138–50

Kremling A, Jahreis K, Lengeler JW et al. (2000) The organization of metabolic reaction networks: a signal-oriented approach to cellular models. Metab Eng 2(3):190–200

Kremling A, Saez-Rodriguez J (2007) Systems biology - an engineering perspective. J Biotechnol 129(2):329–51

Lapin A, Schmid J, Reuss M (2006) Modeling the dynamics of *E. coli* populations in the three-dimensional turbulent field of a stirred-tank bioreactor – A structured-segregated approach. Chem Eng Sci 61(14):4783–4797

Lee SB, Bailey JE (1984a) Genetically structured models for *lac* promoter-operator function in the chromosome and in multicopy plasmids: *lac* promoter function. Biotechnol Bioeng 26(11):1383–9

Lee SB, Bailey JE (1984b) Genetically structured models for *lac* promoter-operator function in the *Escherichia coli* chromosome and in multicopy plasmids: *lac* operator function. Biotechnol Bioeng 26(11):1372–82

Lee SY (1996) High cell-density culture of *Escherichia coli*. Trends Biotechnol 14(3): 98–105

Lemuth K, Hardiman T, Winter S, Pfeiffer D, Keller MA, Lange S, Reuss M, Schmid RD, Siemann-Herzberg M (2008) Global transcription and metabolic flux analysis of *Escherichia coli* in glucose-limited fed-batch cultivations. Appl Environ Microbiol 74:7002–15

Lengeler JW, Drews G, Schlegel HG (1999) Biology of the prokaryotes. Georg Thieme Verlag, Stuttgart

Likhoshvai V, Ratushny A (2007) Generalized hill function method for modeling molecular processes. J Bioinform Comput Biol 5(2B):521–31

McAdams HH, Arkin A (1998) Simulation of prokaryotic genetic circuits. Annu Rev Biophys Biomol Struct 27:199–224

McClure WR (1985) Mechanism and control of transcription initiation in prokaryotes. Annu Rev Biochem 54:171–204

Mechold U, Murphy H, Brown L et al. (2002) Intramolecular regulation of the opposing (p)ppGpp catalytic activities of Rel_{Seq}, the Rel/Spo enzyme from *Streptococcus equisimilis*. J Bacteriol 184(11):2878–88

Neidhardt FC, Savageau MA (1996) Regulation Beyond the Operon. In: Neidhardt FC, et al. (eds) *Escherichia coli* and *Salmonella*: cellular and molecular biology, American Society for Microbiology Press, Washington DC

Patten CL, Kirchhof MG, Schertzberg MR et al. (2004) Microarray analysis of RpoS-mediated gene expression in *Escherichia coli* K-12. Mol Genet Genomics

Postma PW, Lengeler JW, Jacobson GR (1993) Phosphoenolpyruvate: carbohydrate phosphotransferase systems of bacteria. Microbiol Rev 57(3):543–94

Ramseier TM (1996) Cra and the control of carbon flux via metabolic pathways. Res Microbiol 147(6–7):489–93

Reuss M, Luciano A-V, Mauch K (2007) Reconstruction of dynamic network models from metabolite measurements. In: Nielsen J, Jewett MC (eds) Metabolomics, Springer, Berlin, Heidelberg

Rizzi M, Baltes M, Theobald U et al. (1997) *In vivo* analysis of metabolic dynamics in *Saccharomyces cerevisiae*. 2. Mathematical model. Biotechnol Bioeng 55(4):592–608

Roels JA (1983) Energetics and kinetics in biotechnology. Elsevier Biomedical Press, Amsterdam, New York, Oxford

Rosen R (1968) Recent developments in the theory of control and regulation of cellular processes. 3. Int Rev Cytol 23:25–88

Saier MH, Jr., Ramseier TM (1996) The catabolite repressor/activator (Cra) protein of enteric bacteria. J Bacteriol 178(12):3411–7

Saier MH, Jr., Ramseier TM, Reizer J (1996) Regulation of carbon utilization. In: Neidhardt FC, et al. (ed) *Escherichia coli* and *Salmonella*: cellular and molecular biology, American Society for Microbiology Press, Washington DC

Schaub J, Mauch K, Reuss M (2008) Metabolic flux analysis in *Escherichia coli* by integrating isotopic dynamic and isotopic stationary ^{13}C labeling data. Biotechnol Bioeng 99(5):1170–85

Stephanopoulos GN, Aristidou AA, Nielsen J (1998) Metabolic engineering: principles and methodologies. Academic Press, London

Stormo GD (1988) Computer methods for analyzing sequence recognition of nucleic acids. Annu Rev Biophys Biophys Chem 17:241–63

Stormo GD (1990) Consensus patterns in DNA. Methods Enzymol 183:211–21

Stormo GD (2000) DNA binding sites: representation and discovery. Bioinformatics 16(1):16–23

Stormo GD, Fields DS (1998) Specificity, free energy and information content in protein-DNA interactions. Trends Biochem Sci 23(3):109–13

Streichert F, Ulmer H (2005) JavaEvA: a Java based framework for Evolutionary Algorithms. http://tobias-lib.ub.uni-tuebingen.de/volltexte/2005/1702/

Takeda Y, Sarai A, Rivera VM (1989) Analysis of the sequence-specific interactions between Cro repressor and operator DNA by systematic base substitution experiments. Proc Natl Acad Sci USA 86(2):439–43

Teich A, Meyer S, Lin HY, Andersson L, Enfors S, Neubauer P (1999) Growth rate related concentration changes of the starvation response regulators s^S and ppGpp in glucose-limited fed-batch and continuous cultures of *Escherichia coli*. Biotechnol Prog 15:123–9

Traxler MF, Chang DE, Conway T (2006) Guanosine 3′, 5′-bispyrophosphate coordinates global gene expression during glucose-lactose diauxie in *Escherichia coli*. Proc Natl Acad Sci USA 103(7):2374–2379

Uptain SM, Kane CM, Chamberlin MJ (1997) Basic mechanisms of transcript elongation and its regulation. Annu Rev Biochem 66:117–72

Van Dien SJ, Keasling JD (1998) A dynamic model of the *Escherichia coli* phosphate-starvation response. J Theor Biol 190(1):37–49

Visser D, Heijnen JJ (2003) Dynamic simulation and metabolic re-design of a branched pathway using linlog kinetics. Metab Eng 5(3):164–76

Visser D, Schmid JW, Mauch K et al. (2004) Optimal re-design of primary metabolism in *Escherichia coli* using linlog kinetics. Metab Eng 6(4):378–90

Visser D, van der Heijden R, Mauch K et al. (2000) Tendency modeling: a new approach to obtain simplified kinetic models of metabolism applied to *Saccharomyces cerevisiae*. Metab Eng 2(3):252–75

Vohradsky J (2001a) Neural model of the genetic network. J Biol Chem 276(39):36168–73

Vohradsky J (2001b) Neural network model of gene expression. FASEB J 15(3):846–54

Weaver DC, Workman CT, Stormo GD (1999) Modeling regulatory networks with weight matrices. Pacific Symposium on Biocomputing 4:112–123

Wendrich TM, Blaha G, Wilson DN et al. (2002) Dissection of the mechanism for the stringent factor RelA. Mol Cell 10(4):779–88

Wong P, Gladney S, Keasling JD (1997) Mathematical model of the *lac* operon: inducer exclusion, catabolite repression, and diauxic growth on glucose and lactose. Biotechnol Prog 13(2): 132–43

Yagil G (1975) Quantitative aspects of protein induction. Curr Top Cell Regul 9:183–236

Yagil G, Yagil E (1971) On the relation between effector concentration and the rate of induced enzyme synthesis. Biophys J 11(1):11–27

Yee L, Blanch HW (1992) Recombinant protein expression in high cell density fed-batch cultures of *Escherichia coli*. Biotechnology (NY) 10(12):1550–6

Chapter 12
Genome-Scale Models and the Genetic Basis for *E. coli* Adaptation

M. Kenyon Applebee and Bernhard Ø. Palsson

Contents

Abstract In this chapter we describe work stemming from the development of a stoichiometrically-constrained model of *Escherichia coli* metabolism, to experimental evolution of strains, to the analysis of the function of acquired adaptive mutations with the goal of understanding their system-wide effect on phenotype.

12.1 Introduction

E. coli is among the most extensively characterized microorganisms (Feist et al. 2007, Karp et al. 2007), making it a good candidate organism to initiate studies of how systems of biological molecules function and interact to produce an organism's physiological behavior – an endeavor that falls under the umbrella of systems

B.Ø. Palsson (✉)
Department of Bioengineering, University of California – San Diego, 9500 Gilman Drive, La Jolla, CA 92093-0412 USA
e-mail: palsson@ucsd.edu

S.Y. Lee (ed.), *Systems Biology and Biotechnology of Escherichia coli*,
DOI 10.1007/978-1-4020-9394-4_12, © Springer Science+Business Media B.V. 2009

biology. The study of systems with numerous interacting elements is often aided by the development of computational models of the systems, both to facilitate calculations and to discover emergent properties (Cohen and Harel 2007). The models can then be used to predict the cell's simulated response to a perturbation, which can be compared to the system's actual response; failed predictions can be analyzed to determine what may be missing or incorrect in the model and thus direct research towards critical elements or interactions, creating an iterative process of model testing and discovery.

12.2 Metabolic Models

Genome-scale *in silico* metabolic models simulate how nutrients are processed by an organism's metabolic network to harness energy and biomass, based on the set of reactions the organism's complement of enzymes can catalyze. The dynamics of enzyme-catalyzed reactions are traditionally described using kinetic constants and the concentrations of reactants and products. However, the catalytic efficiency of many enzymes is sensitive to differences that exist between most assay conditions and physiological cellular conditions, such as pH and ionic strength, making the values reported in the literature unrepresentative of *in vivo* conditions. Additionally, the concentration of many metabolites and enzymes within the cell are not accurately known, and they are condition-dependent. These factors have hampered efforts to construct robust metabolic models based on kinetic parameters (Pramanik and Keasling 1997, Varma and Palsson 1994).

For this reason, many recent metabolic modeling efforts have focused on simulating flux states of metabolic maps using only stoichiometric constraints. In this approach, metabolism is represented by a set of stoichiometrically-balanced equations of each metabolic reaction. This can be used to calculate all possible steady-state flux distributions that can result from passing a defined supply of simulated nutrients through the metabolic network. Additional algorithms can then be applied to identify solutions that represent physiological behavior. For example, since the exponential growth of bacteria has been hypothesized to optimize biomass (Lenski et al. 1991), the model can be used to make predictions about the exponential growth rate of the *E. coli* under specified growth conditions by searching the solution space for flux distributions that use the simulated nutrients to produce the most biomass (Pramanik and Keasling 1997), using a method known as flux balance analysis (FBA). FBA uses linear programming to find flux states that maximize a determined objective (Reed and Palsson 2003, Varma and Palsson 1994), such as biomass. Many of these methods are discussed in more depth in Chapter 11 of this book, and elsewhere (Feist and Palsson 2008, Reed and Palsson 2003).

FBA with the biomass objective function predicts the maximum growth rates that can be produced given the specified availability of oxygen and other limiting nutrients (usually the carbon source), represented as the oxygen and substrate uptake rate (OUR and SUR, respectively) (Varma and Palsson 1994). The results of this calculation can be visualized as a three-dimensional phenotypic phase plane (PhPP)

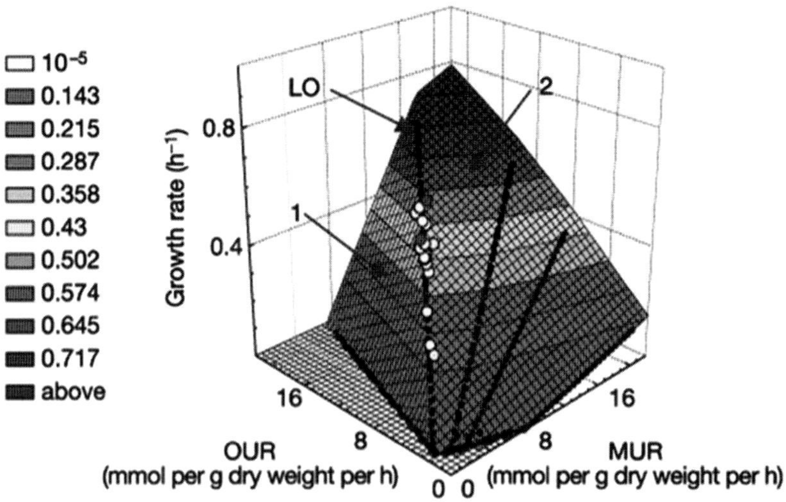

Fig. 12.1 The Phenotype Phase Plane. This figure depicts the maximum growth rate that is predicted to be produced across the range of feasible oxygen and substrate uptake rates. This example shows predicted optimized growth rates of *E. coli* grown aerobically on malate. The line of optimality (LO) is highlighted. White and dark grey circles represent the experimentally-observed growth phenotypes of *E. coli* cultures, where open circles represent wild type cultures grown over a range of malate concentrations (0.25–3g/L) and temperatures (29–37C). The filled circles represent wild type before and after adaptive evolution on malate (2 g/L). Note that all experimentally-derived measurements cluster on the line of optimality. OUR, oxygen uptake rate; MUR, malate uptake rate. Figure originally published in (Ibarra et al. 2002)

(Fig. 12.1) that show the maximum growth rate across the range of allowable uptake rates. These phase planes often have several faces, each of which represents a mode of growth such as aerobic growth or fermentation of a byproduct (Ibarra et al. 2002). The intersection separating oxygen-limited and carbon source-limited growth is known as the line of optimality (LO), and represents the growth mode in which the uptake rate of oxygen and the carbon source are stoichiometrically balanced, producing completely aerobic growth without any energy loss to futile cycles (for example, due to concurrent activity of glycolytic and gluconeogenic pathways). Interestingly, more than one set of uptake rates or even intracellular fluxes can produce the same predicted maximum growth rate (Reed and Palsson 2004).

The growth rate predictions made by genome-scale metabolic models of *E. coli* K12 MG1655 have generally been accurate when compared to growth rates of cultures grown under the relevant conditions during exponential growth (Ibarra et al. 2002). Figure 12.1 shows the results of plotting the measured growth rate, SUR and OUR of growing cultures onto the generated phenotypic phase plane. Cultures were grown under various conditions, including variable temperatures, substrates (glucose, malate, succinate, and acetate), and substrate concentrations in minimal media. When the growth phenotypes of the cultures are plotted on the phenotypic phase plane, most of cultures cluster around the line of optimality as predicted (Ibarra et al. 2002).

12.3 Adaptive Evolution

12.3.1 Adaptation to a Substrate Challenge

Experimentally-measured growth phenotypes do not always fall on line of optimality predicted by biomass-optimized flux balance analysis. For example, *E. coli* K12 MG1655 was found to grow significantly slower than the model-predicted optimum on glycerol and lactate. These prediction failures were not believed to be due to model errors, since the metabolic pathways for utilizing these substrates have been well-characterized. Rather, the discrepancy was hypothesized to be due the organism not being suitably adapted to maximally utilize these substrates.

Given that the metabolic pathways for these substrates exist, the growth capacity could be constrained by factors not included in the metabolic model that regulate flux through the metabolic network, such as enzyme expression, kinetics, or feedback regulation. Such constraints, especially transcriptional regulation, are genetically malleable and are refined by evolution. The short generation time of *E. coli* makes it possible to observe the evolution of adaptive traits in long-term cultures (Lenski et al. 1991). If the prediction failures are due to the wild type strains not being well adapted, subjecting the strains to natural selection by long term exponential growth on the challenging substrate should allow the strains to approach or achieve the FBA-predicted optimal growth rates.

Such experiments, known as laboratory adaptive evolution experiments, were conducted by culturing *E. coli* K12 MG1655 on each of the challenging substrates for a prolonged ($\geq$ 500 generations) period of time (Fong et al. 2003, Ibarra et al. 2002). The resulting strains all had increased growth rates on the targeted substrate (Table 12.1) (Ibarra et al. 2002). Additionally, the growth profiles of each strain migrated towards the predicted line of optimality, and once the growth phenotype aligned with the predicted line of optimality generally it only migrated along it (Table 12.1, Fig. 12.2). Further growth rate increases were accomplished by increasing both uptake rates proportionally so that the phenotype moved up the line of optimality (Edwards et al. 2001, Ibarra et al. 2002). However, it should be noted that several strains eventually increased their sugar uptake rates beyond what could be fully aerobically metabolized, due to physical limits on the oxygen uptake rate – these strains were driven by growth-rate dependent selection to ferment the excess sugar, and consequently migrated off of the predicted line of optimality.

Overall, the outcomes indicate that actual growth limits are captured by biomass-optimized flux-balance analysis of the stoichiometric reconstruction of the *E. coli* metabolic network. Additionally, they suggest that the cellular architecture is efficient at relieving constraints aside from those imposed by chemical or catalytic limitations. Unfortunately, FBA alone cannot be used to determine how the evolved strains metabolize nutrients, since each growth rate can be associated with multiple flux distributions (Reed and Palsson 2004). Quantitative metabolomic profiling is necessary to both identify and validate the flux distribution predictions, which is

Table 12.1 Summary of substrate-challenged evolution experiments

	Growth phenotype near predicted LO					
	Temp. (°C)	Generations evolved	Wild type	Evolved strains	Percent of GR increase	Ref.
Glucose	37	500	Yes	Yes	17	(Fong and Palsson 2004)
Malate	37	500	Yes	Yes	19	(Fong et al. 2005a)
Acetate	37	700	Yes	Yes	20	(Fong et al. 2005b)
Lactate	30	950	No (acetate sec.)	Yes	147	
Lactate	30	950	No (acetate sec.)	Yes	132	
Glycerol	30	1000	No (futile cycles)	Yes	140	
Lactate	37	870	No (futile cycles)	No (O_2 maxed)	80	
Pyruvate	37	1200	No (acetate sec.)	No (O_2 maxed)	69	
Pyruvate	30	1000	Yes	No (O_2 maxed)	115	
α-Ketoglutarate	37	625	Yes	No	41	
	30	440	Yes	No	48	

This table compares outcomes of adaptive evolution experiments that challenged MG1655 wild type *E. coli* to increase growth on a variety of substrates with *in-silico* flux-balance analysis (FBA) growth rate predictions. The growth phenotype of wild type on some substrates did not meet the FBA prediction because of carbon or oxygen uptake imbalance, that resulted in acetate secretion (too much carbon uptake) or futile cycles (too much oxygen uptake). The growth phenotype of some evolved strains did not fall on the LO because they increased their carbon uptake beyond the limits of physiologically-available O_2. LO – Line of Optimality. GR – Growth rate.

still a very challenging experimental endeavor. Additionally, these results do not begin to identify the mechanics of the adaptation, which will be addressed later in this chapter.

12.3.2 Adaptation to Deletion of a Metabolic Gene

In addition to growth on rarely-encountered carbon sources, strains can also be perturbed from the optimal growth state by deletion of a metabolic gene. Deleting a metabolic gene produces a strain with a different potential optimum growth rate that can be predicted by removing the catalyzed reaction from the *in silico* metabolic model, changing the solution space calculated by flux balance analysis (Reed and Palsson 2003). The new line of optimality is composed of solutions that most effectively redistribute the metabolic flux around the lost reaction to produce the most biomass (deletion of essential genes are not considered since they have no

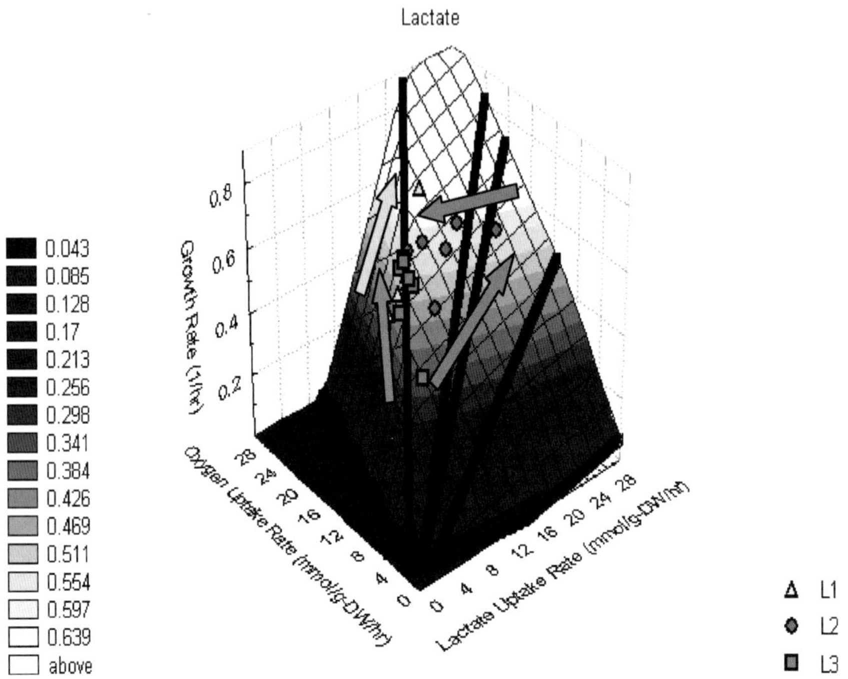

Fig. 12.2 Migration of the growth phenotype of evolving *E. coli* strains towards the line of optimality. This example shows the growth phenotype three replicate experimental evolutions of wild type on lactate (L1-L3), which all eventually migrate towards the line of optimality. Figure originally published in Fong et al. 2003

solution space that supports growth). Strains of *E. coli* with single metabolic gene deletions were adaptively evolved to test whether the predicted growth phenotype reflects the actual potential of *E. coli* to adapt to gene loss (Fong and Palsson 2004). These experiments test the plasticity of cell regulation to allow the redistribution of metabolic fluxes.

One of six genes was deleted from strains of *E. coli*, (acetate kinase A (*ackA*), fumarate reductase (*frd*), phosphoenolpyruvate carboxykinase (*ppc*), phosphoenol pyruvate carboxylase (*ppc*), triosephosphate isomerase (*tpi*), or glucose 6-phospate-1-dehydrogenase (*zwf*)). These genes encode enzymes required for gluconeogenesis, fermentation, or the pentose phosphate pathway (Fig. 12.3). Adaptive evolution experiments were performed with these strains across a set of substrates that enter central metabolism at a variety of points (Fong and Palsson 2004). Approximately 80% of the adaptively evolved gene-deletion strains made gains in growth rate within 10% of their respective FBA prediction. This success rate further validates biomass-optimized flux balance analysis of the stoichiometrically-constrained metabolic model, and indicates that the stoichiometric model captures many of the physiologically-relevant constraints on growth. The results of both these and the substrate-challenged adaptive evolution studies suggests that *E. coli* is fairly adept

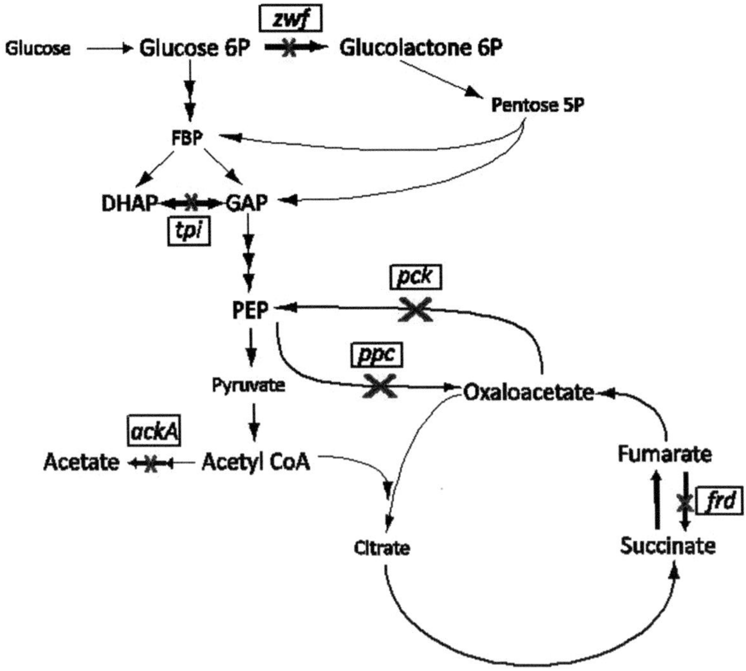

Fig. 12.3 Genes deleted in evolved gene-deletion strains described in Fong and Palsson, 2004. Deleting *zwf* (glucose 6-phosphate dehydrogenase) prevents forward flux through the pentose phosphate pathway. Deleting *tpi* (triosphosphate isomerase) interferes with glycolysis and gluconeogenesis. Deleting *pck* (phosphoenolpyruvate carboxykinase) can interfere with gluconeogenesis from citric acid cycle intermediates. Deletion of *ppc* (phosphoenolpyruvate carboxylase) impedes the ability to replenish oxaloacetate to the citric acid cycle. Deletion of *frd* (fumarate reductase) obstructs utilization of the reductive pathway of the citric acid cycle. Deletion of *ackA* (acetate kinase A) blocks a pathway needed to secrete acetate. These strains were experimentally evolved on a variety of carbon sources to select for increased growth rate

at rerouting its metabolic flux to achieve the optimal growth phenotype available within the limits of its metabolic chemical capacity.

12.3.3 Application of Experimental Evolution for Rational Design of Production Strains (OptKnock)

A practical application for metabolic modeling is to assist the design of strains that secrete desired product(s). A number of algorithms, which can be used to find growth-coupled strains (including OptKnock (Burgard et al. 2003, Pharkya et al. 2003), OptStrain (Pharkya et al. 2004), and OptGene (Patil et al. 2005)), calculate the predicted growth phenotype across all possible gene deletion strains and identify permutations of the metabolic network that maximize both biomass formation and production of the secreted compound. These two objectives arc simultaneously met when metabolic reactions that allow growth by secretion of alternative, more-energetically favorable fermentation products are removed, making growth dependent on secretion of the desired compound. Since these designs involve deleting metabolic genes, adaptive evolution can be employed to drive the generated strains to recover and to optimize their growth rate and secretion rate. Manual examination of the metabolic network can also be used to predict gene deletions that will couple growth to product secretion; however, the computational algorithms can identify designs that are not intuitively obvious.

The accuracy of both intuitive designs and non-intuitive designs predicted by the algorithm OptKnock to optimize the production of lactic acid were tested experimentally by generating the strains with the indicated gene deletions, and the growth rates of the constructs were optimized by adaptive evolution (Fong et al. 2005a). Three strain designs were tested, (1) *pta-adhE* double deletion strain, (2) *pta-pfk* double deletion strain, and (3) *pta-adhE-pfk-glk* quadruple deletion strain, summarized in Fig. 12.4. The first strain design, *pta-adhE,* is an intuitive design as it deletes reactions in the ethanol and acetate fermentation pathways. In the second design, *pta-pfk*, the reason for deleting phosphofructokinase in not intuitively

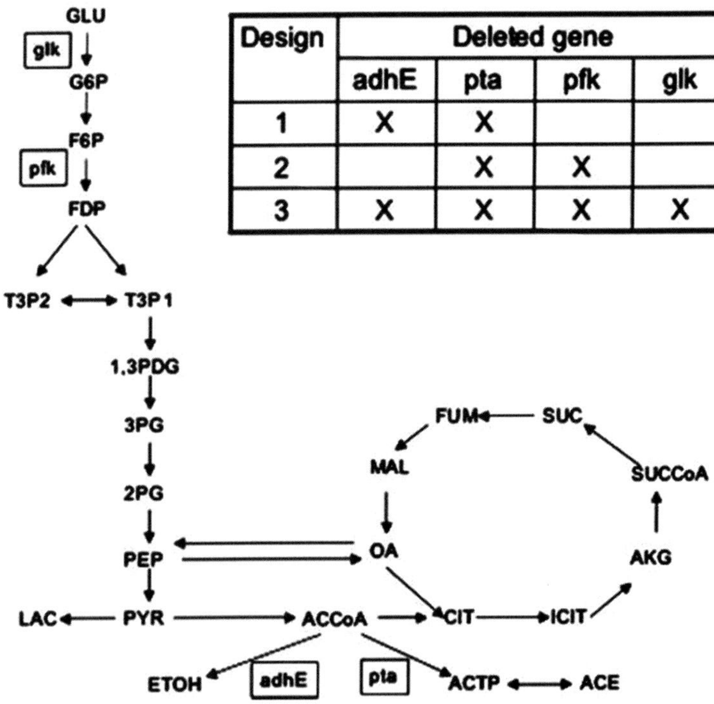

Design	Deleted gene			
	adhE	pta	pfk	glk
1	X	X		
2		X	X	
3	X	X	X	X

Fig. 12.4 OptKnock strain designs. Strain designs 1–3 were generated by deleting the genes indicated on the table. The reactions lost by each gene deletion are indicated on the diagram of central metabolism. Figure originally published in Fong et al. 2005a

obvious, but it promotes lactic acid production by increasing NADH and pyruvate by forcing flux through the Entner-Doudoroff and pentose phosphate pathway. In the third strain, the additional deletion of glucose kinase also contributes to lactate secretion by increasing pyruvate production, via forcing all metabolized glucose to be phosphorylated by the phosphotransferase system.

Adaptive evolution increased the growth rates of all four strain designs interestingly to approximately the same rate (0.24–0.26 hr^{-1}). Lactic acid production increased with growth rate in each case, proving that all of the designs successfully coupled lactate secretion to growth. The two-gene deletion strains agreed well with the predictions made by OptKnock in terms of migration of both the growth rate and lactate secretion rate; the quadruple gene-deletion strain, Δ*pta-adhE-pfk-glk*, actually acquired slightly faster growth rates than predicted. However, even though the lactic acid secretion rates in all three designs increased over the adaptation, the final lactate titer in the media recovered after culturing did not, possibly suggesting the existence of metabolic feedback mechanisms not included in the current model that can arrest secretion beyond some threshold. Additionally, the intuitively-designed strain, *pta-adhE*, produced the most lactic acid in terms of both secretion rate and titer among the three designs, suggesting that the non-intuitive upstream gene deletions (*pfk* and *glk*) may have more complex effects on metabolism than those currently captured by the model – again, this could easily be attributable to metabolic feedback mechanisms.

This study demonstrated that the OptKnock algorithm can successfully be used identify gene deletions that couple a secondary objective, such as fermentation product secretion, to growth. The useable set of solutions is of course limited to those that do not require the loss of an essential gene or set of genes. And even though in this case the intuitive design were the most effective, this may not always be so – indeed, this tool allows researchers to search for designs that produce compounds when there are none that are intuitively obvious. Additionally, the range of compounds that can be produced by modeled organisms like *E. coli* can conceivably be increased by including reactions and pathways carried out by enzymes that can be transferred from other microorganisms, as performed by the OptStrain algorithm (Pharkya et al. 2004). Growth-coupling algorithms have the potential to allow systems-scale models to be used to drive the development of strains for practical applications.

12.4 Characterizing Intracellular Mechanisms of Adaptation

Beyond illustrating that selection during exponential growth produces optimized phenotypes that converge with biomass-optimized FBA predictions, the intracellular changes that facilitate the phenotype shifts are also of inherent interest. Adaptations can act through many intracellular activities, including but not limited to metabolism, transcriptional regulation, protein turnover, and intracellular feedback mechanisms. Thus identifying the mechanism through which adaptive mutations act can require multiple high-throughput experimental methods, including mRNA transcription profiling, global metabolomic and flux profiling, and chromatin immunoprecipitation-on-chip – and requires an integrated analysis of such data sets. Additionally, the timescale required to change the growth phenotype implies the adaptations involve genetic mutations, rather than an adjustment to an existing mode of regulation. Therefore it is necessary to identify the acquired mutations in order to pursue the fundamental goal of understanding the mechanism that underlies the phenotypic change.

12.4.1 Phenotypic Characterization of Replicate Endpoints

As previously touched upon, flux balance analysis of an *in-silico* model cannot directly be used to assess the intracellular state. This is because the line of optimality actually represents a set of flux states that produce the same, maximum biomass (although, the differences between the solutions in the set are often restricted to variation in flux among a small set of reactions (Reed and Palsson 2004)). The existence of multiple metabolic states that can achieve the same optimal biomass objective parallels the frequent observation that replicated adaptive evolution experiments often achieve nearly identical growth rates and nutrient uptake rates in the evolved environment, but have distinct phenotypes such as growth rate on alternative substrates and byproduct secretion rates (Fong et al. 2005b, Fong et al. 2003, Fong and Palsson 2004). Such observations suggest these replicate lineages acquire different adaptive changes that produce the same or similar adjustment to growth in the environment of the evolution, but which have different (pleiotropic) effects on growth under other conditions. To sample the range of genetically-distinct endpoint strains that can be produced by replicate evolutions to the same metabolic challenge, we have extensively characterized replicate endpoint strains adaptively evolved on glycerol, lactate, or after deletion of various single metabolic genes (Fong et al. 2005b, 2006, Fong and Palsson 2004).

The glycerol- and lactate-adapted endpoint strains all achieved growth rates, SURs, and OURs within 12% of each of the other replicates. Additionally, while the replicate evolved strains generally had similar growth rates on other substrates, there was sufficient variability to suggest each endpoint strain had acquired different adaptations. The variation between replicate strains is even more pronounced between the mid-point evolution cultures (day 20) than between the replicate endpoint strains, in terms of growth rate, and the oxygen and substrate uptake rate during growth on the alternative substrates (Fig. 12.5). This variation may indicate that there were multiple adaptive strategies available during the initial stages of adaptation that lead to divergence between replicate cultures, followed by a period of more discriminate selection during the later period that caused the phenotypes of the endpoints to converged towards a single optimal phenotype. Interestingly, the endpoint strains generally grow faster than the wild type strain on many other carbon sources in minimal media, suggesting that some of the acquired changes were generally beneficial.

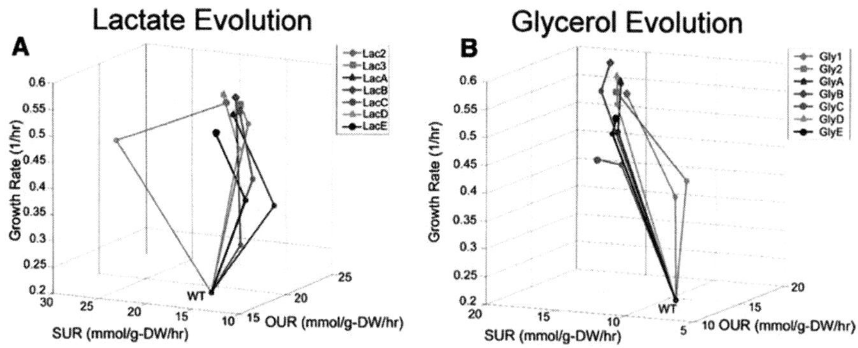

Strain	SUR (mmol/g-DW/hr)	OUR (mmol/g-DW/hr)	GR (1/hr)
Wild-type	13.8 ± 0.4	15.9 ± 2.1	0.23
Lac2 D60	14.9 ± 0.2	19.9 ± 1.0	0.56
Lac3 D60	14.5 ± 0.04	20.2 ± 1.1	0.53
LacA D60	14.6 ± 0.03	18.9 ± 0.5	0.54
LacB D60	16.4 ± 1.4	21.2 ± 0.6	0.55
LacC D60	13.0 ± 0.2	17.7 ± 0.2	0.56
LacD D60	14.4 ± 0.4	17.8 ± 0.5	0.59
LacE D60	16.4 ± 0.1	18.9 ± 0.7	0.50

Strain	SUR (mmol/g-DW/hr)	OUR (mmol/g-DW/hr)	GR (1/hr)
Wild-type	7.6 ± 0.4	12.0 ± 0.7	0.22
Gly1 D44	14.7 ± 0.2	17.3 ± 0.7	0.54
Gly2 D44	14.9 ± 0.2	16.6 ± 0.4	0.55
GlyA D44	15.1 ± 0.2	17.3 ± 0.3	0.56
GlyB D44	17.2 ± 0.1	19.3 ± 0.2	0.58
GlyC D44	13.4 ± 0.1	12.2 ± 0.1	0.46
GlyD D44	14.5 ± 0.2	16.0 ± 0.2	0.58
GlyE D44	15.0 ± 0.2	16.6 ± 0.1	0.50

Fig. 12.5 Phenotypic variability between replicate lactate- and glycerol-evolved strains. (**A**) Lactate Evolution. The top figure shows the trajectory of each replicate as it evolved from wild type, in terms of changing growth, substrate uptake, and oxygen uptake rate. Note that most strains begin to converge to the same growth phenotype. The bottom table lists the values of the measured parameters for the wild type and the final (day 60) endpoint strains. (**B**) Similar to (**A**), except applies to strains evolved to glycerol. Figure originally published in Fong et al. 2005b

12.4.2 Identifying Changes in Metabolic Pathway Utilization

Since increasing the growth rate involves increasing the efficiency with which available nutrients are metabolized, it is informative to determine how flux through various metabolic pathways has changed over the course of adaptation. One method for comparing functional flux states in microorganisms tracks the catabolism of ^{13}C-labeled substrates, and has been successfully used to access *in-vivo* reaction rates (Fischer et al. 2004, Sauer 2004). This technique has been used to track flux changes following adaptation to lactate and deletion of various single metabolic genes.

Results from the ^{13}C-labeling experiments on the lactate evolutions (Hua et al. 2007) showed that the first major flux change was a dramatic (up to 80%) increase in the uptake of lactate, and increased flux capacity through most metabolic reactions, over the first 20 days of evolution. Additionally, though the replicate strains showed significant phenotypic diversity at day 10 that later converged, their flux profiles were relatively similar at day 10 relative to later in the evolution. During this period, flux profiling showed that all of the strains shift more metabolites into the TCA cycle rather than acetate fermentation, allowing the cells to generate more energy and anabolic precursors, and thus achieve a faster growth rate. After day

10 the flux changes among the replicate evolutions diverged, though an important consistency was discovered across the course of all of the replicate evolutions – approximately two-thirds of the metabolized lactate was consistently partitioned to the TCA cycle through pyruvate dehydrogenase, indicating that partitioning between gluconeogenic and catabolic fluxes is tightly regulated – likely by feedback mechanisms that recognize cellular concentrations of phosphoenolpyruvate (Chulavatnatol and Atkinson 1973).

An additional ^{13}C-labeling experiment was performed with adaptively evolved gene deletion strains, though with deletion of different genes than those described in Section 12.3.2. The genes deleted in this study encode metabolic enzymes that catalyze key metabolic branch points in central metabolism, and were chosen because their loss is expected to most dramatically change pathway utilization (Fong et al. 2006) (phosphoglucose isomerase (*pgi*), phosphoenolpyruvate carboxylase (*ppc*), triose-phosphate isomerase (*tpi*), or phosphate transacetylase (*pta*)) (Fig. 12.6). As

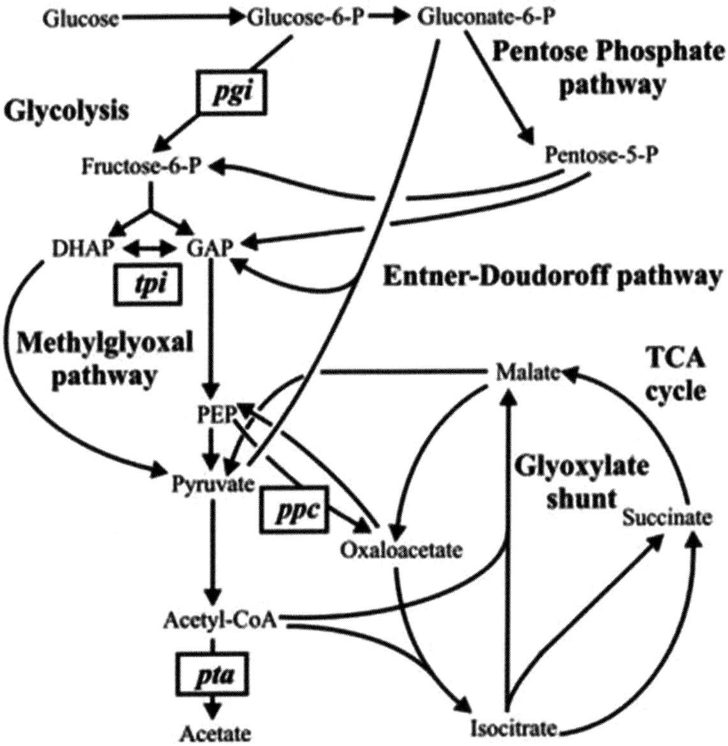

Fig. 12.6 Map of metabolism highlighting the gene deletion strains that were adaptively evolved on glucose in minimal medium. Deleting *pgi* (phosphoglucose isomerase) impedes glycolysis, forcing flux to be rerouted through the pentose phosphate or Entner-Doudoroff pathway. Deleting *pta* (phosphate acetyltransferase) impedes acetate fermentation. The *tpi* and *ppc* deletions are described in the Fig. 12.3 caption. Figure originally published in Fong et al. 2006

in the previous studies, the generated endpoint strains achieved near-wild type growth rates, and replicate endpoints differed in terms of phenotypic traits like byproduct secretion, suggesting they had acquired different specific adaptations.

The outcome of the ^{13}C-labeling experiments revealed that most the flux changes involved activating alternative pathways to locally reroute metabolites around the lost gene function. This rerouting was accomplished by activating pathways typically used for growth under other conditions, such as the pentose-phosphate pathway (*pgi* strains), glyoxylate shunt (*ppc* strains), and the methyl-glyoxyl bypass (*tpi* strains) (Fig. 12.6). No evidence of new enzyme activities or metabolic pathways was observed. The evolved gene deletion strains often utilized the same pathways to circumvent the lost gene activity as those used in pre-evolved deletion strain, and evolution involved increasing the flux capacity through those pathways – similar to increasing lactate uptake in the previously described study. However, the outcome of this study suggests that *E. coli* immediately responds to a breakdown in metabolic capacity by activating repressed pathways to search for a way to redistribute flux through the metabolic network (a process that may favor short routes that cause the least disruption of the wild type flux configuration (Segre et al. 2002)). Thus the adaptation process may often consist of refining the most easily established solutions, rather than searching for.

However, this should not be taken to suggest that every replication of adaptive evolution under the same conditions will eventually converge to the same flux configuration. Although all of the evolved replicate strains circumvented their gene deletions with the same latent pathway, there were several significant differences in how some replicate strains utilized other pathways. For example, one evolved *pgi* strain primarily utilizes the TCA cycle and secretes acetate while another has more flux through the glyoxylate shunt and secretes no acetate. These outcomes suggest that much of the variability between replicate evolutions may stem from variable means of making downstream metabolic adjustments that are necessary to refine usage of the major adaptive flux shift.

12.4.3 Gene Expression Changes

While growth rate increases are dependent on improving flux through the metabolic network, those improvements are a result of refining the activity of metabolic elements. Metabolic activity is regulated at multiple levels, including allosteric control of enzymes, turnover rates, and transcriptional regulation. We used mRNA transcription profiling to identify the adaptive mechanism utilized in each strain, since high throughput methods are not available to screen the other types of regulation, such as feedback regulation.

We measured changes in genome-wide mRNA transcription levels of the replicate glycerol- and lactate-evolved strains over the course of their evolutions, as well as the cultures at day 1 and day 20 of each evolution, and wild type grown on glucose (Fong et al. 2005b). Interestingly, the transcription state of each evolved endpoint

Table 12.2 Number of expression changes across evolutions on glycerol and lactate

		Glycerol evolved		Lactate evolved	
relative to wild type (glucose)	Day 1	39%	1687 genes	18%	756 genes
	Day 20	18%	770 genes	4%	194 genes
	Day 44	11%	498 genes	7%	323 genes

This table shows the average number of significant gene expression changes in adaptively evolved strains relative to wild type grown on glucose at different time points in the evolution experiments. Data from Fong et al. 2005b

was distinct, despite similarity in endpoint growth phenotypes, providing further evidence that endpoint cultures result from acquiring different sets of adaptations.

Analysis of the transcription profiles revealed that the largest number of changes in gene expression were between wild-type grown on glucose and wild-type grown on the challenging substrate (day 1 cultures), and that the course of adaptive evolution returned the expression of many of these genes to pre-evolution levels (Table 12.2). This may be attributable to a large scale carbon-scavenging response intended to allow metabolism of any "low-quality" carbon sources available in the absence of preferred substrates (Liu et al. 2005), in which case the adaptation process may involve refining the regulatory response to only activate genes specific to glycerol or lactate metabolism.

Expression changes that developed over the course of evolution common across replicate strains were also identified. Approximately 70 gene expression changes were identified among the glycerol-evolved strains, and only two among the lactate-evolved strains – a striking difference. The small number of genes identified across multiple lactate strains may be due to their adaptive pathways being more divergent compared to glycerol strains, which is also discussed later in this chapter (Section 12.5). It appears to be very difficult to identify adaptive mechanisms through examination of expression changes alone. Adaptation generally results in changed expression of a large number of genes, and it is difficult without knowledge of changes to other intracellular systems, such as metabolism or of specific mutations, to identify critical expression shifts that mediate the adaptive mechanism rather than result as a down-stream response.

Additionally, the expression profiles of the evolved deletion strains have also been measured, and have been compared to the flux changes discussed in the previous section (Section 12.5.2) to try to identify gene expression changes responsible for the metabolic and phenotypic adaptations. However, as previously stated, not all of the flux changes are caused by expression changes since there are other mechanisms that regulate metabolism, such as post-translational regulation, enzyme kinetics, and allosteric control. Additionally, not all gene expression changes directly alter the phenotype or metabolic flux, or their effects may not be adaptive but rather are a secondary response to other changes. Wide-spread expression levels can also be caused by altered mRNA stability or RNase activities. But both data sets can be combined to identify adaptive flux changes that are caused by changes to gene expression.

Many of the observed flux changes among the evolved deletion strains can be linked to expression changes (Fong et al. 2006). Expression changes correlated well to changes in flux through the glyoxylate shunt, methylglyoxylate shunt, and TCA cycle, suggesting that these pathways are at least partially controlled at the transcriptional level. On the other hand, no expression changes were identified that correlate to observed flux changes through glycolysis or the pentose phosphate pathway, suggesting that other mechanisms may predominantly regulate flux through those pathways. Interestingly, no correlation was found between flux and gene expression changes among the pre-evolved deletion strains. This may indicate that the major, shared flux shifts were initially mostly dependent on other mechanisms of metabolic regulation, and that evolution was necessary to acquire changes involving transcriptional regulation.

Additionally, the flux and expression data collected from the evolved gene-deletion strains has been used to inform the metabolic model to try to identify why some strains failed to reach model-predicted growth rates (Herrgard et al. 2006). The developed method, called optimal metabolic network identification (OMNI), searches for metabolic reactions that, when eliminated, can produce flux distributions that are the closest fit to those actually measured in the relevant strains. The enzymes of these reactions may act as bottlenecks to optimal growth, that for some reason are disadvantageously regulated in a manner that had not been overcome by the point at which the experimental evolution was terminated. The validity of this approach is supported by the fact that expression of many of the genes identified as possibly causing flux bottlenecks can be seen to have reduced expression in the evolved strain compared to wild type.

The previously described studies successfully correlated flux change to shifts in expression of several relevant genes, partially illustrating the adaptive mechanism. It is likely that additional conclusions can be made from the expression profiling data if it were analyzed with a more sophisticated view of cell regulation, possibly facilitated by a comprehensive model of transcription regulation.

12.5 Genome Resequencing

Adaptive changes in the phenotype of evolved strains are ultimately caused by mutations (and possibly epigenetic changes to the DNA), and a comprehensive understanding of the adaptation process requires their identification. Further, the effect of each mutation on the DNA-encoded function must be determined, whether it effects regulation of nearby genes, causes an amino acid change in a protein that affects its function or activity, or some other effect. Ultimately, we would like to know how acquired mutations cause all of the observed changes in metabolic flux, gene expression, and the overall systems dynamics that translate into the observed phenotype. Such an accomplishment would significantly contribute to building an understanding of how genetic structure produces phenotype.

Conclusively identifying all of the mutations acquired during adaptive evolution requires a search of the entire genome that can find changes as small as of a single nucleotide (or a single nucleotide polymorphism – SNP). Six of the glycerol-evolved strains have undergone whole-genome resequencing (Herring et al. 2006). With one exception, the strains acquired two to three SNPs within the coding region of annotated genes, as described in Table 12.3. All five resequenced strains acquired a different mutation within *glpK*, which encodes glycerol kinase which catalyzes the rate-limiting step in glycerol metabolism. Three of the strains also acquired mutations in RNA polymerase subunits β and β', encoded by *rpoB* and *rpoC*, which was surprising given the extensive influence these genes may have on global transcription regulation. Additionally, mutations were acquired by two strains that effect peptidoglycan biosynthesis (within genes *dapF* and *murE*).

Table 12.3 Mutations identified in glycerol-evolved strains

Clone	Gene	Product/Function	Mutation	Gene position nt	Region	Genome Position
GB-1	*glpK*	Glycerol kinase	a-> t	218	Coding	4115028
	rpoC	RNA polymerase	27 bp deletion	3132–3158	Coding	4186504–4186530
GC-1	*glpK*	Glycerol kinase	g-> t	184	Coding	4115062
	n/a	All genes between insC-5 & insD6	1313 kb duplication[1]	n/a	n/a	~3189209–4497523
GD-1	*glpK*	Glycerol kinase	g-> a	816	Coding	4114430
	rpoB	RNA polymerase	a-> t	1685	Coding	4180952
	murE	peptidoglycan biosynthase	a-> c	8	Coding	93173
GE-1	*glpK*	Glycerol kinase	a-> c	113	Coding	4115133
	rpoC	RNA polymerase	c-> t	2249	Coding	4185621
	dapF	Lysine/peptidoglycan biosynthase	c-> a	512	Coding	3993293
G2-1	*glpK*	Glycerol kinase	9 bp duplication	705	Coding	4114541
	rph-pyrE	RNAse PH/pyrimidine synthesis	82 bp deletion	rph: 610-end	Coding + Intergenic	3813882–3813963
	pdxK-crr	Pyridoxal kinase/enzyme IIa glucose	28 bp deletion	pdxK: 833-end	Coding + Intergenic	2534400–2534427

Mutations identified by whole-genome resequencing of *E. coli* strains adaptively-evolved to increase growth on glycerol minimal media. Previously published. 1. Evident in CGS mapping data; not independently validated.

The impact of individual mutations on the fitness phenotype was determined by assessing strains that had been altered to carry one or more of the discovered mutations using site-directed mutagenesis (Herring et al. 2006). Importantly, the set of mutations identified in each strain have been proven to be responsible for the phenotype change, since the endpoint phenotype has been shown to be reproduced by inducing the mutation sets into wild type by site-directed mutagenesis. The phenotype of mutant strains carrying only single mutations indicate that the *rpoB/C* mutations have the greatest impact on growth rate, followed by the *glpK* mutations. Additionally, results of competitions between strains with different induced mutations indicated that some of the *glpK* and *rpoB/C* mutations may have cooperative (epistatic) effects (Applebee et al. 2008); possible mechanisms are still being investigated. Mutations to the peptidoglycan synthesis genes (*murE* & *dapF*) were also only shown to have a significant effect on growth rate in strains that also carried the co-acquired *glpK* and *rpoB/C* mutations.

The *rpoB/C* mutations clearly perform a critical function in optimizing the growth of *E. coli* on glycerol, and this function most likely involves adjusting transcriptional regulation. These mutations have the greatest impact on fitness among those acquired by glycerol-evolved stains, and they may additionally be responsible for the increased growth capacity in minimal media on a wide range of non-glycerol substrates (unpublished results). One hypothesis suggests that these mutations improve growth by reducing the sensitivity of RNA polymerase to stress response signals, particularly ppGpp, that may be induced by the transition from rich to minimal media. This may prevent the expression of unnecessary or detrimental proteins that are associated with the stress response (Liu et al. 2005). The effect of these mutations on RNA polymerase activity and global gene expression is currently being investigated.

Strains evolved on lactate or in response to deletion of the *pgi* gene have also undergone whole-genome resequencing (unpublished data). Interestingly, there was more variation in the number of mutations these strains acquired (0–7 mutations per strain, versus 2–3 in glycerol), and they appeared across a more diverse set of genes than glycerol-evolved strains. It is not yet clear why these evolutions produced more genetically divergent replicates compared to the glycerol-evolved strains, though it suggests there may simply be more adaptive routes that these replicate evolutions can sample.

A developing pattern seen across the different experimental evolutions is that strains frequently acquire mutations to both a metabolic gene and a global transcription factor. The function of mutations to metabolic genes (PEP synthase in lactate-evolved strains, NADPH/NADH transhydrogenase genes in evolved Δ*pgi* strains) is assumed to increase the activity of rate-limiting enzymes under the growth conditions, as has been shown for the *glpK* mutations in the glycerol evolved strains (Herring et al. 2006). The function of mutated global regulatory or transcription factors is likely more complicated. Among the mutations discovered in lactate and Δ*pgi* strains are mutations to *crp, cyaA*, and *rpoS* (unpublished data). This trend suggests that the *E. coli* regulatory network is both robust to mutations that alter the function of major regulatory elements, and that these are more accessible or advantageous than mutations that effect the transcriptional regulation of a smaller, more specific set of genes.

12.6 Summary Remarks

This chapter has covered how stoichiometrically-constrained models of *E. coli* metabolism have been used to predict optimal growth rates on a variety of carbon sources. The physiological relevance of these predictions has been validated by demonstrating they approximate the actual growth phenotype observed on these substrates in exponential growth phase. Not all substrates produce growth in wild type that approximates the model-derived optimum, but adaptive evolution studies have demonstrated that prolonged exposure to these substrates generates adaptations that cause the growth phenotype to converge towards the predicted optimum. Additionally, the models have been successful at predicting the optimum growth phenotype that *E. coli* will adapt to following loss of a metabolic gene function. These outcomes suggest that the model captures the critical constraints on growth capacity before considering metabolic regulation, without requiring kinetic constraints. Additionally, it indicates that metabolic regulation is readily malleable to selection pressure, to allow the optimum growth phenotype to be found in response to a wide range of environmental or metabolic challenges.

Each evolution experiment was performed multiple times, and although each of the replicate strains generally acquired the same growth phenotype (converging towards the predicted optimum), they generally differed in terms of other phenotypic traits such as byproduct secretion or growth capacity on alternative substrates. Significant differences between replicate strains were also validated by differences between their flux and transcription profiles. The basis for those differences have now been genetically identified – no two replicate strains have acquired identical mutations. Thus different genetic changes can produce similar phenotypic changes. The existence of multiple metabolic flux shifts that produce the optimal phenotype is actually predicted from the results of flux balance analysis (Reed and Palsson 2004). The degree of variation among replicate evolved strains may indicate how many adaptive routes exist to produce this phenotype.

Attempts to understand the intracellular changes causing the growth phenotype changes have involved measuring changes in pathway utilization and transcription expression over the course of evolution, and identifying acquired mutations. At this point the challenge involves deducing the mechanism by which the discovered mutations alter metabolic flux through regulatory mechanisms to produce increased growth rate. Among the most intriguing discoveries are the mutations to so-called "global regulators" like RNA polymerase subunits and elements of catabolite repression, that appear to play a significant role in inducing the adapted phenotype. It remains to be proven whether this truly is a general mechanism of adaptation, and if so how it functions.

These studies begin to highlight the potential of using adaptive evolution to discover important cellular dynamics that exist on the systems-biology level. The eventual goal is to be able to predict what regulatory and metabolic changes are necessary to produce a desired phenotype. We will have accomplished a true understanding of the relationship between phenotype and genotype when we understand the adaptive function of acquired mutations – and are able to predict them.

Acknowledgments Thank you to Karsten Zengler, Jan Schellenberger, and Adam Feist for their insightful feedback, to Addiel de Alba for lending his graphics expertise and to Jessica Na for her encouragement during the writing of this chapter.

References

Applebee MK, Herrgard MJ, Palsson BO (2008) Impact of individual mutations on increased fitness in adaptively evolved strains of *Escherichia coli*. J Bacteriol 190(14):5087–94

Burgard AP, Pharkya P, Maranas CD (2003) Optknock: a bilevel programming framework for identifying gene knockout strategies for microbial strain optimization. Biotechnol Bioeng 84(6):647–57

Chulavatnatol M, Atkinson DE (1973) Kinetic competition *in vitro* between phosphoenolpyruvate synthetase and the pyruvate dehydrogenase complex from *Escherichia coli*. J Biol Chem 248(8):2716–21

Cohen IR, Harel D (2007) Explaining a complex living system: dynamics, multi-scaling and emergence. J R Soc Interface 4(13):175–82

Edwards JS, Ibarra RU, Palsson BO (2001) *In silico* predictions of *Escherichia coli* metabolic capabilities are consistent with experimental data. Nat Biotechnol 19(2):125–30

Feist AM, Henry CS, Reed JL et al. (2007) A genome-scale metabolic reconstruction for *Escherichia coli* K-12 MG1655 that accounts for 1260 ORFs and thermodynamic information. Mol Syst Biol 3:121

Feist AM, Palsson BO (2008) The growing scope of applications of genome-scale metabolic reconstructions using *Escherichia coli*. Nat Biotechnol 26(6):659–67

Fischer E, Zamboni N, Sauer U (2004) High-throughput metabolic flux analysis based on gas chromatography-mass spectrometry derived 13C constraints. Anal Biochem 325(2):308–16

Fong SS, Burgard AP, Herring CD et al. (2005a) *In silico* design and adaptive evolution of *Escherichia coli* for production of lactic acid. Biotechnol Bioeng 91(5):643–8

Fong SS, Joyce AR, Palsson BO (2005b) Parallel adaptive evolution cultures of *Escherichia coli* lead to convergent growth phenotypes with different gene expression states. Genome Res 15(10):1365–72

Fong SS, Marciniak JY, Palsson BO (2003) Description and interpretation of adaptive evolution of *Escherichia coli* K-12 MG1655 by using a genome-scale *in silico* metabolic model. J Bacteriol 185(21):6400–8

Fong SS, Nanchen A, Palsson BO et al. (2006) Latent pathway activation and increased pathway capacity enable *Escherichia coli* adaptation to loss of key metabolic enzymes. J Biol Chem 281(12):8024–33

Fong SS, Palsson BO (2004) Metabolic gene-deletion strains of *Escherichia coli* evolve to computationally predicted growth phenotypes. Nat Genet 36(10):1056–8

Herrgard MJ, Fong SS, Palsson BO (2006) Identification of genome-scale metabolic network models using experimentally measured flux profiles. PLoS Comput Biol 2(7):e72

Herring CD, Raghunathan A, Honisch C et al. (2006) Comparative genome sequencing of *Escherichia coli* allows observation of bacterial evolution on a laboratory timescale. Nat Genet 38(12):1406–12

Hua Q, Joyce AR, Palsson BO et al. (2007) Metabolic characterization of *Escherichia coli* strains adapted to growth on lactate. Appl Environ Microbiol 73(14):4639–47

Ibarra RU, Edwards JS, Palsson BO (2002) *Escherichia coli* K-12 undergoes adaptive evolution to achieve *in silico* predicted optimal growth. Nature 420(6912):186–9

Karp PD, Keseler IM, Shearer A et al. (2007) Multidimensional annotation of the *Escherichia coli* K-12 genome. Nucleic Acids Res 35(22):7577–90

Lenski RE, Rose MR, Simpson SC et al. (1991) Long-term experimental evolution in *Escherichia coli*. I. Adaptation and divergence during 2,000 generations. Am Nat 138:1315–1341

Liu M, Durfee T, Cabrera JE et al. (2005) Global transcriptional programs reveal a carbon source foraging strategy by *Escherichia coli*. J Biol Chem 280(16):15921–7

Patil KR, Rocha I, Forster J et al. (2005) Evolutionary programming as a platform for *in silico* metabolic engineering. BMC Bioinformatics 6:308

Pharkya P, Burgard AP, Maranas CD (2003) Exploring the overproduction of amino acids using the bilevel optimization framework OptKnock. Biotechnol Bioeng 84(7):887–99

Pharkya P, Burgard AP, Maranas CD (2004) OptStrain: a computational framework for redesign of microbial production systems. Genome Res 14(11):2367–76

Pramanik J, Keasling JD (1997) Stoichiometric model of *Escherichia coli* metabolism: incorporation of growth-rate dependent biomass composition and mechanistic energy requirements. Biotechnol Bioeng 56(4):398–421

Reed JL, Palsson BO (2003) Thirteen years of building constraint-based *in silico* models of *Escherichia coli*. J Bacteriol 185(9):2692–9

Reed JL, Palsson BO (2004) Genome-scale *in silico* models of *E. coli* have multiple equivalent phenotypic states: assessment of correlated reaction subsets that comprise network states. Genome Res 14(9):1797–805

Sauer U (2004) High-throughput phenomics: experimental methods for mapping fluxomes. Curr Opin Biotechnol 15(1):58–63

Segre D, Vitkup D, Church GM (2002) Analysis of optimality in natural and perturbed metabolic networks. Proc Natl Acad Sci USA 99(23):15112–7

Varma A, Palsson BO (1994) Stoichiometric flux balance models quantitatively predict growth and metabolic by-product secretion in wild-type *Escherichia coli* W3110. Appl Environ Microbiol 60(10):3724–31

Chapter 13
Organisation of Complex *Escherichia coli* Promoters

Douglas F. Browning, David C. Grainger, and Stephen J. W. Busby

Contents

Abstract Promoters are locations on a chromosome that are responsible for the recruitment of RNA polymerase to different transcription units. This recruitment is tightly regulated and this chapter discusses how this regulation is effected at the molecular level in *Escherichia coli*. Transcription factors play a major role in this regulation and the different mechanisms by which they control transcription initiation at simple and complex promoters are outlined. At some promoters, the actions of transcription factors are modulated by nucleoid associated proteins. Two examples where this occurs, the regulatory regions of the *Escherichia coli nir* and *dps* genes, are described in detail.

13.1 Transcriptional Regulation in *Escherichia coli*

Gene expression in *Escherichia coli* is primarily regulated at the level of transcription initiation, the point at which RNA synthesis begins. The enzyme responsible for RNA synthesis is RNA polymerase and, predictably, it is the target for many regulatory factors. This is mainly done by DNA sequence elements at promoters, and by transcription factors that modulate promoter activity. In addition, a range of

S.J.W. Busby (✉)
School of Biosciences, University of Birmingham, Birmingham B15 2TT UK
e-mail: s.j.w.busby@bham.ac.uk

S.Y. Lee (ed.), *Systems Biology and Biotechnology of Escherichia coli*,
DOI 10.1007/978-1-4020-9394-4_13, © Springer Science+Business Media B.V. 2009

RNA polymerase-binding proteins, nucleoid-associated proteins, small molecules and metabolites also intervene (Browning and Busby 2004).

The bacterial RNA polymerase exists in two states. One form, known as the core enzyme, can catalyse RNA synthesis, but is unable to bind to promoter targets in DNA. The second form, the holoenzyme, is capable of both RNA synthesis and promoter recognition. The bacterial RNA polymerase is a multi-subunit enzyme and both forms of RNA polymerase posses two α subunits, and the β and β' and ω subunits. The holoenzyme form contains an additional subunit, σ, and it is this subunit that facilitates promoter DNA recognition directly. Following σ mediated DNA binding, transcription initiation occurs, the σ subunit then dissociates from the RNA polymerase-DNA-mRNA complex, and the core enzyme completes the process of gene transcription (Helmann and Chamberlin 1988). It is estimated that there are $\sim$ 5000 copies of RNA polymerase in growing *Escherichia coli* K-12 cells, which must be distributed between $\sim$ 3000 transcription units. Everything we know about bacterial transcription tells us that this distribution is not even and, thus, the cell has to regulate the binding of RNA polymerase across its chromosome prudently (Ishihama 1997).

Since the σ subunit of RNA polymerase is responsible for DNA recognition, it plays a pivotal role in managing the chromosome-wide distribution of the transcriptional machinery. *Escherichia coli*, like most bacteria, contains one major σ factor (σ^{70}), responsible for the recognition of most promoters, and several alternate σ factors, each present at lower levels. Each of the alternative σ factors is responsible for transcription of a subset of genes, usually in response to a stress. Thus, for example, the stationary phase σ factor (σ^{38}) controls the expression of many proteins needed for the long-term survival of non-growing cells (Ishihama 1997).

Figure 13.1 illustrates our current understanding of how RNA polymerase holoenzyme recognizes promoters (Murakami and Darst 2003). Most bacterial σ factors contain several independently folding domains that recognize different promoter elements. Thus, Domain 2 and Domain 4 recognize promoter -10 and -35 hexamer elements, that are located 10 and 35 base pairs upstream of the transcription start site at most promoters. Other sequence elements recognized by RNA polymerase are the extended -10 TG motif, immediately upstream of the -10 element, and the UP element, located upstream of the -35 element. Extended -10 elements are recognized by Domain 3 of σ subunits, whilst UP elements interact with the C-terminal domains of the RNA polymerase α subunits. Consensus sequences for these elements have been defined and the activity of any promoter is primarily defined by the correspondence of these sequences to the consensus. Note that it is rare for all 4 elements to be functional at a promoter, and hence different bacterial promoters carry different combinations of functional elements (Miroslavova and Busby 2006). Whilst these elements set the promoter strength, regulation requires transcription factors. These are sequence-specific DNA binding proteins that modulate the frequency of transcription initiation at target promoters.

Transcriptional activators and repressors share many properties and often exert their effect in response to environmental cues (Perez-Rueda and Collado-Vides

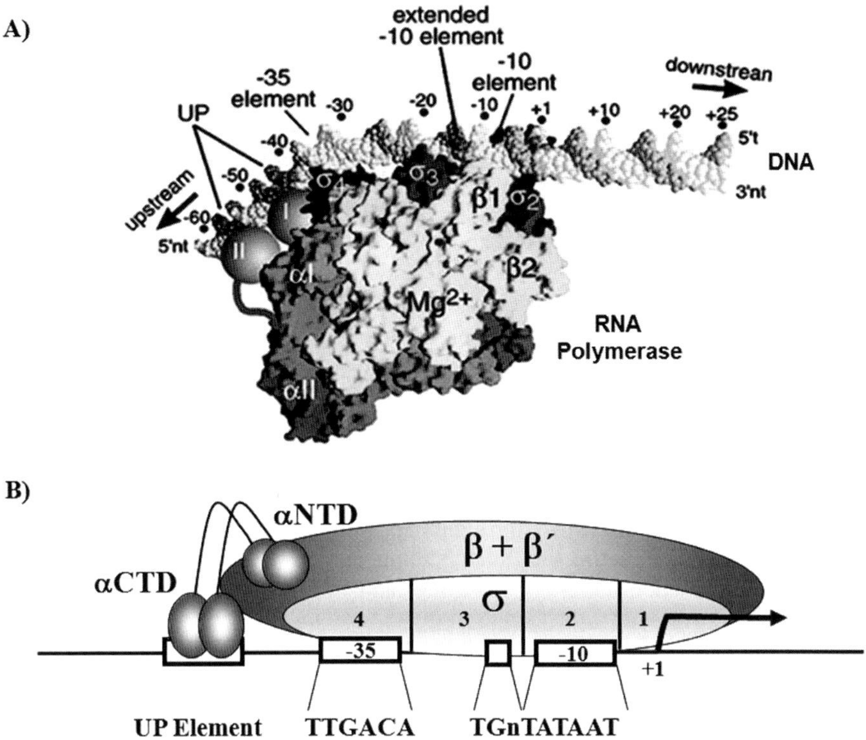

Fig. 13.1 RNA polymerase and its interactions at promoters. (**A**) Model based on crystallographic studies of the initial docking of holo RNA polymerase to a promoter (adapted from Murakami and Darst 2003). The β, β′, σ and α subunits of RNA polymerase are indicated and the different promoter elements are shown. Spheres indicate the binding locations of αCTD. (**B**) Cartoon illustration of the above, showing the different interactions between promoter elements and RNA polymerase. Consensus sequences for the −35 (TTGACA), extended −10 (TGn) and −10 (TATAAT) elements are shown

2000). A small number of transcription factors, termed 'global' regulators, influence the expression of a large number of transcription units. Conversely, a large number of 'specific' transcription factors each affect the expression of a small number of transcription units. The expression of many transcription units is regulated by a combination of both global and specific transcription factors and this allows bacteria to differentially regulate the gene expression in response to combinations of different environmental stimuli.

13.2 Simple Repression and Activation at Bacterial Promoters

Some promoters are active in the absence of additional factors and when the genes under their control are not required, they are silenced by transcription repressors. However, most promoters lack a good match to the consensus elements for RNA polymerase binding, and many of these require ancillary proteins, known as transcription activators, to function.

Repressor proteins reduce transcription initiation at target promoters and the textbook view is that this is simple to understand. Thus, at some promoters, a single repressor is involved and its binding prevents promoter recognition by RNA polymerase (Choy and Adhya 1996). In these instances, the repressor binding site is located at, or close to, the core promoter elements (Fig.13.2A). Note that, in some cases, the repressor may not prevent binding of RNA polymerase but, rather, interferes with post-recruitment steps in transcription initiation (Rojo 2001). At other promoters, multiple repressor molecules bind to promoter-distal sites, and repression may be caused by DNA looping, which shuts off transcription initiation within the looped domain (Fig. 13.2B).

At some promoters, activation of transcription is simple, and involves the action of a single activator (Busby and Ebright 1994, Rhodius and Busby 1998) Three general mechanisms are used for 'simple' activation. In Class I activation (Fig. 13.3A), the activator binds to a target located upstream of the promoter −35 element and recruits RNA polymerase to the promoter by directly interacting with the RNA polymerase αCTD. Because the linker joining the αCTD and αNTD is flexible, activators that function using a Class I mechanism can bind at various locations

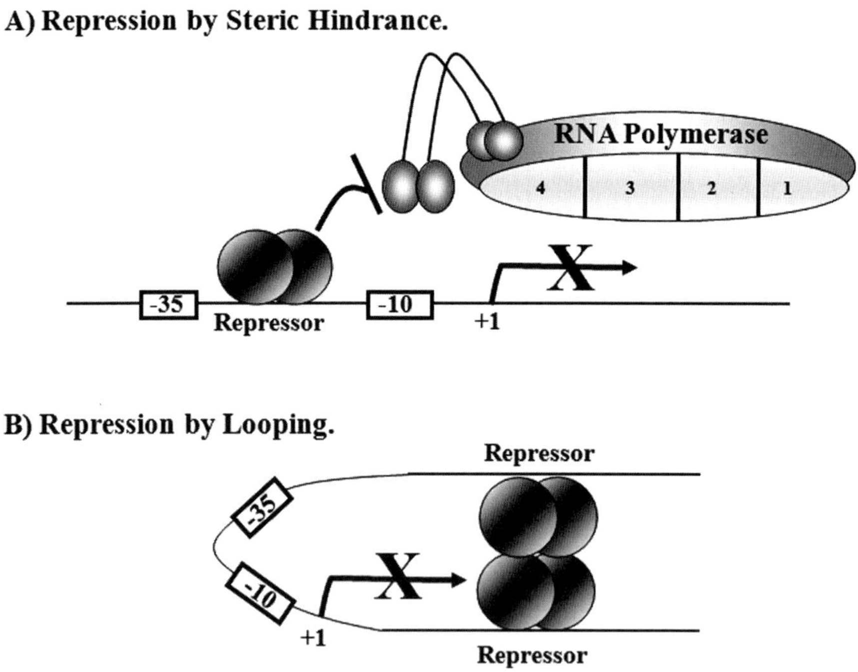

Fig. 13.2 Mechanisms of repression. (**A**) Repression by steric hindrance. The repressor binding site overlaps core promoter elements. (**B**) Repression by looping. Repressors bind to distal sites and interact by looping, repressing the intervening promoter

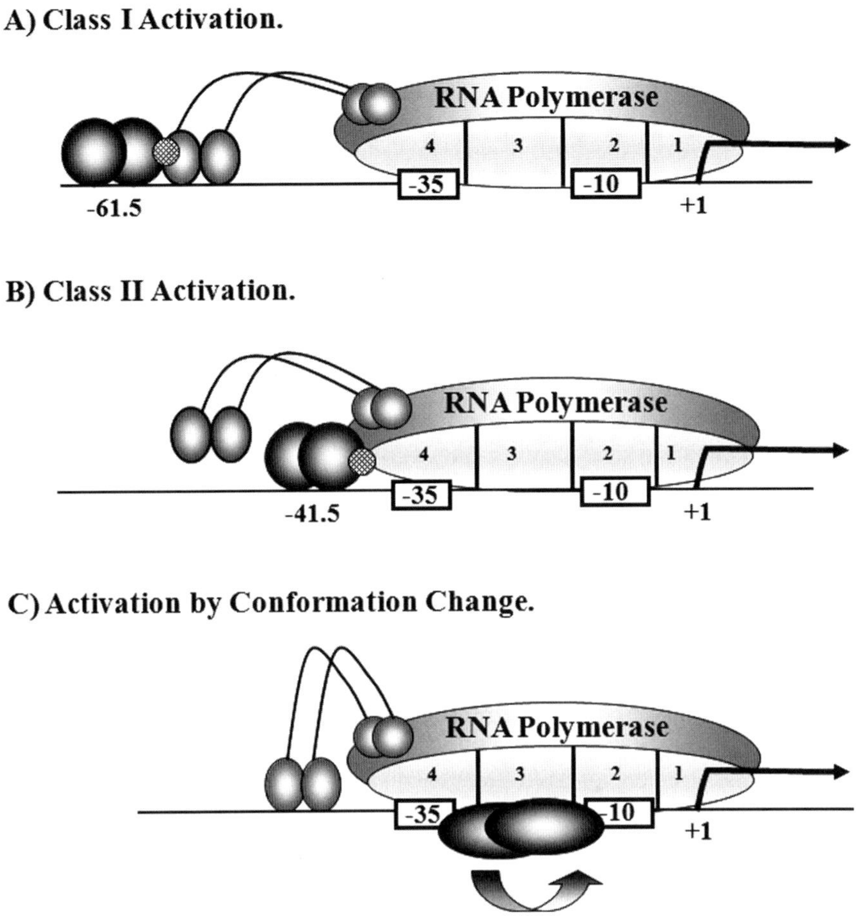

Fig. 13.3 Activation at simple bacterial promoters. The figure illustrates the organisation of RNA polymerase and activator subunits during activation. (**A**) Class I activation. The activator is bound to an upstream site and contacts the αCTD of RNA polymerase, thereby recruiting polymerase to the promoter. (**B**) Class II activation. The bound activator overlaps the promoter −35 element and interacts with Domain 4 of σ^{70}. (**C**) Activation by conformational changes in promoter DNA. The activator realigns the −10 and −35 elements

upstream of promoters. In Class II activation (Fig. 13.3B), the activator binds to a target that overlaps the promoter −35 element and contacts Domain 4 of the RNA polymerase σ subunit. This contact also results in recruitment of RNA polymerase to the promoter, but other steps in initiation can also be affected. The third mechanism for simple activation is found in cases where the activator alters the conformation of the target promoter, to enable the interaction of RNA polymerase with the promoter −10 and −35 elements. This requires the activator to bind at or very near to the promoter elements (Fig. 13.3C). In the case of members of the MerR family, the activator binds between promoter −10 and −35 elements and alters their relative orientation so as to facilitate interaction with holo RNA polymerase (Brown et al. 2003).

13.3 Complex Repression and Activation at Bacterial Promoters

Most bacterial promoters are regulated by more than one transcription factor and this permits regulatory input from multiple environmental cues (Martinez-Antonio and Collado-Vides 2003). At promoters that are co-dependent on two or more activators, complicated mechanisms are brought into play, and the four mechanisms so far discovered are illustrated in Fig. 13.4. These involve the repositioning of one activator by another, independent activator-RNA polymerase contacts, co-operative activator binding and anti-repression by an activator (Barnard et al. 2004, Browning and Busby 2004).

For mechanisms involving repositioning, the role of the secondary activator is to reposition the primary activator from a location where it is unable to activate transcription to a location where it can activate transcription. This repositioning can involve either shifting the primary activator from one DNA site to another or altering the conformation of the DNA to allow the primary activator to interact with RNA polymerase (Fig. 13.4A). A different mechanism operates at promoters where activators must make independent contacts with RNA polymerase for transcription activation. At some complex promoters, both activators function by a Class I mechanism, whilst at others, one activator functions by a Class I mechanism, with the other using a Class II mechanism (Fig. 13.4B). These promoters (often referred to as Class III promoters) contrast with simple Class I and Class II activator-dependent promoters, where the interaction of RNA polymerase with a single activator is sufficient for full activation (Fig. 13.3A,B). In most cases studied to date, where two activators make independent contacts with RNA polymerase, the two activators bind independently at the target promoter. Because, in these cases, the different activators are functioning independently, promoters of this type are easy to evolve and hence are widespread. However, in some cases, activators bind co-operatively, and this provides another mechanism for ensuring co-dependence, since one activator is unable to bind in the absence of the other (Fig. 13.4C). Finally, in some cases,

Fig. 13.4 Mechanisms of promoter co-dependence on two activator proteins. (**A**) Repositioning of the primary activator by a secondary activator. In (i), the secondary activator repositions the primary activator from a location where it is unable to activate transcription to a location where it can activate. In (ii), the secondary activator alters the conformation of the DNA by bending, bringing the primary activator into a position from which it can activate. (**B**) Independent contacts by both activators are required for optimal activation. In (i), both activators function by a Class I mechanism. In (ii), one activator functions by a Class II mechanism, and the second by a Class I mechanism. (**C**) Co-operative binding: the binding of one activator is dependent upon the binding of the second. (**D**) Anti-repression. The binding of the secondary activator is required to counteract the inhibitory effects of a repressor, in order to allow the primary activator to function

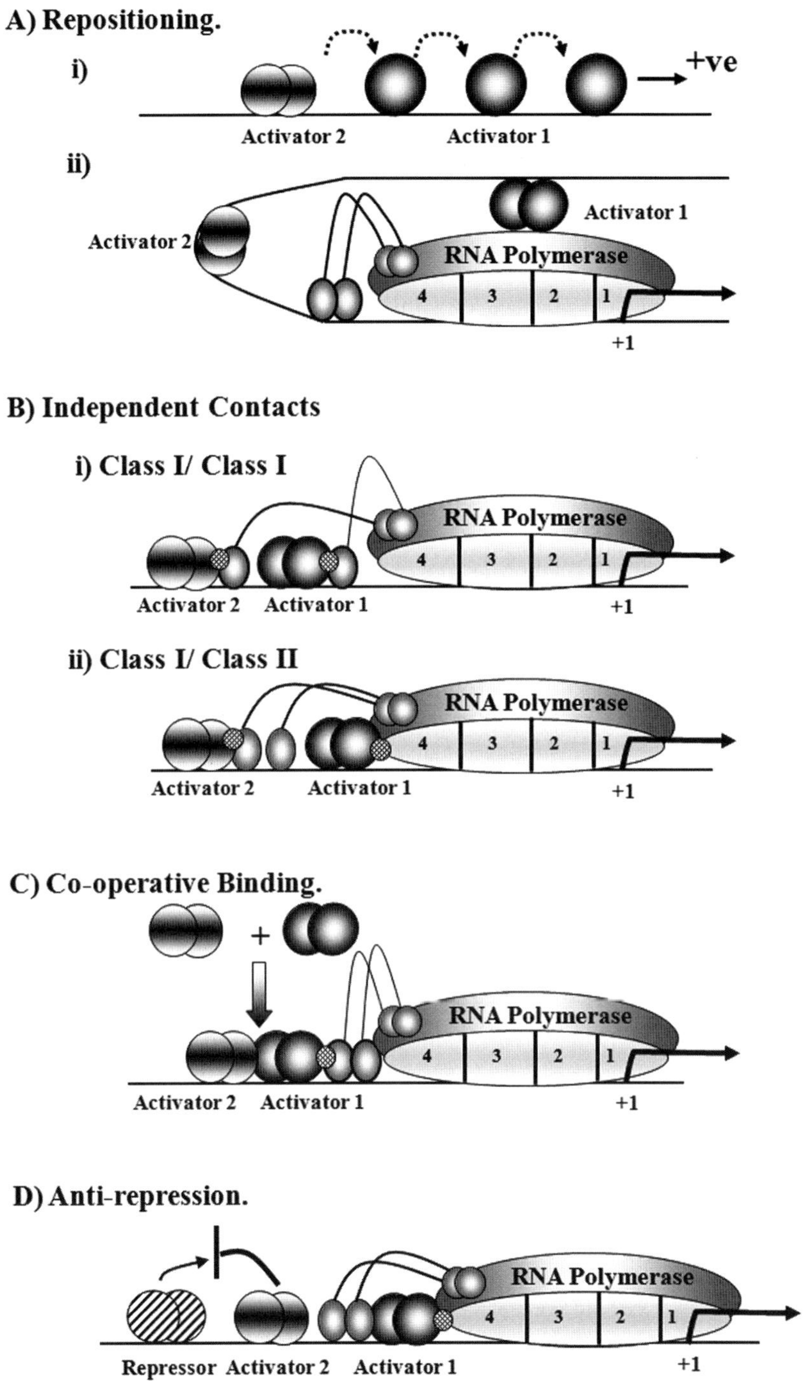

Fig. 13.4 (continued)

the role of the second activator is not to activate directly, but rather to prevent the action of a repressor that is interfering with the function of the primary activator (Fig. 13.4D).

13.4 Nucleoid Associated Proteins can Participate at Complex Promoters

The *Escherichia coli* chromosome folds into a compact structure, the nucleoid, and a set of nucleoid-associated proteins is involved in maintaining this (Dame 2005, Thanbichler et al. 2005). Many of these proteins are present at high levels and sharply bend target DNA upon binding, and it is often thought that they bind across the entire chromosome with little sequence specificity, like eukaryotic histones. However, the activity of some promoters is modulated by nucleoid-associated proteins binding at specific targets (McLeod and Johnson 2001). Also, it is now clear that many of the 'repressors' responsible for conferring codependence of promoters on two activators (Fig. 13.4D) are nucleoid-associated proteins or combinations thereof. In this context, three of the best studied *Escherichia coli* nucleoid proteins are Fis (Factor for Inversion Stimulation), H-NS (Histone-like Nucleoid-structuring protein) and IHF (Integration Host Factor). To gain insight into the global roles of these factors, chromatin immunoprecipitation was exploited to find their binding locations (Grainger et al. 2006). To do this, the sequence composition of immunoprecipitated DNA was analysed using high density microarrays. Figure 13.5 shows a typical set of results illustrating the distribution of Fis, H-NS and IHF across the chromosome of *Escherichia coli* K-12. Each scan shows enrichment (y-axis) for DNA sequences at particular loci (x-axis) in the immunoprecipitated DNA sample. As expected, each of the nucleoid-associated proteins binds at hundreds of targets, but, surprisingly, analysis of the binding locations shows that $\sim 60\%$ of the targets are in intergenic regulatory regions. Since these regions cover less than 10% of the total genome, it is clear that Fis, H-NS and IHF binding is highly focused, that they are unlike eucaryotic histones, and that they must orchestrate DNA folding from regulatory regions. Moreover their binding locations are completely consistent with their known roles in transcriptional regulation. Analysis of the target locations revealed many regulatory regions where Fis and H-NS both interact, compared with smaller numbers of targets for Fis and IHF or H-NS and IHF. Interestingly, *Escherichia coli* contains some targets where Fis, H-NS and IHF all bind together. As a first step to understanding the rationale for this binding, and the molecular mech-

→

Fig. 13.5 Binding of Fis, H-NS and IHF across the *E. coli* chromosome. The figure shows chromosome-wide DNA binding profiles for (**A**) Fis, (**B**) H-NS and (**C**) IHF, generated from chromatin immunoprecipitation experiments in which immunoprecipitated DNA was analysed on high density microarrays (Grainger et al. 2006). The x-axis indicate sequence coordinates on the chromosome of *E. coli* K-12 strain MG1655 and the y-axis indicate the signal intensity at that position in each experiment

A)

B)

C)

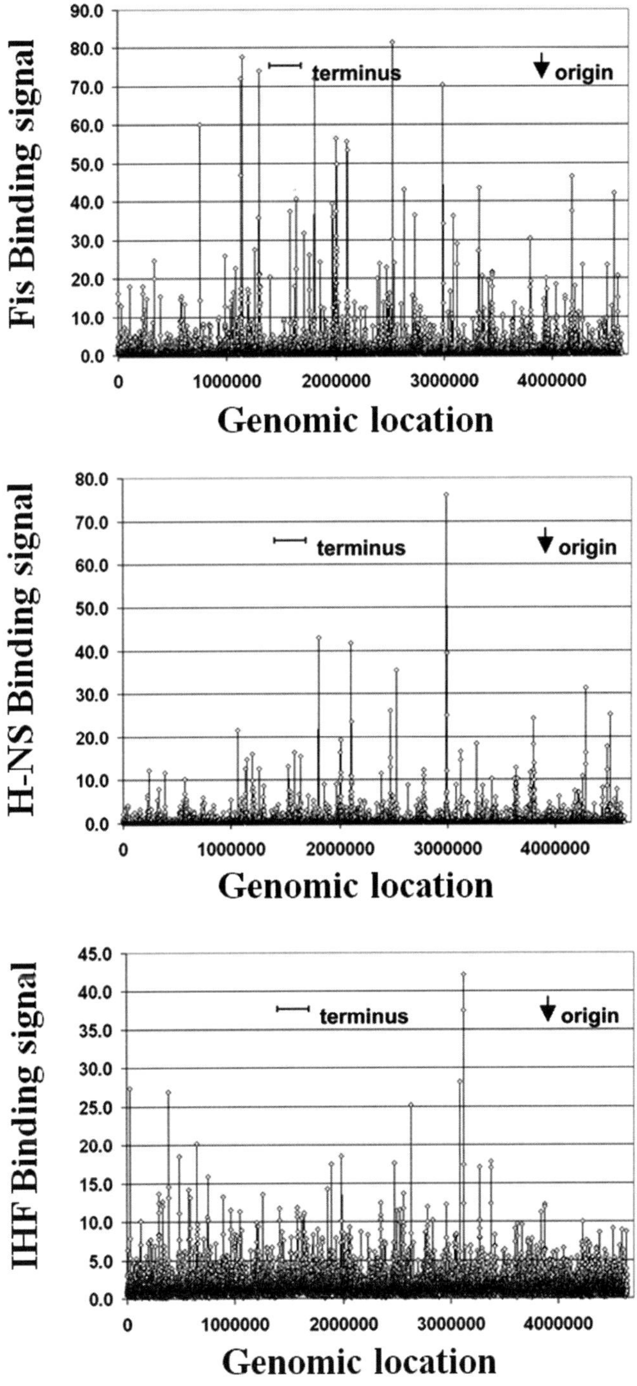

Fig. 13.5 (continued)

anisms that are involved, we have made detailed studies of two of these targets, the regulatory regions of the transcription units covering the *nir* operon and the *dps* gene.

13.5 Regulation of the *Escherichia coli nir* Operon Promoter

The *nir* operon encodes a cytoplasmic nitrite reductase that catalyses the NADH-dependent reduction of nitrite ions to ammonia (Harborne et al. 1992). Transcription from a single startpoint is controlled by a promoter upstream of the *nirB* gene (Jayaraman et al. 1988). At this promoter, H-NS is an overall repressor, whereas Fis and IHF function in concert to confer codependence on two activators (Browning et al. 2000).

Early studies had shown that the *nir* promoter is optimally active when cells are grown in anaerobic conditions in the presence of nitrite or nitrate ions, and also that higher activities are found in rich media (Bell et al. 1990). Induction in anaerobic conditions is due to the activity of FNR, a global transcription activator responsible for the induction of over 100 different transcription units in *Escherichia coli* in response to low oxygen levels (Browning et al. 2002). FNR dimerisation, and hence specific binding at target promoters, requires the formation of an iron-sulphur cluster, and thus its activity is inhibited by oxygen. However as oxygen levels decrease, FNR binds to target sites and activates transcription. In most cases, including the *nir* promoter, the DNA binds to a target near position −41 and functions as a Class II activator (Wing et al. 1995).

In addition to being dependent on FNR, expression from the *nir* promoter is dependent on activation by NarL. NarL is a response regulator, that is activated by nitrate or nitrite ions via sensor kinases, and it is this dependence which ensures that *nir* operon induction is coupled to the presence of nitrate or nitrite ions as well as the lack of oxygen (Tyson et al. 1993). NarL binds as a dimer to a DNA site just upstream of the DNA site for FNR (Fig. 13.6). It was found that NarL binding has no effect on FNR binding, and this raised the puzzle of why NarL is needed, and why FNR, which is perfectly able to function as a Class II activator alone, is unable to function without NarL at the *nir* promoter. The key to this is the observation that the sequences upstream of the DNA site for FNR at the *nir* promoter carry targets for Fis at position −142 and for IHF at positions −88 and −115 (Browning et al. 2000, 2004). Genetic experiments have shown that FNR-dependent activation is suppressed by the binding together of Fis at position −142 and IHF at position −88. *In vitro* experiments with purified components demonstrated that FNR is able to activate open complex formation at the *nir* promoter and that the complex is destabilized by binding of Fis and IHF. The suppression mediated by Fis and IHF is relieved upon binding of NarL, which displaces IHF from position −88 (Fig. 13.6). Thus here, the role of the second activator, NarL, is not to activate directly, but rather to prevent the action of repressors (Fis and IHF) that interfere with the primary activator (FNR). Hence, point mutations that destroy the DNA sites for Fis and

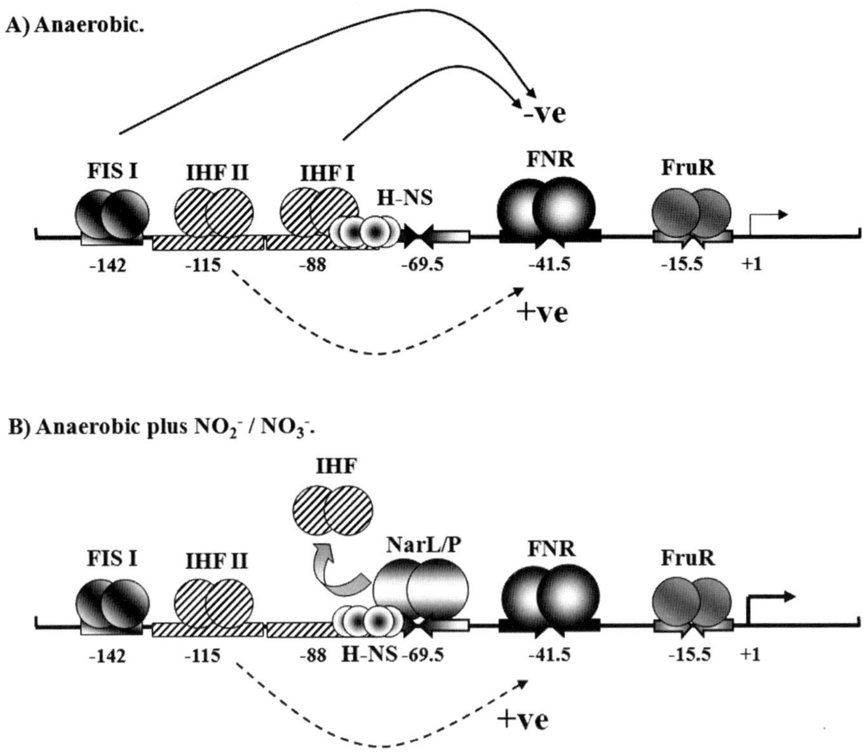

Fig. 13.6 Transcription regulation at the *E. coli nir* promoter. (**A**) Anaerobic conditions. The binding of Fis to Fis I and IHF to IHF I inhibits FNR-dependent transcription (−ve), whilst occupancy of the lower affinity IHF II site stimulates transcription (+ve). The mechanisms by which upstream bound Fis and IHF modulate FNR-dependent transcription initiation at the *nir* promoter are not understood. In addition, the binding of H-NS and FruR can down-regulate transcription initiation at the *nir* promoter. (**B**) Anaerobic conditions plus nitrite or nitrate. The binding of NarL (or its homologue, NarP) displaces IHF from IHF I, thus counteracting the repression mediated by IHF and Fis, and enabling maximal FNR-dependent transcription. In contrast, NarL and NarP have little effect on repression by H-NS and FruR, whose effects are modulated by temperature and nutrient richness respectively

IHF at positions −142 and −88 respectively, release the requirement for NarL, and convert the *nir* promoter into a simple Class II FNR-dependent promoter.

Further analysis has revealed three additional complications. First, IHF binding to a weaker second site at position −115 promotes rather than suppresses FNR-dependent transcription activation (Browning et al. 2004). Thus, IHF bound at two adjacent sites (at positions −88 and −115) have opposite effects on *nir* promoter activity (Fig. 13.6). Interestingly, the relative IHF binding affinities at the two sites differ in diverse enteric bacteria and this sets the basal level of FNR-dependent activation in the absence of NarL. Hence the basal NarL-independent activity of the *nir* promoter is increased by mutations that improve IHF binding at position −115 and decreased by mutations that destroy binding (Browning and Busby, unpublished results). Second, H-NS, binds to a specific single target that overlaps the DNA site for NarL and globally down-regulates *nir* promoter activity (Browning et al. 2000). Finally FruR also represses the *nir* promoter by binding to a site centered at position -15.5 (Tyson et al. 1997). Whilst the significance of repression by H-NS remains unclear, the role of FruR appears to be to down-regulate *nir* expression in poor media. FruR is a *lac* repressor family member that is displaced from its targets by fructose diphosphate and thus it binds and represses *nir* expression when nutrients are in short supply and glycolytic flux is low (Saier and Ramseier 1996).

13.6 Regulation of the *Escherichia coli dps* Promoter

Dps is a nucleoid associated protein that is absent in rapidly growing *Escherichia coli* but accumulates as growth slows and cells enter stationary phase (Almiron et al. 1992). In non-growing cells Dps becomes the most abundant nucleoid-associated protein and this is thought to be a key factor in maintaining the stationary phase folded chromosome. Interestingly, levels of Fis, which is very abundant in growing *Escherichia coli* cells, are reduced to near zero as cell growth slows upon entry into stationary phase (Ali Azam et al. 1999). The expression of Dps depends on a single promoter located just upstream of the *dps* gene and accumulation of Dps requires the stationary phase σ factor, σ^{38}. The observation that the *dps* promoter can be served by RNA polymerase containing either σ^{38} or the major σ factor, σ^{70}, raises the puzzle of what prevents *dps* from being expressed in rapidly growing cells (Altuvia et al. 1994). The key to understanding this is the recent discovery of a DNA site for Fis that overlaps the *dps* promoter -10 region, which suggested that Fis binding might account for repression in rapidly growing cells (Grainger et al. 2008). *In vitro* studies have confirmed this repression but have revealed a novel repression mechanism in which bound Fis jams RNA polymerase containing σ^{70} at the *dps* promter. Thus, in rapidly growing cells, the *dps* promoter is silenced by a ternary repression complex containing RNA polymerase with σ^{70}, Fis and promoter DNA. Remarkably, Fis has little or no effect on the activity of RNA polymerase containing σ^{38}, and thus Fis discriminates between different forms of RNA polymerase.

As well as being repressed by Fis, the *dps* promoter is also regulated by Fis, H-NS, IHF and OxyR (Fig.13.7). Like Fis, H-NS acts as a repressor that discriminates between RNA polymerase containing σ^{70} and σ^{38} (Grainger et al. 2008). H-NS displaces RNA polymerase containing σ^{70} from the *dps* promoter, whilst not interfering with RNA polymerase containing σ^{38}. Thus, together with Fis, H-NS confers σ factor dependence on *dps* expression. In contrast, a third nucleoid-associate protein, IHF, binds upstream of the core *dps* promoter elements and functions as an activator during σ^{38}-dependent transcription in stationary phase (Altuvia et al. 1994, Ohniwa et al. 2006). Finally a second activator, OxyR, which is triggered by oxidative stress, also binds upstream, and is responsible for transient induction of *dps* during oxidative stress in rapidly growing cells (Altuvia et al. 1994, Ohniwa et al. 2006). In these circumstances, the repression by Fis and H-NS must be overcome, but the mechanism for this is unclear at present.

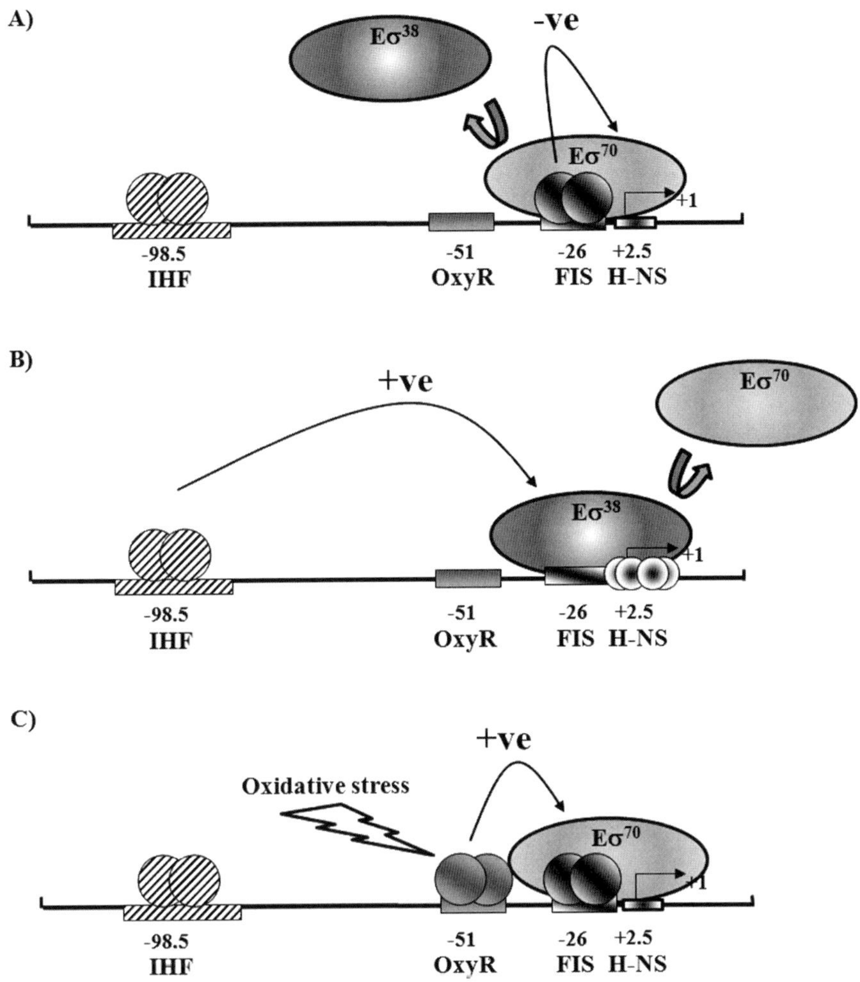

Fig. 13.7 Selective regulation of the *E. coli dps* promoter. (**A**) Selective repression by Fis. During rapid growth, transcription from the *dps* promoter is repressed by Fis, which binds to the promoter in unison with $E\sigma^{70}$ and shuts down the promoter. (**B**) Selective repression by H-NS. Binding of H-NS to the *dps* promoter blocks binding of $E\sigma^{70}$ but permits binding of $E\sigma^{38}$. Transcription by $E\sigma^{38}$ (but not $E\sigma^{70}$) can be stimulated by IHF. (**C**) Activation by OxyR. In response to oxidative stress, transcription from the *dps* promoter by $E\sigma^{70}$ is enhanced by OxyR, which somehow overcomes the negative effects of Fis and H-NS

13.7 Perspectives

Escherichia coli is found in many places, and most of these, such as the guts of animals and aquatic environments, are subject to rapid and frequent fluctuations. As for most bacteria, survival depends on the selective expression of gene products to cope with the environment, and thus, it is no surprise that *Escherichia coli* has evolved sophisticated systems to control transcription. This is most apparent in the high proportion of its gene products that are dedicated to regulating transcription initiation and in the complexity of even the simplest promoter. Thus, the *nir* operon promoter is regulated by four transcription factors: by FNR, by NarL (and its homologue, NarP) and by FruR and their activity is modulated by three nucleoid-associated proteins, IHF, Fis and H-NS. Although we can assume that different combinations of these factors are used in different conditions, most studies have been performed in 'simple' laboratory conditions, and the relative importance of the different factors in 'real' environments is still poorly understood. A quick glance at the *Ecocyc* database will convince anyone that the simplistic models for promoter regulation that appear in the textbooks are misleading. These models are mostly based on a small number of paradigm promoters (such as the *lac* promoter) and were established early in the history of this subject area. We now know that many, if not most, promoters are very complicated, with multiple factors interacting and other factors such as small ligands, the local chromosome landscape and DNA topology intervening. The challenge now is for us to put all the facts together, to produce integrated models, and, most important, to understand how systems are evolving.

Acknowledgments Work in the authors' laboratory has been supported by the Wellcome trust and the UK BBSRC.

References

Ali Azam T, Iwata A, Nishimura A et al. (1999) Growth phase-dependent variation in protein composition of the *Escherichia coli* nucleoid. J Bacteriol 181(20):6361–70

Almiron M, Link AJ, Furlong D et al. (1992) A novel DNA-binding protein with regulatory and protective roles in starved *Escherichia coli*. Genes Dev 6(12B):2646–54

Altuvia S, Almiron M, Huisman G et al. (1994) The *dps* promoter is activated by OxyR during growth and by IHF and sigma S in stationary phase. Mol Microbiol 13(2):265–72

Barnard A, Wolfe A, Busby S (2004) Regulation at complex bacterial promoters: how bacteria use different promoter organizations to produce different regulatory outcomes. Curr Opin Microbiol 7(2):102–8

Bell AI, Cole JA, Busby SJ (1990) Molecular genetic analysis of an FNR-dependent anaerobically inducible *Escherichia coli* promoter. Mol Microbiol 4(10):1753–63

Brown NL, Stoyanov JV, Kidd SP et al. (2003) The MerR family of transcriptional regulators. FEMS Microbiol Rev 27(2–3):145–63

Browning DF, Busby SJ (2004) The regulation of bacterial transcription initiation. Nat Rev Microbiol 2(1):57–65

Browning DF, Cole JA, Busby SJ (2000) Suppression of FNR-dependent transcription activation at the *Escherichia coli nir* promoter by Fis, IHF and H-NS: modulation of transcription initiation by a complex nucleo-protein assembly. Mol Microbiol 37(5):1258–69

Browning DF, Cole JA, Busby SJ (2004) Transcription activation by remodelling of a nucleoprotein assembly: the role of NarL at the FNR-dependent *Escherichia coli nir* promoter. Mol Microbiol 53(1):203–15

Browning DF, Lee DJ, Green J et al. (2002) Secrets of bacterial transcription initiation taucht by the *Escherichia coli* FNR protein. In: Hodgson D, Thomas C (ed) Signals, Switches, Regulons & Cascades: Control of Bacterial Gene Expression, SGM Symposium

Busby S, Ebright RH (1994) Promoter structure, promoter recognition, and transcription activation in prokaryotes. Cell 79(5):743–6

Choy H, Adhya S (1996) Negative Control. In: Neidhardt F (ed) *Escherichia coli* and *Salmonella*, ASM Press, Washington DC

Dame RT (2005) The role of nucleoid-associated proteins in the organization and compaction of bacterial chromatin. Mol Microbiol 56(4):858–70

Grainger DC, Goldberg MD, Lee DJ et al. (2008) Selective repression by Fis and H-NS at the *Escherichia coli dps* promoter. Mol Microbiol 68(6):1366–77

Grainger DC, Hurd D, Goldberg MD et al. (2006) Association of nucleoid proteins with coding and non-coding segments of the *Escherichia coli* genome. Nucleic Acids Res 34(16): 4642–52

Harborne NR, Griffiths L, Busby SJ et al. (1992) Transcriptional control, translation and function of the products of the five open reading frames of the *Escherichia coli nir* operon. Mol Microbiol 6(19):2805–13

Helmann JD, Chamberlin MJ (1988) Structure and function of bacterial sigma factors. Annu Rev Biochem 57:839–72

Ishihama A (1997) Adaptation of gene expression in stationary phase bacteria. Curr Opin Genet Dev 7(5):582–8

Jayaraman PS, Gaston KL, Cole JA et al. (1988) The *nirB* promoter of *Escherichia coli*: location of nucleotide sequences essential for regulation by oxygen, the FNR protein and nitrite. Mol Microbiol 2(4):527–30

Martinez-Antonio A, Collado-Vides J (2003) Identifying global regulators in transcriptional regulatory networks in bacteria. Curr Opin Microbiol 6(5):482–9

McLeod SM, Johnson RC (2001) Control of transcription by nucleoid proteins. Curr Opin Microbiol 4(2):152–9

Miroslavova NS, Busby SJ (2006) Investigations of the modular structure of bacterial promoters. Biochem Soc Symp 73:1–10

Murakami KS, Darst SA (2003) Bacterial RNA polymerases: the wholo story. Curr Opin Struct Biol 13(1):31–9

Ohniwa RL, Morikawa K, Kim J et al. (2006) Dynamic state of DNA topology is essential for genome condensation in bacteria. Embo J 25(23):5591–602

Perez-Rueda E,Collado-Vides J (2000) The repertoire of DNA-binding transcriptional regulators in *Escherichia coli* K-12. Nucleic Acids Res 28(8):1838–47

Rhodius VA, Busby SJ (1998) Positive activation of gene expression. Curr Opin Microbiol 1(2):152–9

Rojo F (2001) Mechanisms of transcriptional repression. Curr Opin Microbiol 4(2):145–51

Saier MH, Jr., Ramseier TM (1996) The catabolite repressor/activator (Cra) protein of enteric bacteria. J Bacteriol 178(12):3411–7

Thanbichler M, Wang SC, Shapiro L (2005) The bacterial nucleoid: a highly organized and dynamic structure. J Cell Biochem 96(3):506–21

Tyson K, Busby S, Cole J (1997) Catabolite regulation of two *Escherichia coli* operons encoding nitrite reductases: role of the Cra protein. Arch Microbiol 168(3):240–4

Tyson KL, Bell AI, Cole JA et al. (1993) Definition of nitrite and nitrate response elements at the anaerobically inducible *Escherichia coli nirB* promoter: interactions between FNR and NarL. Mol Microbiol 7(1):151–7

Wing HJ, Williams SM, Busby SJ (1995) Spacing requirements for transcription activation by *Escherichia coli* FNR protein. J Bacteriol 177(23):6704–10

Chapter 14
Plasmid Regulation and Systems-Level Effects on *Escherichia coli* Metabolism

Dave Siak-Wei Ow, Dong-Yup Lee, Hsiu-Hui Tung, and Sue Lin-Chao

Contents

Abstract ColE1-type plasmids are multicopy extra-chromosomal vectors with wide-spread applications in many areas of genetic engineering and biotechnology. While the regulation of ColE1 replication is primarily effected by plasmid-encoded factors, the continual discovery of new host-encoded factors modulating ColE1 replication such as RNases and exoribonucleases reveals that the *Escherichia coli* host could exert a considerable effect on plasmid replication as well. On the other hand, the presence of plasmids also imposes a metabolic burden impeding host growth and metabolism. The basis of this metabolic burden is multifaceted

D.S.-W. Ow (✉)
Bioprocessing Technology Institute, Agency for Science, Technology and Research (A*STAR), 20 Biopolis Way, #06-01 Centros, Singapore 138668
e-mail: dave_ow@bti.a-star.edu.sg

S. Lin-Chao (✉)
Institute of Molecular Biology, Academia Sinica, Taipei 11529, Taiwan

S.Y. Lee (ed.), *Systems Biology and Biotechnology of Escherichia coli*,
DOI 10.1007/978-1-4020-9394-4_14,

and appears to involve both the plasmid-related drain of cellular resources from the host cell and the perturbation of cellular regulatory state mediated by global transcriptional regulators. Through the systems-level analysis by "omics" tools and *in-silico* modeling, we are gaining better understanding of plasmid-host interactions. This chapter will discuss the interaction of host-encoded factors with the regulation of ColE1-type plasmid replication and the systems-level effects of these multicopy plasmids on metabolism of the *E. coli* host.

14.1 Plasmids and Its Biotechnological Applications

Plasmids are self-replicating extra-chromosomal DNA elements found in many bacteria and yeasts. Initially revealed as the F factor for conjugative gene transfer in *Escherichia coli* (Hayes 1953, Lederberg 1998, Lederberg et al. 1952), later studies on plasmids led to remarkable contributions to the field of molecular biology and biotechnology (Cohen 1993). Figure 14.1 summarizes the major scientific impacts arising from the studies of plasmids. The first replication origins were isolated and characterized from plasmids (Lovett and Helinski 1976, Timmis et al. 1975), providing the foundation for the construction of artificial chromosomes and our present understanding of DNA replication and topology. Subsequent analysis of plasmid-derived genetic elements such as operon and replicon builds further fundamental knowledge on DNA conjugation and fertility, gene expression, genetic recombination, gene transfer and transposable elements (Cohen 1993).

Early investigation of ColE1 type plasmid replication first led to the discovery of antisense RNA and its control on gene expression (Tomizawa et al. 1981). Subsequent studies on antisense RNA of ColE1 plasmid helped to clarify the mechanism of RNA decay (Lin-Chao and Cohen 1991, Xu et al. 1993). In addition, these plasmids also play a critical role in the development of recombinant DNA technology, gene cloning and genome evolution (Cohen 1993). The discovery of

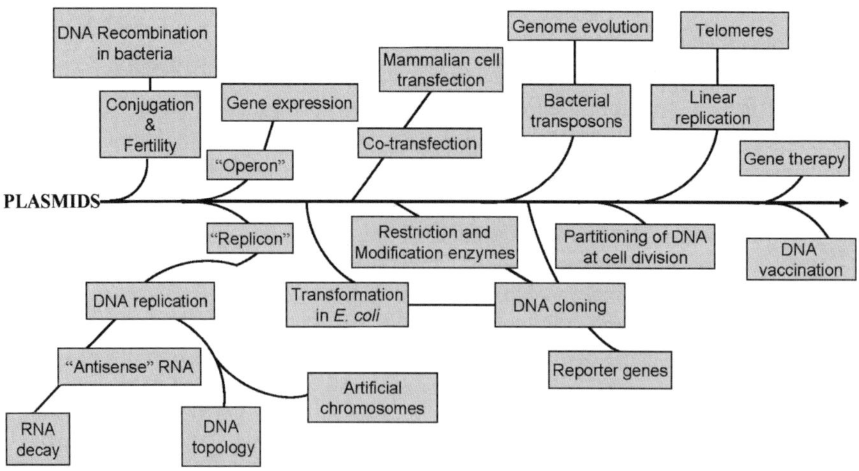

Fig. 14.1 Major scientific impacts rising from the studies of plasmid. (Cohen 1993)

Table 14.1 Commonly-utilized ColE1-type plasmids in biotechnology

Family	Examples	Application
pBR322	pBR322, pBR325, pBR328	Cloning
pBluescript	pBluescript SK, pBluescript KS, pBluescript II SK, pBluescript II KS	Cloning
pUC	pUC18, pUC118, pUC19, pUC119	Cloning/Expression
pET	pET3, pET5, pET7, pET9, pET11, pET12, pET39	Expression
Others	pVAC, pDNAVACCultra, pcDNA series, pCMV series	DNA vaccination

restriction endonucleases and stable transformation of *E. coli* using plasmid DNA led to the invention of recombinant DNA technology and gave rise to the present field of genetic engineering (Cohen 1973, Cohen et al. 1972, Watanabe et al. 1966).

Plasmids used in genetic engineering and biotechnology are commonly known as vectors. Plasmid-based vectors are important tools in biotechnology, where they allow the efficient cloning of genes and expression of desired recombinant proteins in *E. coli* and other microorganisms. Furthermore, there is also emerging interest to apply plasmid DNA as non-viral vectors for delivery of therapeutic or antigenic genes during gene therapy and DNA vaccination (Anderson and Schneider 2007, Ledley 1995, Liu and Huang 2002, Weide et al. 2008). The majority of plasmid vectors used in current recombinant DNA work were high-copy derivatives of ColE1-type plasmids (or its close relative pMB1) (Bolivar et al. 1977, Kahn et al. 1979). Examples of these are high-copy cloning vectors like the well-known pBR322, pBluescript and pUC series (Balbas et al. 1986). Table 14.1 lists some common ColE1-related plasmids currently in use. These high-copy plasmids are usually smaller than low-copy plasmids and, when transformed into the *E. coli* host, are routinely used for gene cloning, recombinant protein production and plasmid DNA production. Despite their widespread utility, we are only beginning to comprehend the complexity of plasmid-host interactions. A better understanding of plasmid-host interactions would allow the design of enhanced plasmid vectors and host strains for biotechnological processes. In view of this, the current review will discuss the interaction of host-encoded molecules with the regulation of ColE1-type plasmid replication, followed by the systems-level effects of these multicopy plasmids on *E. coli* metabolism.

14.2 Host Factor-Mediated Regulation of ColE1 Plasmid Replication

14.2.1 Replication of ColE1 Type Plasmids

ColE1 type plasmids are small circular plasmids naturally occurring in members of the family Enterobacteriaceae and they include plSA, pMB1, RSF1010 (NTP1), CloDF13 (Selzer et al. 1983), NTP16 (Lambert et al. 1987), and other coligenic plasmids (Zverev et al. 1984). The original ColEl is a 6.6 kb *E. coli* plasmid with a copy number of nearly 20 (Chan et al. 1985). It encodes for a 57 kDa colicin E1 toxin which can kill other *E. coli* cells by depolarizing the bacterial membrane and another protein (*Imm*), offering self-immunity to its colicin. The native plasmid contains the following genes: *cea, imm, kil, inc, RNAII, RNAI, rom, mob, cer* and *exc*; it also carries an origin of replication (*oriV*) and a region (*bom*) from which it can be mobilized for transfer to other bacteria.

The fundamental regulation of ColE1 plasmid replication by plasmid-encoded molecules has been extensively studied (Cabello et al. 1976, Davison 1984, Panayotatos 1984, Schmidt and Inselburg 1982). A region of about 600 bp in the ColE1 plasmid and several *E. coli* enzymes are involved in replication of ColE1. The initiation of replication of ColE1 plasmid proceeds from 555 bp upstream of the *oriV* and leads to the transcription of a pre-primer RNAII by RNA polymerase (Itoh and Tomizawa 1979). The nascent RNAII transcript hybridizes with the DNA template of ColEl and forms a DNA-RNA hybrid with a specific secondary structure which can be recognized and cleaved by RNase H, resulting in a free 3′-OH end which serves as a primer for DNA synthesis by DNA Polymerase I (Tomizawa and Som 1984). Plasmid replication proceeds by covalent extension of RNA primer (pRNA) from the *oriV* region (Fig. 14.2).

The repression of Co1E1 plasmid replication depends on the inhibition of the primer precursor, RNAII, by its plasmid-encoded antisense molecule, RNAI. RNAI is a 108-nucleotide molecule which is transcribed from 445 bp upstream of the *oriV*, in the opposite direction to RNAII, from a promoter located between the RNAII promoter and the origin of replication of ColE1 type plasmids. As the sequence of RNAI is complementary to the 5′-end of RNAII, RNAI can bind to RNAII and form a stable RNA-RNA hybrid (Cesareni et al. 1991, Kues and Stahl 1989). The binding of RNAII to RNAI leads to a conformational change in RNAII, preventing the formation of the DNA-RNA hybrid (Masukata and Tomizawa 1986, Polisky et al. 1990). Consequently, RNAII may not be able to function as a replication primer. Thus, RNAI plays a key role in the control of ColE1 plasmid copy number as the inhibitor of plasmid replication.

In addition to RNAI and RNAII, a third plasmid-encoded factor, Rom or Rop, can also negatively control the replication of ColE1. Rom is a small protein which has been proposed to accelerate the binding of RNAI to RNAII (Tomizawa and Som 1984) and inhibit RNAII primer formation (Cesareni et al. 1982). Thus, the expression of Rom protein reduces plasmid copy number and the *rom* gene is absent in many high-copy plasmids like pUC or pET. In line with that, a point mutation in RNAII that suppresses Rom was shown to increase plasmid copy number (Lin-Chao et al. 1992). Figure 14.2 illustrates the mechanism of ColE1 replication by antisense RNAI regulation.

14.2.2 Host-Mediated Regulation of ColE1 Plasmid Replication

Although copy number control of Co1E1-type plasmid in *E. coli* by plasmid-encoded molecules has been studied at length, there is recent evidence that other

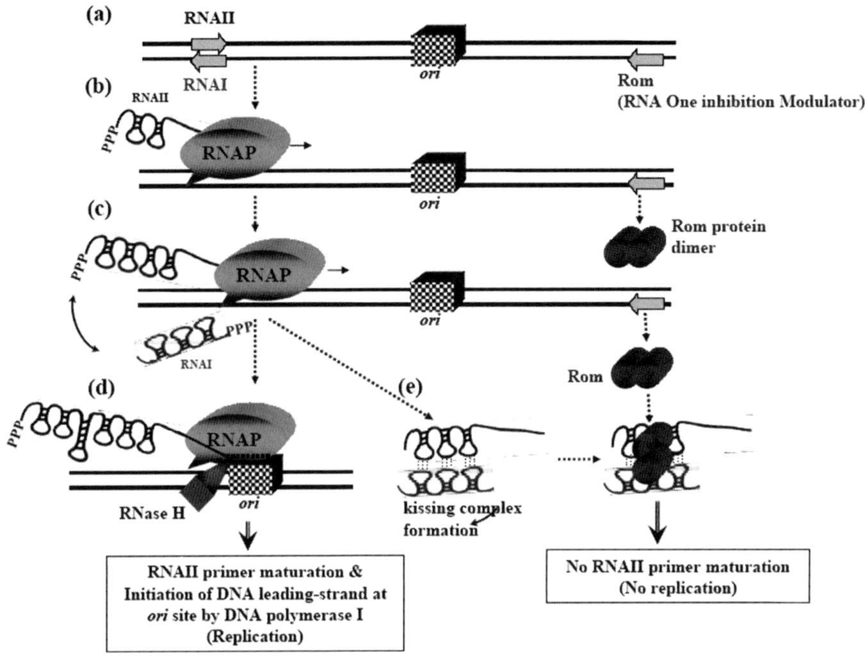

Fig. 14.2 The mechanism of ColE1 plasmid replication by anti-sense RNA regulation. (**a**) A genetic map of ColE1 plasmid replication. The blue arrows ( ) indicate the transcription direction for RNAII (primer RNA,), RNAI (anti-sense RNA,) and Rom (RNA One Modulator). The replication origin (*ori*,) is indicated. (**b–d**) The pre-primer RNAII and anti-sense RNAI are synthesized by the host RNA polymerase (RNAP,). RNase H ( ) cleavage at the DNA-RNA hybrid between RNAII and the DNA template at the origin region generates the 3′-OH end of the RNAII primer (i.e. RNAII primer maturation) for initiation of leading strand synthesis by DNA polymerase I. (**e**) When RNAI interacts with RNAII, the kissing complex is formed and stabilized by the Rom ( ) proteins. This anti-sense and primer RNA interaction inhibits the formation of DNA RNAII hybrid, and prevents maturation of pre-primer RNAII. As a result, no RNAII primer is available for plasmid replication

host-encoded factors also modulate overall plasmid copy number. In principle, any host-encoded factors affecting the stability or secondary structure of RNAI and RNAII will also interfere with the plasmid copy number. It has been shown that some host-encoded enzymes regulate the degradation of RNAI by endo- or exonucleolytic cleavage: (i) RNase E has been shown to have an endonucleolytic activity in RNAI decay (Lin-Chao and Cohen 1991); (ii) RNase III plays a role in turnover of RNAI (Binnie et al. 1999); (iii) polynucleotide phosphorylase is implicated in degradation of RNAI (Xu and Cohen 1995); and (iv) poly(A) polymerase I has been shown to play a role in the regulation of ColE1 plasmid copy number and RNAI decay (Xu et al. 2002). Figure 14.3 shows how host-encoded proteins interfering with the RNA-RNA interaction or regulating RNAI degradation can affect ColE1 plasmid replication (e.g. r-protein L_4; Singh et al. 2008).

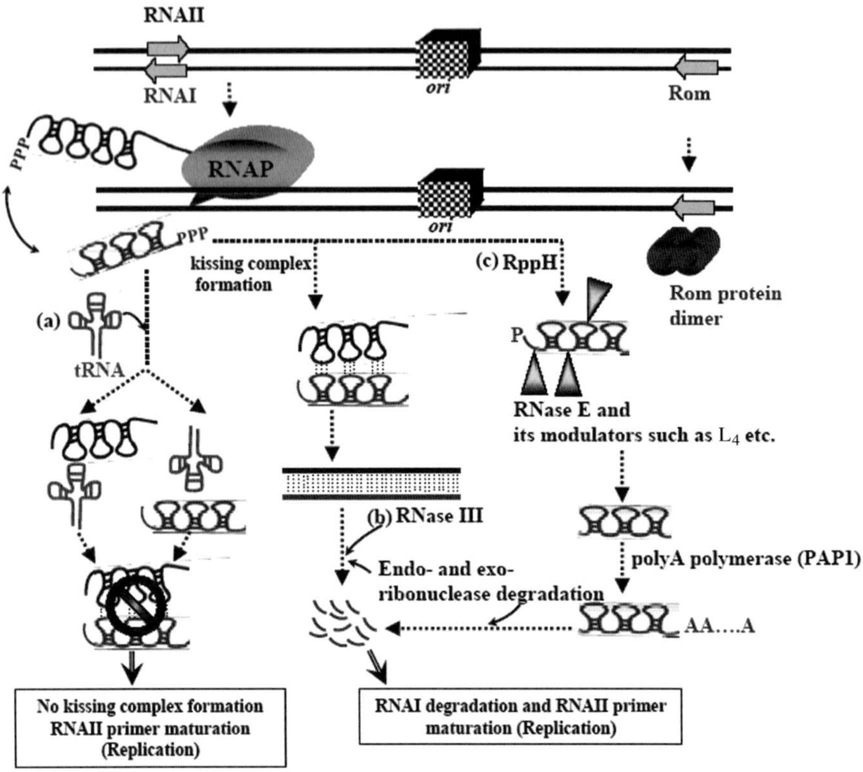

Fig. 14.3 Host-encoded proteins and tRNAs interfering with RNA-RNA interaction or regulating RNAI degradation can affect ColE1 plasmid replication. (**a**) Owing to sequence homologies with RNAII and RNAI, uncharged tRNA ( ) can interact with RNAII or RNAI and block kissing complex formation. Thus, matured primer-RNA is produced which promotes initiation of plasmid DNA replication. (**b**) When RNAII and RNAI form a kissing complex, RNase III can cleave the double-stranded RNAs into single-stranded RNA. These substrates are degraded by endo- and exo-ribonucleases (**c**) When RNAI is transcribed by RNAP, the tri-phosphate RNAI (ppp-RNAI) phosphate groups are removed by RppH (pyrophosphohydrolase) proteins to form monophosphate RNAI (p-RNAI). The p-RNAI substrates are cleaved by RNase E to form $pRNAI_{-5}$. Some RNase E modulators such as inhibitor L_4 are also involved in RNAI metabolism. Poly(A) tails are added to the $pRNAI_{-5}$ substrates by PAPI (polyA polymerase 1) proteins. These intermediates of RNAI are then degraded rapidly by endo- and exo-ribonucleases. When all RNAI intermediates have been degraded, no kissing complex can form and replication starts

14.2.2.1 Role of RNases and Polyadenylation in the Regulation of ColE1 Plasmid Replication

Several RNases encoded by *E. coli* have been shown to be implicated in ColE1 plasmid replication. The endoribonuclease RNase E, which is involved in the maturation of 5S rRNA (Apirion 1978, Mackie 1998), has also been shown to have an endonucleolytic activity in RNAI decay (Lin-Chao and Cohen 1991). It mediates the cleavage of RNAI from the full length triphosphorylated pppRNA I108 to pRNA I105 (Lin-Chao and Cohen 1991). Later, Kaberdin and co-workers also identified multiple cleavage sites in the stem loops of RNAI by RNase E (Kaberdin et al. 1996). RNase H recognizes and cleaves the RNAII–DNA hybrids at the origin with a sequence of AAAAA of RNAII and then generates mature primer RNAII (Naito and Uchida 1986). RNase III, an endonuclease which recognizes and cleaves double-strand RNA, is a third implicated RNase. It is involved in both the processing of rRNA and the degradation or processing of a variety of mRNA (Babitzke et al. 1993). It has been shown that RNase III is involved in the processing or degradation of RNAI and RNAII during the formation of RNAI and RNAII complex (Binnie et al. 1999).

More recently, polyadenylation has also been shown to be involved in regulating the copy number of ColE1 plasmids (Xu et al. 1993). Polynucleotide phosphorylase (PNPase) is an exoribonuclease which is implicated in mRNA degradation by removing nucleotides from the 3′-end of the RNA(Donovan and Kushner 1986) and it can also degrade RNAI (Xu and Cohen 1995). The degradation of RNAI by PNPase is further promoted by its polyadenylation. When a poly(A) tail is added to the 3′-end of RNAI by poly(A) polymerase (encoded by *pcnB*), this facilitates further exonucleolytic cleavage by PNPase after cleavage by RNase E (Xu et al. 2002). Therefore, the addition of a poly(A) tail hastens the decay of RNAI. In contrast, mutation in the *pcnB* gene stabilizes the RNA intermediate and reduces the copy number of ColE1 plasmid (Masters et al. 1990, Sarkar et al. 2002).

14.2.2.2 Role of tRNA on the Regulation of ColE1 Plasmid Replication

Previous studies have reported that the sequence of the loop II of RNAI and the dihydrouridylic loop of tRNA have close homologies (Yavachev and Ivanov 1988). This therefore implies a possible role for tRNA in the regulation of ColE1 plasmid replication. It has been shown that uncharged tRNA can interact with RNAI to regulate the replication of ColE1 plasmid (Wang et al. 2002, Wang et al. 2004, Wrobel and Wegrzyn 1998). It was speculated that tRNA–RNAI interactions may interfere with hybridization between RNAI and RNA II, thus allowing more maturation of the pre-primer RNAII and initiation of plasmid DNA replication (Wegrzyn 1999). Another effect of tRNA in the control of replication of ColE1-type plasmids, based on the interaction of the 3′-CCA sequence of uncharged tRNA with RNAI, has also been proposed (Wang et al. 2004). Understanding the role of uncharged tRNA in regulating the replication of ColE1-type plasmid is important, because amino acid starvation (which leads to the accumulation of uncharged tRNA) has been considered to be a more effective method than temperature shift for up-regulating the replication of ColE1-type plasmids (Wegrzyn 1999). Figure 14.3 illustrates how tRNAs interference with RNA-RNA interaction can affect ColE1 plasmid replication.

14.2.2.3 Other Host Factor-Mediated Regulation of ColE1 Type Plasmid Replication

Other host factors are also involved in regulation of ColE1 plasmid replication. It has been shown that RraA, a regulator of ribonuclease A activity, interacts with RNase E and inhibits RNase E endonucleolytic cleavage (Lee et al. 2003). Inhibition of RNase E by RraA prolongs the half-life of substrates such as RNAI, and thus interferes with the replication of ColE1 plasmids. Moreover, the degradation of most transcripts of *E. coli* proceeds through a 5′-end-dependent pathway and begins with endonucleolytic cleavage. The endonuclease responsible is RNase E, whose cleavage is the initial, rate-limiting step of mRNA degradation in *E. coli*. Previous study reported that the mechanism of the 5′-end dependent pathway for RNA decay is triggered by 5′-pyrophosphate removal (Celesnik et al. 2007). RppH, an RNA pyrophosphohydrolase which belongs to the Nudix protein family, has been reported to initiate the degradation of mRNA by this 5′-end dependent pathway as it removes the phosphates from the 5′-end of a triphosphorylated primary transcript (Deana et al. 2008). Therefore, RppH triggers RNase E cleavage and controls the rate of RNA decay (such as the degradation of RNAI) and also affects the ColE1 plasmid copy number.

14.3 Systems-Level Effects of Plasmid on *E. coli* Host Metabolism

14.3.1 Cellular Metabolic Burden from the Presence of Multicopy Plasmids

Although multicopy derivatives of ColE1 plasmids are widely-used, the introduction of these plasmids to *E. coli* often imposes a metabolic burden causing systems-level perturbation to cellular metabolism (Glick 1995). Phenotypically, the metabolic burden can be directly observed as reduction in cellular growth rate and final biomass (Fig. 14.4). Growth rate of plasmid-bearing (P+) *E. coli* decline with increasing plasmid copy number or size (Bentley et al. 1990, Cheah et al. 1987, Seo and Bailey 1985). Conversely, with increasing growth rate, the ratio of the RNAI inhibitor of plasmid replication over the replication pre-primer RNAII has been shown to increase correspondingly (Lin-Chao and Bremer 1986), suggesting the existence

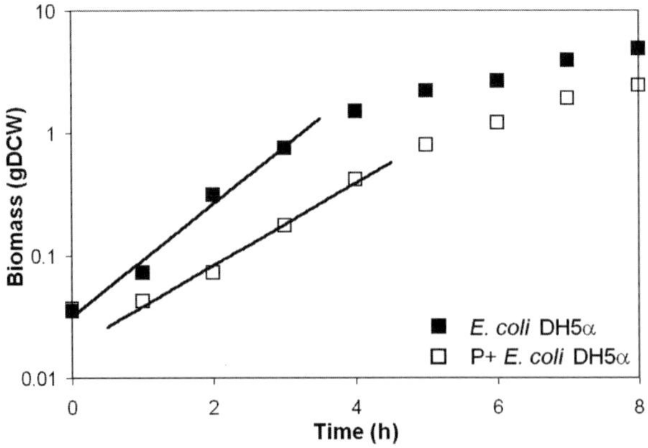

Fig. 14.4 Growth of plasmid-free (■) and plasmid-bearing (□) *E. coli DH5α* cells during 2-L batch cultures. Due to the metabolic burden of maintaining multicopy plasmids, the plasmid-bearing (P+) cells showed a reduced growth rate and final biomass relative to the plasmid-free host cells

of an intricate relationship between plasmid copy regulation and growth rate. The majority of growth-related phenotypic changes from plasmid presence have been extensively covered in two prior reviews (Glick 1995, Ricci and Hernandez 2000). Other than growth retardation, plasmid presence could also incur additional physiological changes, including elevation of oxygen uptake rates (Khosravi et al. 1990), increased glucose uptake and ATP synthesis (Diaz-Ricci et al. 1992), co-localization and presumed interaction with the host replication machinery at the cell membrane (Pogliano 2002, Yao et al. 2007), and loss of viability and cell lysis during fed-batch cultures (Andersson et al. 1996).

Conventionally, the basis of plasmid metabolic burden has been attributed to the metabolic drain of biosynthetic precursors, energy and other cellular resources for the maintenance of multicopy plasmids (Seo and Bailey 1985). As illustrated in Fig. 14.5, plasmid DNA replication and plasmid-encoded mRNA and protein synthesis share the same precursors, energy and enzymatic machinery as the analogous host metabolic processes (Peretti and Bailey 1987). Accordingly, the maintenance of plasmids would inevitably compete with the cellular growth for a limited pool of cellular resources, including biosynthetic precursors like deoxyribonucleotides, ribonucleotides and amino acids and high-energy molecules like ATP, GTP, NADH and NADPH. All these precursors and high-energy molecules required for plasmid maintenance are derived from the distribution of carbon fluxes through the central metabolic pathways (CMP) into assorted branches of biosynthetic and catabolic pathways (Holms 1996).

First proposed by Diaz Ricci and Hernandez (2000), an alternate proposition for the metabolic burden is that the presence of plasmids distresses host metabolism by perturbing the global transcriptional network. In other words, the *E. coli* host could perceive plasmids or plasmid-encoded products as an intracellular stress stimulus,

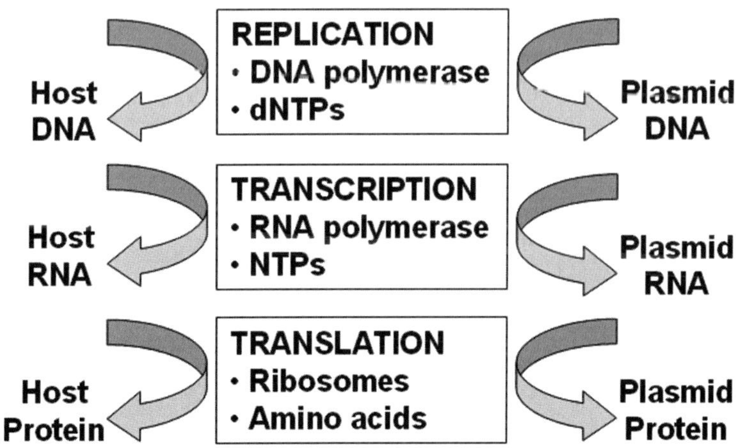

Fig. 14.5 Schematic illustration of competition between plasmid DNA and the bacterial host for cellular resources (Peretti and Bailey 1987). Copyright (1987, John Wiley & Sons, Inc.), reprinted with permission of Wiley-Liss, Inc., a subsidiary of John Wiley & Sons, Inc

triggering a cascade of stress signals affecting the *E. coli* transcriptional network. That in turn could lead to substantial changes in global gene expression and cellular phenotype.

14.3.2 Plasmid Perturbation of the Global Transcriptional Regulatory Network in E. coli

In bacteria, transcription regulation is generally considered the main mode of gene regulation. Figure 14.6 illustrates the multi-layer hierarchical structure of the *E. coli* global transcriptional network (Ma et al. 2004). Of the 4280 transcripts identified in *E. coli*, 267 are known or putative transcriptional regulators (Babu et al. 2004). The overall regulation of transcription in response to environmental and physiological changes is coordinated by a set of specific and global transcriptional regulators. While specific transcriptional regulators mainly regulate single transcriptional units consisting of genes with related functions known as operons, global transcriptional regulators are pleiotropic proteins with the ability to regulate operons belonging to several metabolic pathways or functional classes (Gottesman 1984).

There are now evidences suggesting that global transcriptional regulators play a key role in mediating the plasmid metabolic burden response. The first global

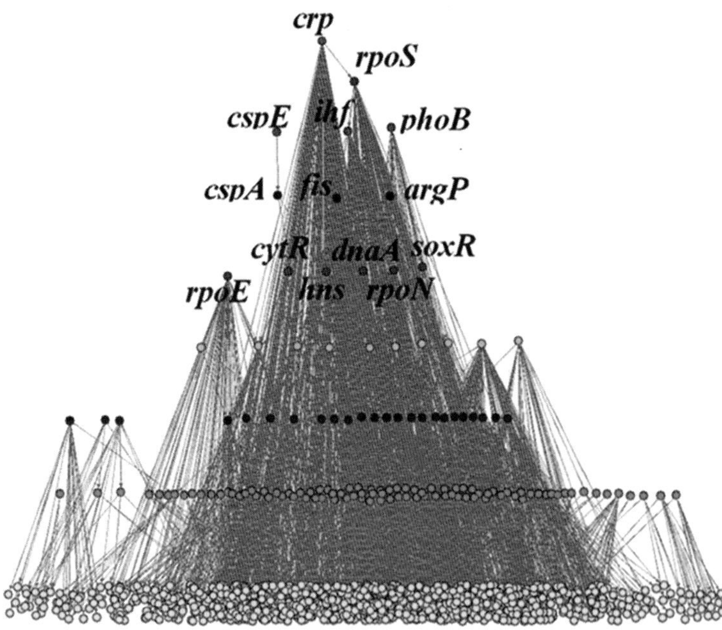

Fig. 14.6 Multi-layer hierarchical structure of the *E. coli* global transcriptional network. In the extended transcriptional regulatory network model containing 1278 genes and 2724 regulatory interactions by Ma and co-workers, the top layer regulators tend to be global transcriptional regulators, while the regulated metabolic enzymes are at the bottom layer (Ma et al. 2004)

transcriptional regulator implicated in plasmid metabolic burden is the cyclic AMP-response protein (CRP). The presence of plasmids is reported to lead to an increase in intracellular concentration of cyclic AMP (cAMP) in three *E. coli* strains (Diaz-Ricci et al. 1995). cAMP is a signal molecule responsible for activating the CRP, a global transcriptional regulator that directly regulates the expression of 197 *E. coli* genes including 22 other transcriptional regulators (Martinez-Antonio and Collado-Vides 2003). The cAMP-CRP complex activates the expression of catabolic operons in response to the availability of glucose and is also involved in cell division, motility, starvation function and anaerobiosis (Botsford and Harman 1992). During the growth of *E. coli* HB101, DH1 and JM109 carrying 3 different plasmids (including the ColE1 derived pUC19), an increase in intracellular cAMP concentration was accompanied by the higher activity of the cAMP-CRP activated β-galactosidase and an elevated rate of glucose uptake relative to the corresponding plasmid-free cells (Diaz-Ricci et al. 1995). Thereafter, the authors proposed, "*plasmids affect host metabolism through the perturbation of the cAMP-CRP complex, which in turn causes the alteration of the regulatory status of host regulations.*" (Diaz Ricci and Hernandez 2000). Despite the finding that plasmid presence is related to higher cellular cAMP levels, the exact mechanisms involved remain unclear.

A second global regulator-like molecule that could be involved in the transcription response to plasmid metabolic burden is guanosine tetraphosphate or ppGpp (Magnusson et al. 2005). ppGpp is the effector of the stringent response to amino acid starvation, widely-observed as the overall down-regulation of rRNA biosynthesis and ribosome production (Stent and Brenner 1961). The intracellular level of ppGpp rises in response to amino acid, carbon or energy depletion(Cashel et al. 1996) and is shown to correlate inversely with cellular growth rate (Joseleau-Petit et al. 1994). Recently, ppGpp is proposed to be the master regulator coordinating the binding of various sigma factors with RNA polymerase core enzyme (Nystrom 2004); in doing so, ppGpp in turn regulates the transcription of various stress-responsive genes mediated by alternative sigma factors, including σ^s (regulator of stationary phase response), σ^{32} (regulator of heat shock response) and σ^{54} (regulator of nutrient limitation and alternative carbon utilization). In recombinant plasmid-bearing cells under significant metabolic stress, ppGpp levels could be elevated. An intracellular ppGpp level of 0.45 μMol/gDCW was reported in uninduced pET11ahSOD plasmid-bearing *E. coli* HMS174(DE3) cells not producing the recombinant protein product (Cserjan-Puschmann et al. 1999). The subtle metabolic burden displayed by these plasmid-bearing cells was attributed to the replication and expression of the plasmid and its marker protein.

At present, the sole global transcriptional regulator shown to directly affect the metabolic burden-related retardation of cellular growth is FruR (fructose repressor, also known as Cra or catabolite activator repressor). FruR is a global transcription regulator of major catabolic enzymes using a cAMP-CRP independent mechanism (Saier 1996). Primarily, FruR represses the transcription of catabolic enzymes involved in the glycolytic pathway (*pfkA, pykF, gapA, pgk, eno*), Entner-Doudoroff pathway (*edd, eda*) and alternative sugar catabolism (*fruBAK, mtlADR*). At the same time, it positively activates genes in glyconeogenesis (*fbp, ppsA*), TCA cycle (*acnA, icdA*), glyoxalate shunt (*aceBA*) and electron transport chain (*cydAB*) (Ramseier 1996). Corresponding to FruR regulation of central metabolic pathways, knockout of the *fruR* gene in *E. coli* was shown to enhance carbon flow though glycolytic pathway and inhibit carbon flow though glyconeogenesis (Ramseier et al. 1995).

The inactivation of FruR by gene knockout was found to significantly improve the growth rates of plasmid-bearing cells relative to the respective wildtype cells (Ow et al. 2007). For *E. coli DH5α* carrying a ColE1-derived pcDNA3.1d/NS3 plasmid, the cellular growth rate during 2-L batch cultures improved from $0.75\,h^{-1}$ to $0.91\,h^{-1}$ after *fruR* knockout. This considerable growth rate recovery from plasmid metabolic burden was accompanied by a corresponding up-regulation of glycolytic enzymes and down-regulation of TCA cycle and stress proteins as revealed from proteomic and transcriptional analyses (Fig. 14.7).

As revealed from these studies, there is mounting evidence that, mediated by the action of global transcriptional regulators, the presence of plasmid leads to alterations in the global transcriptional network. Two implicated global transcriptional regulators, CRP and FruR, are both recognized to be major regulators of central metabolic gene expression. This appears to point towards the prevailing role of central metabolic gene expression in effecting the metabolic burden phenomenon. In all, these findings do not disprove the former proposition of plasmid metabolic drain. It is more likely that both the drain of cellular resources and the perturbation of the cellular regulatory network act synergistically together to contribute to the metabolic burden.

14.3.3 Central Metabolic Gene Expression and Plasmid Metabolic Burden

As metabolic fluxes within pathways have to synchronize with biosynthetic demands for precursors and energy during cell growth, they have evolved to be under tight regulatory control (Nielsen 2003). Regulation of metabolic fluxes can occur at the level of transcription (mRNA synthesis and degradation), translation (protein synthesis and proteolysis) and enzyme activity (allosteric regulation; Table 14.2). Due to the existence of various feedback loops and signaling cascades, these regulatory processes are interlinked to form a complex regulatory network (Vemuri and Aristidou 2005). Despite our extensive knowledge on *E. coli*, the exact nature of the regulatory network and its impact on central metabolism has not been clearly elucidated.

Although many CMP enzymes are constitutively expressed (Fraenkel 1996) and a few key enzymes are regulated by allosteric binding of effector molecules, the transcription of CMP enzymes has been observed to vary in response to different physiological condition (Sabnis et al. 1995). For instance, considerable transcriptional changes were observed within the glycolytic, gluconeogenic and TCA cycle pathways in *E. coli* during growth in acetate versus glucose media (Oh and Liao 2000). These transcriptional changes were found to qualitatively correlate to the actual CMP metabolic fluxes, which indicate the existence of significant regula-

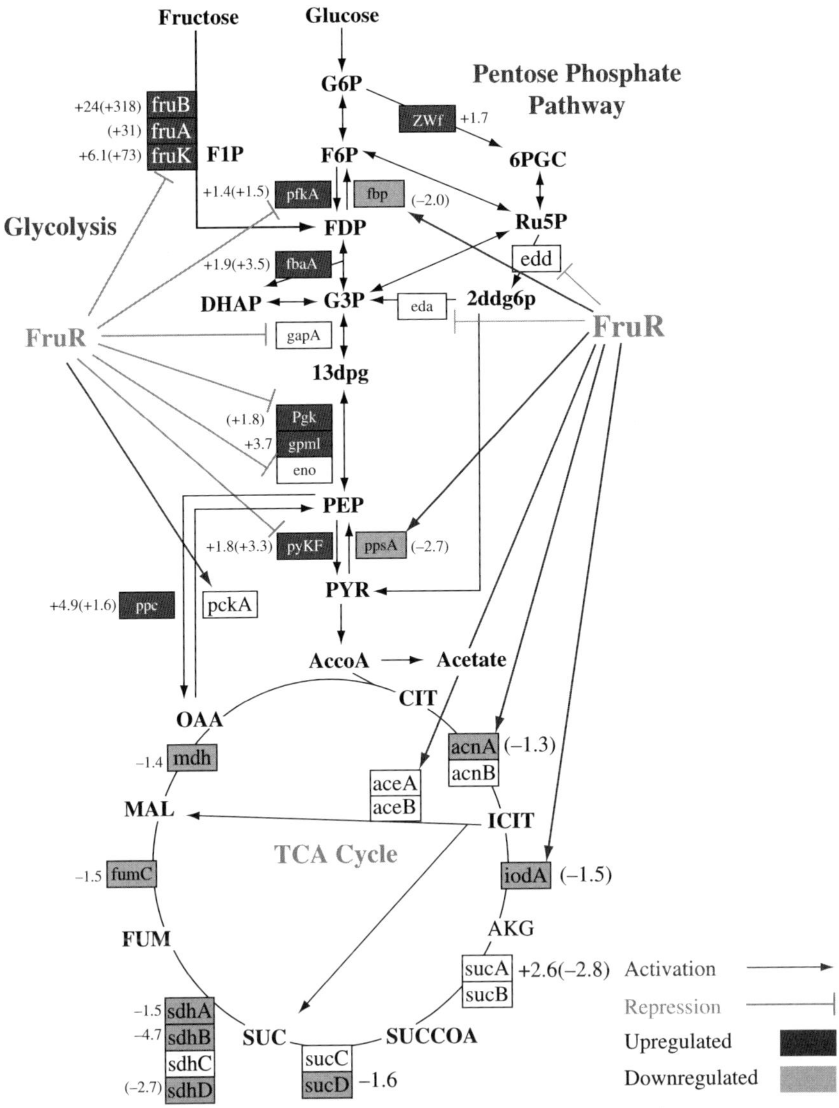

Fig. 14.7 Central metabolic gene expression changes in plasmid-bearing *DH5α* cells after *fruR* gene knockout. Values without brackets are protein expression fold changes, while values in brackets are the transcriptional fold changes. A trend of down-regulated (green boxes) glycolytic genes and up-regulated (red boxes) TCA cycle genes was observed. FruR mediated activation and repression are indicated by → or ⊣ respectively. (Figure from Ow et al. 2007)

tion in these pathways. During high cell density culturing of *E. coli* (Yoon et al. 2003), global transcriptional and proteomic studies showed a pattern of CMP gene expression changes that relates to the various physiological growth phases. These

Table 14.2 Allosteric regulation of enzyme activity in glycolysis and TCA cycle (compiled from EcoCyc database, Keseler et al. 2005)

Enzyme	Activator	Inhibitor
Phosphofructokinase 1	ADP	PEP
Fructose-1,6-biphosphate		AMP
Pyruvate kinase 1	FBP	
Pyruvate kinase 2	AMP	
Citrate synthase		NADH, OAA
Phosphoenolpyruvate carboxylase	FBP, acetyl-CoA	Aspartate, MAL
Phosphoenolpyruvate carboxykinase		NADH

observations indicate that transcriptional regulation of CMP is also of physiological importance (Sabnis et al. 1995).

The apparent role that the central metabolic gene expression plays in affecting plasmid metabolic burden is further supported by another recent study (Flores et al. 2004), whereby the overexpression of the *zwf* gene (encoding for the first enzyme of the pentose phosphate pathway, glucose 6-phosphate dehydrogenase) increased the growth rate of plasmid-bearing *E. coli* JM101 from $0.46\,h^{-1}$ to $0.64\,h^{-1}$ (Fig. 14.8). The growth rate recovery was ascribed to the potential increase in carbon flux to the oxidative branch of the pentose phosphate pathway. Since the PP pathway provides: (1) NADPH, a source of reducing power for many biosynthetic reactions and (2) precursors (ribose-5-phosphate and erythrose-4-phosphate) for nucleotide, histidine, and aromatic amino acids biosynthesis, they hypothesized that the availability

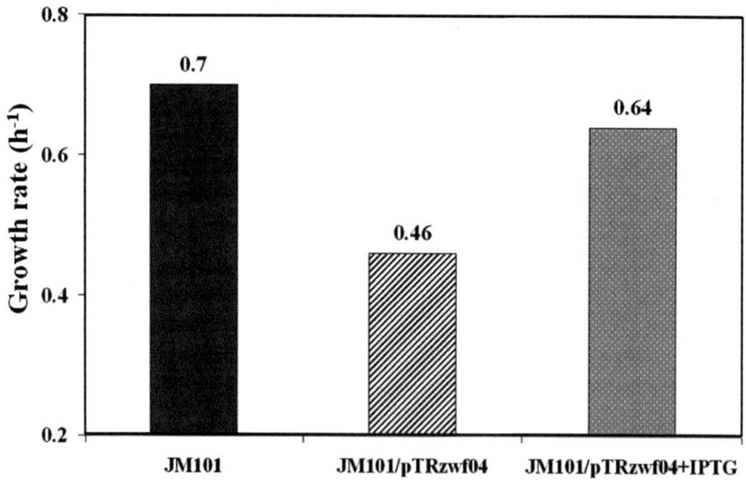

Fig. 14.8 Specific growth rates of *E. coli* JM101 strains over-expressing the *zwf* gene (Flores et al. 2004). The overexpression of *zwf* (encoding for the first enzyme of the pentose phosphate pathway, glucose 6-phosphate dehydrogenase) with IPTG led to an increase in growth rate from $0.46\,h^{-1}$ to $0.64\,h^{-1}$ in cells carrying plasmid pTRzwf04

of some of these metabolites could be limiting for the biosynthesis of plasmids or foreign proteins.

14.3.4 Global Transcriptional and Proteomic Studies of Plasmid Metabolic Burden

The first semi-global proteomic study to investigate protein expression changes due to plasmid presence in exponentially-growing *E. coli* was conducted by Birnbaum and Bailey (Birnbaum and Bailey 1991). In the study, two ColE1-derived plasmids that differ only by a mutated RNAI sequence were used to generate two plasmid copy number mutants with: (1) a mid plasmid copy number of 56 (designated strain P60), and (2) a high plasmid copy number of 240 (designated strain P120). Protein expression trends of the P60 and P120 strains were compared with the plasmid-free HB101 parental strain grown in minimal media supplemented with 20 amino acids. From 93 polypeptides identified, 34 were examined.

It was found that the levels of TCA cycle enzymes increase as the plasmid copy number increases initially from 0 to 56 (Birnbaum and Bailey 1991). Subsequently, at the higher copy number of 240, an increase in the anaplerotic PEP carboxylase expression was accompanied by a corresponding reduction in the expression of pyruvate kinase I, pyruvate dehydrogenase complex and TCA cycle enzymes. This indicates that, when grown on amino acids as the sole carbon source, cells carrying more copies of plasmids replenish TCA intermediates for precursor generation at the expense of TCA cycle flux. In additional, reduced expression of proteins of the protein synthesis machinery (2 elongation factors, 9 ribosomal subunits, asparyl-tRNA synthetase) and increased expression of 4 heat shock proteins were also seen. Together with a decrease in total cellular RNA and ribosome content as revealed from sucrose gradient profiles, the results denote a reduced translational capacity and elevated metabolic stress for cells carrying more plasmid.

Subsequently, a global transcriptome-proteome study was conducted to examine gene expression changes from plasmid presence in *E. coli* $DH5\alpha$ grown on glucose-containing complex media (Ow et al. 2006). The ColE1-type pcDNA3.1d/NS3 plasmid used was a DNA vaccine carrying a non-expressing antigenic gene against Dengue virus and has a copy number of approximately 100–150 during exponential phase (Lee et al. 2006). In the study, pcDNA3.1d/NS3 plasmid-bearing cells showed a 25% drop in growth rate over the plasmid-free host cells. Comparison of the exponentially growing plasmid-bearing cells over the plasmid-free host cells identified 364 genes and 18 proteins with more than 1.2 fold changes in gene expression.

A general downregulation of biosynthetic and key aerobic respiratory genes was observed (Ow et al. 2006). The downregulation of NADH dehydrogenase II (*ndh*) and several aerobic terminal oxidases (*cydA*, *cydB*, *cyoA*, *cyoB*) indicated an overall repression of major respiratory energy pathways in the plasmid-bearing cells. Among the upregulated genes were 6 stress-response heat shock proteins (*lon*, *mopA, clpB, hslV, ibpB, ibpA*). In particular, the upregulation of the heat shock *clpB* chaperone and *hslV* protease have not been previously associated with plasmid presence. Consistent with reports of higher cAMP-CRP activity in cells carrying plasmids, most upregulated carbon transporters are activated by the cAMP-CRP global regulator. Moreover, the downregulation of two key glycolytic genes, *pfkA* and *pykF* was seen. Interestingly, the only known transcriptional repressor of *pfkA* and *pykF* is FruR, another global regulator implicated with plasmid metabolic burden.

In the previous two studies, the comparison of gene expression was made on plasmid-bearing and the host cells showing evident variations in growth rates during exponential phase. As any variations in growth rates or physiological conditions could affect the interpretation of gene expression studies, Wang and colleagues used glucose-limited chemostat cultures to equilibrate the growth rates of two BL21 strains carrying a mid or a high copy ColE1 plasmid (copy numbers of 80 and 420) with the plasmid-free host (Wang et al. 2006). At the identical steady-state growth rate of $0.20\,h^{-1}$, glucose consumption rates for the plasmid-bearing cells were higher by approximately 2.5–3 fold relative to the host. Correspondingly, the acetate excretion rates for the plasmid-bearing cells were higher by 5–11 fold.

Microarray transcriptional analysis on these plasmid-bearing cells over the host showed a clear gene expression trend of an increase in glycolysis and TCA cycle and decrease in pentose phosphate pathway (Wang et al. 2006). These central metabolic gene expression trends were found to be consistent with the corresponding data from enzyme activity assays and metabolic flux analysis. In line with the experimental observations of higher glucose consumption and acetate excretion rates, the upregulation of *ptsG* for glucose uptake and acetate metabolic genes (*ackA*, *pta*) was also reported. In contrast with previous studies, only subtle changes in expression of genes related to cellular structure, DNA replication, and transcription/translation processes were observed. The authors reported that only the CMP related genes showed the largest expression changes. Hence, it appeared that, when growing at the same physiological growth rate as the host, CMP gene expression changes in the plasmid-bearing cells are dominant over other functional changes.

14.3.5 In-silico Simulation of Plasmid Metabolic Burden

It is now widely accepted that mathematical modeling and simulation of complex biological systems play a pivotal role in further improving our understanding of systems-level characteristics and functions. To date, various quantitative models have been presented for describing host-plasmid interactions in *E. coli*. Once predictive models are developed, various simulations under different conditions can be conducted by changing relevant parameters, thus allowing us to explore effects of plasmid presence on metabolic burden. As discussed previously, the presence of plasmids in the cell results in metabolic burden effects leading to retarded growth, changes in gene regulation, enzyme activities and metabolic flux. Major mechanistic processes involved in host-vector systems compose of plasmid replication, mRNA transcription, and plasmid-encoded mRNA translation of the foreign protein. Thus, detailed kinetics and control structures describing those processes along with key factors affecting host physiology can be formulated within plasmid-host interaction models. Experimentally observed metabolic burden is then characterized through *in silico* simulation of the models.

Plasmid replication is the first step in any plasmid-host interactions related to plasmid metabolic burden. In order to understand the underlying strategy of plasmid replication control, many researchers have presented mathematical models for describing the mechanism of ColE1 plasmid replication by anti-sense RNA regulation (Brendel and Perelson 1993). Paulsson and Ehrenberg determined optimal ColE1 copy number to achieve increased segregational stability, thereby reducing metabolic burden to the host cell (Paulsson and Ehrenberg 1998). Peretti and Bailey presented a mechanistically detailed single-cell model for *E. coli*, considering various competitive interactions found in plasmid-host systems (Peretti and Bailey 1987). They simulated recombinant cell growth by changing the relevant factors to metabolic activity, including plasmid copy number, promoter strength and ribosome binding strength. From these simulation experiments, strategies for enhancing cloned-gene productivity or reducing metabolic burden could be evaluated.

The second principal factor affecting metabolic burden is the expression of plasmid-encoded protein, which is commonly the antibiotic resistance marker protein. A simple mathematical model was developed for guiding stable target protein production and excretion (Togna et al. 1993). In the latter study, the empirical expression for the specific rate of plasmid production and less structured model of the *lac* operon induction was included in the dynamic model formulation. Bentley and co-workers presented a metabolically-structured kinetic model where expression of foreign proteins such as chloramphenicol-acetyl-transferease (CAT) and resistance marker protein (b-lactamase) was described based on plasmid content in addition to replication and mRNA transcription (Bentley et al. 1990). They observed the close corelationship between growth rate and foreign protein expression while the effect of plasmid replication on the growth rate was negligible. This implies that the metabolic drain of precursors and energy associated with the expression of proteins prevail over that for the replication of plasmid DNA.

Most plasmid-host interaction models are kinetics-based dynamic models which require extensive kinetic and regulatory information for modeling. Most often than not, experimental measurements are not easily obtained for determining a large number of kinetic parameters (e.g. intracellular reaction rates). The stationary modeling approach is, therefore, a good alternative to the kinetic model for the simulation of plasmid-bearing host metabolism. Assuming the pseudo-steady state, the kinetic model can be simplified into static representation, taking into account the network's connectivity and capacity as time-invariant properties of the metabolic system. To investigate the effect of plasmid-directed synthesis on metabolic stoichiometry, da Silva and Bailey calculated additional energetic and material requirements caused by plasmids (da Silva and Bailey 1986), hence deriving an early stoichiometric model for plasmid synthesis. Subsequently, Ozkan and colleagues developed a metabolic model for cell growth and recombinant protein overproduction in *E. coli* that included precursor balances and energetic requirement for plasmid replication, and protein expression within the metabolic balance model (Ozkan et al. 2005). Using some of these stoichiometric models for plasmid synthesis, the physiological effect of plasmid metabolic burden on *E. coli* metabolism can be further explored by constraints-based flux modeling (Ow et al. 2009). exploited a genome-scale *E. coli* model using various linear-programming cellular objective functions to identify the most plausible describer of the physiological state within the plasmid-bearing cells. The study demonstrated that flux simulations by maximizing maintenance energy expenditure showed good consistency with experimental data, suggesting that the plasmid-bearing cells are less energetically-efficient and could require more maintenance energy.

Current models are still limited by insufficient knowledge on global regulation and kinetic information. As there are now evidences of global regulatory changes in plasmid-bearing cells, future modeling approaches should systematically combine dynamic and stationary models with regulatory information and high-throughput "omics" data analysis to characterize the metabolic burden, thereby identifying engineering strategies for overcoming plasmid metabolic burden in *E. coli*.

14.4 Conclusions and Future Prospects

ColE1-type plasmids have been extensively characterized and widely applied in biotechnology. Early studies of plasmids and the successive development of ColE1-type plasmid vectors have contributed extensively to the present progress in molecular biology and recombinant DNA technology. Studies on the basic regulation of ColE1 replication have been initialized more than three decades ago. Although it is now well known that the control of ColE1 replication is primarily regulated by plasmid-encoded factors, the continual discovery of new host-encoded factors modulating ColE1 replication reveals that the *E. coli* host could exert a considerable effect on the replication of ColE1 plasmids as well.

While *E. coli* host produces several factors modulating ColE1 replication, plasmids also impose a metabolic burden impeding host growth and metabolism. The basis of this metabolic burden is complex and appears to involve both the plasmid-related drain of cellular resources from central metabolism and the perturbation of cellular regulatory state mediated by global transcriptional regulators. Through the application of systems level "omics" tools and *in-silico* modeling, we are beginning to gain better understanding of plasmid-host interactions. From the initial discovery of plasmids, to successive vector construction and emerging applications like genetic therapy and vaccination, it is anticipated that the current trend towards systems-level studies of plasmid-host interactions will give rise to even more knowledge and further biotechnological applications.

Acknowledgments The authors wish to thank: The Agency for Science, Technology and Research (A*STAR) for their funding; Pei-Hsun Lin for her excellent assistance with the figures and writing; Victor Wong, Andrea Camattari, Ken Chan and Wen-Yu Lin for reviewing and giving their valuable comments on the manuscript.

References

Anderson RJ, Schneider J (2007) Plasmid DNA and viral vector-based vaccines for the treatment of cancer. Vaccine 25 Suppl 2:B24–34

Andersson L, Yang S, Neubauer P et al. (1996) Impact of plasmid presence and induction on cellular responses in fed batch cultures of *Escherichia coli*. J Biotechnol 46(3):255–63

Apirion D (1978) Isolation, genetic mapping and some characterization of a mutation in *Escherichia coli* that affects the processing of ribonuleic acid. Genetics 90(4):659–71

Babitzke P, Granger L, Olszewski J et al. (1993) Analysis of mRNA decay and rRNA processing in *Escherichia coli* multiple mutants carrying a deletion in RNase III. J Bacteriol 175(1):229–39

Babu MM, Luscombe NM, Aravind L et al. (2004) Structure and evolution of transcriptional regulatory networks. Curr Opin Struct Biol 14(3):283–91

Balbas P, Soberon X, Merino E et al. (1986) Plasmid vector pBR322 and its special-purpose derivatives–a review. Gene 50(1–3):3–40

Bentley WE, Mirjalili N, Andersen DC et al. (1990) Plasmid-encoded protein: The principal factor in the "metabolic burden" associated with recombinant bacteria. Biotechnol Bioeng 35(7): 668–81

Binnie U, Wong K, McAteer S et al. (1999) Absence of RNASE III alters the pathway by which RNAI, the antisense inhibitor of ColE1 replication, decays. Microbiology 145(Pt 11): 3089–100

Birnbaum S, Bailey JE (1991) Plasmid presence changes the relative levels of many host cell proteins and ribosome components in recombinant *Escherichia coli*. Biotechnol Bioeng 37(8):736–45

Bolivar F, Rodriguez RL, Greene PJ et al. (1977) Construction and characterization of new cloning vehicles. II. A multipurpose cloning system. Gene 2(2):95–113

Botsford JL, Harman JG (1992) Cyclic AMP in prokaryotes. Microbiol Rev 56(1):100–22

Brendel V, Perelson AS (1993) Quantitative model of ColE1 plasmid copy number control. J Mol Biol 229(4):860–72

Cabello F, Timmis K, Cohen SN (1976) Replication control in a composite plasmid constructed by *in vitro* linkage of two distinct replicons. Nature 259(5541):285–90

Cashel M, Gentry DR, Hernandez VJ et al. (1996) The stringent response. In: (ed) In *Escherichia coli* and *Salmonella*, Cellular and molecular biology. Neidhardt FC

Celesnik H, Deana A, Belasco JG (2007) Initiation of RNA decay in *Escherichia coli* by 5′ pyrophosphate removal. Mol Cell 27(1):79–90

Cesareni G, Helmer-Citterich M, Castagnoli L (1991) Control of ColE1 plasmid replication by antisense RNA. Trends Genet 7(7):230–5

Cesareni G, Muesing MA, Polisky B (1982) Control of ColE1 DNA replication: the *rop* gene product negatively affects transcription from the replication primer promoter. Proc Natl Acad Sci USA 79(20):6313–7

Chan PT, Ohmori H, Tomizawa J et al. (1985) Nucleotide sequence and gene organization of ColE1 DNA. J Biol Chem 260(15):8925–35

Cheah UE, Weigand WA, Stark BC (1987) Effects of recombinant plasmid size on cellular processes in *Escherichia coli*. Plasmid 18(2):127–34

Cohen HJ (1973) Peroxide detoxification affecting the production of immunoglobulin by mouse myeloma tumor cells *in vitro*. Experientia 29(10):1285–7

Cohen SN (1993) Bacterial plasmids: their extraordinary contribution to molecular genetics. Gene 135(1–2):67–76

Cohen SN, Chang AC, Hsu L (1972) Nonchromosomal antibiotic resistance in bacteria: genetic transformation of *Escherichia coli* by R-factor DNA. Proc Natl Acad Sci USA 69(8):2110–4

Cserjan-Puschmann M, Kramer W, Duerrschmid E et al. (1999) Metabolic approaches for the optimisation of recombinant fermentation processes. Appl Microbiol Biotechnol 53(1):43–50

da Silva NA, Bailey JE (1986) Theoretical growth yield estimates for recombinant cells. Biotechnol Bioeng 28(5):741–6

Davison J (1984) Mechanism of control of DNA replication and incompatibility in ColE1-type plasmids–a review. Gene 28(1):1–15
Deana A, Celesnik H, Belasco JG (2008) The bacterial enzyme RppH triggers messenger RNA degradation by 5′ pyrophosphate removal. Nature 451(7176):355–8
Diaz-Ricci JC, Bode J, Rhee JI et al. (1995) Gene expression enhancement due to plasmid maintenance. J Bacteriol 177(22):6684–7
Diaz-Ricci JC, Tsu M, Bailey JE (1992) Influence of expression of the *pet* operon on intracellular metabolic fluxes of *Escherichia coli*. Biotechnol Bioeng 39(1):59–65
Diaz Ricci JC, Hernandez ME (2000) Plasmid effects on *Escherichia coli* metabolism. Crit Rev Biotechnol 20(2):79–108
Donovan WP, Kushner SR (1986) Polynucleotide phosphorylase and ribonuclease II are required for cell viability and mRNA turnover in *Escherichia coli K-12*. Proc Natl Acad Sci USA 83(1):120–4
Flores S, de Anda-Herrera R, Gosset G et al. (2004) Growth-rate recovery of *Escherichia coli* cultures carrying a multicopy plasmid, by engineering of the pentose-phosphate pathway. Biotechnol Bioeng 87(4):485–94
Fraenkel DG (1996) Glycolysis. In: Neidhardt F (ed) *Escherichia coli* and *Salmonella*, American Society for Microbiology, Washington, D.C
Glick BR (1995) Metabolic load and heterologous gene expression. Biotechnol Adv 13(2):247–61
Gottesman S (1984) Bacterial regulation: global regulatory networks. Annu Rev Genet 18:415–41
Hayes W (1953) Observations on a transmissible agent determining sexual differentiation in Bacterium coli. J Gen Microbiol 8(1):72–88
Holms H (1996) Flux analysis and control of the central metabolic pathways in *Escherichia coli*. FEMS Microbiol Rev 19(2):85–116
Itoh T, Tomizawa J (1979) Initiation of replication of plasmid ColE1 DNA by RNA polymerase, ribonuclease H, and DNA polymerase I. Cold Spring Harb Symp Quant Biol 43(Pt 1):409–17
Joseleau-Petit D, Thevenet D, D'Ari R (1994) ppGpp concentration, growth without PBP2 activity, and growth-rate control in *Escherichia coli*. Mol Microbiol 13(5):911–7
Kaberdin VR, Chao YH, Lin-Chao S (1996) RNase E cleaves at multiple sites in bubble regions of RNA I stem loops yielding products that dissociate differentially from the enzyme. J Biol Chem 271(22):13103–9
Kahn M, Kolter R, Thomas C et al. (1979) Plasmid cloning vehicles derived from plasmids ColE1, F, R6K, and RK2. Methods Enzymol 68:268–80
Khosravi M, Ryan W, Webster DA et al. (1990) Variation of oxygen requirement with plasmid size in recombinant *Escherichia coli*. Plasmid 23(2):138–43
Kues U, Stahl U (1989) Replication of plasmids in gram-negative bacteria. Microbiol Rev 53(4):491–516
Lambert CM, Wrighton CJ, Strike P (1987) Characterization of the drug resistance plasmid NTP16. Plasmid 17(1):26–36
Lederberg J (1998) Plasmid (1952–1997). Plasmid 39(1):1–9
Lederberg J, Cavalli LL, Lederberg EM (1952) Sex Compatibility in *Escherichia coli*. Genetics 37(6):720–30
Ledley FD (1995) Nonviral gene therapy: the promise of genes as pharmaceutical products. Hum Gene Ther 6(9):1129–44
Lee CL, Ow DS, Oh SK (2006) Quantitative real-time polymerase chain reaction for determination of plasmid copy number in bacteria. J Microbiol Methods 65(2):258–67
Lee K, Zhan X, Gao J et al. (2003) RraA. a protein inhibitor of RNase E activity that globally modulates RNA abundance in *E. coli*. Cell 114(5):623–34
Lin-Chao S, Bremer H (1986) Effect of the bacterial growth rate on replication control of plasmid pBR322 in *Escherichia coli*. Mol Gen Genet 203(1):143–9
Lin-Chao S, Chen WT, Wong TT (1992) High copy number of the pUC plasmid results from a Rom/Rop-suppressible point mutation in RNA II. Mol Microbiol 6(22):3385–93
Lin-Chao S, Cohen SN (1991) The rate of processing and degradation of antisense RNAI regulates the replication of ColE1-type plasmids *in vivo*. Cell 65(7):1233–42

Liu F, Huang L (2002) Development of non-viral vectors for systemic gene delivery. J Control Release 78(1–3):259–66
Lovett MA, Helinski DR (1976) Method for the isolation of the replication region of a bacterial replicon: construction of a mini-F'kn plasmid. J Bacteriol 127(2):982–7
Ma HW, Kumar B, Ditges U et al. (2004) An extended transcriptional regulatory network of *Escherichia coli* and analysis of its hierarchical structure and network motifs. Nucleic Acids Res 32(22):6643–9
Mackie GA (1998) Ribonuclease E is a 5′-end-dependent endonuclease. Nature 395(6703):720–3
Magnusson LU, Farewell A, Nystrom T (2005) ppGpp: a global regulator in *Escherichia coli*. Trends Microbiol 13(5):236–42
Martinez-Antonio A, Collado-Vides J (2003) Identifying global regulators in transcriptional regulatory networks in bacteria. Curr Opin Microbiol 6(5):482–9
Masters M, March JB, Oliver IR et al. (1990) A possible role for the *pcnB* gene product of *Escherichia coli* in modulating RNA: RNA interactions. Mol Gen Genet 220(2):341–4
Masukata H, Tomizawa J (1986) Control of primer formation for ColE1 plasmid replication: conformational change of the primer transcript. Cell 44(1):125–36
Naito S, Uchida H (1986) RNase H and replication of ColE1 DNA in *Escherichia coli*. J Bacteriol 166(1):143–7
Nielsen J (2003) It is all about metabolic fluxes. J Bacteriol 185(24):7031–5
Nystrom T (2004) Growth versus maintenance: a trade-off dictated by RNA polymerase availability and sigma factor competition? Mol Microbiol 54(4):855–62
Oh MK, Liao JC (2000) Gene expression profiling by DNA microarrays and metabolic fluxes in *Escherichia coli*. Biotechnol Prog 16(2):278–86
Ow DS, Lee RM, Nissom PM et al. (2007) Inactivating FruR global regulator in plasmid-bearing *Escherichia coli* alters metabolic gene expression and improves growth rate. J Biotechnol 131(3):261–9
Ow DSW, Lee DY, Yap MGS et al. (2009) Identification of Cellular Objective for Elucidating the Physiological State of Plasmid-Bearing *Escherichia coli* Using Genome-Scale *in silico* Analysis. Biotechnol Prog 25(1)
Ow DSW, Nissom PM, Philp R et al. (2006) Global transcriptional analysis of metabolic burden due to plasmid maintenance in *Escherichia coli* $DH5\alpha$ during batch fermentation. Enzyme Microb Technol 39(3):391–398
Ozkan P, Sariyar B, Utkur FO et al. (2005) Metabolic flux analysis of recombinant protein overproduction in *Escherichia coli*. Biochem Eng J 22(2):167–195
Panayotatos N (1984) DNA replication regulated by the priming promoter. Nucleic Acids Res 12(6):2641–8
Paulsson J, Ehrenberg M (1998) Trade-off between segregational stability and metabolic burden: a mathematical model of plasmid ColE1 replication control. J Mol Biol 279(1):73–88
Peretti SW, Bailey JE (1987) Simulations of host-plasmid interactions in *Escherichia coli*: Copy number, promoter strength, and ribosome binding site strength effects on metabolic activity and plasmid gene expression. Biotechnol Bioeng 29(3):316–28
Pogliano J (2002) Dynamic cellular location of bacterial plasmids. Curr Opin Microbiol 5(6): 586–90
Polisky B, Zhang XY, Fitzwater T (1990) Mutations affecting primer RNA interaction with the replication repressor RNA I in plasmid ColE1: potential RNA folding pathway mutants. Embo J 9(1):295–304
Ramseier TM (1996) Cra and the control of carbon flux via metabolic pathways. Res Microbiol 147(6–7):489–93
Ramseier TM, Bledig S, Michotey V et al. (1995) The global regulatory protein FruR modulates the direction of carbon flow in *Escherichia coli*. Mol Microbiol 16(6):1157–69
Sabnis NA, Yang H, Romeo T (1995) Pleiotropic regulation of central carbohydrate metabolism in *Escherichia coli* via the gene *csrA*. J Biol Chem 270(49):29096–104
Saier MH, Jr. (1996) Cyclic AMP-independent catabolite repression in bacteria. FEMS Microbiol Lett 138(2–3):97–103

Sarkar N, Cao GJ, Jain C (2002) Identification of multicopy suppressors of the *pcnB* plasmid copy number defect in *Escherichia coli*. Mol Genet Genomics 268(1):62–9

Schmidt L, Inselburg J (1982) ColE1 copy number mutants. J Bacteriol 151(2):845–54

Selzer G, Som T, Itoh T et al. (1983) The origin of replication of plasmid p15A and comparative studies on the nucleotide sequences around the origin of related plasmids. Cell 32(1):119–29

Seo JH, Bailey JE (1985) Effects of recombinant plasmid content on growth properties and cloned gene product formation in *Escherichia coli*. Biotechnol Bioeng 27(12):1668–74

Singh D, Chang S-J, Lin P-H, Averina OV, Kaberdin VR, Sue Lin-Chao (2008) Regulation of Ribonuclease E activity by the L4 ribosomal protein of *Escherichia coli*. Proc Natl Acad Sci USA (In press)

Stent GS, Brenner S (1961) A genetic locus for the regulation of ribonucleic acid synthesis. Proc Natl Acad Sci USA 47:2005–14

Timmis K, Cabello F,Cohen SN (1975) Cloning, isolation, and characterization of replication regions of complex plasmid genomes. Proc Natl Acad Sci USA 72(6):2242–6

Togna AP, Shuler ML, Wilson DB (1993) Effects of plasmid copy number and runaway plasmid replication on overproduction and excretion of beta-lactamase from *Escherichia coli*. Biotechnol Prog 9(1):31–9

Tomizawa J, Itoh T, Selzer G et al. (1981) Inhibition of ColE1 RNA primer formation by a plasmid-specified small RNA. Proc Natl Acad Sci USA 78(3):1421–5

Tomizawa J, Som T (1984) Control of ColE1 plasmid replication: enhancement of binding of RNA I to the primer transcript by the Rom protein. Cell 38(3):871–8

Vemuri GN, Aristidou AA (2005) Metabolic engineering in the -omics era: elucidating and modulating regulatory networks. Microbiol Mol Biol Rev 69(2):197–216

Wang Z, Le G, Shi Y et al. (2002) A model for regulation of ColE1-like plasmid replication by uncharged tRNAs in amino acid-starved *Escherichia coli* cells. Plasmid 47(2):69–78

Wang Z, Xiang L, Shao J et al. (2006) Effects of the presence of ColE1 plasmid DNA in *Escherichia coli* on the host cell metabolism. Microb Cell Fact 5:34

Wang Z, Yuan Z, Hengge UR (2004) Processing of plasmid DNA with ColE1-like replication origin. Plasmid 51(3):149–61

Watanabe T, Takano T, Arai T et al. (1966) Episome-mediated Transfer of Drug Resistance in Enterobacteriaceae X. Restriction and Modification of Phages by fi R Factors. J Bacteriol 92(2):477–486

Wegrzyn G (1999) Replication of plasmids during bacterial response to amino acid starvation. Plasmid 41(1):1–16

Weide B, Garbe C, Rammensee HG et al. (2008) Plasmid DNA- and messenger RNA-based anticancer vaccination. Immunol Lett 115(1):33–42

Wrobel B, Wegrzyn G (1998) Replication regulation of ColE1-like plasmids in amino acid-starved *Escherichia coli*. Plasmid 39(1):48–62

Xu F, Cohen SN (1995) RNA degradation in *Escherichia coli* regulated by 3′ adenylation and 5′ phosphorylation. Nature 374(6518):180–3

Xu F, Lin-Chao S, Cohen SN (1993) The *Escherichia coli pcnB* gene promotes adenylylation of antisense RNAI of ColE1-type plasmids *in vivo* and degradation of RNAI decay intermediates. Proc Natl Acad Sci USA 90(14):6756–60

Xu FF, Gaggero C, Cohen SN (2002) Polyadenylation can regulate ColE1 type plasmid copy number independently of any effect on RNAI decay by decreasing the interaction of antisense RNAI with its RNAII target. Plasmid 48(1):49–58

Yao S, Helinski DR, Toukdarian A (2007) Localization of the naturally occurring plasmid ColE1 at the cell pole. J Bacteriol 189(5):1946–53

Yavachev L, Ivanov I (1988) What does the homology between *E. coli* tRNAs and RNAs controlling ColE1 plasmid replication mean? J Theor Biol 131(2):235–41

Yoon SH, Han MJ, Lee SY et al. (2003) Combined transcriptome and proteome analysis of *Escherichia coli* during high cell density culture. Biotechnol Bioeng 81(7):753–67

Zverev VV, Kuzmin NP, Zuyeva LA et al. (1984) Regions of homology in small colicinogenic plasmids. Plasmid 12(3):203–5

Chapter 15
Systems-Level Analysis of Protein Quality in Inclusion Body-Forming *Escherichia coli* Cells

Elena García-Fruitós, Nuria González-Montalbán, Mónica Martínez-Alonso, Ursula Rinas and Antonio Villaverde

Contents

Abstract Recombinant proteins produced in *Escherichia coli* often aggregate as amorphous masses of insoluble material known as inclusion bodies. Being quite homogeneous in their composition, inclusion bodies display amyloid-like properties such as sequence-dependent protein-protein interactions, seeding-driven deposition of their components and β-sheet intermolecular architecture. However, inclusion bodies formed by different proteins and enzymes also show important extents of native-like secondary structure and include significant proportions of properly folded, functional protein, which makes them suitable to be used in catalytic processes. Inclusion bodies are formed as a result of the incapability of the quality

A. Villaverde (✉)
Institut de Biotecnologia i de Biomedicina, Universitat Autònoma de Barcelona, Bellaterra, Barcelona, 08193 Spain
e-mail: antoni.villaverde@uab.es

S.Y. Lee (ed.), *Systems Biology and Biotechnology of Escherichia coli*,
DOI 10.1007/978-1-4020-9394-4_15,

control cell system to cope with the non physiological amounts of misfolding-prone proteins produced upon recombinant gene expression. Multiple cellular proteins involved in the quality control, namely chaperones and proteases, participate in their formation and co-ordinately determine the amount of aggregated protein, the size of aggregates and the main structural and functional properties of the embedded polypeptides, such as their inner molecular organization.

15.1 Recombinant Protein Production: An Historical Overview

The discovery of restriction enzymes in the 70s offered one of the most powerful tools in molecular biology, dramatically fuelling the progress of recombinant DNA technologies. Before the systematic use of restriction enzymes, genetic manipulation was restricted to poorly controlled genetic modifications such as those caused in bacterial genomes by bacteriophages and plasmids. Restriction enzymes permitted the isolation and cloning of genes and their regulated expression in heterologous cell hosts, such as bacteria. This allowed the production of polypeptides that, being of interest for scientific, pharmaceutical or industrial purposes, occurred in low amounts in their natural sources and therefore, were difficult to obtain. This simple gene-cloning-and-expression strategy offered a solid methodological background on which the modern biotechnology fully developed. The use of cells (mainly microbial) as biological systems for the regulated production of recombinant proteins (and also of natural substances of biotechnological interest) originated the "Cell Factory" concept. This notion, underlying any man-driven, cell-mediated production process in single cells, refers to the engineering of the cell's biosynthetic machinery and the supporting genetic programme for applied purposes.

In early DNA recombinant times, it was believed that recombinant protein production in microbial cells would be the source of any relevant protein of pharmacological interest with high added value (such as immunogens, hormones, enzymes and complex molecular assemblies such as virus-like particles) as well as enzymes of straightforward industrial applicability (such as lipases, glycosidases, proteases, etc.). Therefore, the implementation of recombinant DNA technologies was predicted to result into a dramatic positive impact in biotechnology and biomedicine, expanding the spectrum of protein products available in the market. However, those expectations were rapidly frustrated since generally the quality of recombinant proteins produced in bacteria was not comparable to that of those obtained from natural sources, and therefore those recombinant proteins were not suitable for use. Essentially, the major bottlenecks encountered during recombinant protein production are proteolytic digestion by cell proteases (Enfors 1992) and aggregation as insoluble protein deposits known as inclusion bodies (IBs) (Georgiou and Valax 1996, Marston 1986). Human insulin was among the first proteins for which the accumulation in morphologically discrete aggregates in the bacterial cytoplasm was shown

(Paul et al. 1983, Williams et al. 1982), which delayed its further development as a pharmaceutical in human therapy.

The majority of proteins deposited as inclusion bodies are produced in non-functional conformation, in particular if they are of eukaryotic origin and contain disulfide bonds and, thus, require solubilisation and refolding for generation of the biologically active version of the protein (Clark 2001, Fahnert et al. 2004, Jungbauer and Kaar 2007, Middelberg 2002, Vallejo and Rinas 2004). Although inclusion body formation in general leads to additional down-stream steps during protein production and purification, inclusion body based production processes involving solubilization and refolding are economically viable options for many biopharmaceuticals. For example, human insulin, nowadays produced as recombinant protein in tons per year quantities, is produced using two major routes (Walsh 2005). One route involves the production of proinsulin in form of inclusion bodies using *E. coli* as expression host with subsequent solubilization and refolding procedures. The other route involves the utilization of yeast-based expression systems leading to the secretion of a soluble proinsulin into the culture supernatant. Both routes are economically viable.

Both proteolysis and IB formation result from the unability of many recombinant proteins to reach their native conformation in recombinant cells, and the efforts addressed to minimize them have only resulted partially successful. Therefore, the number of recombinant proteins that have entered the biotechnological market represents only a very minor fraction of those that have ever been produced in heterologous cells. At least partially, the incomplete exploitation of the recombinant DNA technologies for protein production can be attributed to a limited understanding of the cell physiology under the non-physiological cellular conditions associated to recombinant gene expression.

15.1.1 Protein Production in Escherichia coli

Since the first experiments of gene cloning and expression, the gram-negative bacterium *Escherichia coli* has been universally used as a convenient host for protein production. Despite other heterologous hosts have been progressively incorporated into production processes (namely gram-positive bacteria, yeast, insect cells, mammalian cells, filamentous fungi and others) (Gasser et al. 2008), *E. coli* is still a main cell factory for protein production, essentially because of its high growth rate, the high cell densities reached in fed-batch cultures, the relatively inexpensive growth media, the deep knowledge of its genetics and the availability of diverse genetic tools, such as plasmids, transposons and viruses acting on this species. Although recombinant proteins can be obtained in the cell periplasm if fused to secretion peptides, *E. coli* is mostly used to produce proteins in the cytoplasm, which need to be recovered from cell extracts after cell disruption.

However, many recombinant proteins produced in *E. coli* are unable to reach their native conformation, especially if they have eukaryotic origins or require posttranslational modifications such as disulfide bridge formation for their folding, and are rapidly degraded by cell proteases. Among the set of *E. coli* proteases, the ATP-dependent proteases Lon and ClpP are responsible for the degradation of most of the recombinant polypeptides (Maurizi 1992). Although the use of protease-deficient mutants as cell hosts has been explored as a method to enhance the stability of recombinant proteins (Baneyx and Georgiou 1991, Gottesman et al. 1997, Tomoyasu et al. 2001b), the issue appears progressively more complex as the physiology of *in vivo* protein folding is better understood. In this regard, cell proteases are an important arm of the quality control system, in which they act in cooperation with folding assistant proteins to survey conformational quality (Fig. 15.1). Therefore, minimizing proteolysis in protease deficient mutants leads to the accumulation of misfolded protein species as IBs (Garcia-Fruitos et al. 2007a, Rosen et al. 2002, Vera et al. 2005). During the growth of non recombinant *E. coli* cells at 37 °C, protein aggregation affects only a background fraction of cell proteins (Gonzalez-Montalban et al. 2006), being rather irrelevant from a quantitative point of view. However, both cell growth at high temperatures and the production of recombinant proteins cause aggregation of cell proteins or recombinant species, respectively, and trigger the expression of heat-shock genes (many of which encode chaperones and proteases

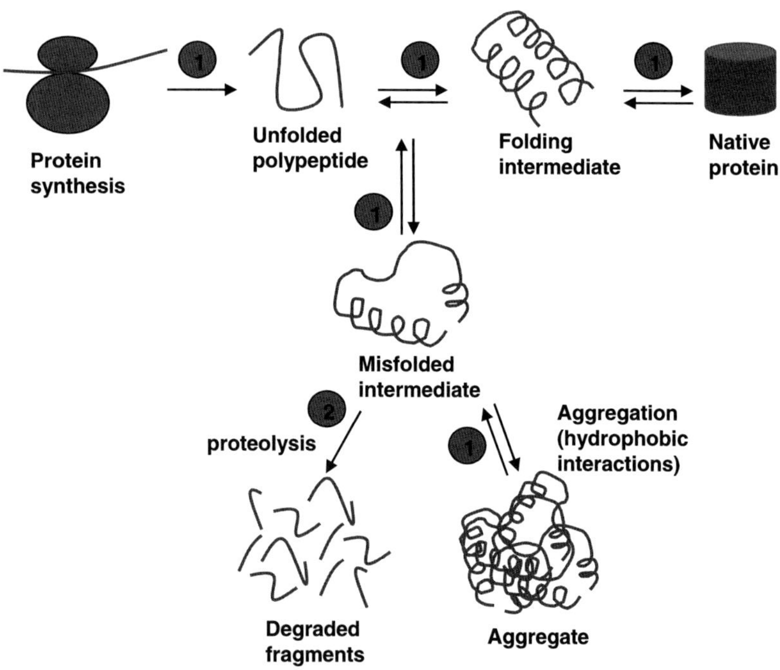

Fig. 15.1 Conventional model of protein folding, aggregation and proteolysis. A chain newly synthesized on a ribosome may fold to a native state, can aggregate or can be proteolysed. In living systems, environmental conditions and the quality control system highly regulate the transition between the different states. (1) Chaperones assist protein intermediates and misfolded proteins to reach their native state. (2) Proteases proteolyse misfolded proteins that have failed to reach a native conformation

for quality control). In recombinant cells, significant fractions of the recombinant protein are often found as IBs. This indicates that the quality control is inefficient in the processing of non-physiological amounts of heterologous proteins, what results in production processes rendering insoluble and biologically unuseful material. Recombinant protein misfolding and aggregation is one of the major concerns when facing *in vivo* protein production processes.

15.2 Molecular Basis of Protein Folding

Proteins have multiple and critical roles in all organisms, being the most abundant molecules in biological systems other than water. Protein folding is the process through which unfolded, nascent polypeptide chains convert into tightly folded compact structures with biological functions. Pioneering studies on protein folding by Anfinsen showed that the amino acid sequence of a protein encodes its functional three-dimensional structure (Anfinsen 1973). The underlying mechanism by which this complex process takes place is becoming progressively understood, because of the development of both physicochemical techniques and computational methods. In fact, understanding protein folding is not only relevant for biotechnological purposes but also to solve the molecular mechanisms responsible for conformational diseases such as Alzheimer, type II diabetes and Creutzfeldt-Jakob, among others.

Apparently, there is a common mechanism for the folding of the enormous spectrum of proteins in nature, irrespective of their native structure or amino acid sequence (Snow et al. 2002), in which the necessary information to reach a unique native state in a finite time is defined (Karplus 1997). Among the total number of possible conformations that a polypeptide could reach, finding a particular structure would take a length of time many orders of magnitude greater than the real time required for proteins to fold. This inconsistency, known as Levinthal paradox (Karplus 1997), has been solved with the development of the so called "new view" (Yon 2001), in which folding is described as a stochastic search of conformational space rather than as a series of mandatory structural transitions (Baldwin 1994, Dill and Chan 1997, Matagne and Dobson 1998, Wolynes et al. 1995). In essence, the inherent fluctuations in the conformation of an incompletely folded polypeptide enable the contact even of residues located at very different positions in the amino acid sequence. Therefore, as correct (native-like) interactions are more stable than non-native ones, this search mechanism is able to find the structure with the lowest energy (Baldwin 1994, Dinner et al. 2000). As the native state is approached, the conformational space accessible to the polypeptide chain is reduced (Wolynes et al. 1995). The fundamental mechanism of protein folding involves the formation of a folding-nucleus of residues in the protein, around which the remainder structure rapidly condenses (Otzen and Fersht 1998).

Small, single-domain proteins do not require many partially folded intermediates to reach a native conformation, and usually only extreme conditions unfold them (Jackson 1998). In contrast, folding of large, multidomain proteins involve several intermediates prior to the formation of the completely folded native state. They usually fold in modules that finally interact to conform the fully native structure (Khan et al. 2003, Panchenko et al. 1996, Vendruscolo et al. 2003) but often require the assistance of folding modulators, namely isomerases and foldases. The requirement of such cell elements dramatically increases in the context of recombinant protein production, in which the host cell receives an extremely dramatically high input of *de novo* synthesized polypeptides. In fact, chaperones are considered limiting factors in recombinant cells.

The term "misfolding" is used to describe the process that results in a protein acquiring a sufficient number of persistent non-native interactions to affect its overall architecture and/or its properties in a biologically significant manner (Dobson 2004). Misfolded and incompletely folded molecules are susceptible to aggregate, due to the exposure of hydrophobic regions that are buried in the native state (Fink 1998) (Fig. 15.1). To avoid aggregation, cells of living organisms have auxiliary factors, including folding catalysts that accelerate rate-limiting steps, and molecular chaperones that assist protein folding (Gething and Sambrook 1992, Hartl and Hayer-Hartl 2002). Moreover, such cell quality control mechanism targets for proteolytic destruction any protein molecule that has not folded correctly (Fig. 15.1). Protein misfolding in recombinant bacteria and other microbial cell factories is a major concern in Biotechnology, as misfolding not only results in protein degradation and/or aggregation but also in a global conformational stress status that triggers a set of cell responses.

15.3 The *Escherichia coli* Quality Control System

The protein quality control machinery is mainly based on the activity of chaperones and proteases that co-ordinately act assisting protein folding, preventing accumulation of misfolded species, removing protein from aggregates and degrading folding-reluctant species (Bukau et al. 2006). Therefore, this system's coordinated activity promotes protein solubility by minimizing the amount of aggregated species. In the biotechnological context, solubility is the parameter commonly used to evaluate the quality of a recombinant protein in a production process (de Marco et al. 2007, Schultz et al. 2006), and it is given as the amount of recombinant protein present in the soluble cell fraction relative to the total amount of recombinant protein occurring in the cell (usually expressed as percentage). In *E. coli*, this system is composed by periplasmic and cytoplasmic arms, which control polypeptides secreted and retained in the cytoplasm respectively. Periplasmic quality control has been extensively reviewed elsewhere (Miot and Betton 2004) and the next sections will mainly focus on the cytoplasmic regulators of protein folding and quality.

15.3.1 Chaperones and Proteases

The term chaperone was first used to describe an activity associated with nucleoplasmin in *Xenopus* oocytes (Laskey et al. 1978). Since then, the term has been expanded to include more than 20 protein families with a central role in the conformational quality control of the proteome (Bukau et al. 2006, Ellis 1987, Young et al. 2004). Specifically, molecular chaperones are a group of structurally diverse proteins highly conserved in all kingdoms of life which form a complex network to assist proper protein folding, prevent their deposition and dissolve deposits of misfolded proteins (Kazemi-Esfarjani and Benzer 2000, Krobitsch and Lindquist 2000, Mogk et al. 1999, Muchowski et al. 2000, Warrick et al. 1999). Even though chaperones are constitutively expressed under physiological conditions, many of them are upregulated under conformational stress conditions. In *E. coli*, such regulation is mainly controlled by the sigma factor σ^{32}, encoded by *rpoH* gene (Straus et al. 1987). Since chaperone abundance increases in cells upon thermal stress, these molecules have been traditionally named heat shock proteins (Hsps) (Lemaux et al. 1978), although not all chaperones are heat shock proteins and not all heat shock proteins are chaperones. Molecular chaperones can be divided into three functional subclasses based on their mechanism of action:

"Folding" chaperones mediate the folding of their substrates in an ATP-dependent process. These cell molecules increase the yield of properly folded proteins but not the folding rate. In the *E. coli* cytoplasm the three chaperone systems involved in this process are trigger factor (TF) (Bukau et al. 2000), DnaK-DnaJ-GrpE and GroEL-GroES (Grantcharova et al. 2001).

"Holding" chaperones maintain proteins partially folded on their surface to await availability of folding chaperones upon stress conditions, preventing polypeptides from aggregation (Ehrnsperger et al. 1997, Mogk et al. 1999, Veinger et al. 1998). The most extensively characterized bacterial holdases are IbpA and IbpB, both belonging to the small Hsp family (Narberhaus 2002) and commonly found within IBs (Allen et al. 1992) with a suspected role in its physiological disintegration (Lethanh et al. 2005). Hsp31 (Sastry et al. 2002) and Hsp33 (Graf and Jakob 2002) are also classified as holdases; while Hsp31 binds early unfolding intermediates in times of severe stress, thereby preventing overload of the DnaK-DnaJ-GrpE system (Malki et al. 2003, Mujacic et al. 2004), Hsp33 manages oxidative protein misfolding (Graf and Jakob 2002).

Finally, "disaggregating" chaperones promote protein removal from IBs and other aggregates (Rinas et al. 2007). Among them, ClpB is the best characterized. It has a secondary role, assisting refolding and promoting the solubilisation of proteins that have become aggregated as a result of stress (Ben-Zvi and Goloubinoff 2001, Schirmer et al. 1996). This chaperone acts in cooperation with DnaK and IbpAB chaperones (Mogk et al. 2003, Schlieker et al. 2004, Thomas and Baneyx 2000).

Moreover, at least several chaperones, including DnaK, ClpA and ClpX, work in cooperation with proteases (Garcia-Fruitos et al. 2007b, Hoskins et al. 1998, Matouschek 2003).

15.3.1.1 Trigger Factor

The ribosome-associated trigger factor (TF) is a three-domain protein that binds to the large subunit of the ribosomes, in the vicinity of the peptide exit site, to interact with nascent polypeptides and protect them (Bukau et al. 2000). Trigger factor exhibits both peptidyl-prolyl *cis/trans* isomerase (PPIase) and chaperone activity (Hoffmann and Rinas 2004, Huang et al. 2000, Nishihara et al. 2000). Therefore, this molecular chaperone supports the *de novo* folding by binding to nascent chains. Once the substrate is released, trigger factor can cycle back to the ribosome, waiting for the next substrate molecule (Bukau et al. 2000).

15.3.1.2 The Hsp70 System: DnaK, DnaJ, GrpE

Hsp70 family proteins are encoded in all living organisms' genomes, being one of the most conserved family in the evolution (Gupta and Singh 1994, Hunt and Morimoto 1985, Lindquist and Craig 1988). In *E. coli*, there are three Hsp70 proteins (namely DnaK, HscA and HscC), being DnaK the best characterized. DnaK is a key element of the multichaperone network, having different recognized roles: (1) it mediates ATP-dependent unfolding, (2) prevents aggregation, (3) stabilises the substrates for refolding by GroELS (Goloubinoff et al. 1999, Gupta and Singh 1994, Hoffmann and Rinas 2004, Hunt and Morimoto 1985, Lindquist and Craig 1988, Nishihara et al. 1998, Thomas and Baneyx 1996a), (4) participates in proteolysis (Bukau 1993, Yura and Nakahigashi 1999), cooperating in some cases with Lon protease, (5) folds newly synthesized polypeptides (Hartl and Hayer-Hartl 2002, Teter et al. 1999), (6) solubilises protein aggregates in cooperation with ClpB and Ibps (Ben-Zvi and Goloubinoff 2001, Carrio and Villaverde 2002, Glover and Tkach 2001, Goloubinoff et al. 1999, Mogk and Bukau 2004, Mogk et al. 1999, Zolkiewski 1999), (7) protects proteins against oxidative damages (Echave et al. 2002, Fredriksson et al. 2005) and (8) negatively regulates the heat shock response (Nagai et al. 1994) minimizing the expression of the heat shock σ^{32} regulon, which encodes the main chaperones and proteases, including DnaK itself (Morita et al. 2000, Tomoyasu et al. 2001a, Tomoyasu et al. 1998).

DnaK has an N-terminal ATPase domain of 44 kDa, two β-sheets forming a substrate binding site and a C-terminal domain of 27 kDa that can interact with partner proteins to modulate chaperone function (Genevaux et al. 2007, Genevaux et al. 2001). DnaK partners are a J-domain protein (JDP) co-chaperone, belonging to the Hsp40 family, termed DnaJ (Hennessy et al. 2005) and a nucleotide exchange factor (NEF) named GrpE. When ATP is bound, DnaK binds the substrate through weak, hydrophobic interactions and hydrogen bonds (Zhu et al. 1996). Upon ATP hydrolysis, there is a conformational change that stabilises substrate binding (Hoffmann and Rinas 2004). In this process, the co-chaperone DnaJ has an important role accelerating the ATP hydrolysis rate, while the co-chaperone GrpE accelerates the exchange of ADP with ATP, leading to substrate ejection. The released polypeptide may reach a native conformation, undergo additional cycles in the chaperone system until it folds, or be transferred to GroEL-GroES (Ewalt et al. 1997). The system formed by DnaK chaperone and DnaJ and GrpE co-chaperones is usually abbreviated as KJE.

15.3.1.3 ClpB

ClpB is an ATP-dependent molecular chaperone, member of Hsp100 family. Specifically, ClpB is a "disagregase" that works in cooperation with DnaK-DnaJ-GrpE reverting aggregation (Carrio and Villaverde 2001, Hoffmann and Rinas 2004,

Mogk et al. 2003, Parsell et al. 1994). This molecular chaperone has an important role, in cooperation with DnaK, in dissolving protein aggregates, reducing the aggregate size and exposing hydrophobic surfaces (Ben-Zvi and Goloubinoff 2001, Goloubinoff et al. 1999, Zolkiewski 1999). However, the full recovery of the native state cannot be achieved until the partially unfolded substrate is transferred from ClpB to DnaK (Glover and Lindquist 1998, Goloubinoff et al. 1999, Mogk et al. 1999, Motohashi et al. 1999, Zolkiewski 1999).

15.3.1.4 Hsp60 System: GroEL and GroES

GroEL is a bacterial chaperonin of approximately 60 kDa that belongs to the Hsp60 family. This molecular chaperone, essential for growth at all temperatures (Fayet et al. 1989), prevents aggregation (Kedzierska et al. 1999), acting as the main folder element in the chaperone network (Grantcharova et al. 2001). GroEL is formed by two stacked homoheptameric rings which define a central cavity in which incompletely folded polypeptides up to around 60 kDa (Sakikawa et al. 1999) can properly fold. When ATP is bound, a conformational change takes place (Ranson et al. 2001) rendering GroEL competent to bind the 10 kDa accessory protein GroES (Hartl and Hayer-Hartl 2002). The GroES-bound GroEL protein undergoes a second conformational modification, allowing the folding of the non-native polypeptide. If the protein has not reached the native state, a further round of binding and attempted folding follows.

15.3.1.5 Small Heat Shock Proteins

The best defined small heat shock proteins (sHsps) in bacteria have been Inclusion Body Proteins (Ibps), which are regularly associated to inclusion bodies (Allen et al. 1992) and commonly organized in large oligomeric structures (Haslbeck 2002, Narberhaus 2002). There are two different types of Ibps encoded on a single-operon (Allen et al. 1992, Chuang et al. 1993), IbpA and IbpB of 14 and 16 kDa size, respectively. Although IbpA is insoluble and IbpB is mainly soluble, IbpB comigrates to the insoluble fraction when produced with IbpA (Kuczynska-Wisnik et al. 2002). Even though Ibps function is not well understood, they seem to recognize hydrophobic patches in unfolded proteins, remaining bound to these polypeptides and protecting them from aggregation until they are transferred to DnaK or GroEL for refolding (Kitagawa et al. 2002, Kuczynska-Wisnik et al. 2002, Shearstone and Baneyx 1999, Thomas and Baneyx 1998). Moreover, it has been recently described that IbpA and IbpB facilitate the disaggregation and refolding activity of ClpB (Mogk et al. 2003).

15.3.1.6 Proteases

Proteolysis of misfolded proteins that have failed to reach a native conformation plays a crucial role in the quality control system, preventing the aggregation of abnormal polypeptides as well as allowing the amino acid recycling within the cell. The main proteases of *E. coli* cytoplasm are ClpP and Lon (Gottesman et al. 1997,

Maurizi 1992, Wickner et al. 1999). These heat-shock ATP-dependent proteases recognize hydrophobic surfaces, as chaperones do (Wickner et al. 1999). Moreover, these cell proteases degrade not only unprotected, misfolded polypeptides localized in the soluble cell fraction (Carrio et al. 1999, Maurizi 1992), but also those found embedded in protein aggregates (Corchero et al. 1997, Vera et al. 2005).

Lon is a tetrameric serine protease of 87 kDa subunits containing three functional domains. Its N-terminus is involved in substrate recognition and binding, its central domain is responsible for ATPase activity and its C-terminus domain has proteolytic activity. In addition to being responsible for bulk protein degradation (Missiakas et al. 1996, Tomoyasu et al. 2001b), Lon also exerts a regulatory function by degrading a class of proteins that are designed to be unstable.

ClpP is a protein organized as two stacked heptamers of 23 kDa each. Their substrates are folded, misfolded or incompletely synthesized proteins that are targeted for degradation. This protease forms a complex with two members of the Hsp100 family of ATPases (ClpA and ClpX) (Hoskins et al. 1998, Levchenko et al. 1995, Wickner et al. 1994) to form a fully-competent degrading machinery. ClpA and ClpX, which are flanking the rings of ClpP, act as molecular chaperones, unfolding proteins in an ATP-dependent manner and translocating substrates into the ClpP central channel (Matouschek 2003).

15.4 Composition of Inclusion Bodies

In general, the major component of IBs is the recombinant protein itself that can reach up to around 95% of the deposited protein material (Villaverde and Carrio 2003). However, in addition to the target protein other plasmid or host cell derived proteins or other cell components coprecipitate during IB recovery, adsorb to IBs or can get even entrapped *in vivo* during IB construction. For instance, lipids, DNA and outer membrane proteins are not integral IB components but coprecipitate after mechanical cell breakage with the aggregates during sedimentation by centrifugation (Bowden et al. 1991). The outer membrane proteins, for example, are also found in the particulate fraction after cell breakage prior to induction of recombinant protein synthesis and in cells not producing recombinant proteins (Hart et al. 1990, Rinas and Bailey 1992, Rinas and Bailey 1993, Rinas et al. 1993, Schmidt et al. 1999). These outer membrane proteins can be removed from IB preparations by detergent washing and other procedures that do not unfold proteins but solubilise membrane proteins (Estapé and Rinas 1996, Hart et al. 1990). Other non-integral macromolecular host cell contaminants of crude IB preparations, e.g. nucleic acids, phospholipids, and lipopolysaccharides are also removed by washing procedures using buffers composed of detergents, EDTA as well as cell wall- and DNA-degrading enzymes (Harris et al. 1986, Hartley and Kane 1988, Marston 1986, Marston and Hartley 1990, Marston et al. 1984, Schoemaker et al. 1985, Sugrue et al. 1990). In addition to the outer membrane protein OmpA, which constitutes the major portion of contaminating proteins in crude IB preparations (Hart et al. 1990, Rinas and

Bailey 1992, Rinas et al. 1993), other host cell or plasmid-encoded proteins also coprecipitate after mechanical cell breakage of IB-containing cells but also in corresponding control cells not producing the recombinant protein. Examples include the other outer membrane proteins OmpF and OmpC (Hart et al. 1990, Rinas and Bailey 1992, Rinas et al. 1993), other membrane proteins such as the flavoprotein subunit of succinate dehydrogenase SdhA, and ribosomal subunit proteins L7/L12 (Rinas and Bailey 1992). Moreover, the plasmid-encoded cI857 repressor, a thermolabile protein used for controlling temperature-inducible lambda promoter based expression systems, has been found in the insoluble cell fraction of IB-containing cells but also in respective control cells suggesting that its aggregation is not related to target protein production (Rinas et al. 2007).

However, there are also other proteins specifically associated with the aggregated fraction of inclusion body producing cells which are not found in the aggregated fraction of respective control cells. For instance, truncated versions of the recombinant target protein, other plasmid-encoded proteins e.g. those conferring resistance to antibiotics, and defined host cell proteins have been found entrapped within bacterial IBs (Hart et al. 1990, Jurgen et al. 2000, Neubauer et al. 2007, Rinas and Bailey 1992, Rinas and Bailey 1993, Rinas et al. 1993, Wagner et al. 2007). In particular, putative DnaK substrates such as the elongation factor Tu (ET-Tu) and the metabolic enzymes dihydrolipoamide dehydrogenase (LpdA), tryptophanase (TnaA), and D-tagatose-1,6-bisphosphate aldolase (GatY) have been identified only in the aggregated fraction of inclusion body producing cells (Rinas et al. 2007). GatY, in particular, a notoriously insoluble protein depending on GroEL (Chapman et al. 2006, Kerner et al. 2005) and DnaK for proper folding (Mogk et al. 1999), has also been found in other inclusion body preparations (Josef Lengeler and Peter Neubauer, personal communication). In some cases entrappment of precursors of membrane and periplasmic proteins into cytoplasmic IBs has been reported (Rinas and Bailey 1993, Wagner et al. 2007)

The most prominent host cell derived protein contaminants of IBs were identified as members of the heat-shock protein family (Allen et al. 1992). As their function was completely unknown at that time these IB contaminants were named inclusion body proteins (IbpA and IbpB). Since then, their presence within bacterial IBs has been further reported (Lethanh et al. 2005, Wagner et al. 2007) but also their absence in IB preparations has been noted (Rinas et al. 2007). Today, their function is still not completely understood. *In vitro* studies on thermal aggregates indicate that both together efficiently stabilize thermally aggregated proteins in a disaggregation competent state and allow more effective reactivation through the disaggregating chaperones ClpB and DnaK (Kuczynska-Wisnik et al. 2002, Laskowska et al. 2004, Lewandowska et al. 2007, Matuszewska et al. 2005, Veinger et al. 1998). Moreover, *in vivo* studies revealed that the presence of IbpA and IbpB renders the aggregated polypeptides in a conformationally more native state, with higher enzymatic activity compared to IBs produced in *ibpAB* deletion strains (Kuczynska-Wisnik et al. 2004). Other members of the heat shock protein family, namely the chaperones DnaK and GroEL, have also been found associated with IBs (Carrio and Villaverde 2002). DnaK is localized preferentially on the surface of inclusion bodies (Carrio and Villaverde 2005) and, together with ClpB, is recovered with low density protein aggregates during sucrose density centrifugation (Schrodel and de Marco 2005). The presence of DnaK has also been verified in other inclusion body preparations (Rinas et al. 2007, Wagner et al. 2007). GroEL, on the other hand, is homogeneously distributed in the cytosol, absent from the IB surface, but found in minor amounts also inside the aggregates (Carrio and Villaverde 2005). During *in vitro* recovery by sucrose density centrifugation, GroEL is recovered together with IbpB with high density protein aggregates (Schrodel and de Marco 2005). However, other researchers report absence or at most very small quantities of GroEL in IB preparations (Bowden et al. 1991, Carrio and Villaverde 2002, Rinas and Bailey 1992, Rinas et al. 1993, Rinas et al. 2007, Wagner et al. 2007).

It has been long debated if intracellular protein aggregation as IBs is a specific process between identical protein chains or is a process where different protein chains interact with each other forming mixed aggregates. *In vitro* mixed refolding studies using the P22 tailspike and coat proteins revealed that the two proteins did not coaggregate with each other but only with themselves, suggesting that aggregation is caused by specific interactions among protein chains (Speed et al. 1996). Moreover, *in vitro* seeding of pure soluble protein solutions with purified IBs revealed that aggregation of the soluble protein was only induced when seeding occurred with IBs composed of the same protein but not with IBs composed of unrelated proteins (Carrio et al. 2005). Also, recent experiments using Fluorescence Resonance Energy Transfer (FRET) indicated that coproduction of two different aggregation prone proteins in *E. coli* does not lead to mixed intermolecular interactions between the different protein chains (Morell et al. 2008). In this line, *in vivo* studies using a human cell line (HEK293) revealed that coexpression of two unrelated aggregation prone proteins did not lead to coaggregation but to deposition in separate aggregates in the same cell, suggesting strong specificity of protein aggregation (Rajan et al. 2001). On the other hand, kanamycin phosphotransferase, a plasmid encoded protein conferring resistance to kanamycin, can only be solubilized under conditions that also solubilize the plasmid-encoded target protein, bovine growth hormone, strongly suggesting that both proteins are tightly associated within IBs (Schoner et al. 1985). However, tight association of proteins within bacterial IBs does not necessarily imply interactions between unrelated peptide chains but could simply indicate colocalization of small protein aggregates within inclusion bodies. This is not unexpected having in mind that the cellular environment is a very crowded space (Ellis 2001), with protein concentrations in the cytoplasm in the order of 200 g/L (Neidhardt and Umbarger 1996) and concentrations of all macromolecules together reaching more than 340 g/L (Zimmerman and Minton 1993, Zimmerman and Trach 1991). Moreover, protein diffusion experiments suggest that in solutions containing proteins at concentrations comparable to those found in biological fluid media, the diffusive transport of larger proteins and aggregates may be slower than in dilute solution by several orders of magnitude (Muramatsu and Minton 1988). By using very strong expression systems, induction leads to almost exclusive synthesis of the target protein (Schoner et al. 1985). For example, temperature-induction of recombinant protein synthesis can increase total protein synthesis rates four fold (with 60% of protein synthesis dedicated to the synthesis of the target protein) but leading only to 10% target protein accumulation (Hoffmann and Rinas 2000, Hoffmann and Rinas 2001). Thus, during a limited period of time very high protein synthesis rates can occur in protein producer cells which can explain entrappment of normally soluble host cell or plasmid derived proteins into bacterial inclusion bodies. In this line, high level expression of an aggregation prone target protein can also lead to entrappment of another aggregation prone plasmid-encoded protein into inclusion bodies although this other plasmid-encoded protein is produced at lower rates during target protein synthesis compared to respective control conditions without target protein overexpression (Neubauer et al. 2007, Rinas and Bailey 1993). Thus, non-target but aggregation prone proteins might directly aggregate where they are synthesized due to diffusive transport limitations during high level target proteins synthesis thereby leading to inclusion bodies of mixed micro aggregates.

15.5 Structural Properties of Bacterial Inclusion Bodies

IBs are protein aggregates with spherical or ovoid shapes, formed either in the cytoplasm or the periplasm, and that are observed as refractile particles (usually one or two per cell) by optical microscopy (Bowden et al. 1991, Carrio et al. 2005) and as electrodense masses by transmission electron microscopy (Bowden et al. 1991). Soluble polypeptides can be extracted *in vitro* from IBs by denaturation and refolding sequential procedures (Rudolph and Lilie 1996, Vallejo and Rinas 2004), that permit to obtain soluble protein species through protein-tailored protocols. Interestingly, the arrest of protein synthesis in recombinant bacteria promotes the fast disintegration of IBs (Carrio et al. 1999) proving that they result from an unbalanced equilibrium between protein deposition and cell mediated protein removal, in which both chaperones and proteases are involved. This fact is also being considered when designing *in vivo* protein recovery protocols (de Marco et al. 2007).

The inner molecular organization of bacterial IBs has been a matter of deep scientific discussion. To date, some spectroscopic techniques have been developed or fitted to the analysis of bacterial IBs, namely Circular Dichroism (Chiti et al. 1998, Lewandowska et al. 2007, Plakoutsi et al. 2005, Umetsu et al. 2004, Umetsu et al. 2005), Raman Spectroscopy (Przybycien et al. 1994), Dynamic Light Scattering (Grudzielanek et al. 2007, Plakoutsi et al. 2005), and Nuclear Magnetic Resonance (Umetsu et al. 2004). However, Fourier-Transform Infrared Spectroscopy (FTIR) has proven to be the most useful and powerful tool for this purpose (Ami et al. 2005, Jevsevar et al. 2005), especially Attenuated Total Reflection-FTIR (ATR-FTIR) (Gonzalez-Montalban et al. 2006, Gonzalez-Montalban et al. 2007b, Vera et al. 2007). FTIR, in contrast to other optical spectroscopic methods, resolves measurements which are essentially unaffected by light scattering on residual protein-lipid interactions or contaminant membrane fragments. For this reason, this technique was originally developed for the study of structural characterization of membrane or lipid-associated proteins (Chapman and Haris 1989, Surewicz and Mantsch 1988, Surewicz et al. 1988).

Infrared spectroscopy is a form of vibrational spectroscopy which reports directly on the secondary structure of the proteins. For this purpose, the major areas of interest in the spectra are Amide I and Amide II bands. Amide I band arises predominantly (about 80%) from the C = O stretching vibration of the amide functional group which absorbs basically in the 1600–1700cm^{-1} region. The Amide II band arises from N-H bending and C-N stretching vibrations which absorb in the 1500–1600cm^{-1} region. However, structural studies on protein aggregation are usually based on evaluations of Amide-I-band contour, since only this band is a sensitive marker of secondary structure, being the analysis of Amide II band, in general, less relevant.

Aggregation as IBs has been long thought to be an unspecific process driven by the random interactions of hydrophobic patches, thus rendering protein aggregates with no specific internal molecular architecture. However, more recently, evidences against this view have been rapidly increasing (Mozell et al. 2008, Wang et al. 2008, Ami et al. 2005, Ami et al. 2006, Carrio et al. 2005, Oberg et al. 1994, Przybycien et al. 1994, Umetsu et al. 2004), picturing IBs as highly ordered structures. As FTIR analysis reveals, IBs seem to build up through a constant type of intermolecular protein interactions, resulting in a molecular architecture characterized by the formation of new β-sheet structures (Ami et al. 2003, Carrio et al. 2005, Garcia-Fruitos et al. 2007b, Gonzalez-Montalban et al. 2007b, Umetsu et al. 2005) at expenses of α-helical structures (Fink 1998, Przybycien et al. 1994), even common to rich-β-sheet native proteins (Oberg et al. 1994, Vera et al. 2007). In cases where the aggregation-prone protein is an all-α-protein, as it happens with interleukin-4(IL-4) (Umetsu et al. 2004), IBs are characterized by a sharp increment in β-sheet content and by an almost undetectable α-helical moieties signal. These structural data suggest that the new formed β-sheet structures may be interacting in a different way from native β-sheet conformation, probably by a network of hydrogen bonds between different chains creating a tightly packed extended, intermolecular β-sheet conformation (Fink 1998). Altogether, these observations seem to point out that the interactions leading to IB formation and the molecular reorganization that aggregated proteins undergo within the deposit are not likely to be unspecific interactions.

Table 15.1 Common structural features of amyloid fibrils and inclusion bodies

Structural characteristics	References
Structural homogeneity	(Ami et al. 2005, Carrio et al. 2005, Fink 1998, Garcia-Fruitos et al. 2007b, Vera et al. 2007)
Intermolecular, cross β-sheet organization or enrichement of β-structures	(Carrio et al. 2005, Garcia-Fruitos et al. 2007b, Gonzalez-Montalban et al. 2006, Przybycien et al. 1994)
Amyloid-tropic dye binding	(Carrio et al. 2005)
Cytotoxicity linked to amyloid-like structures	(Gonzalez-Montalban et al. 2007b)

Interestingly, all these secondary structural features greatly resemble to those that have been proven to characterize amyloid fibril formation (Table 15.1). In the case of amyloid fibrils, sequence determinants acting as aggregating "hot-spots" seem to modulate the specific nucleation of amyloid proteins (Ivanova et al. 2004, Ventura 2005, Ventura and Villaverde 2006, Ventura et al. 2004). In fact, in recent years it has been shown that IBs formation is a highly specific process since this kind of protein aggregates are essentially formed by the recombinant protein (Carrio et al. 1998, Garcia-Fruitos et al. 2005a, Gonzalez-Montalban et al. 2006) and organized in a very homogeneous architecture (Ami et al. 2005, Fink 1998). Furthermore, preformed IBs can seed specifically misfolded counterparts promoting the deposition of homologous but not heterologous domains (Carrio et al. 2005). As in the case of amyloid fibrils, whose formation seems to be preceded by the formation of intermediate amyloid-like species linked to cellular toxicity (Bucciantini et al. 2002, 2004, Stefani and Dobson 2003), IB structure is reported to be deleterious for mammalian cells in a structural-dependent manner (Gonzalez-Montalban et al. 2007b).

Intriguingly, the increase in the newly formed, non-native β-sheet content does not necessarily involve the full unfolding of the protein sequestered in IBs. In fact, there is a significant number of reports indicating the occurrence of native-like secondary structure of IB polypeptides (Table 15.2).

The native structure of the soluble IL-2 and its IB counterpart is almost identical, with only packing the degree and the nature of molecular β-sheet interaction being the main differences (Oberg et al. 1994). In fact, the little variations in FTIR signals seem to reflect subtle rather than significant changes in the secondary structure. TEM β-lactamase IBs seem to retain about 60% of the native secondary structure of the soluble protein (Georgiou et al. 1994). A particular example is represented by recombinant hyperthermophilic archaeon proteins. At least 3 proteins (namely PH0979, PH0628 and PH1830), when embedded in IBs, maintained some degree of a native-like rigid secondary structure (Umetsu et al. 2004). These observations were also made for IBs formed by *Pseudomonas fragi* lipase, human growth hormone and interferon-alpha-2b (Ami et al. 2005, Ami et al. 2006). VP1LAC, a re-

Table 15.2 Coincidence of native-like structure and amyloid-like aggregation pattern in inclusion bodies

IB-forming protein	Amyloid-like structure	Native-like structure	References
K97V 1L-1β	Yes	Yes	(Oberg et al. 1994)
hDHFR	Yes	Yes	(Garcia-Fruitos et al. 2005b)
VP1GFP	Yes	Yes	(Garcia-Fruitos et al. 2005b)
Aβ42(F19D)-BFP	Yes	Yes	(Garcia-Fruitos et al. 2005b)
hG-CSF	Yes	Yes	(Jevsevar et al. 2005)
LPF	Yes	Yes	(Ami et al. 2005)
h-GH	Yes	Yes	(Ami et al. 2006)
IFN-alpha-2b	Yes	Yes	(Ami et al. 2006)
S65T GFP	Yes	Yes	(Vera et al. 2007)
VP1LAC	Yes	Yes	(Gonzalez-Montalban et al. 2006)

combinant *E. coli* β-galactosidase, retained a great amount of native-like structure when forming IBs in a mutant strain lacking a fully functional chaperone GroEL (Gonzalez-Montalban et al. 2006). This trait is also common to recombinant fluorescent proteins. The structural analysis by ATR-FTIR and fluorescence measurements showed that green fluorescent protein (GFP) (Vera et al. 2007), VP1GFP (a GFP fused to a foot-and-mouth disease virus capsid protein) and Aβ42(F19D)-BFP (an amyloid peptide fused to blue fluorescent protein (BFP)) (Garcia-Fruitos et al. 2005b) aggregated as IBs maintain native-like structure. This feature seems to permit an easier solubilization of the embedded protein. In this regard, L-arginine can easily disaggregate GFP (Tsumoto et al. 2003) and β2 microglobulin (Umetsu et al. 2005) from IBs due to the retention of native-like structure of the embedded polypeptides. Human granulocyte-colony stimulating factor (hG-CSF) produced in *E. coli* at low temperatures enables the formation of "non-classical" IBs, which contain high amounts of correctly folded hG-CSF. HG-CSF can be readily extracted from these "non-classical" IBs by nondenaturing conditions and low concentrations of polar solvents (Jevsevar et al. 2005).

15.6 Strategies to Minimize Inclusion Body Formation

In general, the refolding processes required to recover the protein in a native form are complex, expensive and not always convenient from an industrial point of view (Vallejo and Rinas 2004). For this reason, much effort has been invested to minimize IB formation during the production process itself, aiming to improve the yield of soluble protein species. Recombinant protein can account up to around 30% of the total cell protein, producing an enormous metabolic load on the *E. coli* biosynthetic machinery (Sahdev et al. 2008). Thus, as summarized below, some of the strategies devised to minimize aggregation are based on a tight control of the *E. coli* cellular milieu, while others are addressed to favour protein folding by either physicochemical or biological approaches.

15.6.1 Media Composition

The folding of certain proteins requires the presence of specific cofactors in the growth media, such as metal ions (e.g., iron-sulphur) or polypeptide-cofactors (e.g., flavin-mononucleotide). By adding these factors to the growth media, both protein solubility and folding rates can be enhanced (Apiyo and Wittung-Stafshede 2002, Bruser et al. 2003). The composition of growth media also affects the levels of soluble protein. By optimizing media composition, reduced expression times, increased soluble fraction yield and enhanced biological activity of enzymes have been achieved. These modifications have been recently reviewed (Sahdev et al. 2008).

15.6.2 Protein Production at Low Temperatures

A number of proteins have been successfully produced in a soluble form in *E. coli* by lowering the growth temperature of the culture (Chesshyre and Hipkiss 1989, Niiranen et al. 2007, Schein and Noteborn 1988, Vera et al. 2007). As the hydrophobic interactions that determine IB formation are temperature-dependent, protein production at temperatures below the optimal of 37 °C for *E. coli* growth usually leads to increased stability and correct folding (Sahdev et al. 2008). Moreover, the increased production of a number of chaperones also accounts for the better protein quality obtained at lower growth rates (Ferrer et al. 2003). In addition, some of the heat shock proteases induced during recombinant protein production are poorly active at low temperatures. This accounts for the reduced degradation of recombinant protein observed within a temperature range of 15–23 °C (Hunke and Betton 2003, Spiess et al. 1999).

However, disadvantages are also present in the use of this strategy, as low temperatures lead to reduced transcription and translation rates, which results in low yields and poor turnover of the recombinant protein.

15.6.3 *Genetic Modification of Producing* **Escherichia coli** *Strains*

Genetic background largely affects recombinant protein production. Ideally, host strains should be deficient in the most harmful proteases, confer a stable maintenance of the expression plasmid and be compatible with the expression system chosen by providing the genetic elements required (e.g., DE3 strain for the pET expression system) (Sorensen and Mortensen 2005a).

E. coli BL21 (Novagen, USA) is one among the most common hosts. The non-pathogenic *E. coli* B strains can grow in minimal media and are deficient in *ompT* and Lon proteases, providing increased protein stability. The most important BL21 derivatives include:

- *BLR* RecA$^-$ for stabilization of target plasmids containing repetitive sequences.
- *trxB/gor mutants* for enhancement of cytoplasmic disulfide bond formation (Novagen Origami and AD494 strains).
- *Rosetta-gami strains* for overcoming codon bias associated problems through the overexpression of a rare tRNA expression vector, in addition to the *trxB*/gor mutation described above.
- *lac ZY* deletion mutants for uniform and adjustable protein expression in all the cells (Novagen Tuner series).
- Origami-B strains derived from a *lac ZY* mutant of BL21, also including *trxB* and *gor* mutations and OmpT and Lon deficiencies of *BL21*.
- Avidis C1(DE3) and C43(DE3) strains, for soluble expression of IB prone and membrane proteins.

15.6.4 Co-production of Folding Modulators

Molecular chaperones important for the control of protein quality are believed to be limiting in bacterial cell factories. Therefore, co-production strategies have been widely tested to overcome limitations due to IB formation during recombinant protein expression, but to date the results obtained are in general controversial and inconsistent (Baldwin 1986, Baneyx 2004, Thomas et al. 1997). Some successful examples of improved solubility by coproduction of some of the major cytosolic chaperones (namely the DnaK-DnaJ-GrpE system or the GroEL-ES complex) are human ORP150, human lysozyme, p50csk protein tyrosine kinase, phosphomannose isomerase, endostatin, transglutaminase and fusion protein PreS2-S′-β-galactosidase (Amrein et al. 1995, Dale et al. 1994, de Marco et al. 2000, Nishihara et al. 2000, Proudfoot et al. 1996, Thomas and Baneyx 1996a, Thomas and Baneyx 1996b, Yokoyama et al. 1998). However, although a trial and error approach is still needed to determine the best set of chaperones for a determined target protein, so far the best results have been obtained by coexpression of several sets of folding modulators.

Recently, a systematic analysis of the combined power of the major cytosolic chaperone systems of *E. coli* (KJE, ELS, ClpB and IbpAB) was performed (de Marco et al. 2007). Of the 50 proteins tested, the solubility of around 50% of them was improved by chaperone co-overproduction, being KJE, ClpB and ELS the most successful combination. The study also suggested an enhancement of the native state acquisition due to chaperone overproduction.

Optimization of the procedure was done by allowing chaperone-assisted folding in absence of protein synthesis, which was blocked by either inducer withdrawal or chloramphenicol addition. Solubility yields increased in comparison to the one-step procedure, with some of the proteins requiring the two-step procedure for any solubilisation. Coproduction of IbpAB also improved solubility, even being the only combination that solubilised some of the proteins tested.

15.6.5 Fusion Tags

A different strategy consists of fusion protein technology, in which a solubility "tag" is fused to the target protein (Sahdev et al. 2008). Tags are proteins or peptides that upon fusion, help to the proper folding of their fusion partners and lead to enhanced solubility of the protein (Esposito and Chatterjee 2006). Some tags can also be used for affinity purification, and provide advantages such as protection from proteolysis or being expression reporters (GFP). When solubility tags do not double as affinity tags, they may be combined with another hexahistidine (His6) tag, this way allowing for purification. The use of small peptide tags called SET tags has also been successful for some proteins (Zhang et al. 2004). The small size of these tags ($<$ 30 amino acids) may lead to less folding interferences, making the protein suitable for structural studies without the need of removing the tag.

Table 15.3 Commonly used tags for solubility enhancement

Tag	Protein	Solubility enhancement	Affinity purification
MBP	Maltose-binding protein	Yes	Yes
GST	Glutathione-S-transferase	Yes	Yes
Trx	Thioredoxin	Yes	No
NusA	N-Utilization substance	Yes	No
SUMO	Small ubiquitin-modifier	Yes	No
SET	Solubility-enhancing tag (synthetic)	Yes	No
DsbC	Disulfide bond C	Yes	No
Skp	Seventeen kilodalton protein	Yes	No
T7PK	Phage T7 protein kinase	Yes	No
GB1	Protein G B1 domain	Yes	No
ZZ	Protein A IgG ZZ repeat domain	Yes	No
His6	Hexahistidine tag	No	Yes
FLAG	FLAG tag peptide	No	Yes
BAP	Biotin acceptor peptide	No	Yes
Strep-II	Streptavidin-binding peptide	No	Yes
CBP	Calmodulin-binding peptide	No	Yes

Table adapted from reference (Esposito and Chatterjee 2006)

This technique poses some technical disadvantages, such as the need for tag removal and the question of whether the protein of interest remains in its native state and active once the tag has been removed. Nevertheless, if the target protein is linked to its fusion partner through a protease-specific recognition sequence, this will allow for an easy separation of the purified recombinant protein by cleavage with the specific protease. Because of its high specificity and ease of production, one of the most commonly used proteases is TEV, from tobacco etch virus (Kapust et al. 2002, Kapust et al. 2001).

Some commonly used tags, either for solubility enhancement or combined affinity purification, are listed in Table 15.3.

15.7 Conformational Quality and Biological Activity of Recombinant Proteins in Inclusion Bodies

Although it has been historically believed that proteins deposited as IBs were devoid of any biological activity, independent studies of unrelated aggregating enzymes and fluorescent proteins have demonstrated that IBs are enzymatically active or fluorescent respectively (de Groot and Ventura 2006, Garcia-Fruitos et al. 2005a, Garcia-Fruitos et al. 2005b, Kuczynska-Wisnik et al. 2004, Tokatlidis et al. 1991, Vera et al. 2007, Worrall and Goss 1989). The functional protein species do not occur in the IB interface but in the core of the aggregates, indicating that active polypeptides are not mere contaminants from the soluble cell fraction but true structural components (Garcia-Fruitos et al. 2007a). This is in agreement with the observation of native-like secondary structure in IBs as discussed above, and indicates that solubility and biological activity are not linked parameters. Therefore, aggregation of recombinant proteins as IBs does not split the population of recombinant polypeptides into functional and non functional (Gonzalez-Montalban et al. 2007a), and aggregation determinants must then be defined stretches instead of large protein segments, and not necessarily linked to active sites or fluorophors. Probably, aggregation patches coexist in a single polypeptide molecule, with properly folded regions and conformational quality of protein embedded in IBs depending on how fast the aggregation occurs after protein synthesis (de Groot and Ventura 2006, Waldo et al. 1999). On the other hand, the occurrence in recombinant cells of "soluble aggregates", namely protein deposits present in the soluble fraction (Schrodel and de Marco 2005, Sorensen and Mortensen 2005b, Ventura and Villaverde 2006) is another indicator that solubility is not matching conformational quality, and strongly suggests that there is a wide spectrum of protein conformations in both soluble and insoluble cell fractions (Ventura and Villaverde 2006). The fact that the biological activity of both soluble and insoluble recombinant protein versions is favoured or impaired in

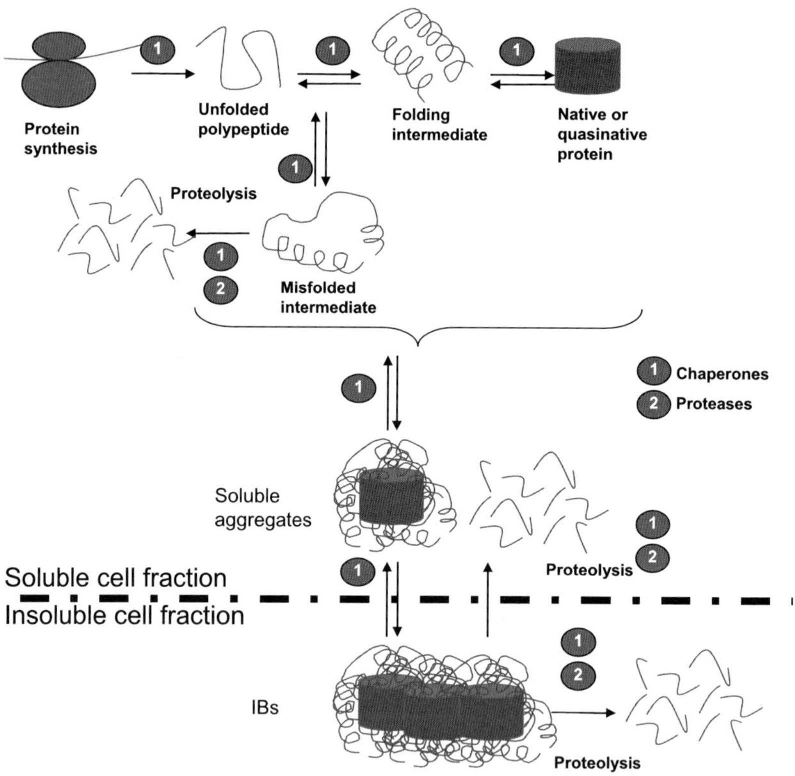

Fig. 15.2 Novel model of protein folding, aggregation and proteolysis in the *E. coli* cytoplasm. Several conformational versions of newly synthesized polypeptides, including those reaching native or native-like forms, can interact to form soluble aggregates, the putative precursors of inclusion bodies. Both soluble aggregates and inclusion bodies are then expected to be heterogeneous regarding protein folding status. The formation of insoluble inclusion bodies is highly favoured at high concentrations of recombinant protein. Chaperones (1) regulate aggregation and disaggregation but also protease (2)-mediated digestion of both soluble and insoluble protein versions

Table 15.4 Inclusion bodies used as biocatalysers

Inclusion bodies as biocatalysers	References
β-Galactosidase	Garcia-Fruitos et al. 2007a
Polyphosphate kinase	Nahalka et al. 2006
D-amino acid oxidase fusion protein	Nahalka and Nidetzky 2007
Maltodextrin phosphorylase fusion protein	Nahalka 2008
Sialic acid aldolase fusion protein	Nahalka et al. 2008

parallel by experimental conditions such as growth temperature (Vera et al. 2007) or availability of chaperones (Martinez-Alonso et al. 2007), indicates that IBs are not excluded from quality control but fully integrated in the cell processing of aberrant proteins (Fig. 15.2).

Regarding practical issues, functional IBs (such as those formed by enzymes) have been proposed as useful catalysts in bioprocesses without the need of protein removal and *in vitro* refolding (Garcia-Fruitos et al. 2007a, Garcia-Fruitos et al. 2005b). This principle has recently been proven with a diversity of aggregating recombinant enzymes such as D-amino oxidase from *Trigonopis variabilis* (Nahalka and Nidetzky 2007), polyphosphate kinase (Nahalka et al. 2006), maltodextrin phosphorylase from *Pyrococcus furious* (Nahalka 2008) and sialic acid aldolase (Nahalka et al. 2008), and opens new and challenging possibilities in the biotechnological market of recombinant proteins (Table 15.4).

15.8 Complex Systems Control of Protein Quality, Aggregation and IB Formation

Interestingly, the dramatic impact that different mutations in chaperone and protease genes have on IB disintegration (Carrio and Villaverde 2003, Vera et al. 2005) indicates that many components of the cell quality coordinately regulate the biology of these aggregates. In this context, a recent study shows that the total or partial intactivation of different genes of the *E. coli* quality control apparatus (including *dnaK, groEL, groES, clpA, clpP* and *lon*) results, as expected, in less solubility, but, surprisingly, in much more functional proteins in both soluble and insoluble populations (Garcia-Fruitos et al. 2007b). In particular, a deficiency in the chaperone DnaK, which is essentially found on the IBs surface in wild type recombinant cells (Carrio and Villaverde 2005), promotes the accumulation of high amounts of highly fluorescent GFP in IBs. This and other intriguing recent findings, such as the DnaK-inhibited activation and folding of β-galactosidase within IBs (Gonzalez-Montalban et al. 2008), the negative effect of DnaK on GFP folding and fluorophore activation (Garcia-Fruitos et al. 2007b, Martinez-Alonso et al. 2007), the DnaK-mediated stimulation of Lon- and Clp-mediated recombinant protein degradation (Garcia-Fruitos et al. 2007b) and the impact of DnaK on the partitioning of recombinant proteins into soluble and insoluble cell fractions (Garcia-Fruitos et al. 2005a, Gonzalez-Montalban et al. 2006) show that the quality control in general and the particular role of DnaK as a chaperone might have been largely misunderstood, specially regarding IB-forming recombinant cells.

By using GFP as a reporter recombinant protein, it has been determined that, in wild type cells, proteolysis acts on aggregation-prone but functional (or suitable to be activated) polypeptides. However, in cells deficient in chaperones such as ClpB, DnaK or GroEL or proteases such as Lon and ClpP, protein stability significantly increases. It seems that Lon and ClpP, in cooperation with DnaK, ClpB and others, proteolyse polypeptides on the IB surface (Garcia-Fruitos et al. 2007b), probably associated to their release during the continuous *in vivo* IB reconstruction (Carbonell and Villaverde 2002, Carrio et al. 1999, Carrio and Villaverde 2001, Carrio and Villaverde 2002, Corchero et al. 1997, Cubarsi et al. 2005). On the contrary, IbpA and IbpB play an antagosistic role, protecting recombinant proteins from proteolysis (Garcia-Fruitos et al. 2007b, Han et al. 2004). Interestingly, the combination of all these events and in particular, the unexpected role of DnaK in promoting proteolytic digestion of functional protein species and impairing *in situ* IB protein folding, results in a negative correlation between solubility and biological activity (and therefore, conformational quality) of recombinant proteins (Martínez-Alonso et al. 2008, and (Garcia-Fruitos et al. 2007b)). These observations point out that solubility is not a parameter representative of protein quality, since in recombinant cells conformational quality and solubility show a divergent genetic control. At least under recombinant protein production conditions, the bacterial quality control system tends to promote solubility at expenses of conformational quality, what could partially explain the inconsistent results found under coexpression of particular chaperones as discussed above. Also, the soluble and insoluble fractions, as virtual cell compartments, do not have much biological sense regarding protein quality and activity (Fig. 15.2).

To sum up, IBs, rather than being mere molecular "dust-balls" of the protein folding pipeline, are transient but highly dynamic protein reservoirs, fully integrated in the protein quality system, and whose formation and maintenance implies the complex activities of multigenetic networks. From a functional side, IB formation involves a tight cell control of protein folding and proteolytic stability.

Acknowledgments The authors appreciate the finantial support received for their research in recombinant protein production and aggregation, through grants BIO2005-23732-E and BIO2007-61194 MEC, Spain, and 2005SGR-00956 AGAUR, Spain.

Abbreviations

IBs:	inclusion bodies
TF:	trigger factor
KJE:	DnaK-DnaJ-GrpE system
Hsp:	heat shock protein
sHsps:	small heat shock proteins
Ibps:	inclusion bodies proteins
FTIR:	Fourier-transform infrared spectroscopy
ATR-FTIR:	Attenuated Total Reflection-FTIR
GFP:	Green fluorescent protein
BFP:	Blue fluorescent protein
hG-CSF:	Human granulocyte-colony stimulating factor
PPIase:	peptidyl-prolyl *cis/trans* isomerase
NEF:	nucleotide exchange factor

References

Allen SP, Polazzi JO, Gierse JK et al. (1992) Two novel heat shock genes encoding proteins produced in response to heterologous protein expression in *Escherichia coli*. J Bacteriol 174(21):6938–47

Ami D, Bonecchi L, Cali S et al. (2003) FT-IR study of heterologous protein expression in recombinant *Escherichia coli* strains. Biochim Biophys Acta 1624(1–3):6–10

Ami D, Natalello A, Gatti-Lafranconi P et al. (2005) Kinetics of inclusion body formation studied in intact cells by FT-IR spectroscopy. FEBS Lett 579(16):3433–6

Ami D, Natalello A, Taylor G et al. (2006) Structural analysis of protein inclusion bodies by Fourier transform infrared microspectroscopy. Biochim Biophys Acta 1764(4):793–9

Amrein KE, Takacs B, Stieger M et al. (1995) Purification and characterization of recombinant human p50csk protein-tyrosine kinase from an *Escherichia coli* expression system overproducing the bacterial chaperones GroES and GroEL. Proc Natl Acad Sci USA 92(4):1048–52

Anfinsen CB (1973) Principles that govern the folding of protein chains. Science 181(96):223–30

Apiyo D, Wittung-Stafshede P (2002) Presence of the cofactor speeds up folding of *Desulfovibrio desulfuricans* flavodoxin. Protein Sci 11(5):1129–35

Baldwin RL (1986) Temperature dependence of the hydrophobic interaction in protein folding. Proc Natl Acad Sci USA 83(21):8069–72

Baldwin RL (1994) Protein folding. Matching speed and stability. Nature 369(6477):183–4

Baneyx F (2004) Keeping up with protein folding. Microb Cell Fact 3(1):6

Baneyx F, Georgiou G (1991) Construction and characterization of *Escherichia coli* strains deficient in multiple secreted proteases: protease III degrades high-molecular-weight substrates *in vivo*. J Bacteriol 173(8):2696–703

Ben-Zvi AP, Goloubinoff P (2001) Review: mechanisms of disaggregation and refolding of stable protein aggregates by molecular chaperones. J Struct Biol 135(2):84–93

Bowden GA, Paredes AM, Georgiou G (1991) Structure and morphology of protein inclusion bodies in *Escherichia coli*. Biotechnology (N Y) 9(8):725–30

Bruser T, Yano T, Brune DC et al. (2003) Membrane targeting of a folded and cofactor-containing protein. Eur J Biochem 270(6):1211–21

Bucciantini M, Calloni G, Chiti F et al. (2004) Prefibrillar amyloid protein aggregates share common features of cytotoxicity. J Biol Chem 279(30):31374–82

Bucciantini M, Giannoni E, Chiti F et al. (2002) Inherent toxicity of aggregates implies a common mechanism for protein misfolding diseases. Nature 416(6880):507–11
Bukau B (1993) Regulation of the *Escherichia coli* heat-shock response. Mol Microbiol 9(4): 671–80
Bukau B, Deuerling E, Pfund C et al. (2000) Getting newly synthesized proteins into shape. Cell 101(2):119–22
Bukau B, Weissman J, Horwich A (2006) Molecular chaperones and protein quality control. Cell 125(3):443–51
Carbonell X, Villaverde A (2002) Protein aggregated into bacterial inclusion bodies does not result in protection from proteolytic digestion. Biotechnol Lett 24(23):1939–1944
Carrio M, Gonzalez-Montalban N, Vera A et al. (2005) Amyloid-like properties of bacterial inclusion bodies. J Mol Biol 347(5):1025–37
Carrio MM, Corchero JL, Villaverde A (1998) Dynamics of *in vivo* protein aggregation: building inclusion bodies in recombinant bacteria. FEMS Microbiol Lett 169(1):9–15
Carrio MM, Corchero JL, Villaverde A (1999) Proteolytic digestion of bacterial inclusion body proteins during dynamic transition between soluble and insoluble forms. Biochim Biophys Acta 1434(1):170–6
Carrio MM, Villaverde A (2001) Protein aggregation as bacterial inclusion bodies is reversible. FEBS Lett 489(1):29–33
Carrio MM, Villaverde A (2002) Construction and deconstruction of bacterial inclusion bodies. J Biotechnol 96(1):3–12
Carrio MM, Villaverde A (2003) Role of molecular chaperones in inclusion body formation. FEBS Lett 537(1–3):215–21
Carrio MM, Villaverde A (2005) Localization of chaperones DnaK and GroEL in bacterial inclusion bodies. J Bacteriol 187(10):3599–601
Chapman D, Haris PI (1989) Biomembrane structures. Fourier transform infrared spectroscopy and biomembrane technology. Biochem Soc Trans 17(6):951–3
Chapman E, Farr GW, Usaite R et al. (2006) Global aggregation of newly translated proteins in an *Escherichia coli* strain deficient of the chaperonin GroEL. Proc Natl Acad Sci USA 103(43):15800–5
Chesshyre JA, Hipkiss AR (1989) Low temperatures stabilize interferon α-2 against proteolysis in *Methylophilus methylotrophus* and *Escherichia coli*. Appl Microbiol Biotechnol 31(2): 158–162
Chiti F, Taddei N, van Nuland NA et al. (1998) Structural characterization of the transition state for folding of muscle acylphosphatase. J Mol Biol 283(4):893–903
Chuang SE, Burland V, Plunkett G, 3rd et al. (1993) Sequence analysis of four new heat-shock genes constituting the *hslTS/ibpAB* and *hslVU* operons in *Escherichia coli*. Gene 134(1):1–6
Clark ED (2001) Protein refolding for industrial processes. Curr Opin Biotechnol 12(2):202–7
Corchero JL, Cubarsi R, Enfors S et al. (1997) Limited *in vivo* proteolysis of aggregated proteins. Biochem Biophys Res Commun 237(2):325–30
Cubarsi R, Carrio MM, Villaverde A (2005) A mathematical approach to molecular organization and proteolytic disintegration of bacterial inclusion bodies. Math Med Biol 22(3):209–26
Dale GE, Schonfeld HJ, Langen H et al. (1994) Increased solubility of trimethoprim-resistant type S1 DHFR from *Staphylococcus aureus* in *Escherichia coli* cells overproducing the chaperonins GroEL and GroES. Protein Eng 7(7):925–31
de Groot NS, Ventura S (2006) Protein activity in bacterial inclusion bodies correlates with predicted aggregation rates. J Biotechnol 125(1):110–3
de Marco A, Deuerling E, Mogk A et al. (2007) Chaperone-based procedure to increase yields of soluble recombinant proteins produced in *E. coli*. BMC Biotechnol 7:32
de Marco A, Volrath S, Bruyere T et al. (2000) Recombinant maize protoporphyrinogen IX oxidase expressed in *Escherichia coli* forms complexes with GroEL and DnaK chaperones. Protein Expr Purif 20(1):81–6
Dill KA, Chan HS (1997) From Levinthal to pathways to funnels. Nat Struct Biol 4(1):10–9

Dinner AR, Sali A, Smith LJ et al. (2000) Understanding protein folding via free-energy surfaces from theory and experiment. Trends Biochem Sci 25(7):331–9

Dobson CM (2004) Principles of protein folding, misfolding and aggregation. Semin Cell Dev Biol 15(1):3–16

Echave P, Esparza-Ceron MA, Cabiscol E et al. (2002) DnaK dependence of mutant ethanol oxidoreductases evolved for aerobic function and protective role of the chaperone against protein oxidative damage in *Escherichia coli*. Proc Natl Acad Sci USA 99(7):4626–31

Ehrnsperger M, Graber S, Gaestel M et al. (1997) Binding of non-native protein to Hsp25 during heat shock creates a reservoir of folding intermediates for reactivation. Embo J 16(2):221–9

Ellis J (1987) Proteins as molecular chaperones. Nature 328(6129):378–9

Ellis RJ (2001) Macromolecular crowding: obvious but underappreciated. Trends Biochem Sci 26(10):597–604

Enfors SO (1992) Control of *in vivo* proteolysis in the production of recombinant proteins. Trends Biotechnol 10(9):310–5

Esposito D, Chatterjee DK (2006) Enhancement of soluble protein expression through the use of fusion tags. Curr Opin Biotechnol 17(4):353–8

Estapé D, Rinas U (1996) Optimized procedures for purification and solubilization of basic fibroblast growth factor inclusion bodies. Biotechnol. Tech 10(7):481–484

Ewalt KL, Hendrick JP, Houry WA et al. (1997) *In vivo* observation of polypeptide flux through the bacterial chaperonin system. Cell 90(3):491–500

Fahnert B, Lilie H, Neubauer P (2004) Inclusion bodies: formation and utilisation. Adv Biochem Eng Biotechnol 89:93–142

Fayet O, Ziegelhoffer T, Georgopoulos C (1989) The *groES* and *groEL* heat shock gene products of *Escherichia coli* are essential for bacterial growth at all temperatures. J Bacteriol 171(3): 1379–85

Ferrer M, Chernikova TN, Yakimov MM et al. (2003) Chaperonins govern growth of *Escherichia coli* at low temperatures. Nat Biotechnol 21(11):1266–7

Fink AL (1998) Protein aggregation: folding aggregates, inclusion bodies and amyloid. Fold Des 3(1):R9–23

Fredriksson A, Ballesteros M, Dukan S et al. (2005) Defense against protein carbonylation by DnaK/DnaJ and proteases of the heat shock regulon. J Bacteriol 187(12):4207–13

Garcia-Fruitos E, Aris A, Villaverde A (2007a) Localization of functional polypeptides in bacterial inclusion bodies. Appl Environ Microbiol 73(1):289–94

Garcia-Fruitos E, Carrio MM, Aris A et al. (2005a) Folding of a misfolding-prone beta-galactosidase in absence of DnaK. Biotechnol Bioeng 90(7):869–75

Garcia-Fruitos E, Gonzalez-Montalban N, Morell M et al. (2005b) Aggregation as bacterial inclusion bodies does not imply inactivation of enzymes and fluorescent proteins. Microb Cell Fact 4.27

Garcia-Fruitos E, Martinez-Alonso M, Gonzalez-Montalban N et al. (2007b) Divergent genetic control of protein solubility and conformational quality in *Escherichia coli*. J Mol Biol 374(1):195–205

Gasser B, Saloheimo M, Rinas U et al. (2008) Protein folding and conformational stress in microbial cells producing recombinant proteins: a host comparative overview. Microb Cell Fact 7:11

Genevaux P, Georgopoulos C, Kelley WL (2007) The Hsp70 chaperone machines of *Escherichia coli*: a paradigm for the repartition of chaperone functions. Mol Microbiol 66(4):840–57

Genevaux P, Schwager F, Georgopoulos C et al. (2001) The *djlA* gene acts synergistically with *dnaJ* in promoting *Escherichia coli* growth. J Bacteriol 183(19):5747–50

Georgiou G, Valax P (1996) Expression of correctly folded proteins in *Escherichia coli*. Curr Opin Biotechnol 7(2):190–7

Georgiou G, Valax P, Ostermeier M et al. (1994) Folding and aggregation of TEM beta-lactamase: analogies with the formation of inclusion bodies in *Escherichia coli*. Protein Sci 3(11):1953–60

Gething MJ, Sambrook J (1992) Protein folding in the cell. Nature 355(6355):33–45

Glover JR, Lindquist S (1998) Hsp104, Hsp70, and Hsp40: a novel chaperone system that rescues previously aggregated proteins. Cell 94(1):73–82

Glover JR, Tkach JM (2001) Crowbars and ratchets: hsp100 chaperones as tools in reversing protein aggregation. Biochem Cell Biol 79(5):557–68

Goloubinoff P, Mogk A, Zvi AP et al. (1999) Sequential mechanism of solubilization and refolding of stable protein aggregates by a bichaperone network. Proc Natl Acad Sci USA 96(24): 13732–7

Gonzalez-Montalban N, Garcia-Fruitos E, Ventura S et al. (2006) The chaperone DnaK controls the fractioning of functional protein between soluble and insoluble cell fractions in inclusion body-forming cells. Microb Cell Fact 5:26

Gonzalez-Montalban N, Garcia-Fruitos E, Villaverde A (2007a) Recombinant protein solubility – does more mean better? Nat Biotechnol 25(7):718–20

Gonzalez-Montalban N, Natalello A, Garcia-Fruitos E et al. (2008) In situ protein folding and activation in bacterial inclusion bodies. Biotechnol Bioeng 100(4):797–802

Gonzalez-Montalban N, Villaverde A, Aris A (2007b) Amyloid-linked cellular toxicity triggered by bacterial inclusion bodies. Biochem Biophys Res Commun 355(3):637–42

Gottesman S, Wickner S, Maurizi MR (1997) Protein quality control: triage by chaperones and proteases. Genes Dev 11(7):815–23

Graf PC, Jakob U (2002) Redox-regulated molecular chaperones. Cell Mol Life Sci 59(10): 1624–31

Grantcharova V, Alm EJ, Baker D et al. (2001) Mechanisms of protein folding. Curr Opin Struct Biol 11(1):70–82

Grudzielanek S, Velkova A, Shukla A et al. (2007) Cytotoxicity of insulin within its self-assembly and amyloidogenic pathways. J Mol Biol 370(2):372–84

Gupta RS, Singh B (1994) Phylogenetic analysis of 70 kD heat shock protein sequences suggests a chimeric origin for the eukaryotic cell nucleus. Curr Biol 4(12):1104–14

Han MJ, Park SJ, Park TJ et al. (2004) Roles and applications of small heat shock proteins in the production of recombinant proteins in *Escherichia coli*. Biotechnol Bioeng 88(4): 426–36

Harris TJ, Patel T, Marston FA et al. (1986) Cloning of cDNA coding for human tissue-type plasminogen activator and its expression in *Escherichia coli*. Mol Biol Med 3(3):279–92

Hart RA, Rinas U, Bailey JE (1990) Protein composition of Vitreoscilla hemoglobin inclusion bodies produced in *Escherichia coli*. J Biol Chem 265(21):12728–33

Hartl FU, Hayer-Hartl M (2002) Molecular chaperones in the cytosol: from nascent chain to folded protein. Science 295(5561):1852–8

Hartley DL, Kane JF (1988) Properties of inclusion bodies from recombinant *Escherichia coli*. Biochem Soc Trans 16(2):101–2

Haslbeck M (2002) sHsps and their role in the chaperone network. Cell Mol Life Sci 59(10): 1649–57

Hennessy F, Nicoll WS, Zimmermann R et al. (2005) Not all J domains are created equal: implications for the specificity of Hsp40-Hsp70 interactions. Protein Sci 14(7):1697–709

Hoffmann F, Rinas U (2000) Kinetics of heat-shock response and inclusion body formation during temperature-induced production of basic fibroblast growth factor in high-cell-density cultures of recombinant *Escherichia coli*. Biotechnol Prog 16(6):1000–7

Hoffmann F, Rinas U (2001) On-line estimation of the metabolic burden resulting from the synthesis of plasmid-encoded and heat-shock proteins by monitoring respiratory energy generation. Biotechnol Bioeng 76(4):333–40

Hoffmann F, Rinas U (2004) Roles of heat-shock chaperones in the production of recombinant proteins in *Escherichia coli*. Adv Biochem Eng Biotechnol 89:143–61

Hoskins JR, Pak M, Maurizi MR et al. (1998) The role of the ClpA chaperone in proteolysis by ClpAP. Proc Natl Acad Sci USA 95(21):12135–40

Huang GC, Li ZY, Zhou JM et al. (2000) Assisted folding of D-glyceraldehyde-3-phosphate dehydrogenase by trigger factor. Protein Sci 9(6):1254–61

Hunke S, Betton JM (2003) Temperature effect on inclusion body formation and stress response in the periplasm of *Escherichia coli*. Mol Microbiol 50(5):1579–89

Hunt C, Morimoto RI (1985) Conserved features of eukaryotic hsp70 genes revealed by comparison with the nucleotide sequence of human hsp70. Proc Natl Acad Sci USA 82(19): 6455–9

Ivanova MI, Sawaya MR, Gingery M et al. (2004) An amyloid-forming segment of beta2-microglobulin suggests a molecular model for the fibril. Proc Natl Acad Sci USA 101(29):10584–9

Jackson SE (1998) How do small single-domain proteins fold? Fold Des 3(4):R81–91

Jevsevar S, Gaberc-Porekar V, Fonda I et al. (2005) Production of nonclassical inclusion bodies from which correctly folded protein can be extracted. Biotechnol Prog 21(2):632–9

Jungbauer A, Kaar W (2007) Current status of technical protein refolding. J Biotechnol 128(3):587–96

Jurgen B, Lin HY, Riemschneider S et al. (2000) Monitoring of genes that respond to overproduction of an insoluble recombinant protein in *Escherichia coli* glucose-limited fed-batch fermentations. Biotechnol Bioeng 70(2):217–24

Kapust RB, Tozser J, Copeland TD et al. (2002) The P1′ specificity of tobacco etch virus protease. Biochem Biophys Res Commun 294(5):949–55

Kapust RB, Tozser J, Fox JD et al. (2001) Tobacco etch virus protease: mechanism of autolysis and rational design of stable mutants with wild-type catalytic proficiency. Protein Eng 14(12): 993–1000

Karplus M (1997) The Levinthal paradox: yesterday and today. Fold Des 2(4):S69–75

Kazemi-Esfarjani P, Benzer S (2000) Genetic suppression of polyglutamine toxicity in Drosophila. Science 287(5459):1837–40

Kedzierska S, Staniszewska M, Wegrzyn A et al. (1999) The role of DnaK/DnaJ and GroEL/GroES systems in the removal of endogenous proteins aggregated by heat-shock from *Escherichia coli* cells. FEBS Lett 446(2–3):331–7

Kerner MJ, Naylor DJ, Ishihama Y et al. (2005) Proteome-wide analysis of chaperonin-dependent protein folding in *Escherichia coli*. Cell 122(2):209–20

Khan F, Chuang JI, Gianni S et al. (2003) The kinetic pathway of folding of barnase. J Mol Biol 333(1):169–86

Kitagawa M, Miyakawa M, Matsumura Y et al. (2002) *Escherichia coli* small heat shock proteins, IbpA and IbpB, protect enzymes from inactivation by heat and oxidants. Eur J Biochem 269(12):2907–17

Krobitsch S, Lindquist S (2000) Aggregation of huntingtin in yeast varies with the length of the polyglutamine expansion and the expression of chaperone proteins. Proc Natl Acad Sci USA 97(4):1589–94

Kuczynska-Wisnik D, Kedzierska S, Matuszewska E et al. (2002) The *Escherichia coli* small heat-shock proteins IbpA and IbpB prevent the aggregation of endogenous proteins denatured *in vivo* during extreme heat shock. Microbiology 148(Pt 6):1757–65

Kuczynska-Wisnik D, Zurawa-Janicka D, Narkiewicz J et al. (2004) *Escherichia coli* small heat shock proteins IbpA/B enhance activity of enzymes sequestered in inclusion bodies. Acta Biochim Pol 51(4):925–31

Laskey RA, Honda BM, Mills AD et al. (1978) Nucleosomes are assembled by an acidic protein which binds histones and transfers them to DNA. Nature 275(5679):416–20

Laskowska E, Bohdanowicz J, Kuczynska-Wisnik D et al. (2004) Aggregation of heat-shock-denatured, endogenous proteins and distribution of the IbpA/B and Fda marker-proteins in *Escherichia coli* WT and grpE280 cells. Microbiology 150(Pt 1):247–59

Lemaux PG, Herendeen SL, Bloch PL et al. (1978) Transient rates of synthesis of individual polypeptides in *E. coli* following temperature shifts. Cell 13(3):427–34

Lethanh H, Neubauer P, Hoffmann F (2005) The small heat-shock proteins IbpA and IbpB reduce the stress load of recombinant *Escherichia coli* and delay degradation of inclusion bodies. Microb Cell Fact 4(1):6

Levchenko I, Luo L, Baker TA (1995) Disassembly of the Mu transposase tetramer by the ClpX chaperone. Genes Dev 9(19):2399–408

Lewandowska A, Matuszewska M, Liberek K (2007) Conformational properties of aggregated polypeptides determine ClpB-dependence in the disaggregation process. J Mol Biol 371(3):800–11

Lindquist S, Craig EA (1988) The heat-shock proteins. Annu Rev Genet 22:631–77

Malki A, Kern R, Abdallah J et al. (2003) Characterization of the *Escherichia coli* YedU protein as a molecular chaperone. Biochem Biophys Res Commun 301(2):430–6

Marston FA (1986) The purification of eukaryotic polypeptides synthesized in *Escherichia coli*. Biochem J 240(1):1–12

Marston FA, Hartley DL (1990) Solubilization of protein aggregates. Methods Enzymol 182: 264–76

Marston FAO, Lowe PA, Doel MT et al. (1984) Purification of Calf Prochymosin(Prorennin) Synthesized in *Escherichia coli*. Bio/Technology 2(9):800–804

Martinez-Alonso M, Vera A, Villaverde A (2007) Role of the chaperone DnaK in protein solubility and conformational quality in inclusion body-forming *Escherichia coli* cells. FEMS Microbiol Lett 273(2):187–95

Martinez-Alonso M, González-Montalbán N, Garcia-Fruitós E, Villaverde A. (2008) The functional quality of soluble recombinant polypeptides produced in *Escherichia coil* is defined by a wide conformational spectrum. Appl Environ Microbiol 74(23):7431–3

Matagne A, Dobson CM (1998) The folding process of hen lysozyme: a perspective from the 'new view'. Cell Mol Life Sci 54(4):363–71

Matouschek A (2003) Protein unfolding–an important process *in vivo*? Curr Opin Struct Biol 13(1):98–109

Matuszewska M, Kuczynska-Wisnik D, Laskowska E et al. (2005) The small heat shock protein IbpA of *Escherichia coli* cooperates with IbpB in stabilization of thermally aggregated proteins in a disaggregation competent state. J Biol Chem 280(13):12292–8

Maurizi MR (1992) Proteases and protein degradation in *Escherichia coli*. Experientia 48(2): 178–201

Middelberg AP (2002) Preparative protein refolding. Trends Biotechnol 20(10):437–43

Miot M, Betton JM (2004) Protein quality control in the bacterial periplasm. Microb Cell Fact 3(1):4

Missiakas D, Schwager F, Betton JM et al. (1996) Identification and characterization of HsIV HsIU (ClpQ ClpY) proteins involved in overall proteolysis of misfolded proteins in *Escherichia coli*. Embo J 15(24):6899–909

Mogk A, Bukau B (2004) Molecular chaperones: structure of a protein disaggregase. Curr Biol 14(2):R78–80

Mogk A, Deuerling E, Vorderwulbecke S et al. (2003) Small heat shock proteins, ClpB and the DnaK system form a functional triade in reversing protein aggregation. Mol Microbiol 50(2):585–95

Mogk A, Tomoyasu T, Goloubinoff P et al. (1999) Identification of thermolabile *Escherichia coli* proteins: prevention and reversion of aggregation by DnaK and ClpB. Embo J 18(24): 6934–49

Morell M, Bravo R, Espargaro A et al. (2008): Inclusion bodies: Specificity in their aggregation process and amyloid-like structure. Biochim Biophys Acta 1783(10):1815–25

Morita MT, Kanemori M, Yanagi H et al. (2000) Dynamic interplay between antagonistic pathways controlling the sigma 32 level in *Escherichia coli*. Proc Natl Acad Sci U S A 97(11):5860–5

Motohashi K, Watanabe Y, Yohda M et al. (1999) Heat-inactivated proteins are rescued by the DnaK.J-GrpE set and ClpB chaperones. Proc Natl Acad Sci USA 96(13):7184–9

Muchowski PJ, Schaffar G, Sittler A et al. (2000) Hsp70 and hsp40 chaperones can inhibit self-assembly of polyglutamine proteins into amyloid-like fibrils. Proc Natl Acad Sci USA 97(14):7841–6

Mujacic M, Bader MW, Baneyx F (2004) *Escherichia coli* Hsp31 functions as a holding chaperone that cooperates with the DnaK-DnaJ-GrpE system in the management of protein misfolding under severe stress conditions. Mol Microbiol 51(3):849–59

Muramatsu N, Minton AP (1988) Tracer diffusion of globular proteins in concentrated protein solutions. Proc Natl Acad Sci USA 85(9):2984–8

Nagai H, Yuzawa H, Kanemori M et al. (1994) A distinct segment of the sigma 32 polypeptide is involved in DnaK-mediated negative control of the heat shock response in *Escherichia coli*. Proc Natl Acad Sci USA 91(22):10280–4

Nahalka J (2008) Physiological aggregation of maltodextrin phosphorylase from *Pyrococcus furiosus* and its application in a process of batch starch degradation to alpha-D-glucose-1-phosphate. J Ind Microbiol Biotechnol 35(4):219–23

Nahalka J, Gemeiner P, Bucko M et al. (2006) Bioenergy beads: a tool for regeneration of ATP/NTP in biocatalytic synthesis. Artif Cells Blood Substit Immobil Biotechnol 34(5):515–21

Nahalka J, Nidetzky B (2007) Fusion to a pull-down domain: a novel approach of producing Trigonopsis variabilisD-amino acid oxidase as insoluble enzyme aggregates. Biotechnol Bioeng 97(3):454–61

Nahalka J, Vikartovska A, Hrabarova E (2008) A crosslinked inclusion body process for sialic acid synthesis. J Biotechnol 134(1–2):146–53

Narberhaus F (2002) Alpha-crystallin-type heat shock proteins: socializing minichaperones in the context of a multichaperone network. Microbiol Mol Biol Rev 66(1):64–93; table of contents

Neidhardt FC, Umbarger HE (1996) Chemical composition of *Escherichia coli*. In: (ed) *Escherichia coli* and *Salmonella*, American Society for Microbiology Press, Washington, D.C.

Neubauer A, Soini J, Bollok M et al. (2007) Fermentation process for tetrameric human collagen prolyl 4-hydroxylase in *Escherichia coli*: improvement by gene optimisation of the PDI/beta subunit and repeated addition of the inducer anhydrotetracycline. J Biotechnol 128(2): 308–21

Niiranen L, Espelid S, Karlsen CR et al. (2007) Comparative expression study to increase the solubility of cold adapted *Vibrio* proteins in *Escherichia coli*. Protein Expr Purif 52(1):210–8

Nishihara K, Kanemori M, Kitagawa M et al. (1998) Chaperone coexpression plasmids: differential and synergistic roles of DnaK-DnaJ-GrpE and GroEL-GroES in assisting folding of an allergen of Japanese cedar pollen, Cryj2, in *Escherichia coli*. Appl Environ Microbiol 64(5):1694–9

Nishihara K, Kanemori M, Yanagi H et al. (2000) Overexpression of trigger factor prevents aggregation of recombinant proteins in *Escherichia coli*. Appl Environ Microbiol 66(3):884–9

Oberg K, Chrunyk BA, Wetzel R et al. (1994) Nativelike secondary structure in interleukin-1 beta inclusion bodies by attenuated total reflectance FTIR. Biochemistry 33(9):2628–34

Otzen DE, Fersht AR (1998) Folding of circular and permuted chymotrypsin inhibitor 2: retention of the folding nucleus. Biochemistry 37(22):8139–46

Panchenko AR, Luthey-Schulten Z, Wolynes PG (1996) Foldons, protein structural modules, and exons. Proc Natl Acad Sci USA 93(5):2008–13

Parsell DA, Kowal AS, Singer MA et al. (1994) Protein disaggregation mediated by heat-shock protein Hsp104. Nature 372(6505):475–8

Paul DC, Van Frank RM, Muth WL et al. (1983) Immunocytochemical demonstration of human proinsulin chimeric polypeptide within cytoplasmic inclusion bodies of *Escherichia coli*. Eur J Cell Biol 31(2):171–4

Plakoutsi G, Bemporad F, Calamai M et al. (2005) Evidence for a mechanism of amyloid formation involving molecular reorganisation within native-like precursor aggregates. J Mol Biol 351(4):910–22

Proudfoot AE, Goffin L, Payton MA et al. (1996) *In vivo* and *in vitro* folding of a recombinant metalloenzyme, phosphomannose isomerase. Biochem J 318 (Pt 2):437–42

Przybycien TM, Dunn JP, Valax P et al. (1994) Secondary structure characterization of beta-lactamase inclusion bodies. Protein Eng 7(1):131–6

Rajan RS, Illing ME, Bence NF et al. (2001) Specificity in intracellular protein aggregation and inclusion body formation. Proc Natl Acad Sci USA 98(23):13060–5

Ranson NA, Farr GW, Roseman AM et al. (2001) ATP-bound states of GroEL captured by cryo-electron microscopy. Cell 107(7):869–79

Rinas U, Bailey JE (1992) Protein compositional analysis of inclusion bodies produced in recombinant *Escherichia coli*. Appl Microbiol Biotechnol 37(5):609–14

Rinas U, Bailey JE (1993) Overexpression of bacterial hemoglobin causes incorporation of pre-beta-lactamase into cytoplasmic inclusion bodies. Appl Environ Microbiol 59(2):561–6

Rinas U, Boone TC, Bailey JE (1993) Characterization of inclusion bodies in recombinant *Escherichia coli* producing high levels of porcine somatotropin. J Biotechnol 28(2–3):313–20

Rinas U, Hoffmann F, Betiku E et al. (2007) Inclusion body anatomy and functioning of chaperone-mediated *in vivo* inclusion body disassembly during high-level recombinant protein production in *Escherichia coli*. J Biotechnol 127(2):244–57

Rosen R, Biran D, Gur E et al. (2002) Protein aggregation in *Escherichia coli*: role of proteases. FEMS Microbiol Lett 207(1):9–12

Rudolph R, Lilie H (1996) *In vitro* folding of inclusion body proteins. Faseb J 10(1):49–56

Sahdev S, Khattar SK, Saini KS (2008) Production of active eukaryotic proteins through bacterial expression systems: a review of the existing biotechnology strategies. Mol Cell Biochem 307(1–2):249–64

Sakikawa C, Taguchi H, Makino Y et al. (1999) On the maximum size of proteins to stay and fold in the cavity of GroEL underneath GroES. J Biol Chem 274(30):21251–6

Sastry MS, Korotkov K, Brodsky Y et al. (2002) Hsp31, the *Escherichia coli yedU* gene product, is a molecular chaperone whose activity is inhibited by ATP at high temperatures. J Biol Chem 277(48):46026–34

Schein CH, Noteborn (1988) Formation of soluble recombinant proteins in *Escherichia coli* is favored by lower growth temperature. Bio/Technology 6(3):291–294

Schirmer EC, Glover JR, Singer MA et al. (1996) HSP100/Clp proteins: a common mechanism explains diverse functions. Trends Biochem Sci 21(8):289–96

Schlieker C, Tews I, Bukau B et al. (2004) Solubilization of aggregated proteins by ClpB/DnaK relies on the continuous extraction of unfolded polypeptides. FEBS Lett 578(3):351–6

Schmidt M, Babu KR, Khanna N et al. (1999) Temperature-induced production of recombinant human insulin in high-cell density cultures of recombinant *Escherichia coli*. J Biotechnol 68(1):71–83

Schoemaker JM, Brasnett AH, Marston FA (1985) Examination of calf prochymosin accumulation in *Escherichia coli*: disulphide linkages are a structural component of prochymosin-containing inclusion bodies. Embo J 4(3):775–80

Schoner RG, Ellis LF, Schoner BE (1985) Isolation and purification of protein granules from *Escherichia coli* cells overproducing bovine growth hormone. Bio/Technology 3:151–154

Schrodel A, de Marco A (2005) Characterization of the aggregates formed during recombinant protein expression in bacteria. BMC Biochem 6:10

Schultz T, Martinez L, de Marco A (2006) The evaluation of the factors that cause aggregation during recombinant expression in *E. coli* is simplified by the employment of an aggregation-sensitive reporter. Microb Cell Fact 5:28

Shearstone JR, Baneyx F (1999) Biochemical characterization of the small heat shock protein IbpB from *Escherichia coli*. J Biol Chem 274(15):9937–45

Snow CD, Nguyen H, Pande VS et al. (2002) Absolute comparison of simulated and experimental protein-folding dynamics. Nature 420(6911):102–6

Sorensen HP, Mortensen KK (2005a) Advanced genetic strategies for recombinant protein expression in *Escherichia coli*. J Biotechnol 115(2):113–28

Sorensen HP, Mortensen KK (2005b) Soluble expression of recombinant proteins in the cytoplasm of *Escherichia coli*. Microb Cell Fact 4(1):1

Speed MA, Wang DI, King J (1996) Specific aggregation of partially folded polypeptide chains: the molecular basis of inclusion body composition. Nat Biotechnol 14(10):1283–7

Spiess C, Beil A, Ehrmann M (1999) A temperature-dependent switch from chaperone to protease in a widely conserved heat shock protein. Cell 97(3):339–47

Stefani M, Dobson CM (2003) Protein aggregation and aggregate toxicity: new insights into protein folding, misfolding diseases and biological evolution. J Mol Med 81(11):678–99

Straus DB, Walter WA, Gross CA (1987) The heat shock response of *E. coli* is regulated by changes in the concentration of sigma 32. Nature 329(6137):348–51

Sugrue R, Marston FA, Lowe PA et al. (1990) Denaturation studies on natural and recombinant bovine prochymosin (prorennin). Biochem J 271(2):541–7

Surewicz WK, Mantsch HH (1988) New insight into protein secondary structure from resolution-enhanced infrared spectra. Biochim Biophys Acta 952(2):115–30

Surewicz WK, Stepanik TM, Szabo AG et al. (1988) Lipid-induced changes in the secondary structure of snake venom cardiotoxins. J Biol Chem 263(2):786–90

Teter SA, Houry WA, Ang D et al. (1999) Polypeptide flux through bacterial Hsp70: DnaK cooperates with trigger factor in chaperoning nascent chains. Cell 97(6):755–65

Thomas JG, Ayling A, Baneyx F (1997) Molecular chaperones, folding catalysts, and the recovery of active recombinant proteins from *E. coli*. To fold or to refold. Appl Biochem Biotechnol 66(3):197–238

Thomas JG, Baneyx F (1996a) Protein folding in the cytoplasm of *Escherichia coli*: requirements for the DnaK-DnaJ-GrpE and GroEL-GroES molecular chaperone machines. Mol Microbiol 21(6):1185–96

Thomas JG, Baneyx F (1996b) Protein misfolding and inclusion body formation in recombinant *Escherichia coli* cells overexpressing Heat-shock proteins. J Biol Chem 271(19):11141–7

Thomas JG, Baneyx F (1998) Roles of the *Escherichia coli* small heat shock proteins IbpA and IbpB in thermal stress management: comparison with ClpA, ClpB, and HtpG *In vivo*. J Bacteriol 180(19):5165–72

Thomas JG, Baneyx F (2000) ClpB and HtpG facilitate de novo protein folding in stressed *Escherichia coli* cells. Mol Microbiol 36(6):1360–70

Tokatlidis K, Dhurjati P, Millet J et al. (1991) High activity of inclusion bodies formed in *Escherichia coli* overproducing *Clostridium thermocellum* endoglucanase D. FEBS Lett 282(1):205–8

Tomoyasu T, Arsene F, Ogura T et al. (2001a) The C terminus of sigma(32) is not essential for degradation by FtsH. J Bacteriol 183(20):5911–7

Tomoyasu T, Mogk A, Langen H et al. (2001b) Genetic dissection of the roles of chaperones and proteases in protein folding and degradation in the *Escherichia coli* cytosol. Mol Microbiol 40(2):397–413

Tomoyasu T, Ogura T, Tatsuta T et al. (1998) Levels of DnaK and DnaJ provide tight control of heat shock gene expression and protein repair in *Escherichia coli*. Mol Microbiol 30(3):567–81

Tsumoto K, Umetsu M, Kumagai I et al. (2003) Solubilization of active green fluorescent protein from insoluble particles by guanidine and arginine. Biochem Biophys Res Commun 312(4):1383–6

Umetsu M, Tsumoto K, Ashish K et al. (2004) Structural characteristics and refolding of *in vivo* aggregated hyperthermophilic archaeon proteins. FEBS Lett 557(1–3):49–56

Umetsu M, Tsumoto K, Nitta S et al. (2005) Nondenaturing solubilization of beta2 microglobulin from inclusion bodies by L-arginine. Biochem Biophys Res Commun 328(1):189–97

Vallejo LF, Rinas U (2004) Strategies for the recovery of active proteins through refolding of bacterial inclusion body proteins. Microb Cell Fact 3(1):11

Veinger L, Diamant S, Buchner J et al. (1998) The small heat-shock protein IbpB from *Escherichia coli* stabilizes stress-denatured proteins for subsequent refolding by a multichaperone network. J Biol Chem 273(18):11032–7

Vendruscolo M, Paci E, Dobson CM et al. (2003) Rare fluctuations of native proteins sampled by equilibrium hydrogen exchange. J Am Chem Soc 125(51):15686–7

Ventura S (2005) Sequence determinants of protein aggregation: tools to increase protein solubility. Microb Cell Fact 4(1):11

Ventura S, Villaverde A (2006) Protein quality in bacterial inclusion bodies. Trends Biotechnol 24(4):179–85

Ventura S, Zurdo J, Narayanan S et al. (2004) Short amino acid stretches can mediate amyloid formation in globular proteins: the Src homology 3 (SH3) case. Proc Natl Acad Sci USA 101(19):7258–63

Vera A, Aris A, Carrio M et al. (2005) Lon and ClpP proteases participate in the physiological disintegration of bacterial inclusion bodies. J Biotechnol 119(2):163–71

Vera A, Gonzalez-Montalban N, Aris A et al. (2007) The conformational quality of insoluble recombinant proteins is enhanced at low growth temperatures. Biotechnol Bioeng 96(6):1101–6

Villaverde A, Carrio MM (2003) Protein aggregation in recombinant bacteria: biological role of inclusion bodies. Biotechnol Lett 25(17):1385–95

Wagner S, Baars L, Ytterberg AJ et al. (2007) Consequences of membrane protein overexpression in *Escherichia coli*. Mol Cell Proteomics 6(9):1527–50

Waldo GS, Standish BM, Berendzen J et al. (1999) Rapid protein-folding assay using green fluorescent protein. Nat Biotechnol 17(7):691–5

Walsh G (2005) Therapeutic insulins and their large-scale manufacture. Appl Microbiol Biotechnol 67(2):151–9

Wang L, Maji SK, Sawaya MR et al. (2008): Bacterial inclusion bodies contain amyloid-like strucure. PLoS Biol 6(8):e195

Warrick JM, Chan HY, Gray-Board GL et al. (1999) Suppression of polyglutamine-mediated neurodegeneration in *Drosophila* by the molecular chaperone HSP70. Nat Genet 23(4):425–8

Wickner S, Gottesman S, Skowyra D et al. (1994) A molecular chaperone, ClpA, functions like DnaK and DnaJ. Proc Natl Acad Sci USA 91(25):12218–22

Wickner S, Maurizi MR, Gottesman S (1999) Posttranslational quality control: folding, refolding, and degrading proteins. Science 286(5446):1888–93

Williams DC, Van Frank RM, Muth WL et al. (1982) Cytoplasmic inclusion bodies in *Escherichia coli* producing biosynthetic human insulin proteins. Science 215(4533):687–9

Wolynes PG, Onuchic JN, Thirumalai D (1995) Navigating the folding routes. Science 267(5204):1619–20

Worrall DM, Goss NH (1989) The formation of biologically active beta-galactosidase inclusion bodies in *Escherichia coli*. Aust J Biotechnol 3(1):28–32

Yokoyama K, Kikuchi Y, Yasueda H (1998) Overproduction of DnaJ in *Escherichia coli* improves *in vivo* solubility of the recombinant fish-derived transglutaminase. Biosci Biotechnol Biochem 62(6):1205–10

Yon JM (2001) Protein folding: a perspective for biology, medicine and biotechnology. Braz J Med Biol Res 34(4):419–35

Young JC, Agashe VR, Siegers K et al. (2004) Pathways of chaperone-mediated protein folding in the cytosol. Nat Rev Mol Cell Biol 5(10):781–91

Yura T, Nakahigashi K (1999) Regulation of the heat-shock response. Curr Opin Microbiol 2(2):153–8

Zhang YB, Howitt J, McCorkle S et al. (2004) Protein aggregation during overexpression limited by peptide extensions with large net negative charge. Protein Expr Purif 36(2):207–16

Zhu X, Zhao X, Burkholder WF et al. (1996) Structural analysis of substrate binding by the molecular chaperone DnaK. Science 272(5268):1606–14

Zimmerman SB, Minton AP (1993) Macromolecular crowding: biochemical, biophysical, and physiological consequences. Annu Rev Biophys Biomol Struct 22:27–65

Zimmerman SB, Trach SO (1991) Estimation of macromolecule concentrations and excluded volume effects for the cytoplasm of *Escherichia coli*. J Mol Biol 222(3):599–620

Zolkiewski M (1999) ClpB cooperates with DnaK, DnaJ, and GrpE in suppressing protein aggregation. A novel multi-chaperone system from *Escherichia coli*. J Biol Chem 274(40):28083–6

Chapter 16
Mechanistic Challenges and Engineering Applications of Protein Export in *E. coli*

Eva-Maria Strauch and George Georgiou

Contents

Abstract Protein secretion and subcellular localization in *E. coli* has been under investigation for more than 60 years. While many details about the molecular mechanisms of these processes have been revealed, several facets of protein translocation still remain unclear. Bacteria secrete numerous proteins such as pathogenicity factors, toxins or degradative enzymes (Fernandez and Berenguer 2000). Six different secretion mechanisms for extruding proteins into the extracellular environment have been identified to-date. In Gram-negative bacteria such as *E. coli*, secretion into the extracellular medium requires crossing of two biological membranes, the inner and outer membranes of the cell. However, systems for protein translocation into the extracellular medium are generally highly protein-specific and with very few exceptions have not yet been engineered for the efficient export of recombinant proteins. More relevant from a technical and engineering standpoint, is the translocation of polypeptides from the cytoplasm into the periplasmic space, the main secretory compartment which is equivalent to the endoplasmic reticulum of eukaryotic cells.

G. Georgiou (✉)
Department of Chemical Engineering, Biomedical Engineering, Molecular Genetics and Microbiology and the Institute for Cellular and Molecular Biology, University of Texas at Austin, Austin, TX, USA
e-mail: gg@che.utexas.edu

S.Y. Lee (ed.), *Systems Biology and Biotechnology of Escherichia coli*,
DOI 10.1007/978-1-4020-9394-4_16, © Springer Science+Business Media B.V. 2009

In the first part of this chapter, we discuss export via the general Sec pathway and the Twin-Arginine Translocase (Tat) pathway. Compartmentalized molecular chaperones facilitate folding, impose a quality control step on the maturation of certain secreted proteins, especially those exported via Tat, and further facilitate the decision which protein export route should be chosen. The second part of this chapter focuses on the design of genetic screens or selections that capitalize on protein secretion to aid the screening of libraries of protein variants for molecular recognition or catalysis. We will briefly summarize the major *E. coli*-based display technologies and introduce new methodologies particularly those utilizing the Twin-Arginine Translocase pathway.

16.1 Protein Transport in *E. coli*

Proteins are charged, bulky heteropolymers of which transport across, or insertion into, the low dielectric barrier of a lipid bilayer membrane is thermodynamically highly unfavorable. Hence, there are several different transport pathways that expend metabolic energy to overcome this physical barrier. The Sec protein translocase utilizes energy mainly generated by the hydrolysis of nucleoside triphosphates. However, there are several different transport processes across biological membranes which solely rely on ion gradients.

In gamma-proteobacteria, the main route for protein transport across the cytoplasmic membrane is through the Sec translocon, a set of transmembrane proteins which form a hydrophilic channel. The SecYEG translocon is the bacterial homologue of the Sec61$\alpha\beta\gamma$ in eukaryotic cells. Sec protein translocation can be summarized in a basic set of rules (Schatz and Dobberstein 1996): A precursor protein containing a targeting sequence which is typically N-terminal, is maintained in an unfolded state prior to export. This unfolded, export-competent state of the precursor protein is either achieved by cytoplasmic chaperones (post-translational) or by an immediate association of the protein-synthesizing ribosome with a receptor in the membrane (co-translational export). For Sec substrates, folding and cofactor assembly occur in the periplasm. Folding is assisted by periplasmic chaperones such as DegP which functions as either protease or chaperone depending on the growth temperature, Skp that binds to non-native forms of periplasmic or outer membrane proteins preventing their aggregation, and four peptidyl-proline *cis-trans* isomerases, PpiA, PpiD, SurA and FkpA. The periplasm is an oxidizing environment mainly due to the action of DsbA which oxidizes free cysteines, whereas the isomerase DsbC rearranges disulfide bridges to their native conformation (Georgiou and Segatori 2005).

In contrast, proteins exported via the Tat pathway first fold within the cytoplasm (DeLisa et al. 2003). The cytoplasmic folding environment contains the general chaperones GroEL, DnaK/DnaJ/GrpE, the trigger factor, ClpB and the small heat-shock chaperones (IbpAB) among others. DnaK and possibly other chaperons play a role in the folding of some Tat substrate proteins prior to export (Graubner et al.

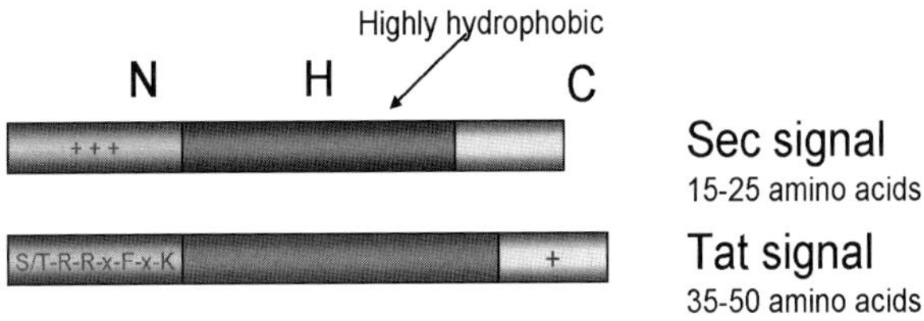

Fig. 16.1 Signal peptides. Comparison between a Sec and Tat signal peptides. N, H and C region are indicated. The positive charge within the C-region of a Tat signal peptide marks the Sec avoidance signal

2007, Perez-Rodriguez et al. 2007). Further, for the incorporation of metal cofactors into Tat substrates, complex protein maturation pathways involving several maturation enzymes are necessary. For example, the incorporation of iron-sulfur cluster requires at least 8 proteins (Tokumoto et al. 2002).

N-terminal signal peptides are evolutionarily well conserved. Signal peptides have three distinct regions (Fig. 16.1). The N-region harbors a positive charge, whereas the hydrophobic H-region comprises the center and the longest part of the signal peptide. The H-region of Tat signal peptides has a less hydrophobic character and is typically longer than the one of Sec signal peptides. The N-terminal positively charged region of Tat signal peptides contains the hallmark twin-arginine amino acids with the consensus sequence S/T-R-R-x-F-x-K. The C-region of both Sec and Tat signal peptides bears the signal peptidase cleavage site which is recognized by the type I signal peptidase. However, signal peptides of lipoproteins in *E. coli*, which so far have been only found to be exported by the Sec pathway, are cleaved by the type II signal peptidase (Paetzel et al. 2002). In general, Tat signal peptides typically contain a lysine or arginine residue within the C-terminal region which serves as Sec avoidance signal (Blaudeck et al. 2003). The more positively charged the C-region together with the beginning part of the mature protein is, the lower the likelihood that the precursor protein will be targeted to the Sec pathway (Tullman-Ercek et al. 2007).

16.1.1 The General Pathway for Secretion: The Sec Pathway

The Sec apparatus transports substrate polypeptides in an unfolded state through a narrow pore of about 5–8 Å minimally established by the SecY/E/G membrane proteins. Translocation can occur in two different ways which have been referred to as the co-translational and the post-translational export modes. Co-translational export involves the signal recognition particle (SRP) which is composed of the protein Ffh (fifty four homologue, based on its similarity to the eukaryotic SRP version in the endoplasmatic reticulum) and the 4.5S RNA unit (Luirink et al. 1992). The co-translational mode is often also referred to as the SRP pathway and ubiquitously found in all three kingdoms of life. The SRP complex binds to either a signal peptide or to a highly hydrophobic peptide stretch corresponding to a transmembrane domain of a integral membrane protein, as it exits the ribosome (Kim et al. 2001) (Fig. 16.2A). The loaded SRP complex then binds to the membrane-bound receptor

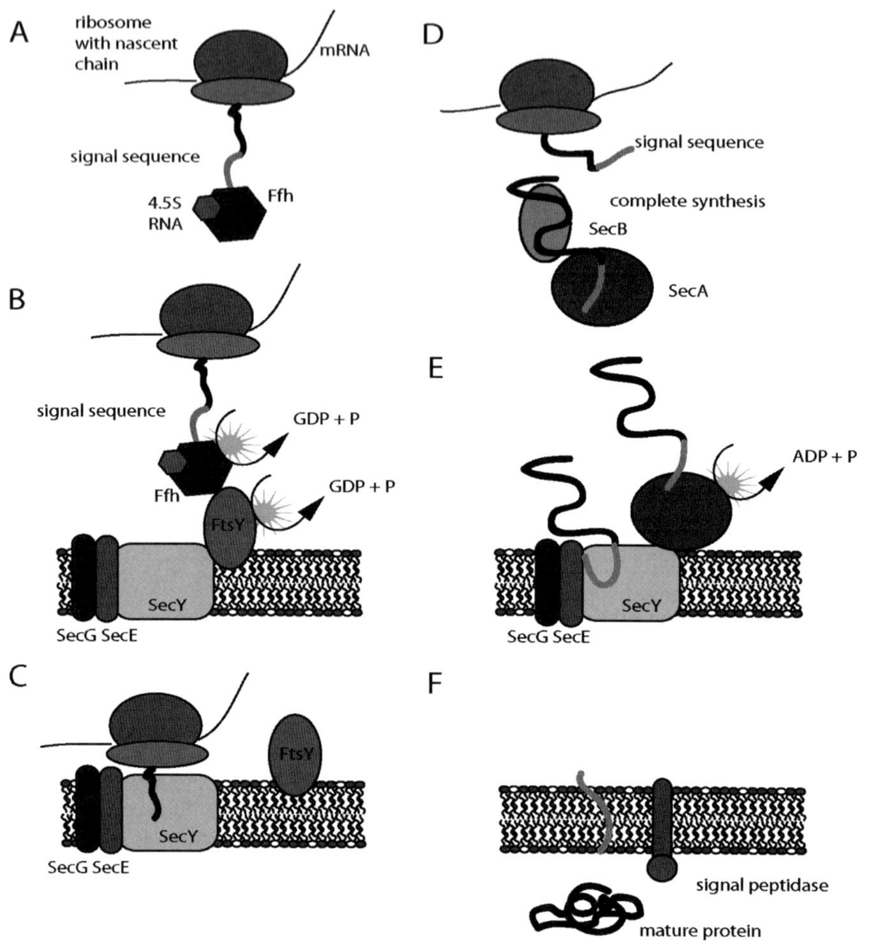

Fig. 16.2 Overview of protein targeting to the Sec translocon via the co-translational (**A–C**) and post-translational route (**D–E**). (**A**) A hydrophobic signal peptide or transmembrane domain of the nascent polypeptide chain is recognized by SRP (Ffh protein and 4.5S RNA unit). (**B**) SRP guides the ribosome with the nascent chain to the membrane-embedded receptor FtsZ which ensures the transfer of the nascent chain to the Sec apparatus. GTP hydrolysis is required for the release of SRP and the receptor. (**C**) The membrane-associated ribosome proceeds to synthesize the protein directly into the Sec system. (**D**) The signal peptide exiting the ribosome as a nascent chain is recognized by SecB which prevents its folding. SecA associates with SecB. (**E**) SecB transfers the preprotein to SecA and dissociates since it is not necessary for the translocation step. SecA associates with the Sec apparatus and proceeds to insert around 20 amino acids at a time into the translocation machinery; ATP hydrolysis is necessary for this motion. (**F**) Upon completion of translocation signal peptidase I cleaves off the signal peptide

FtsY which mediates the interaction of Ffh with the Sec translocon (Fig. 16.2B); GTP binding to both SRP and the receptor FtsY is a prerequisite for their interaction. GTP hydrolysis is precisely timed to transfer the ribosome nascent chain to the Sec translocon releasing the SRP from its receptor (Bange et al. 2007). In this manner the ribosome can resumes protein synthesis (Fig. 16.2C). Proteins following the co-translational export route are typically inner membrane proteins, but a few soluble proteins utilize this route as well. Probably the best studied SRP substrate is the disulfide oxidase DsbA. DsbA appears to fold too rapidly to be maintained in an unfolded state which is required for post-translational Sec transport (Schierle et al. 2003). Other proteins that utilize the cotranslational route include TorT, TolB or FlgI (Huber et al. 2005a). It is conceivable that the co-translational route could be utilized for the expression of any protein which is otherwise prone to aggregation in the cytoplasm.

The post-translational secretion mode normally involves the tetrameric chaperone SecB which binds to a nascent chain exiting the ribosome to prevent its immediate folding (Fig. 16.2D). In a SecB mutant, other general chaperones such as GroEL and/or DnaK can compensate for the loss of SecB (Kumamoto 1991). The association with SecB maintains the protein in a transport-compatible state since only the unfolded protein can be threaded through the membrane. SecB typically transfers its substrate directly to SecA to which it binds asymmetrically in its dimeric form. Binding of SecB presumably results in the dissociation of one SecA monomer, which may be important for the transfer step of the precursor protein to SecA (Fekkes et al. 1997). The translocation through the Sec pore is a step-wise event in which ATP hydrolysis by SecA allows the threading of around 20–30 amino acids of the polypeptide at a time through the SecYEG pore (van der Wolk et al. 1997), (Fig. 16.2E). However, once the preprotein is inserted into the membrane, translocation can be completed in the presence of solely a electrochemical potential, even without SecA (Duong and Wickner 1997, Schiebel et al. 1991).

16.1.2 Transporting Folded Proteins: The Twin-Arginine Translocase

The Tat pathway was discovered in bacteria 12 years ago (Berks 1996), and in the thylakoids membrane of plants 16 years ago (Cline et al. 1992). Little is known about the detailed molecular mechanism of protein translocation via Tat. The most remarkable feature of the Tat pathway is that it exports completely folded and assembled protein substrates. An unknown step in the translocation process serves as proof-reading function allowing only native proteins to be exported (DeLisa et al. 2003). The energy for translocation is derived solely from the proton motive force (Alder and Theg 2003, Bageshwar and Musser 2007) and so far no ATP requirement has been demonstrated. The Tat pathway is highly conserved in archaea, in most bacteria and in the chloroplasts of plants. Tat signal sequences can be found in most organisms and are partially interchangeable. Only a few protozoa encode proteins with homology to Tat components in their mitochondrial genome (Gray et al. 2004). The majority of the protein substrates for this pathway function in alternative anaerobic respiration pathways and catalyze redox reactions. They typically require the assembly of a set of complex cofactors and are often composed of multiple polypeptide subunits. The incorporation of these cofactors often necessitates specialized chaperones which are only available in the cytoplasm. For example, trimethylamine N-oxide reductase (TorA) contains a Fe-S cluster and a bis-molybdopterin guanine dinucleotide (MGM) cofactor (Mejean et al. 1994). Next to the advantage of having cytoplasmic chaperones assisting their folding maturation, iron-sulfur clusters are sensitive to oxidants which can be easier avoided in the reducing environment of the cytoplasm. Particularly, folding in the cytoplasm is of great advantage for halophilic organisms, which had they relied on the Sec pathway would have to fold their extracytoplasmic proteins under higher salt concentrations that favor protein aggregation. Hence these organisms often solely rely on Tat-mediated export (Dilks et al. 2005, Rose et al. 2002).

The minimal composition of the Tat translocon consists of the membrane proteins TatA, TatB and TatC. TatB is dispensable in some Gram-positive bacteria, and most archaea or can be replaced by mutated TatA variants (Blaudeck et al. 2005). Tat components can be found in two distinct subcomplexes in resting membranes (Orriss et al. 2007): a receptor complex composed of stoichiometric amounts of TatB and TatC, which is responsible for the recognition of Tat signal peptides (Alami et al. 2003, Kreutzenbeck et al. 2007, Strauch and Georgiou 2007b) and a second subcomplex containing high-molecular weight complexes of TatA. TatA forms pore-like structures of varying sizes in certain detergents, leading to the hypothesis that it mediates the actual translocation step (Gohlke et al. 2005). Whether TatA actually forms a channel or whether it is involved in lipid rearrangements that in turn mediate translocation has yet to be clarified. TatC interacts with the twin-arginine motif, whereas TatB associates with the hydrophobic stretch within the signal peptide. Currently, the most favored model for Tat export proposes a "handing-over" mechanism in which the signal peptide is first recognized by the TatB/C complex followed by the recruitment of several TatA oligomers; the substrate is then "handed over" to the TatA complex which mediates the actual translocation event (Fig. 16.3).

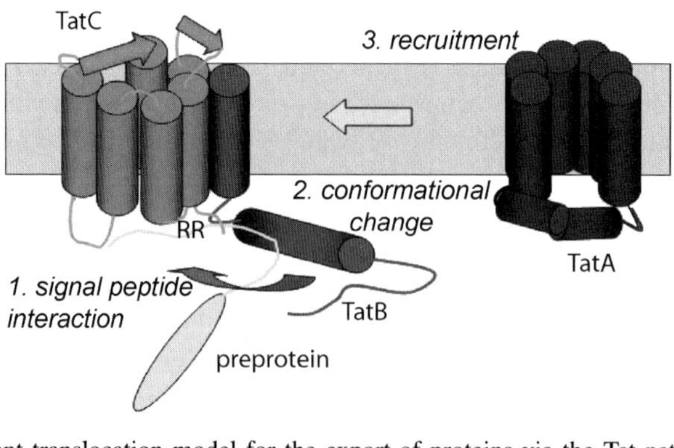

Fig. 16.3 Current translocation model for the export of proteins via the Tat pathway. TatB and TatC establish the signal peptide recognition complex. Upon interaction with the signal peptide (1), a possible conformational change occurs (2), followed by the recruitment of several TatA oligomers (3). TatA oligomers assemble in complexes of variable sizes which might depend on the dimensions of the substrate. TatA putatively mediates the actual translocation event

The division of the translocation step into two separate events is reflected in the biophysical properties of the translocation event (Bageshwar and Musser 2007). The question remains how does the receptor complex signal its interaction with the substrate to TatA? A conformational change is probably the most plausible explanation for this event. Gerard and coworkers suggested a pulling mechanism of the precursor by TatC based on the observation that proteins can be exported when covalently linked to the plant TatC homologue in the thylakoid Tat pathway (Gerard and Cline 2006). The fact that the translocon remains functional even when TatC is fused to its substrate could indicate that TatC at least remains in close proximity to the translocation step. Conformational changes could thus either serve as a trigger for the onset of the translocation event or could provide an actual pulling mechanism.

The fact that the Tat pathway exports folded proteins, begs the question whether there is a relationship between folding quality and export competence. Does the pathway discriminate between folded and unfolded proteins? DeLisa and coworkers (2003) showed that alkaline phosphatase, in which two intramolecular disulfide bridges must form to assume its active dimer conformation, cannot be exported when expressed in cells with a reducing cytoplasm that prevents disulfide bond formation (DeLisa et al. 2003). Deletions of *gor* and *trxB* which inactivate the thioredoxin and the glutathione reduction pathways that normally maintain the cytoplasm under reducing conditions allow disulfide bond formation in alkaline phosphatase and result in export via Tat. Similarly, Fisher et al. (2006) reported that the export rates of maltose binding protein variants (MBP-G32D, MBP-I33P, and MalE31-G32D/I33P) correlates with their solubility and the *in vitro* folding kinetics. These observations further support the notion that some step in the Tat pathway functions as a filter to prevent the export of misfolded proteins. Richter and coworkers proposed that it is the exposure of hydrophobic patches in unfolded proteins which allows the pathway to determine whether a protein is folded or misfolded (Richter et al. 2007). Notably they showed that an intrinsically disordered protein could be translocated though Tat but insertion of short hydrophobic stretches in this protein abolished export.

Several cofactor-containing Tat substrates have their own dedicated chaperone that are referred to as redox enzyme maturation proteins (REMPs). REMPs behave as specific proofreading chaperones escorting various oxido-reductases to the Tat apparatus. The enzyme trimethylamine N-oxide reducatase, TorA, for instance, has its own chaperone, TorD, which greatly facilitates the incorporation of its cofactors and retards the export process (Pommier et al. 1998). TorD binds specifically to the core region of the TorA signal peptide, but also to some parts of the mature enzyme (Hatzixanthis et al. 2005). Once TorD is bound to the signal peptide, its affinity for GTP increases. No GTP hydrolysis could be detected *in vitro*, indicating that the role of GTP might be more regulatory than catalytic. Similar to TorA, the DMSO reductase DmsA and the nitrate reductase NarG contain a molybdenum cofactor and an iron-sulfur cluster, respectively. The insertion of the cofactor and folding maturation is assisted by the small chaperones DmsD and NarJ (Chan et al. 2006, Oresnik et al. 2001). Based on phylogenetic analyses, TorD and NarJ have been classified as belonging to one group of maturation chaperones, whereas DmsD and NapD, which assists the folding of the nitrate reductase NapA (Maillard et al. 2007) have been assigned to a second group. At least for NapD, it was recently demonstrated that its molecular role is not only to camouflage the signal peptide, but also to actively inhibit transport before folding maturation has been completed (Maillard et al. 2007). A third group for Tat chaperones, the small chaperones HyaE and HybE assist the folding maturation of the [NiFe]-containing hydrogenase 1 (HyaA) and hydrogenase 2 (HybO/HybC), respectively (Dubini and Sargent 2003).

16.2 Expression and Folding of Exported Recombinant Proteins in *E. coli*

E. coli is widely used as the host organism for preparative protein expression in the laboratory and in the biotechnology industry (Baneyx and Mujacic 2004, de Marco 2007). Expression of heterologous proteins in secreted form is desirable when the heterologous protein contains disulfide bonds or otherwise cannot fold in the cytoplasm and when periplasmic localization confers protection against proteolysis or provides an advantage for downstream processing. Only a small set of *E. coli* proteins are secreted into the extracellular space primarily by pathogenic strains (Lawley et al. 2003, Pallen et al. 2003, Sandkvist 2001). Most naturally transported proteins of non-pathogenic *E. coli* are localized either in the periplasmic space, or associate with the outer membrane. Heterologous proteins secreted via Sec can be expressed at very high levels in the periplasmic space of bacteria (Choi and Lee 2004, Mergulhao et al. 2005). Early reports suggested that Tat-mediated transport results in lower protein yields than Sec transport (Berks et al. 2003, Sargent et al. 1998). However, it now appears that the efficiency of expression via Tat is dependent on the protein of interest. Fisher et al. (2008) reported that the periplasmic accumulation of different proteins such as alkaline phosphatase, GFP and a scFv antibody fragment fused to MBP were comparable or at most two fold lower for Tat-mediated export. Interestingly, the purity and activity levels of Tat exported proteins in the osmotic shock fraction were higher than those exported via Sec (Fisher et al. 2008). On the other hand, Tat export resulted in higher periplasmic yields of thioredoxin variants compared to export via the Sec pathway (Masip et al. 2008). This is presumably because the rapid folding kinetics of thioredoxin render it incompetent for Sec export, but favor secretion through Tat. Upon overexpression of a protein, misfolding leading to polypeptide degradation, aggregation or cell toxicity can occur. Precursor proteins that are exported slowly or that jam the Sec translocon result in cell toxicity and accumulation of the precursor protein in the cytoplasm (Feilmeier et al. 2000, Kiino and Silhavy 1984). However, this does not seem to be the case for export via Tat. Unlike the Sec translocon, the Tat apparatus does not appear to be prone to jamming, since the components of the translocon dissociate from any stuck precursor polypeptides (Cline and McCaffery 2007). Ultimately, in order to achieve higher yields of active proteins, the export pathway has to be chosen carefully. Export rates, yields, purity and yield of active proteins depend strictly on the amino acid sequence of the protein of interest.

Protein folding is a central issue in the expression of secreted proteins. As was discussed above, proteins secreted via Tat must attain a native-like conformation in the cytoplasm whereas proteins exported by the Sec apparatus have to fold within the periplasmic space. The co-expression of the proper set of endogenous chaperones can facilitate the expression of secreted proteins for both transport pathways (de Marco 2007). Cytosolic chaperones are often classified as folding, holding and disaggregation chaperones. The first class includes the ribosome-associated trigger factor (TF), the DnaK system (DnaK with its DnaJ and GrpE co-chaperones; KJE), and the GroEL system (GroEL with its GroES co-chaperone; ELS). Collectively these chaperones assist *de novo* protein folding. Both DnaK and GroEL are capable of refolding host proteins that become unfolded under environmental stress. The second class of cytosolic chaperones comprises of the holdases (the small heat-shock chaperones IbpA and IbpB), the redox-regulated Hsp33 and the "emergency" chaperone Hsp31. Holdases are active during severe stress and bind to early folding intermediates to prevent overloading of the KJE and ELS system (Mujacic et al. 2004). If folding and holding of proteins fail to deter protein aggregation, the third class of chaperones kicks in. Chaperones of this class promote aggregate solubilization and include ClpB ClpA, ClpX and ClpY with the latter three being involved in directing proteins to degradation. Disaggregation chaperones do not participate in the refolding of solubilized proteins, but rather transfer them to DnaK.

Several cytoplasmic chaperones and other cytosolic factors have been shown to increase the efficiency of export via the Tat pathway. For example the general chaperone DnaK aids the folding of several Tat substrates (Graubner et al. 2007, Perez-Rodriguez et al. 2007), resulting in increased export. Improved export of fusions to the TorA signal peptide has been observed upon overexpression of the chaperone TorD (Hatzixanthis et al. 2005, Jack et al. 2004, Li et al. 2006). In addition, overexpression of proteins that do not have a chaperone function, including the Tat pathway components TatABCE (Alami et al. 2003) and PspA (DeLisa et al. 2004) that possibly affects the electron gradient, have been shown to enhance Tat export. On the other hand, proteins translocated via the Sec pathway are released into the periplasm in an unfolded conformation and must attain their native state in that compartment. One of the major folding chaperones in the periplasmic space is DegP, which exhibits two functions: At lower growth temperatures, this protein typically acts as a molecular chaperone whereas at elevated temperatures its function as a degrading enzyme becomes more pronounced (Spiess et al. 1999). It recognizes unfolded proteins presumably via its PDZ domain (Iwanczyk et al. 2007, Wilken et al. 2004). Further, protein degradation in the periplasm can also involve the protease III and Tsp and thus may be alleviated in strains carrying deletions of the respective genes (Meerman and Georgiou 1994).

Many secreted proteins contain disulfide-bridges which need to be correctly formed for the polypeptide to attain its native conformation. In *E. coli* periplasmic protein thiol oxidation is catalyzed by the enzyme DsbA whereas isomerization of misfolded disulfide bonds is mediated by DsbC and to a lesser extend by DsbG (Bessette et al. 1999, Rietsch et al. 1996). Overexpression of DsbA and DsbC can result in a marked increase in the yield of complex recombinant proteins such as the human plasminogen activator (Bessette et al. 1999, Qiu et al. 1998), human nerve growth factor (Kurokawa et al. 2001), insulin-like growth factor-I (Joly et al. 1998) or horseradish peroxidase (Kurokawa et al. 2000). Additionally, overexpression of periplasmic chaperones such as Skp or the peptidyl-proline *cis-trans* isomerases, PpiA, PpiD, SurA or FkpA (Arie et al. 2001, Bothmann and Pluckthun 2000, Missiakas et al. 1996) have been shown to enhance the soluble yield of numerous proteins including antibody fragments (Choi and Lee 2004, Hayhurst et al. 2003). Combinations of overexpressed periplasmic chaperones have been shown to aid the folding of human plasma retinol-binding protein and of the extracellular carbohydrate recognition domain of the dendritic cell membrane receptor DC-SIGN (Schlapschy et al. 2006); for review (Choi and Lee 2004) (Fig. 16.4).

16.2.1 Protein Secretion and Display in Combinatorial Library Screening

16.2.1.1 Phage Display

The display of proteins on the surface of viral particles or cells constitutes the foundation of high throughput screening technologies for protein engineering purposes. Display technologies describe a variety of methodologies for the presentation of biomolecules onto a virus or cell. Protein display allows the screening of large combinatorial protein libraries for the isolation of ligand binding proteins, the engineering of protein stability (Kotz et al. 2004) and catalytic activity (Fernandez-Gacio et al. 2003), the detection of interacting proteins, determining the substrate specificity of proteases (Matthews and Wells 1993) and for several other applications (Hwang et al. 2007, Li et al. 2008, Matthews and Wells 1993). Viral, cell-based and *in vitro* display systems, such as ribosome display (Lipovsek and Pluckthun 2004) have been developed, but for the purposes of the present review we will focus only on viral (bacteriophage) and bacterial cell display methodologies.

Phage display is the first genetic strategy developed for the isolation of ligand-binding proteins from combinatorial libraries (Smith 1985). For filamentous phage display, the phage particles harboring the protein of interest and the gene that encodes it are continuously secreted into the growth medium. The protein of interest is typically displayed as a fusion to one of the coat proteins. Normally, the displayed protein fusion is secreted via the Sec pathway and is incorporated onto the virion during phage assembly in the periplasm. While lytic phages (such as T7 or lambda) have been used for display, non-lytic filamentous phages such as f1, M13 or fd are much more commonly employed. The displayed protein is either encoded in a phagemid, a plasmid containing both an *E. coli* and a phage origin of replications or it is directly integrated into the phage genome. With filamentous phage, the protein of interest is typically fused to the N-terminus of protein pIII (Fig. 16.5A) allowing the presentation of up to 5 copies, or to the major coat protein pVIII (Fig. 16.5A) which allows more than 2700 copies to be displayed on the phage particle (Glucksman et al. 1992, Malik et al. 1996, Zwick et al. 2000).

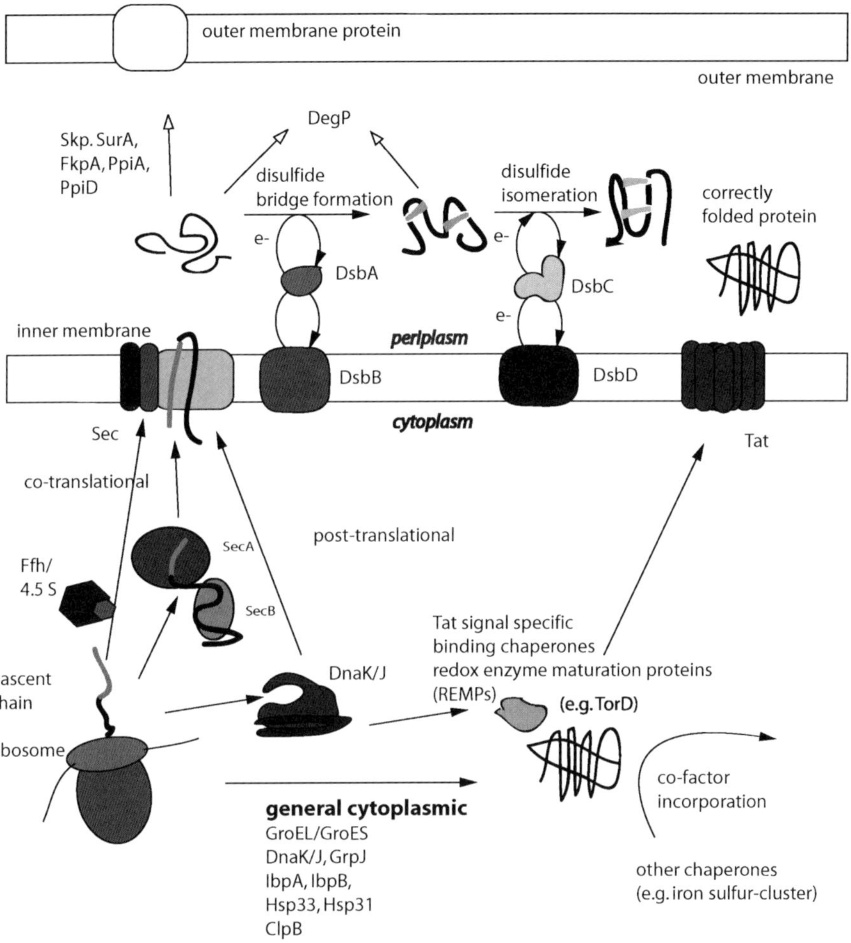

Fig. 16.4 Compartmental chaperones and their contribution to different export pathways. Cytoplasmic chaperones facilitate the export process of most precursor proteins. SecB and SecA bind to the fully synthesized polypeptide to maintain it in its export-competent, unfolded state and the guide it to the Sec translocon (post-translational). On the other hand, the signal recognition particle, composed of Ffh and 4.5S RNA, associates with the nascent chain extruding from the ribosome. The latter allows the recruitment of the translating ribosome to the membrane where it resumes synthesis of the polypeptide directly into the Sec pore (co-translational Sec transport). The presence of the general chaperone DnaK/J can be beneficial for the export of both Sec and Tat substrates. Tat substrates fold prior to export and therefore their folding maturation can be improved by the overexpression of various cytoplasmic chaperones. Additionally, the folding of many Tat substrates requires the participation of specialized cytoplasmic redox enzyme maturation proteins (REMPs). Tat substrates are translocated while in their correctly folded state, whereas folding of Sec substrates takes place in the periplasmic environment. General periplasmic chaperones, such as Skp, SurA etc. improve the solubility or folding of the Sec secreted polypeptides. DegP switches between its protease and foldase function depending on the temperature. DsbA and DsbC catalyze oxidative protein folding in the periplasm. DsbA introduces disulfide bridges, whereas DsbC re-shuffles their confirmation. DsbB and DsbD maintain these two crucial proteins in their appropriate oxidation state.

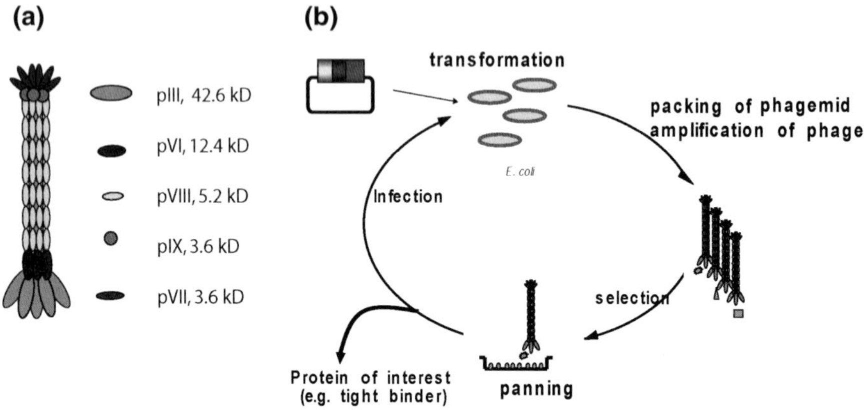

Fig. 16.5 (**A**) Filamentous phage particle, (**B**) panning cycle for enrichment of binders. A phagemid encoding the genetic information for the protein of interest is transformed into *E. coli* cells, which amplify the virion. After purification of the phage particles from the culture supernatant, protein displaying phage particles are applied to the immobilized ligand. Non-binding phage particles are eliminated by washing. Bound phage is eluted, pooled and used to infect new *E. coli* cells allowing a repetition of the panning cycle until interesting ligand binders are identified

Phage displaying polypeptides that bind to a desired ligand are enriched by several rounds of panning onto immobilized ligand. The ligand can be immobilized either directly by adsorption onto a plastic surface or indirectly, e.g. by using a biotin conjugate together with streptavidin-coated beads (Blazek et al. 2004).

Normally, proteins displayed on filamentous phage are secreted by the post-translational Sec pathway (Rapoza and Webster 1993). Often however, limitations associated with the post-translational Sec apparatus restrict the ability to display certain kinds of polypeptides, especially proteins that fold quickly in the cytoplasm or the export of which can block the early steps in the secretion process, e.g. by tight binding to SecA. Employing co-translational Sec export using an appropriate signal peptide such as the one for DsbA can alleviate these problems. For example, the fast-folding designed ankyrin-repeat proteins (DARPins) can be transported with high efficiency when switching the export signal to the co-translational DsbA signal peptide resulting in a 700-fold increase in their display on filamentous phage (Steiner et al. 2006). Export following this route prevents premature cytoplasmic folding prior translocation.

For reasons that probably relate to the fact that coat proteins are embedded in the membrane before they assemble onto the phage particle, it is not possible to export pIII or pV fusions via the Tat apparatus. Export via Tat is desirable for the display of proteins that require the incorporation of cytoplasmic cofactors, for proteins that are unable to fold into the periplasm for other reasons (Feilmeier et al. 2000) or for those that might fold too fast and cannot be maintained in a Sec competent state by the cytoplasmic chaperone machinery. A system that capitalizes on the Tat pathway for protein display is shown in Fig. 16.6 (Paschke and Hohne 2005; Strauch and

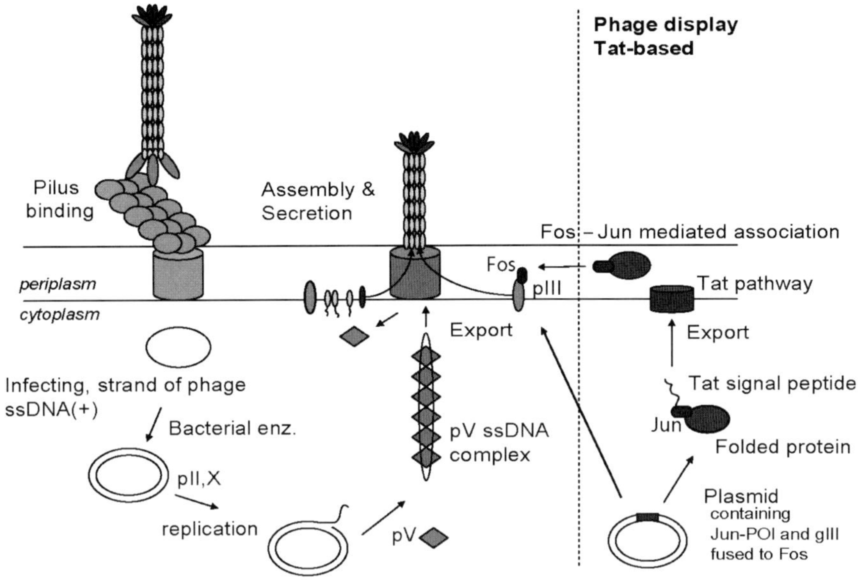

Fig. 16.6 Display of proteins exported via Tat on phage. The phage particle binds to the F-pilus and inserts its single stranded DNA into the bacterium where it uses bacterial enzymes and its own proteins for second strand synthesis and replication. The phage pV protein sequesters the + single strand away to enable its packaging into the phage particle that is assembled in its own secretion apparatus and extruded through the outer membrane via a phage encoded channel formed by the pIV protein. In the Tat-based phage display, pIII is expressed as a fusion of a leucine zipper domain (here Fos) and exported via Sec. The protein of interest (POI) is expressed a fusion to the complementary leucine zipper (here Jun) and a Tat signal peptide. After folding in the cytoplasm, the POI is exported via Tat and associates with the phage by non-covalently binding to the heterodimerizing leucine zipper sequence fused to pIII

Tullman-Ercek, unpublished results). pIII with a N-terminal Sec signal peptide is fused to half of a heterodimerizing leucine zipper sequence whereas the protein of interest is fused to an N-terminal Tat signal peptide and to the complementary leucine zipper sequence. The two gene constructs are expressed from a bicistronic operon. The pIII fusion and the protein of interest are exported via separate routes, namely Sec and Tat respectively, but once in the periplasm, their association is ensured by the leucine zipper dimerization and thus the target protein becomes non-covalently attached to pIII on the surface of phage. Cysteine residues may be placed at the ends of the leucine zipper halves to allow covalent disulfide linkages within the oxidizing environment of the periplasm. This strategy has been successfully employed to display fluorescent GFP (Paschke and Hohne 2005, Strauch 2007) that necessitates folding in the cytoplasm to form the active chromophore.

16.2.1.2 *E. coli*-Based Protein Display

Bacterial display offers several distinct advantages relative to phage: (i) it is possible to display many more protein copies on a bacterium compared to a phage particle:

(ii) complex proteins consisting of multiple polypeptides or proteins containing co-factors are more easy to display on cells; (iii) components of cells surfaces can be exploited for the retention of fluorescent products of enzymatic reactions and (iv) finally but most importantly, because of their larger size, bacteria are compatible with methodologies that utilize fluorescent activated cell sorting (FACS). Using multi-color fluorescence labeling strategies, it is possible to interrogate every clone in a library for the level of expression of a target protein, ligand binding or catalytic activity in a quantitative fashion. The advantages of FACS as a library screening tool have been instrumental in the isolation of very high affinity (picomolar) binding polypeptides and enzymes with high catalytic activity and selectivity from libraries displayed on bacteria.

In *E. coli* proteins can be displayed either on the surface or on a subcellular location that can be made accessible to extracellularly added fluorescently conjugated molecules following chemical treatment. A variety of protein fusions have been used for protein display on the surface of *E. coli* and the topic has been reviewed recently (Daugherty 2007, Lee et al. 2003, Samuelson et al. 2002). Several native outer membrane proteins (OMPs) such as FhuA, OmpA, OmpS, OmpX, and its circular permutated variant CPX, have been utilized for the display of short peptides with varying sizes typically between 12 and 28 aa long. In addition flagellar proteins, such as the commercially available recombinant constructs FliTrx (Lu et al. 1995, Westerlund-Wikstrom 2000) have been used for peptide display whereas Lpp-OmpA fusions and autotransporter proteins from pathogenic *E. coli* have been exploited for the display of several small proteins for ligand binding and enzymatic activity selections (Becker et al. 2007, 2004, Jose et al. 2005, Wentzel et al. 1999). However, surface display of intact proteins is often accompanied by changes in outer membrane permeability and loss of viability. In addition, the display of multi subunit proteins or proteins that contain multiple disulfide bonds is problematic since there is no folding machinery on the surface of the cell (Stathopoulos et al. 1996). Finally, it is not known whether large heterologous polypeptides fused to outer membrane protein targeting sequences can engage the periplasmic folding chaperones (Adams et al. 2005, Bos et al. 2007, Veiga et al. 2002) and the YaeT outer membrane protein localization machinery that might be required for surface display (Kim et al. 2007).

Proteins anchored on the inner membrane or expressed in the periplasm are of course not exposed to the extracellular fluid because the outer membrane of *E. coli* presents a formidable diffusion barrier that excludes molecules larger than 600 kDa. However, various chemical treatments can be used to increase the permeability of, or to completely remove, the outer membrane thus allowing access to periplasmic proteins with externally added ligands. Typically such ligands are fluorescently labeled so that upon binding to an *E. coli* displayed protein they render the cell fluorescent allowing its isolation by flow cytometry (Chen et al. 2001, Harvey et al. 2004). Libraries of scFv antibodies expressed in soluble form in the periplasmic space have been screened for binding to fluorescently labeled low molecular weight ligands that gain access into that compartment by incubating the cells in a high salt environment. Incubation in a hypertonic solution allowed molecules up to 10–15 KDa to diffuse into the cell without the release of the scFv proteins from the periplasm

(Chen et al. 2001). Using the same approach (Ribnicky et al. 2007) isolated a mutant scFv antibody that exhibited improved export via the Tat pathway, leading to greater accumulation of functional protein and therefore increased binding of fluorescently labeled antigen. Interestingly, the selected scFv variant exhibited faster folding kinetics *in vitro*, indicating that the rate of folding within the cytoplasm correlates with competence for translocation via the Tat pathway (Ribnicky et al. 2007).

Access of larger ligands into the periplasm requires rupture of the outer membrane and can be accomplished easily by treating the cells with a combination of chelating agents and lysozyme. However, under these conditions soluble secreted proteins, including the proteins to be displayed, are released from the periplasm either partially or completely. To avoid this problem Harvey et al developed the Anchored Periplasmic Expression (APEx) display system, in which the protein of interest is tethered to the inner membrane by fusing it genetically to an appropriate anchoring sequence. In principle any transmembrane α-helix can be used as an anchoring sequence (Ki et al. 2004). However, fusion to a targeting sequence comprised of a signal sequence followed by the first few amino acids of an inner membrane lipoprotein can be employed to convert the protein of interest into a lipoprotein. This is advantageous because the fusion tag required for display is very short and the expression of lipoproteins is better tolerated by the cell compared to integral membrane proteins. APEx has been used for the isolation of proteins that bind to extracellular ligands, for the engineering of variants that express better in the periplasm and for the detection of protein-protein interactions (Jeong et al. 2007). For the latter application, a bait protein is expressed in membrane-tethered form whereas the prey is expressed solubly in the periplasm. Following permeabilization of the outer membrane, the prey is released from the cell unless it captured by the inner membrane-tethered bait. The resulting complex can be detected by fluorescent anti-prey antibodies allowing isolation of the respective cell by FACS. Recently, a variation of APEx was employed to screen libraries of full length IgG antibodies in bacteria (Mazor et al. 2007). The ability to isolate and express full length IgG in bacteria may allow the rapid generation of antibodies for many therapeutic and diagnostic purposes. Finally our lab recently demonstrated that APEx can be carried out in *dsbA* strains where the formation of disulfide bonds is compromised. In this manner we were able to isolate mutant scFv antibody fragments that are stable and can fold in the absence of disulfide bonds (Seo et al. unpublished). Such antibody fragments are desired for gene therapy applications in which they would be expressed in the cytoplasm (where disulfide bonds cannot normally form) and can be used to disrupt the function of proteins associated with disease (Fig. 16.7).

16.2.2 Exploiting the Secretion Machinery as a Solubility and Folding Filter

As already mentioned, the folding state of a polypeptide is a major determinant of export competence. Huber and coworkers demonstrated that thioredoxin, which

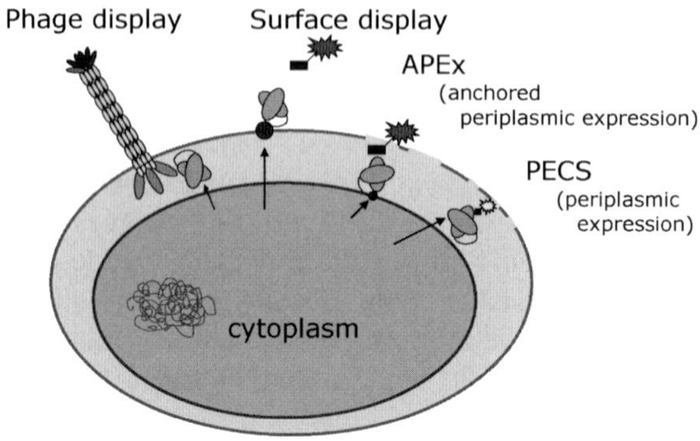

Fig. 16.7 *E. coli*-based display techniques. For simplicity lipopolysaccharides (LPS) and the peptidoglycan layer are not shown. For filamentous phage display, the protein of interest (POI) is exported into the periplasm of *E. coli* before it can be assembled on the phage particle. For surface display, outer membrane proteins or autotransporters can be utilized as carrier proteins for the polypeptide of interest (POI). For anchored periplasmicexpression (APEx), the POI can be either tethered to the inner membrane by a N-terminal NlpA signal peptide fusion or by a fusion to a transmembrane helix. In soluble periplasmic expression (PECS), any transport pathway can be used to secrete the POI into the periplasm. The latter two display technologies require the fracture of the outer membrane via lysozyme-EDTA treatment (APEx) or high salt concentrations (PECS)

folds very rapidly *in vitro*, cannot be secreted via the post-translational Sec pathway but is efficiently translocated into the periplasm when fused to a signal peptide that mediates co-translational export (Huber et al. 2005a). Huber et al. then selected for thioredoxin variants that can be exported post-translationally and showed that these proteins exhibit up to 30 fold slower rates in one of the critical steps of folding (Huber et al. 2005b). In effect, this is the opposite selection to that of Ribnicky who isolated faster folding proteins based on Tat export competence (Ribnicky et al. 2007). In other studies, the quality control feature of the Tat pathway was exploited to select for variant proteins displaying greater solubility (Fisher et al. 2006). In that system the protein of interest is expressed as a tripartite fusion with an N-terminal Tat signal peptide and β-lactamase fused to the C-terminus. Growth on ampicillin is employed to select for mutations in the protein of interest that allow export of the fusion and localization of the β-lactamase moiety in the periplasm. Since the export rates correlate with the solubility of a protein, fusion constructs that conferred ampicillin resistance carried variants with better folding abilities. Utilizing this approach, Fisher and coworkers were able to isolate higher solubility variants of the aggregation-prone amyloid precursor protein Aβ42, a primary constituent of the toxic plaques in Alzheimer disease (Fisher et al. 2006).

The Tat pathway is capable of co-transporting at least two folded protein subunits at a time, only one of which has a signal sequence, via a "hitchhiker export" mechanism (Rodrigue et al. 1999). This observation was recently exploited to develop a Tat based 2-hybrid system in which one protein (bait) is expressed as a fusion to a Tat signal peptide whereas the second protein (prey) is fused to a protein reporter that can confer a phenotype only after export into the bacterial periplasmic space. Since the prey-reporter fusion lacks a signal peptide, it can only be exported as a complex with the bait-signal peptide fusion which is capable of targeting the Tat translocon. Using maltose-binding protein as the reporter, clones expressing interacting proteins could be identified on maltose minimal media or on MacConkey plates. Alternatively, using cysteine disulfide oxidase DsbA as reporter, export of a signal peptide-prey:bait-DsbA complex into the periplasm allowed complementation of $dsbA^-$ mutants. The prey:bait-DsbA complex was able to restore the formation of active alkaline phosphatase, an enzyme that can be easily detected by a chromogenic assay (Strauch and Georgiou 2007a). The Tat two-hybrid system can be used as a new tool to identify protein-protein interaction on a genomic scale by including two libraries as bait and prey, or it can be utilized to identify protein interaction partners for a protein of interest. Additionally, a 2-hybrid assay may be utilized as a tool for the *in vivo* co-evolution of interacting protein pairs as has been demonstrated by DeLisa and coworkers (private communication).

16.3 Conclusions

Protein secretion in *E. coli* is of enormous significance in biotechnology for applications ranging from preparative protein production to combinatorial library screening. After more than 30 years of study many of the mechanistic details of Sec protein translocation have been elucidated. In contrast, the sequence of events that lead to the export of proteins via the Tat pathway is not completely understood. A significant difference between Sec and Tat pathways is that the former exports proteins that are unfolded whereas the latter accepts only proteins that have attained a native-like conformation. The Sec pathway has been used to express recombinant proteins at high levels that can exceed 5 g/L. Periplasmic expression allows the engagement of the post-translational modification apparatus enabling the introduction and refinement of disulfide bridges which can be additionally optimized by the overexpression of endogenous chaperones. Recent evidence suggests that high yields may also be attained with the Tat pathway provided that the protein is compatible for export via this route. However, g/L expression of Tat proteins needs to be demonstrated.

Protein secretion is an essential step for display and the screening of combinatorial libraries. The export pathway inflicts an additional filter step onto the general screening or selection scheme. For example the use of signal peptide that target different export pathways can lead to the isolation of distinct pools of protein variants that exhibit different folding characteristics (Table 16.1).

Despite the fact that little is known about the actual molecular process of targeting to and transport through the Twin-Arginine Translocase, it is a promising candidate for periplasmic expression of heterologous proteins, including those that otherwise would be incompatible with the Sec transport pathway. The use of the Tat pathway does not only allow increased periplasmic yields of actual active proteins, specifically those that fold fast, it further enables the refinement of existing screens or selections, and even allows the launching of novel protein engineering platforms

Table 16.1 Comparisons of current applications of different routes across the cytoplasmic membrane

	Post-translational Sec export	Co-translational	Tat export
Typical signal sequence	ssPelB, ssPhoA, ssOmpA,	ssDsbA, ssTorT	ssTorA,
Folding preference	slow folding	fast folding, aggregation-prone proteins	fast folding, containing cytoplasmically inserted cofactors,
Maturation	disulfide bridges, fatty acylation (lipoproteins), heme insertion (requires periplasmic reduction)	disulfide bridges	cofactors, other protein subunits, disulfide bridges[1]
Current applications	phage display, bacterial display technologies, protein expression, screen for slower folding variants, protein expression	phage display, cell display on the inner membrane	screen for solubility increase or faster folders, Tat two-hybrid, Tat-based phage display, protein expression

[1] Non-native protein substrates containing disulfide bonds must be expressed in strains having an oxidizing cytoplasm.

that capitalize on its proofreading mechanism and the possibility to fold the protein of interest in the cytoplasmic environment of the cell.

References

Adams TM, Wentzel A, Kolmar H (2005) Intimin-mediated export of passenger proteins requires maintenance of a translocation-competent conformation. J Bacteriol 187(2):522–33

Alami M, Luke I, Deitermann S et al. (2003) Differential interactions between a twin-arginine signal peptide and its translocase in *Escherichia coli*. Mol Cell 12(4):937–46

Alder NN, Theg SM (2003) Energetics of protein transport across biological membranes. a study of the thylakoid DeltapH-dependent/cpTat pathway. Cell 112(2):231–42

Arie JP, Sassoon N, Betton JM (2001) Chaperone function of FkpA, a heat shock prolyl isomerase, in the periplasm of *Escherichia coli*. Mol Microbiol 39(1):199–210

Bageshwar UK, Musser SM (2007) Two electrical potential dependent steps are required for transport by the *Escherichia coli* Tat machinery. J Cell Biol 179(1):87–99

Baneyx F, Mujacic M (2004) Recombinant protein folding and misfolding in *Escherichia coli*. Nat Biotechnol 22(11):1399–408

Bange G, Wild K, Sinning I (2007) Protein translocation: checkpoint role for SRP GTPase activation. Curr Biol 17(22):R980–2

Becker S, Michalczyk A, Wilhelm S et al. (2007) Ultrahigh-throughput screening to identify *E. coli* cells expressing functionally active enzymes on their surface. Chembiochem 8(8):943–9

Becker S, Schmoldt HU, Adams TM et al. (2004) Ultra-high-throughput screening based on cell-surface display and fluorescence-activated cell sorting for the identification of novel biocatalysts. Curr Opin Biotechnol 15(4):323–9

Berks BC (1996) A common export pathway for proteins binding complex redox cofactors? Mol Microbiol 22(3):393–404

Berks BC, Palmer T, Sargent F (2003) The Tat protein translocation pathway and its role in microbial physiology. Adv Microb Physiol 47:187–254
Bessette PH, Aslund F, Beckwith J et al. (1999) Efficient folding of proteins with multiple disulfide bonds in the *Escherichia coli* cytoplasm. Proc Natl Acad Sci USA 96(24):13703–8
Blaudeck N, Kreutzenbeck P, Freudl R et al. (2003) Genetic analysis of pathway specificity during posttranslational protein translocation across the *Escherichia coli* plasma membrane. J Bacteriol 185(9):2811–9
Blaudeck N, Kreutzenbeck P, Muller M et al. (2005) Isolation and characterization of bifunctional *Escherichia coli* TatA mutant proteins that allow efficient tat-dependent protein translocation in the absence of TatB. J Biol Chem 280(5):3426–32
Blazek D, Celer V, Navratilova I et al. (2004) Generation and characterization of single-chain antibody fragments specific against transmembrane envelope glycoprotein gp46 of maedi-visna virus. J Virol Methods 115(1):83–92
Bos MP, Robert V, Tommassen J (2007) Biogenesis of the gram-negative bacterial outer membrane. Annu Rev Microbiol 61:191–214
Bothmann H, Pluckthun A (2000) The periplasmic *Escherichia coli* peptidylprolyl cis,trans-isomerase FkpA. I. Increased functional expression of antibody fragments with and without cis-prolines. J Biol Chem 275(22):17100–5
Chan CS, Howell JM, Workentine ML et al. (2006) Twin-arginine translocase may have a role in the chaperone function of NarJ from *Escherichia coli*. Biochem Biophys Res Commun 343(1):244–51
Chen G, Hayhurst A, Thomas JG et al. (2001) Isolation of high-affinity ligand-binding proteins by periplasmic expression with cytometric screening (PECS). Nat Biotechnol 19(6):537–42
Choi JH, Lee SY (2004) Secretory and extracellular production of recombinant proteins using *Escherichia coli*. Appl Microbiol Biotechnol 64(5):625–35
Cline K, Ettinger WF, Theg SM (1992) Protein-specific energy requirements for protein transport across or into thylakoid membranes. Two lumenal proteins are transported in the absence of ATP. J Biol Chem 267(4):2688–96
Cline K, McCaffery M (2007) Evidence for a dynamic and transient pathway through the TAT protein transport machinery. Embo J 26(13):3039–49
Daugherty PS (2007) Protein engineering with bacterial display. Curr Opin Struct Biol 17(4): 474–80
de Marco A (2007) Protocol for preparing proteins with improved solubility by co-expressing with molecular chaperones in *Escherichia coli*. Nat Protoc 2(10):2632–9
DeLisa MP, Lee P, Palmer T et al. (2004) Phage shock protein PspA of *Escherichia coli* relieves saturation of protein export via the Tat pathway. J Bacteriol 186(2):366–73
DeLisa MP, Tullman D, Georgiou G (2003) Folding quality control in the export of proteins by the bacterial twin-arginine translocation pathway. Proc Natl Acad Sci USA 100(10):6115–20
Dilks K, Gimenez MI, Pohlschroder M (2005) Genetic and biochemical analysis of the twin-arginine translocation pathway in halophilic archaea. J Bacteriol 187(23):8104–13
Dubini A, Sargent F (2003) Assembly of Tat-dependent [NiFe] hydrogenases: identification of precursor-binding accessory proteins. FEBS Lett 549(1–3):141–6
Duong F, Wickner W (1997) The SecDFyajC domain of preprotein translocase controls preprotein movement by regulating SecA membrane cycling. Embo J 16(16):4871–9
Feilmeier BJ, Iseminger G, Schroeder D et al. (2000) Green fluorescent protein functions as a reporter for protein localization in *Escherichia coli*. J Bacteriol 182(14):4068–76
Fekkes P, van der Does C, Driessen AJ (1997) The molecular chaperone SecB is released from the carboxy-terminus of SecA during initiation of precursor protein translocation. Embo J 16(20):6105–13
Fernandez-Gacio A, Uguen M, Fastrez J (2003) Phage display as a tool for the directed evolution of enzymes. Trends Biotechnol 21(9):408–14
Fernandez LA, Berenguer J (2000) Secretion and assembly of regular surface structures in Gram-negative bacteria. FEMS Microbiol Rev 24(1):21–44

Fisher AC, Kim J-Y, Perez-Rodriguez R et al. (2008) Exploration of twin-arginine translocation for the expression and purification of correctly folded proteins in *Escherichia coli*. Microbial Biotechnol 1(5):403–415.

Fisher AC, Kim W, DeLisa MP (2006) Genetic selection for protein solubility enabled by the folding quality control feature of the twin-arginine translocation pathway. Protein Sci 15(3):449–58

Georgiou G, Segatori L (2005) Preparative expression of secreted proteins in bacteria: status report and future prospects. Curr Opin Biotechnol 16(5):538–45

Gerard F, Cline K (2006) Efficient twin arginine translocation (Tat) pathway transport of a precursor protein covalently anchored to its initial cpTatC binding site. J Biol Chem 281(10):6130–5

Glucksman MJ, Bhattacharjee S, Makowski L (1992) Three-dimensional structure of a cloning vector. X-ray diffraction studies of filamentous bacteriophage M13 at 7 A resolution. J Mol Biol 226(2):455–70

Gohlke U, Pullan L, McDevitt CA et al. (2005) The TatA component of the twin-arginine protein transport system forms channel complexes of variable diameter. Proc Natl Acad Sci USA 102(30):10482–6

Graubner W, Schierhorn A, Bruser T (2007) DnaK plays a pivotal role in Tat targeting of CueO and functions beside SlyD as a general Tat signal binding chaperone. J Biol Chem 282(10): 7116–24

Gray MW, Lang BF, Burger G (2004) Mitochondria of protists. Annu Rev Genet 38:477–524

Harvey BR, Georgiou G, Hayhurst A et al. (2004) Anchored periplasmic expression, a versatile technology for the isolation of high-affinity antibodies from *Escherichia coli*-expressed libraries. Proc Natl Acad Sci USA 101(25):9193–8

Hatzixanthis K, Clarke TA, Oubrie A et al. (2005) Signal peptide-chaperone interactions on the twin-arginine protein transport pathway. Proc Natl Acad Sci USA 102(24):8460–5

Hayhurst A, Happe S, Mabry R et al. (2003) Isolation and expression of recombinant antibody fragments to the biological warfare pathogen Brucella melitensis. J Immunol Methods 276(1–2):185–96

Huber D, Boyd D, Xia Y et al. (2005a) Use of thioredoxin as a reporter to identify a subset of *Escherichia coli* signal sequences that promote signal recognition particle-dependent translocation. J Bacteriol 187(9):2983–91

Huber D, Cha MI, Debarbieux L et al. (2005b) A selection for mutants that interfere with folding of *Escherichia coli* thioredoxin-1 in vivo. Proc Natl Acad Sci USA 102(52):18872–7

Hwang BY, Varadarajan N, Li H et al. (2007) Substrate specificity of the *Escherichia coli* outer membrane protease OmpP. J Bacteriol 189(2):522–30

Iwanczyk J, Damjanovic D, Kooistra J et al. (2007) Role of the PDZ domains in *Escherichia coli* DegP protein. J Bacteriol 189(8):3176–86

Jack RL, Buchanan G, Dubini A et al. (2004) Coordinating assembly and export of complex bacterial proteins. Embo J 23(20):3962–72

Jeong KJ, Seo MJ, Iverson BL et al. (2007) APEx 2-hybrid, a quantitative protein-protein interaction assay for antibody discovery and engineering. Proc Natl Acad Sci USA 104(20): 8247–52

Joly JC, Leung WS, Swartz JR (1998) Overexpression of *Escherichia coli* oxidoreductases increases recombinant insulin-like growth factor-I accumulation. Proc Natl Acad Sci USA 95(6):2773–7

Jose J, Betscheider D, Zangen D (2005) Bacterial surface display library screening by target enzyme labeling: Identification of new human cathepsin G inhibitors. Anal Biochem 346(2):258–67

Ki JJ, Kawarasaki Y, Gam J et al. (2004) A periplasmic fluorescent reporter protein and its application in high-throughput membrane protein topology analysis. J Mol Biol 341(4):901–9

Kiino DR, Silhavy TJ (1984) Mutation prlF1 relieves the lethality associated with export of beta-galactosidase hybrid proteins in *Escherichia coli*. J Bacteriol 158(3):878–83

Kim J, Rusch S, Luirink J et al. (2001) Is Ffh required for export of secretory proteins? FEBS Lett 505(2):245–8

Kim S, Malinverni JC, Sliz P et al. (2007) Structure and function of an essential component of the outer membrane protein assembly machine. Science 317(5840):961–4

Kotz JD, Bond CJ, Cochran AG (2004) Phage-display as a tool for quantifying protein stability determinants. Eur J Biochem 271(9):1623–9

Kreutzenbeck P, Kroger C, Lausberg F et al. (2007) *Escherichia coli* twin arginine (Tat) mutant translocases possessing relaxed signal peptide recognition specificities. J Biol Chem 282(11):7903–11

Kumamoto CA (1991) Molecular chaperones and protein translocation across the *Escherichia coli* inner membrane. Mol Microbiol 5(1):19–22

Kurokawa Y, Yanagi H, Yura T (2000) Overexpression of protein disulfide isomerase DsbC stabilizes multiple-disulfide-bonded recombinant protein produced and transported to the periplasm in *Escherichia coli*. Appl Environ Microbiol 66(9):3960–5

Kurokawa Y, Yanagi H, Yura T (2001) Overproduction of bacterial protein disulfide isomerase (DsbC) and its modulator (DsbD) markedly enhances periplasmic production of human nerve growth factor in *Escherichia coli*. J Biol Chem 276(17):14393–9

Lawley TD, Klimke WA, Gubbins MJ et al. (2003) F factor conjugation is a true type IV secretion system. FEMS Microbiol Lett 224(1):1–15

Lee SY, Choi JH, Xu Z (2003) Microbial cell-surface display. Trends Biotechnol 21(1):45–52

Li HX, Hwang BY, Laxmikanthan G et al. (2008) Substrate specificity of human kallikreins 1 and 6 determined by phage display. Protein Sci 17(4):664–72

Li SY, Chang BY, Lin SC (2006) Coexpression of TorD enhances the transport of GFP via the TAT pathway. J Biotechnol 122(4):412–21

Lipovsek D, Pluckthun A (2004) In-vitro protein evolution by ribosome display and mRNA display. J Immunol Methods 290(1–2):51–67

Lu Z, Murray KS, Van Cleave V et al. (1995) Expression of thioredoxin random peptide libraries on the *Escherichia coli* cell surface as functional fusions to flagellin: a system designed for exploring protein-protein interactions. Biotechnology (NY) 13(4):366–72

Luirink J, High S, Wood H et al. (1992) Signal-sequence recognition by an *Escherichia coli* ribonucleoprotein complex. Nature 359(6397):741–3

Maillard J, Spronk CA, Buchanan G et al. (2007) Structural diversity in twin-arginine signal peptide-binding proteins. Proc Natl Acad Sci USA 104(40):15641–6

Malik P, Terry TD, Gowda LR et al. (1996) Role of capsid structure and membrane protein processing in determining the size and copy number of peptides displayed on the major coat protein of filamentous bacteriophage. J Mol Biol 260(1):9–21

Masip L, Klein-Marcuschamer D, Quan S et al. (2008) Laboratory evolution of *Escherichia coli* thioredoxin for enhanced catalysis of protein oxidation in the periplasm reveals a phylogenetically conserved substrate specificity determinant. J Biol Chem 283(2):840–8

Matthews DJ, Wells JA (1993) Substrate phage: selection of protease substrates by monovalent phage display. Science 260(5111):1113–7

Mazor Y, Van Blarcom T, Mabry R et al. (2007) Isolation of engineered, full-length antibodies from libraries expressed in *Escherichia coli*. Nat Biotechnol 25(5):563–5

Meerman HJ, Georgiou G (1994) Construction and characterization of a set of *E. coli* strains deficient in all known loci affecting the proteolytic stability of secreted recombinant proteins. Biotechnology (NY) 12(11):1107–10

Mejean V, Iobbi-Nivol C, Lepelletier M et al. (1994) TMAO anaerobic respiration in *Escherichia coli*: involvement of the tor operon. Mol Microbiol 11(6):1169–79

Mergulhao FJ, Summers DK, Monteiro GA (2005) Recombinant protein secretion in *Escherichia coli*. Biotechnol Adv 23(3):177–202

Missiakas D, Betton JM, Raina S (1996) New components of protein folding in extracytoplasmic compartments of *Escherichia coli* SurA, FkpA and Skp/OmpH. Mol Microbiol 21(4):871–84

Mujacic M, Bader MW, Baneyx F (2004) *Escherichia coli* Hsp31 functions as a holding chaperone that cooperates with the DnaK-DnaJ-GrpE system in the management of protein misfolding under severe stress conditions. Mol Microbiol 51(3):849–59

Oresnik IJ, Ladner CL, Turner RJ (2001) Identification of a twin-arginine leader-binding protein. Mol Microbiol 40(2):323–31

Orriss GL, Tarry MJ, Ize B et al. (2007) TatBC, TatB, and TatC form structurally autonomous units within the twin arginine protein transport system of *Escherichia coli*. FEBS Lett 581(21): 4091–7

Paetzel M, Karla A, Strynadka NC et al. (2002) Signal peptidases. Chem Rev 102(12):4549–80

Pallen MJ, Chaudhuri RR, Henderson IR (2003) Genomic analysis of secretion systems. Curr Opin Microbiol 6(5):519–27

Paschke M, Hohne W (2005) A twin-arginine translocation (Tat)-mediated phage display system. Gene 350(1):79–88

Perez-Rodriguez R, Fisher AC, Perlmutter JD et al. (2007) An essential role for the DnaK molecular chaperone in stabilizing over-expressed substrate proteins of the bacterial twin-arginine translocation pathway. J Mol Biol 367(3):715–30

Pommier J, Mejean V, Giordano G et al. (1998) TorD, a cytoplasmic chaperone that interacts with the unfolded trimethylamine N-oxide reductase enzyme (TorA) in *Escherichia coli*. J Biol Chem 273(26):16615–20

Qiu J, Swartz JR, Georgiou G (1998) Expression of active human tissue-type plasminogen activator in *Escherichia coli*. Appl Environ Microbiol 64(12):4891–6

Rapoza MP, Webster RE (1993) The filamentous bacteriophage assembly proteins require the bacterial SecA protein for correct localization to the membrane. J Bacteriol 175(6):1856–9

Ribnicky B, Van Blarcom T, Georgiou G (2007) A scFv antibody mutant isolated in a genetic screen for improved export via the twin arginine transporter pathway exibits faster folding. J Mol Biol 369(3):631–9

Richter S, Lindenstrauss U, Lucke C et al. (2007) Functional Tat transport of unstructured, small, hydrophilic proteins. J Biol Chem 282(46):33257–64

Rietsch A, Belin D, Martin N et al. (1996) An in vivo pathway for disulfide bond isomerization in *Escherichia coli*. Proc Natl Acad Sci USA 93(23):13048–53

Rodrigue A, Chanal A, Beck K et al. (1999) Co-translocation of a periplasmic enzyme complex by a hitchhiker mechanism through the bacterial tat pathway. J Biol Chem 274(19):13223–8

Rose RW, Bruser T, Kissinger JC et al. (2002) Adaptation of protein secretion to extremely high-salt conditions by extensive use of the twin-arginine translocation pathway. Mol Microbiol 45(4):943–50

Samuelson P, Gunneriusson E, Nygren PA et al. (2002) Display of proteins on bacteria. J Biotechnol 96(2):129–54

Sandkvist M (2001) Type II secretion and pathogenesis. Infect Immun 69(6):3523–35

Sargent F, Bogsch EG, Stanley NR et al. (1998) Overlapping functions of components of a bacterial Sec-independent protein export pathway. Embo J 17(13):3640–50

Schatz G, Dobberstein B (1996) Common principles of protein translocation across membranes. Science 271(5255):1519–26

Schiebel E, Driessen AJ, Hartl FU et al. (1991) Delta mu H+ and ATP function at different steps of the catalytic cycle of preprotein translocase. Cell 64(5):927–39

Schierle CF, Berkmen M, Huber D et al. (2003) The DsbA signal sequence directs efficient, cotranslational export of passenger proteins to the *Escherichia coli* periplasm via the signal recognition particle pathway. J Bacteriol 185(19):5706–13

Schlapschy M, Grimm S, Skerra A (2006) A system for concomitant overexpression of four periplasmic folding catalysts to improve secretory protein production in *Escherichia coli*. Protein Eng Des Sel 19(8):385–90

Smith GP (1985) Filamentous fusion phage: novel expression vectors that display cloned antigens on the virion surface. Science 228(4705):1315–7

Spiess C, Beil A, Ehrmann M (1999) A temperature-dependent switch from chaperone to protease in a widely conserved heat shock protein. Cell 97(3):339–47

Stathopoulos C, Georgiou G, Earhart CF (1996) Characterization of *Escherichia coli* expressing an Lpp'OmpA(46–159)-PhoA fusion protein localized in the outer membrane. Appl Microbiol Biotechnol 45(1–2):112–9

Steiner D, Forrer P, Stumpp MT et al. (2006) Signal sequences directing cotranslational translocation expand the range of proteins amenable to phage display. Nat Biotechnol 24(7):823–31

Strauch EM, (2007) Characterization and Applications of the Twin-Arginine Transporter pathway. Department of Biochemistry, pp. 188. University of Texas at Austin

Strauch EM,Georgiou G (2007a) A bacterial two-hybrid system based on the twin-arginine transporter pathway of *E. coli*. Protein Sci 16(5):1001–8

Strauch EM, Georgiou G (2007b) *Escherichia coli* tatC mutations that suppress defective twin-arginine transporter signal peptides. J Mol Biol 374(2):283–91

Tokumoto U, Nomura S, Minami Y et al. (2002) Network of protein-protein interactions among iron-sulfur cluster assembly proteins in *Escherichia coli*. J Biochem (Tokyo) 131(5): 713–9

Tullman-Ercek D, DeLisa MP, Kawarasaki Y et al. (2007) Export pathway selectivity of *Escherichia coli* twin arginine translocation signal peptides. J Biol Chem 282(11):8309–16

van der Wolk JP, de Wit JG, Driessen AJ (1997) The catalytic cycle of the *Escherichia coli* SecA ATPase comprises two distinct preprotein translocation events. Embo J 16(24):7297–304

Veiga E, Sugawara E, Nikaido H et al. (2002) Export of autotransported proteins proceeds through an oligomeric ring shaped by C-terminal domains. Embo J 21(9):2122–31

Wentzel A, Christmann A, Kratzner R et al. (1999) Sequence requirements of the GPNG beta-turn of the *Ecballium elaterium* trypsin inhibitor II explored by combinatorial library screening. J Biol Chem 274(30):21037–43

Westerlund-Wikstrom B (2000) Peptide display on bacterial flagella: principles and applications. Int J Med Microbiol 290(3):223–30

Wilken C, Kitzing K, Kurzbauer R et al. (2004) Crystal structure of the DegS stress sensor: How a PDZ domain recognizes misfolded protein and activates a protease. Cell 117(4):483–94

Zwick MB, Shen J,Scott JK (2000) Homodimeric peptides displayed by the major coat protein of filamentous phage. J Mol Biol 300(2):307–20

Chapter 17
Engineering *E. coli* Central Metabolism for Enhanced Primary Metabolite Production

George N. Bennett and Ka-Yiu San

Contents

Abstract In engineering of *Escherichia coli* for the production of chemicals derived from the central metabolic pathway and in using *E. coli* as a biocatalyst for reactions involving externally supplied specific substrates, there is a need to consider the redox balance and cofactor availability for optimization of the process. Several examples of taking into account the systems biology complexity of redox processes through consideration of gene expression effects, protein level and activity effects, and the role of small molecule effectors of enzyme activity, as well as the role of activation and deactivation of sensitive active site structures are described in the chapter. The manipulation of the availability of reduced cofactors through genetic means and the application of such altered strains for metabolic engineering purposes for the improved production of specific reduced molecules for biofuels, chiral pharmaceutical intermediates, unconjugated colored compounds, and other valuable chemicals is presented.

G.N. Bennett (✉)
Department of Biochemistry and Cell Biology, Rice University, Houston, TX 77005-1892, USA
e-mail: gbennett@rice.edu

S.Y. Lee (ed.), *Systems Biology and Biotechnology of Escherichia coli*,
DOI 10.1007/978-1-4020-9394-4_17, © Springer Science+Business Media B.V. 2009

17.1 The Central Pathway of *E. coli* Metabolism, a Systems View of the Network and Cofactor Considerations

17.1.1 Aerobic Considerations

Under aerobic conditions the level of reduced cofactors formed in glycolysis and in the TCA cycle can be largely converted to energy for the cell via the electron transport chain and the associated oxidative phosphorylation events. The removal of excess reductant under partial aerobic conditions by an NADH oxidase, particularily an enzyme that forms water, has been demonstrated and can aid flow to more oxidized products in lactic acid bacteria (Lopez de Felipe et al. 1998, Neves et al. 2002). The expression of an NADH oxidase from *Streptococcus pneumoniae* was studied. Results showed that expression of NADH oxidase altered the NADH/NAD+ ratio. In an *arcA* host acetate formation was reduced and the biomass yield increased (Vemuri et al. 2006) suggesting that if the NADH level can be kept low, then the TCA cycle can function efficiently even at a high glucose concentration to process the carbon feedstock without build up of intermediates that generate acetate.

In aerobic processes however, if a redox process for the formation of the desired product is required, the cofactor can be recycled and reduced through the metabolism of a suitable precursor. One also has to consider that the possible utilization of the reduced cofactor through the electron transfer system can compete and limit the availability of the reductant for the desired reaction. In this biocatalyst mode the cells are usually held in a non-growing state, and the aerobically generated reductant can be used more fully in a desired microbial conversion reaction.

The contribution of microaerobic conditions to aid cell energetics and growth properties while allowing more efficient use of carbon for products has also been observed. Early enzyme analysis pointed to factors in the transition (Doelle and Hollywood 1978, Thomas et al. 1972) that were important. It appears the ability to respire oxygen under microaerobic conditions aids *E. coli* in intestinal growth and colonization (Jones et al. 2007). In metabolic engineering practice, a similar strategy is used in the formation of partially oxidized products or where the redox balance would not be appropriate for complete anaerobic metabolism. A number of studies have focused on the contribution that the presence of various oxygen binding proteins such as *Vitreoscilla* hemoglobin can make to enhanced respiration under microaerobic conditions and the effects on cell physiology, productivity, and metabolic pattern (Andersson et al. 2000, Frey et al. 2000, Kallio et al. 1996). Studies of the relative expression of genes (Liu and De Wulf 2004, Overton et al. 2006, Salmon et al. 2003) and metabolite patterns under conditions of limited oxygen have been made with wild type and various metabolic and regulatory mutant strains under defined oxygen conditions (Alexeeva et al. 2000, 2002, 2003, Becker et al. 1997, Partridge et al. 2007, Shalel-Levanon et al. 2005a,b,c, Zhu et al. 2006, 2007a). Such measurements have allowed models of the shift between aerobic and anaerobic conditions to be formulated and their general features to be evaluated (Govantes et al. 2000, Peercy et al. 2006, Schramm et al. 2007). The metabolite pattern of products derived from pyruvate arising in various mutant strains under conditions of low oxygen is complicated by many factors influencing the *in vivo* activities of the various enzymes around this node. For example the activities of Pdh and Pfl are affected by gene expression levels, the NADH level and the relative amounts of activation, deactivation of Pfl as well as the YfiD interaction with Pfl. The levels of other enzymes acting around the pyruvate node and the TCA cycle and cytochrome oxidase enzymes also influence the level of small molecules that can affect *in vivo* activity and metabolic flux through the competing routes. Some discussion of these influences is given in (Peercy et al. 2006, Shalel-Levanon et al. 2005b, Zhu et al. 2007a) (Fig. 17.1).

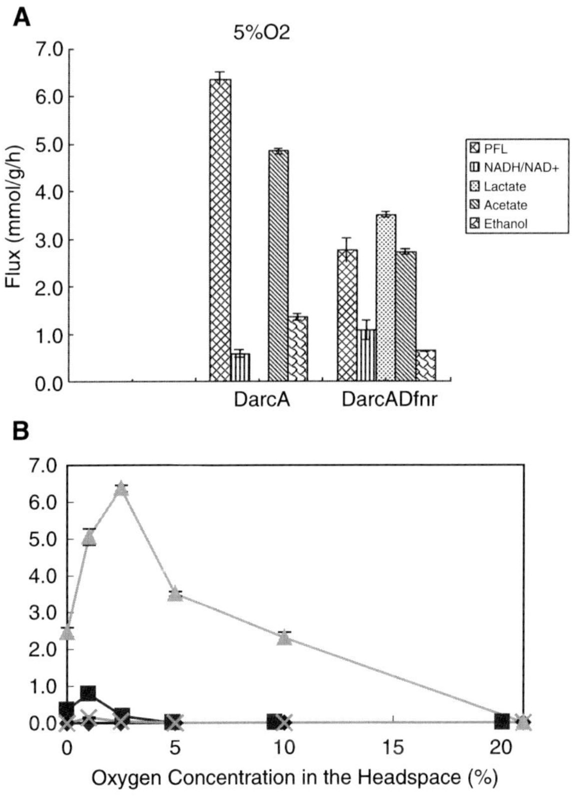

Fig. 17.1 (**a**) Comparison of metabolites and fluxes of cultures of MG1655 DarcA (*arcA* disruption) and MG1655 DarcA, Dfnr strains grown in chemostat under 5% oxygen in the headspace. The difference in lactate flux is most apparent. Other fluxes are shown as indicated. The NADH/NAD+ ratio is also shown. In the parent, MG1655 the other metabolites were not observed see Fig. 17.1b. (**b**) Metabolite fluxes as a function of the oxygen concentration in the headspace at steady state. PFL, lactate, ethanol, and succinate. The fluxes in the individual strains are indicated: (purple, dark gray diamond) MG1655, (red,dark gray squares) MG1655 [DarcA], (green, light gray x) MG1655 [Dfnr], (blue, light gray triangles) MG1655 [DarcA, Dfnr]. The error bars indicate the standard deviation of three samples taken after 7, 7.5, and 8 residence times

17.1.2 Anaerobic Considerations

Under anaerobic growth the reductant formed in glycolysis must be recycled by reactions using available substrates. This process generates the reduced metabolites derived from pyruvate in many bacterial species and the reduced products of the mixed acid fermentation in *E. coli*. By limiting the alternative pathways for cofactor recycling, the metabolic course of the flux into the downstream parts of the central pathway is affected. The dissipation of the reducing equivalents can also be handled through the formation of hydrogen either directly or through the release of a compound such as formate which can easily be converted to hydrogen and carbon dioxide. Bacteria have elaborate sensing mechanisms for oxygen and regulate the specific cytochrome oxidases as well as many other genes through transcriptional regulators such as ArcA and Fnr. The area of aerobic/anaerobic gene regulation mechanisms will not be covered here as it is reviewed elsewhere in this volume and in other reports (Gunsalus and Park 1994, Sawers 1999). The various electron carriers; flavins, nucleotide cofactors, quinones and ferredoxins, act with specific enzymes and while there is interconversion among the reduced compounds the redox potential and relative quantity of each within the cell suggests a distinct role for the individual carriers in the cell. The efficiency of rapid equilibrium among the pools of reduced electron carriers is dependent on a number of factors including the relative location in the cell, association of key molecules with other cell components, and specific binding constants and kinetic parameters of competing reactions. These factors can be adjusted by engineering but the physiological response of the cell is often complicated.

17.2 Strategies for Engineering Metabolic Outputs from Specific Branches

17.2.1 Multiple Deletions in Alternative Pathways

17.2.1.1 Pyruvate and Acetate

Pyruvate is formed under aerobic conditions when it is desired to produce it in high quantity (Causey et al. 2004, Sakai et al. 2007, Tomar et al. 2003, Zelic et al. 2006, 2004a,b). Some similar features have been implemented in the high production of acetate by *E. coli* (Causey et al. 2003). The general strategy for high production of these compounds involves high glycolytic fluxes and the removal of competing pathways, either for the carbon or for the reductant in order to minimize the potential formation of further metabolism of the compound (e.g. pyruvate conversion to lactate). In the case of acetate production, the elimination of reactions involving a key precursor (e.g. pyruvate conversion to other products) can affect the yield and culture performance. Since these compounds are dealt with elsewhere in this volume the specifics of metabolic engineering of *E. coli* for production of these products will not be discussed here.

One area of interest related to industrial production is the reduction of acetate formation that can inhibit growth and limit productivity in a variety of processes including recombinant protein production. Several strategies have been investigated to avoid acetate formation. In cultures, limited glucose addition can avoid some of the problems but requires careful control of the culture. These engineering strategies have become widely practiced as computer controls and sensors have become more sophisticated but are still a concern for optimization and reproducibility in large scale processes. The reduction of glucose uptake and the avoidance of build-up of the glycolytic intermediate, pyruvate, can be accomplished via genetic changes affecting the glucose transport system (Backlund et al. 2008, Chen et al. 1997, De Anda et al. 2006, Hernandez-Montalvo et al. 2003, Lara et al. 2008, Picon et al. 2005, Wong et al. 2008, Yi et al. 2003) or the presence of modified sugars (Aristidou et al. 1999, Chou et al. 1994). A large number of studies on the effects of *ptsG* mutations on production of acetate and other compounds, recombinant proteins, and growth have shown the importance of coordination of glucose uptake with downstream metabolism to avoid excessive acetate production and performance limitations.

The inactivation of genes that encode the major acetate formation pathway enzymes (acetate kinase, *ack* and phosphotransacetylase, *pta*; and pyruvate oxidase, *poxB*) can relieve acetate formation (De Mey et al. 2007) although such mutations may reduce the growth rate under some conditions or in certain genetic backgrounds (Abdel-Hamid et al. 2001, Flores et al. 2004, Vemuri et al. 2005). The effects of fluctuations in oxygen on the formation of acetate and recombinant proteins has been examined with the observation that the genes of fermentative metabolism can be removed with accompanying improved performance (Lara et al. 2006). The differences in *E. coli* strains have been studied and the flux through the glyoxyate pathway, acetate uptake and synthesis, and gluconeogenesis were different among some widely used laboratory strains and accounted for the differences in acetate formation in cultures of *E. coli* B and JM109 (Phue et al. 2005) and the extent of flux through anaplerotic pathways influences acetate excretion (Farmer and Liao 1997). Acetate formation can also be addressed through diversion of the precursor, pyruvate, to a non-toxic compound such as acetoin by incorporation of the gene encoding an acetolactate synthase from another organism (Yang et al. 1999).

17.2.1.2 Lactate

While lactate is readily formed by lactic acid bacteria and other microbes, it is formed naturally in differing amounts by various *E. coli* strains. Lactate formation in *E. coli* has been engineered, with either stereoisomer being formed depending on the particular characteristics of the lactate dehydrogenase employed (Chang et al. 1999, Dien et al. 2001, Fong et al. 2005, Hua et al. 2006, Zhou et al. 2003a,b, Zhou et al. 2005, Zhu et al. 2007b). In this case the fermentation is anaerobic and the other pathways that could use the reduced cofactor formed in glycolysis are removed (e.g. pyruvate conversion to acetyl-CoA and subsequently on to ethanol). Efficient natural production of this compound by other organisms is available and several engineered *E. coli* strains also perform well.

17.2.1.3 Succinate

In the case of succinate production, the conversion of glycolytic intermediates to oxaloacetate is a key step and in order to obtain high conversion enzymes capable of forming OAA or malate (Hong and Lee 2001, Kim et al. 2004, Lin et al. 2004, 2005e, Sanchez et al. 2005b, Stols et al. 1997, Vemuri et al. 2002) are usually overexpressed either through recombinant techniques or by enhancement of the natural system. There is a two fold problem in attaining the highest possible molar yield from glucose; one is the limitation of reductant (Hong and Lee 2002), if the 2 molecules of NADH formed in glycolysis are used to reduce the oxaloacetate, only one molecule of succinate can be formed. There is an alternative way to form succinate that does not consume NADH, i.e. through the glyoxylate route of the TCA cycle. This route also is limited to production of one molecule of succinate from one glucose due to loss of carbon in this normally aerobic pathway (Lin et al. 2005a,c,d). The correct partitioning of oxaloacetate between the reductive and oxidative routes can increase the overall yield while maintaining the NADH balance (Cox et al. 2006, Sanchez et al. 2006).

A variety of mutations to route the products of glycolysis to succinate have been investigated with the effects of redox systems (Yun et al. 2005) and the sugar uptake system (Chatterjee et al. 2001, Wang et al. 2006) showing a significant effect in some backgrounds due to the effects on pyruvate formation. Performance on various hexose and pentose sugars have been studied with glucose generally offering the highest yield compared to fructose or xylose (Andersson et al. 2007, Lin et al. 2005b). Computational methods have also been employed to identify high yielding strains (Lee et al. 2005) or model the immediate metabolic network (Cox et al. 2006). Strains made with an idea of optimal succinate production have included those with a number of defined mutations (Sanchez et al. 2006) and evolved strains derived from a defined parent (Jantama et al. 2008). In the studies various experimental conditions have been examined with the key factors of yield from feedstock, rate of production, productivity per cell mass, and final titer being components of the calculation of the potential of the process.

17.2.2 Alteration of Cofactor Availability (NADH)

Efforts to modify the NADH availability for cell metabolism have been undertaken for many years and have been based on observations of differing metabolic products formed using similar sugars with different oxidation levels such as glucuronic acid, glucose, and sorbitol. In these cultures the pattern of products formed, acetate, ethanol, formate, lactate, and succinate changes with the more oxidized products dominating in the culture from the oxidized substrates and the more reduced products being enhanced upon culture growth on sorbitol, a more reduced substrate. More oxidized products can be formed by depleting the NADH by an NADH oxidase as mentioned above. Here we will consider the changes in metabolites when an effort is made to augment the normal amount of NADH produced by the wild type *E. coli* strain.

Manipulation of the conversion reaction of pyruvate to acetyl-CoA and the subsequent release of formate, if formed, can alter the amount of reductant available to the cell. The production of NADH will favor the formation of more reduced products. There are three general enzymes that can catalyze this reaction each giving its own reduced product; pyruvate dehydrogenase that forms NADH and acetyl-CoA (Cassey et al. 1998, Guest et al. 1981, 1989, Guest and Stephens 1980, Haydon et al. 1993), pyruvate ferredoxin oxidoreductase that forms reduced ferredoxin or flavodoxin and acetyl-CoA (Blaschkowski et al. 1982, Reed et al. 2003, Serres et al. 2001), and pyruvate formate lyase that forms formate and acetyl-CoA (Birkmann et al. 1987, Knappe and Blaschkowski 1975, Knappe et al. 1984, Knappe and Sawers 1990, Pecher et al. 1982, Sauter and Sawers 1990, Sawers and Bock 1988, Varenne et al. 1975) with a number of articles describing the free-radiacal enzyme and its activation under anaerobic conditions and inactivation under aerobic conditions and the participation of proteins such as AdhE, YfiD and PflA in defining the activity of the protein (Becker et al. 1997, 1999, Chase and Rabinowitz 1968, Hoover and Ludwig 1997, Knappe and Wagner 1995, Kulzer et al. 1998, Nnyepi et al. 2007, Reddy et al. 1998, Sawers et al. 1998, Sawers and Watson 1998, Wagner et al. 2001, Zhang et al. 2001, Zhu et al. 2007a). If the reaction gives rise to NADH directly the reduced nucleotide cofactor can be used for production of a desired reduced product. The pyruvate dehydrogenase is generally the active enzyme under aerobic conditions and it is replaced by the pyruvate formate lyase under limiting oxygen conditions. The pyruvate dehydrogenase can still operate under anaerobic conditions, however high NADH is often inhibitory to the reaction (Snoep et al. 1993, Zhu et al. 2007a). The role of PdhR in regulating the Pdh system and effects of mutations of *pdhR* on expression and metabolism have been studied (Haydon et al. 1993, Kim et al. 2007, Ogasawara et al. 2007, Quail and Guest 1995, Zhou et al. 2008). As an added feature, the production of formate, while the final step under neutral pH conditions by Pfl, formate is further hydrolyzed to hydrogen and carbon dioxide under acidic conditions by the formate hydrogen lyase system (Bagramyan and Trchounian 2003, Birkmann et al. 1987). This reaction thereby removes the acidic metabolite formate but does not generate any useful reductant or energy for the cell but could reduce some acid stress due to formate accumulation. The effect of formate hydrogen lyase and other hydrogenases has been studied with regard to hydrogen production (Maeda et al. 2007a, Redwood et al. 2008, Yoshida et al. 2005, 2007). In some cases an uptake hydrogenase can recapture a portion of the hydrogen released and it can thereby affect the pattern of metabolites (Francis et al. 1990, Laurinavichene and Tsygankov 2001, Maeda et al. 2007b, Redwood et al. 2008).

The reducing equivalents available in formate can be recaptured to NADH rather than be released to hydrogen by incorporation of a NADH-dependent formate dehydrogenase (Berrios-Rivera et al. 2002a,b, Galkin et al. 1997, Sanchez et al. 2005a, Slusarczyk et al. 2000). Such NADH coupled enzymes are known in a number of organisms and those of Candida have been used *in vitro* and *in vivo* for regeneration of the NADH pool. Optimal enzymes from *Candida boidinii* and *Mycobacterium vaccae* that are more stable have been generated by mutation (Slusarczyk et al. 2000, Tishkov and Popov 2006, Yamamoto et al. 2005) and NADH-dependent formate

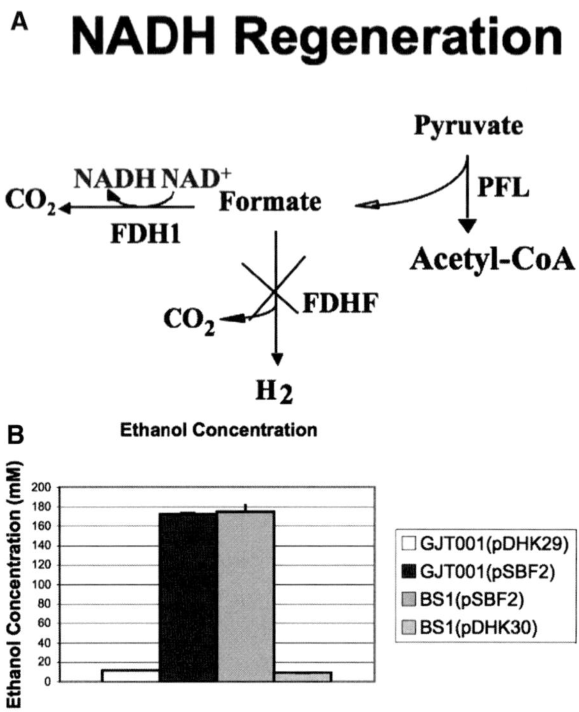

Fig. 17.2 (**a**) NADH coupled formate dehydrogenase pathway. The native NAD independent formate hydrogen lyase pathway uses (FDHF: formate dehydrogenase, NAD independent) to convert formate to hydrogen and carbon dioxide. The newly added NAD+ dependent pathway (in blue, light gray) uses (FDH1: NAD+ dependent formate dehydrogenase, FDH1 encoded by *fdh1* from *Candida boidinii*) to convert formate to carbon dioxide and the reduced cofactor NADH. (**b**) Effects on ethanol formation of expression of a NADH-dependent formate dehydrogenase in *E. coli*. The *E. coli* strain GJT001 is a W3110 derivative parental strain and BS1 has an inactivated *fdhF* gene. The plasmid pDHK29 is the vector and pSBF2 contains the *fdh1* gene from *Candida boidinii*. Growth was in L-broth plus 20 g/L glucose

dehydrogenases from other organisms have been isolated (Nanba et al. 2003a,b). Such enzymes are used to recycle NADH for use in formation of valuable compounds such as the pharmaceutical precursor, ethyl (S)-4-chloro-3-hydroxybutanoate (Yamamoto et al. 2005). The formation of chiral pharmaceutical intermediates using NADH regeneration has been reviewed (Patel 2000) (Fig. 17.2).

17.2.2.1 Ethanol

The capture of all available reducing power from glycolysis and present in pyruvate is needed for optimal formation of 2 molecules of ethanol from glucose. In *E. coli* such high formation of ethanol has been achieved through the addition of the *pdc* and *adh* genes from *Zymomonas mobilis* (Ingram et al. 1987, 1999, Jarboe et al. 2007). The recapture of the reductant in formate via a NADH-dependent formate dehydrogenase can also give essentially complete conversion of glucose to ethanol (Berrios-Rivera et al. 2002b, 2004) and chemostat cultures have shown the effect on metabolites using different carbon sources (Sanchez et al. 2005a).

17.2.2.2 *E. coli* Cells as Single Step Biocatalysts

The use of regenerated NADH to carry out a reduction by a whole cell biocatalyst has some advantage over using a purified enzyme in that the cell takes care of the recycling step and the cofactor is confined within the cell. Several papers have used such recycling systems in roles as cellular biocatalysts for amino acid (Galkin et al. 1997) and mannitol production (Kaup et al. 2003, 2004, 2005).

17.2.3 Alteration of Cofactor Availability (NADPH)

The pentose phosphate pathway, *zwf* and isocitrate dehydrogenase, *icd* are generally considered to be the major sources of reductant NADPH which is used in many biosynthetic reactions. The preference for NADPH can limit the production of the desired product since the NADPH pool is considerably smaller than the pool of NADH. Efforts to enhance the equilibration between the two reduced nucleotide cofactors has been investigated. There are two transhydrogenases in *E. coli*, *udhA* (*sthA*) and *pntAB*. The proton-translocating transhydrogenase PntAB was identified as the major source of NADPH under aerobic growth with the pentose phosphate pathway contributing almost as much and isocitriate dehydrogenase making up most of the remainder. While the energy-independent transhydrogenase UdhA (SthA), seemed to be essential under metabolic conditions with excess NADPH formation suggesting it played more of a role in dissipating NADPH to NADH (Sauer et al. 2004). Alterations of the transhydrogenase do indeed increase the level of NADPH-dependent products that are formed (Weckbecker and Hummel 2004). Another strategy to produce more NADPH for a conversion is to use a biocatalyst with a special system and substrate for producing NADPH based on the oxidation of the specific exogenous added substrate by a NADPH-dependent redox enzyme and the use of the NADPH for synthesis of the desired reduced product (e.g. a chiral alcohol). Another approach is to guide more metabolism through the pentose phosphate pathway where NADPH is formed in an early step. Several papers have analyzed the effects of mutations affecting glycolytic enzymes or overexpression of glucose-6-phosphate-1-dehydrogenase, *zwf*, in the context of NADPH usage. A discussion in consideration of the effects on PHB production is given below.

A more recent strategy is to incorporate a NADPH-utilizing step to replace a natural NADH-dependent step in glycolysis. This approach of using an NADPH-utilizing enzyme from another organism can provide additional NADPH for use by an added pathway that consumes high amounts of the cofactor (a NADPH sink). Several pathways utilize NADPH in high amount such as those for the biodegradable polymer, polyhydroxybutyrate and many unsaturated colored compounds and terpeniod compounds derived from the isoprenyl pyrophosphate pathway. Naturally

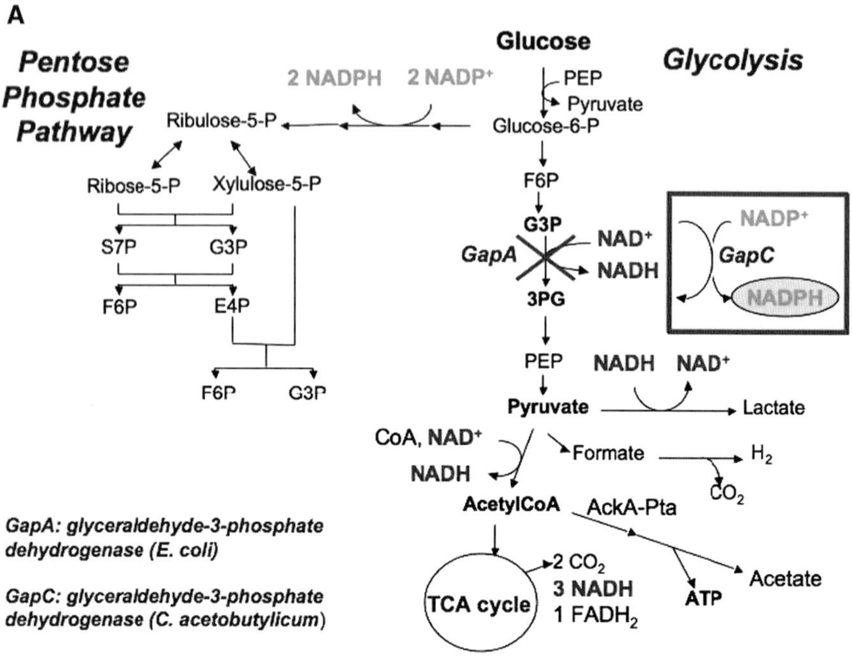

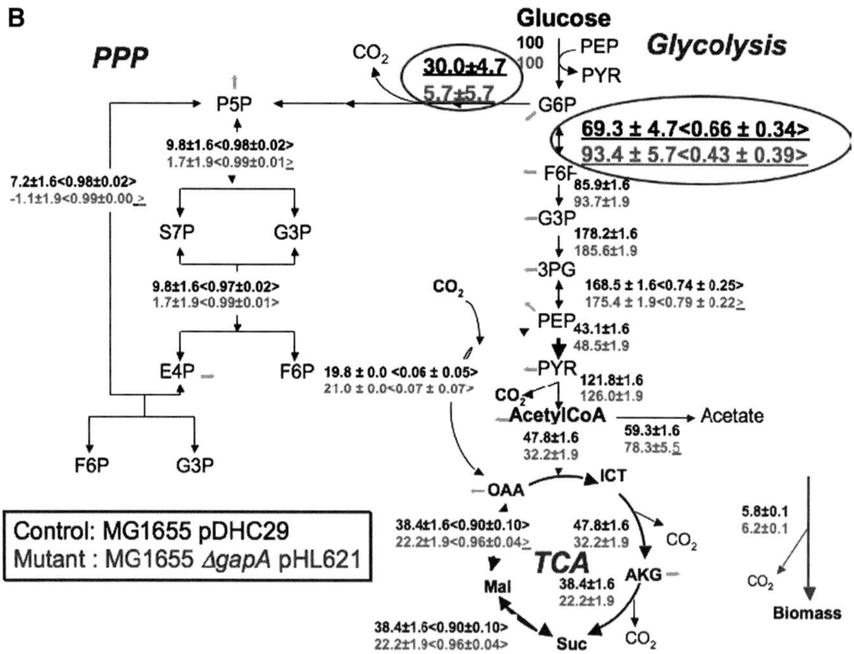

Fig. 17.3 (continued)

existing pathways in *E. coli* or specialized pathways can be introduced to assess the effects of manipulation of NADPH on the production of these compounds. Frequently NADPH is used as a recycling compound in combination with oxidative metabolism, such as with P450 type enzymes and monooxygenases, and studies can examine the efficiency of NADPH recycling systems on processes catalyzed by such enzymes (Fig. 17.3).

17.2.3.1 PHB

The pathway to PHB and other polyhydroxylalkanoates uses NADPH in the reduction step of the individual monomers (Saito et al. 1977) and since a large amount of this product can be formed in engineered *E. coli*, it can serve as a useful test system for accessing the effects of attempts to alter NADPH availability. There have been many studies of the production of PHB type molecules in *E. coli* (Fidler and Dennis 1992, Lee et al. 1994, Peoples and Sinskey 1989, Schubert et al. 1988, Slater et al. 1988, 1992, Timm and Steinbuchel 1992) and recent reviews have appeared (Dias et al. 2006, Keenan et al. 2006, Nomura and Taguchi 2007, Rehm 2007, Steinbuchel 2005, Steinbuchel and Hein 2001). The influences of various approaches are discussed below.

The inactivation of the *talA* gene increased PHB content and effect was thought to arise from effects on supplies of the intermediates NADPH and acetyl-CoA (Song et al. 2006) and a similar effect was noted upon overexpression of the *tktA* gene (Jung et al. 2004). Directly overexpressing *zwf* encoding glucose-6-phosphate dehydrogenase increased PHB accumulation (Lim et al. 2002). These alterations of the pentose pathway would promote increases in the major precursors. Efforts have been made to engineer additional NADPH availability by processing more of the glucose through glucose 6-phosphate dehydrogenase by using a mutation causing *pgi* gene inactivation. NADPH overproduction through the pentose phosphate pathway in the *pgi* mutant strain causes some reducing power imbalance that ultimately can affect the cell growth (Kabir and Shimizu 2003a,b). Experiments analyzing the concentrations of intermediates and coenzyme ratios acetyl-CoA/CoA, total CoA, and NADPH/NADP ratios showed that the PHB flux was highly sensitive to the acetyl-CoA/CoA ratio (response coefficient 0.8), total acetyl-CoA + CoA concentration (response coefficient 0.7), and pH (response coefficient −1.25) (van Wegen et al. 2001). It was less sensitive (response coefficient 0.25) to the NADPH/NADP ratio. The total NADP(H) concentration (NADPH + NADP) had a negligible effect.

Fig. 17.3 (**a**) The pathway diagram shows the formation of NADPH in the pentose phosphate pathway and the modification of the glycolytic pathway by replacement of the normal *gapA* by a *gapC* gene from *C. acetobutylicum*. The GapC can form NADPH and lead to increased availability of NADPH. (**b**) Metabolic flux distribution in control and modified *E. coli* strains. The data in the figure indicate the net flux values in *E. coli* strains calculated from steady state cultures and C-13 labeling experiments. In the top row is shown the values for *E. coli* MG1655 (pDHC29, the vector) and the corresponding values from cultures of the *E. coli gapA* mutant strain harboring the plasmid pHL621 containing *gapC* from *Clostridium acetobutylicum* are shown in the second row. The values in brackets represent the exchange coefficients of the fluxes (Martinez et al. 2008)

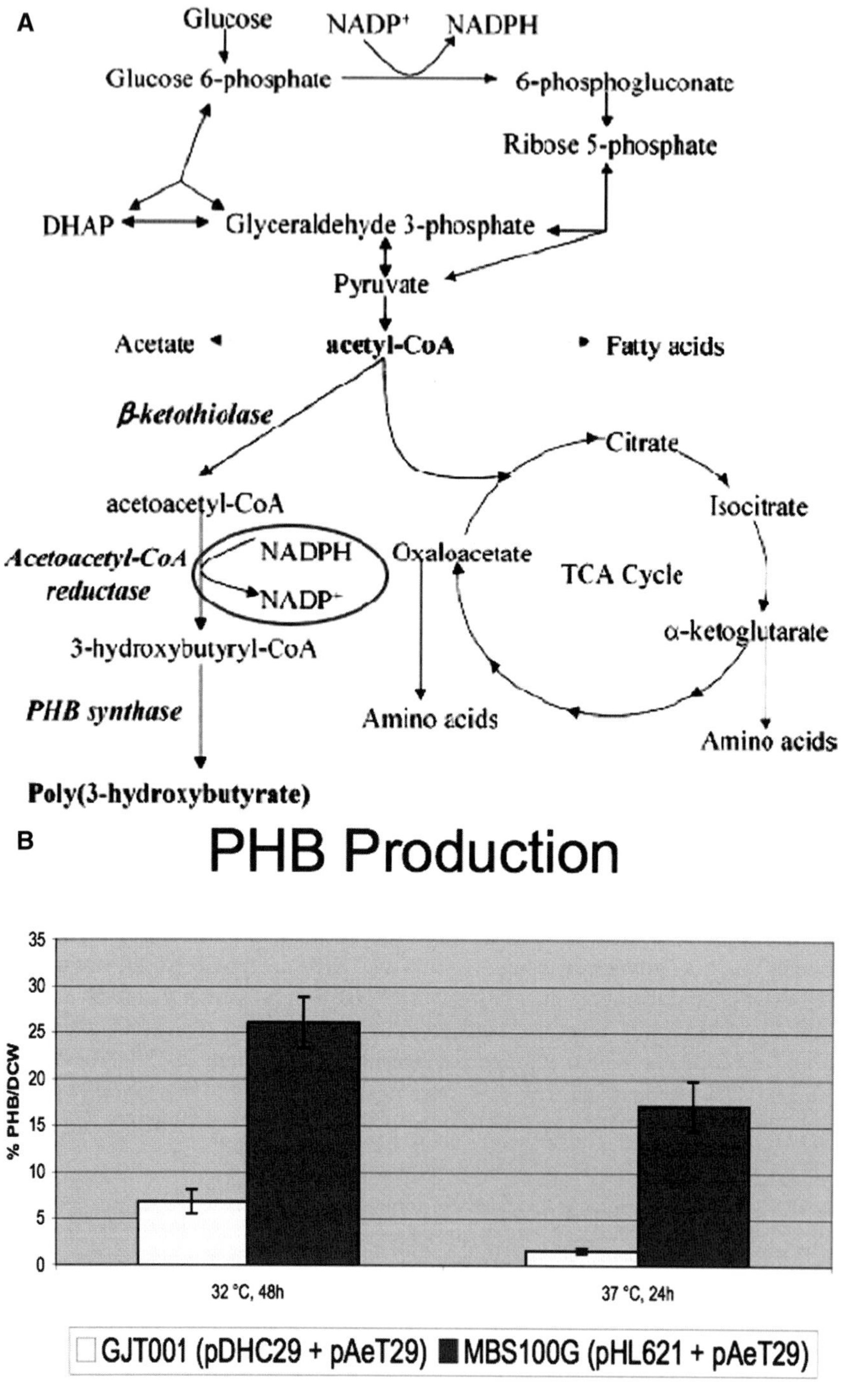

Fig. 17.4 (continued)

The effect of *pta* inactivation on PHB synthesis was studied in cultures grown on several media with the observation that a decrease in Pta activity probably causes some increase in acetyl-CoA as substrate for the PHB synthesis pathway, resulting in increased PHB accumulation (Miyake et al. 2000). The effects of *ack-pta* and *pgi* mutations on PHB synthesis was studied (Shi et al. 1999) and the improved performance of the strain with the *pgi* mutation was observed, however the effect of the alteration of acetyl-CoA suggested it was not so important in that situation.

A strain with altered NADPH availability was tested for PHB production. In this strain the normal NADH-utilizing *E. coli* GAPDH was replaced with a NADPH-utilizing enzyme from *Clostridium acetobutylicum* (Martinez et al. 2008). PHB experiments were performed at 32 °C and 37 °C until glucose was exhausted. Cells grew slower at 32 °C but higher amounts of PHB were produced. After 48h, the modified *E. coli* produced 26% of PHB/DCW compared to 6.8% of PHB/DCW of the control, showing an increase of 3.8-fold. The mutant strain of *E. coli* also produced a significantly higher amount of PHB at 37 °C compared to the control (11-fold) but the final concentration was lower than at 32 °C. These results showed that the *gapA* mutation and introduction of the *gapC* gene did increase the PHB production and further indicated the key role of NADPH availability in allowing high PHB production (Fig. 17.4).

17.2.3.2 Lycopene

Lycopene, a highly unsaturated compound of interest for its color and food ingredient properties, consumes a large amount of NADPH during its biosynthesis. Lycopene synthesis has been studied in *E. coli* with overexpression and engineering of genes of the pathway (Alper et al. 2006, Cunningham et al. 1994, Kim et al. 2008, Kim and Keasling 2001, Linden et al. 1991, Misawa et al. 1990, Misawa and Shimada 1997, Sandmann et al. 1990, Vadali et al. 2005, Wang et al. 2000, Yoon et al. 2006, 2007a,b) and chemical variations of the basic carotenoid compounds have also been formed in *E. coli* (Gallagher et al. 2003, Kajiwara et al. 1997, Lee et al. 2003, Schmidt-Dannert et al. 2000). A variety of approaches have been used to improve production. These include the overexpression of chromosomal genes of *E. coli* by the insertion of strong promoters to direct high level of expression of selected genes (Alper et al. 2005a) or addition of plasmids bearing these genes (Kang et al. 2005). The idea of balance among the levels of various gene products in generating high flux through the pathway while avoiding build-up of any toxic intermediates is a factor in this sort of pathway (Farmer and Liao

Fig. 17.4 (**a**) The pathway for production of PHB. The diagram shows the requirement for NADPH in reduction of the intermediate for polymerization of the PHB precursor. (**b**) Aerobic PHB production by control and *gapA* mutant *E. coli* strain overexpressing the *gapC* gene from *C. acetobutylicum*, both control and modified strains harbor the *phb* operon from *Alcaligenes eutrophus* for PHB synthesis. Control strain: GJT001 (pDHC29 + pAeT29); *gapC* containing mutant strain: MBS100G (pHL621 + pAeT29)

2000, Farmer and Liao 2001, Matthews and Wurtzel 2000, Smolke et al. 2001). As an example Farmer and Liao (Farmer and Liao 2000) manipulated precursor availability to increase lycopene production, they showed the G3P pool could be a limiting factor in their system. The effects of mutations on the synthesis of lycopene have been investigated by computational and experimental approaches (Alper et al. 2005b, Alper and Stephanopoulos 2008, Hemmi et al. 1998, Jin and Stephanopoulos 2007). In recent studies, a large number of individual mutations were screened and several genes were overexpressed in the host. Then combinations of mutations with improved performance were genetically combined to generate a strain with substantially greater production. This type of survey of the metabolic landscape identified the best-engineered strain (T5(P)-*dxs*, T5(P)-*idi*, *rrnB*(P)-*yjiD*-*ycgW*, delta *gdh* delta *aceE* delta *fdhF*, pACLYC), Further study

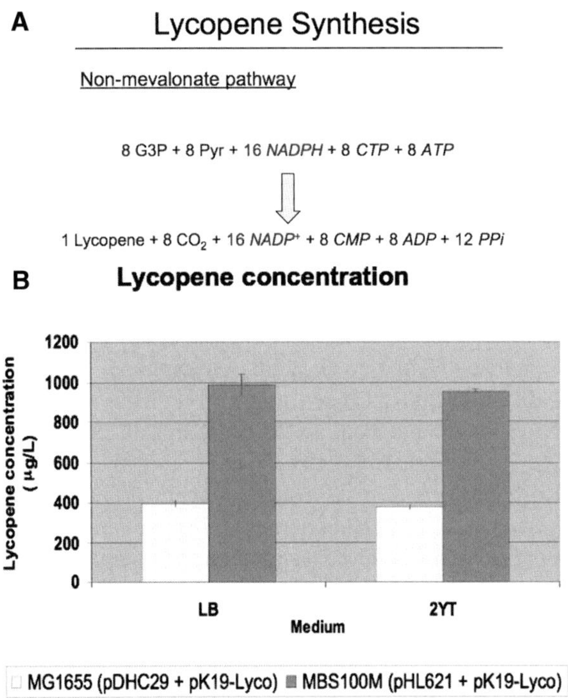

Fig. 17.5 (**a**) Lycopene synthesis by the non-mevalonate pathway requires a high amount of NADPH. (**b**) Effect of increased NADPH availability on lycopene production. The final lycopene concentration of control and modified *E. coli* strains after aerobic culture is shown. The cultures were grown in LB or 2YT medium supplemented with 20 g/L of glucose for 24h at 30 °C and 250 rpm. The data shown are the average of three replicate experiments where the error bars represent the standard deviation. Control strain: MG1655 (pDHC29, + pK19-Lyco); modified strain: MSM (pHL621 + pK19-Lyco). pDHC29 is the control vector and pHL621 carries the *gapC* gene coupling NADPH formation to the glycolytic pathway. pK19-Lyco carries the lycopene biosynthetic pathway genes (Cunningham et al. 1994)

with a large number of mutations demonstrated the complexity of mapping only one genotype to one phenotype. The investigation of combinations identified a particularly interesting mutant, the $\Delta hnr \Delta yliE$ genotype, that exhibited a drastically improved lycopene production (Jin and Stephanopoulos 2007, Alper and Stephanopoulos 2008).

The effects of the above manipulation of NADPH forming pathway, GAPDH alteration, on the levels and productivity of the strains has also been explored. The cell growth of the altered *E. coli* strain was comparable to the parental control and no growth impairment was detected. A significant difference was found in lycopene production between the two strains. The NADPH altered strain produced

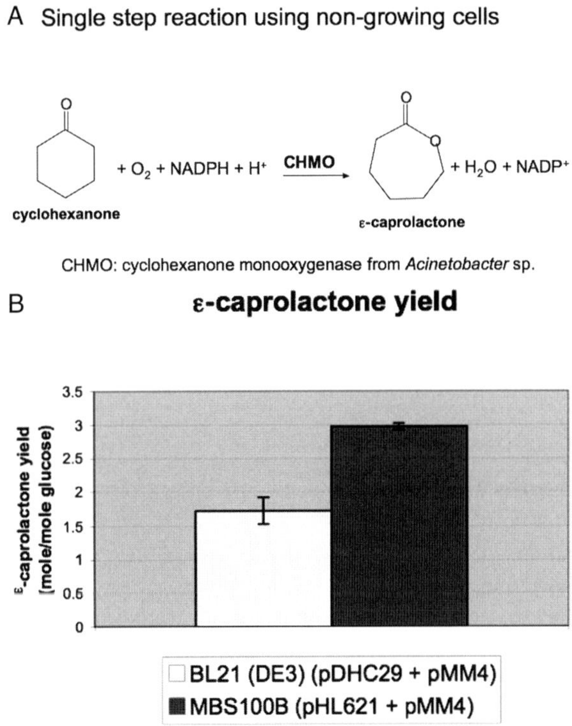

Fig. 17.6 (**a**) Synthesis of ε-caprolactone in recombinant *Escherichia coli* expressing cyclohexanone monooxygenase (CHMO) from *Acinetobacter* sp. (**b**) Effect of increased NADPH availability on conversion of cyclohexanone to the lactone. The final lactone concentration of control and modified *E. coli* strains after aerobic culture is shown. The cultures were grown in LB medium and the expression of CHMO was induced with IPTG. After reaching stationary phase the cells were was re-suspended in 20 ml of non-growing medium containing glucose and 30mM cyclohexanone and incubated for 20 h. Concentrations of cyclohexanone and ε-caprolactone were analyzed. Control strain: BL21 (DE3) contains (pDHC29, + pMM4); the BL21 gapC modified strain: MBS 100B contains (pHL621 + pMM4). pDHC29 is the control vector and pHL621 carries the *gapC* gene coupling NADPH formation to the glycolytic pathway. pMM4 carries the cyclohexanone monooxygenase (CHMO) from *Acinetobacter* sp (Walton and Stewayt 2002)

lycopene equivalent to 2.5-fold that of the control in concentration. The overexpression of the NADPH-utilizing GAPDH from *C. acetobutylicum* together with the knockout of the native NADH-dependent GAPDH improved lycopene synthesis confirming that cofactor availability is a limiting factor for the system (Fig. 17.5).

17.2.3.3 Single Step Biocatalyst

In the area of using engineered *E. coli* as a whole cell biocatalyst for a specific conversion, the emphasis has been on placement of an oxidizing step into the cell and supplying the cell with the substrate. In optimal cases the product of the oxidation step is easily removed from the reaction. The NADPH formed in this step is then used to provide the reductant for the synthesis of the desired product. A useful example of this has been studied using the recycling of mono-oxygenases to form lactones particularly chiral derivatives. In this kind of test system using a strain in which the replacement of a normal glycolytic step using NAD with one capable of using NADP, a positive effect was seen on the production rate and the amount of desired compound formed per mole of glucose consumed. The mutant host strain containing the clostridial GAPDH gene showed a higher ε-caprolactone yield that of the control strain, 2.97 compared to 1.72 mole ε-caprolactone/mole glucose. One mole of NADPH is consumed per mole of ε -caprolactone produced; therefore the mutant strain produced 73% more NADPH than the control strain under the conditions examined (Fig. 17.6).

17.3 Conclusions

In a variety of studies, it has been shown that considerable changes in metabolic pattern can be achieved by manipulation of the availability of the oxidation-reduction cofactors, NADH and NADPH. The alteration in availability of CoA compounds has also indicated that this approach can offer improvements in the synthesis of compounds derived from central pathway CoA containing intermediates. The addition of these cofactor manipulations to the arsenal of metabolic engineering tools should expand the sophistication of cell engineering as well as allow a greater understanding of the role of the various redox carrier systems and activated carriers in cell metabolism and physiology.

Acknowledgments Work in the authors' laboratories was supported by grants from the National Research Initiative Biobased Products and Bioenergy Production Research of the U. S. D. A. Cooperative State Research, Education and Extension Service USDA (2002-35505-11638), (2006-35504-17294), USDA SBIR program (2004-0503) and from the National Science Foundation (BES-0118815, BES-0222691, BES-0000303, BES-0331324, BES-0418289, CBET-0828516 BES-0420840), and NIH RO1 GM071888.

References

Abdel-Hamid AM, Attwood MM, Guest JR (2001) Pyruvate oxidase contributes to the aerobic growth efficiency of *Escherichia coli*. Microbiology 147(Pt 6):1483–98

Alexeeva S, de Kort B, Sawers G et al. (2000) Effects of limited aeration and of the ArcAB system on intermediary pyruvate catabolism in *Escherichia coli*. J Bacteriol 182(17):4934–40

Alexeeva S, Hellingwerf KJ, Teixeira de Mattos MJ (2002) Quantitative assessment of oxygen availability: perceived aerobiosis and its effect on flux distribution in the respiratory chain of *Escherichia coli*. J Bacteriol 184(5):1402–6

Alexeeva S, Hellingwerf KJ, Teixeira de Mattos MJ (2003) Requirement of ArcA for redox regulation in *Escherichia coli* under microaerobic but not anaerobic or aerobic conditions. J Bacteriol 185(1):204–9

Alper H, Fischer C, Nevoigt E et al. (2005a) Tuning genetic control through promoter engineering. Proc Natl Acad Sci USA 102(36):12678–83

Alper H, Jin YS, Moxley JF et al. (2005b) Identifying gene targets for the metabolic engineering of lycopene biosynthesis in *Escherichia coli*. Metab Eng 7(3):155–64

Alper H, Miyaoku K, Stephanopoulos G (2006) Characterization of lycopene-overproducing *E. coli* strains in high cell density fermentations. Appl Microbiol Biotechnol 72(5):968–74

Alper H, Stephanopoulos G (2008) Uncovering the gene knockout landscape for improved lycopene production in *E. coli*. Appl Microbiol Biotechnol 78(5):801–10

Andersson C, Hodge D, Berglund KA et al. (2007) Effect of different carbon sources on the production of succinic acid using metabolically engineered *Escherichia coli*. Biotechnol Prog 23(2):381–8

Andersson CI, Holmberg N, Farres J et al. (2000) Error-prone PCR of *Vitreoscilla* hemoglobin (VHb) to support the growth of microaerobic *Escherichia coli*. Biotechnol Bioeng 70(4): 446–55

Aristidou AA, San KY, Bennett GN (1999) Improvement of biomass yield and recombinant gene expression in *Escherichia coli* by using fructose as the primary carbon source. Biotechnol Prog 15(1):140–5

Backlund E, Markland K, Larsson G (2008) Cell engineering of *Escherichia coli* allows high cell density accumulation without fed-batch process control. Bioprocess Biosyst Eng 31(1):11–20

Bagramyan K, Trchounian A (2003) Structural and functional features of formate hydrogen lyase, an enzyme of mixed-acid fermentation from *Escherichia coli*. Biochemistry (Mosc) 68(11):1159–70

Becker A, Fritz-Wolf K, Kabsch W et al. (1999) Structure and mechanism of the glycyl radical enzyme pyruvate formate-lyase. Nat Struct Biol 6(10):969–75

Becker S, Vlad D, Schuster S et al. (1997) Regulatory O_2 tensions for the synthesis of fermentation products in *Escherichia coli* and relation to aerobic respiration. Arch Microbiol 168(4): 290–6

Berrios-Rivera SJ, Bennett GN, San KY (2002a) The effect of increasing NADH availability on the redistribution of metabolic fluxes in *Escherichia coli* chemostat cultures. Metab Eng 4(3): 230–7

Berrios-Rivera SJ, Bennett GN, San KY (2002b) Metabolic engineering of *Escherichia coli*: increase of NADH availability by overexpressing an NAD(+)-dependent formate dehydrogenase. Metab Eng 4(3):217–29

Berrios-Rivera SJ, Sanchez AM, Bennett GN et al. (2004) Effect of different levels of NADH availability on metabolite distribution in *Escherichia coli* fermentation in minimal and complex media. Appl Microbiol Biotechnol 65(4):426–32

Birkmann A, Zinoni F, Sawers G et al. (1987) Factors affecting transcriptional regulation of the formate-hydrogen-lyase pathway of *Escherichia coli*. Arch Microbiol 148(1):44–51

Blaschkowski HP, Neuer G, Ludwig-Festl M et al. (1982) Routes of flavodoxin and ferredoxin reduction in *Escherichia coli*. CoA-acylating pyruvate: flavodoxin and NADPH: flavodoxin oxidoreductases participating in the activation of pyruvate formate-lyase. Eur J Biochem 123(3):563–9

Cassey B, Guest JR, Attwood MM (1998) Environmental control of pyruvate dehydrogenase complex expression in *Escherichia coli*. FEMS Microbiol Lett 159(2):325–9

Causey TB, Shanmugam KT, Yomano LP et al. (2004) Engineering *Escherichia coli* for efficient conversion of glucose to pyruvate. Proc Natl Acad Sci USA 101(8):2235–40

Causey TB, Zhou S, Shanmugam KT et al. (2003) Engineering the metabolism of *Escherichia coli* W3110 for the conversion of sugar to redox-neutral and oxidized products: homoacetate production. Proc Natl Acad Sci USA 100(3):825–32

Chang DE, Jung HC, Rhee JS et al. (1999) Homofermentative production of D- or L-lactate in metabolically engineered *Escherichia coli* RR1. Appl Environ Microbiol 65(4):1384–9

Chase T, Jr., Rabinowitz JC (1968) Role of pyruvate and S-adenosylmethioine in activating the pyruvate formate-lyase of *Escherichia coli*. J Bacteriol 96(4):1065–78

Chatterjee R, Millard CS, Champion K et al. (2001) Mutation of the *ptsG* gene results in increased production of succinate in fermentation of glucose by *Escherichia coli*. Appl Environ Microbiol 67(1):148–54

Chen R, Hatzimanikatis V, Yap WM et al. (1997) Metabolic consequences of phosphotransferase (PTS) mutation in a phenylalanine-producing recombinant *Escherichia coli*. Biotechnol Prog 13(6):768–75

Chou CH, Bennett GN, San KY (1994) Effect of modulated glucose uptake on high-level recombinant protein production in a dense *Escherichia coli* culture. Biotechnol Prog 10(6):644–7

Cox SJ, Shalel Levanon S, Sanchez A et al. (2006) Development of a metabolic network design and optimization framework incorporating implementation constraints: a succinate production case study. Metab Eng 8(1):46–57

Cunningham FX, Jr., Sun Z, Chamovitz D et al. (1994) Molecular structure and enzymatic function of lycopene cyclase from the cyanobacterium *Synechococcus* sp strain PCC7942. Plant Cell 6(8):1107–21

De Anda R, Lara AR, Hernandez V et al. (2006) Replacement of the glucose phosphotransferase transport system by galactose permease reduces acetate accumulation and improves process performance of *Escherichia coli* for recombinant protein production without impairment of growth rate. Metab Eng 8(3):281–90

De Mey M, Lequeux GJ, Beauprez JJ et al. (2007) Comparison of different strategies to reduce acetate formation in *Escherichia coli*. Biotechnol Prog 23(5):1053–63

Dias JM, Lemos PC, Serafim LS et al. (2006) Recent advances in polyhydroxyalkanoate production by mixed aerobic cultures: from the substrate to the final product. Macromol Biosci 6(11): 885–906

Dien BS, Nichols NN, Bothast RJ (2001) Recombinant *Escherichia coli* engineered for production of L-lactic acid from hexose and pentose sugars. J Ind Microbiol Biotechnol 27(4):259–64

Doelle HW, Hollywood NW (1978) Transitional steady-state investigations during aerobic-anaerobic transition of glucose utilization by *Escherichia coli* K-12. Eur J Biochem 83(2): 479–84

Farmer WR, Liao JC (1997) Reduction of aerobic acetate production by *Escherichia coli*. Appl Environ Microbiol 63(8):3205–10

Farmer WR, Liao JC (2000) Improving lycopene production in *Escherichia coli* by engineering metabolic control. Nat Biotechnol 18(5):533–7

Farmer WR, Liao JC (2001) Precursor balancing for metabolic engineering of lycopene production in *Escherichia coli*. Biotechnol Prog 17(1):57–61

Fidler S, Dennis D (1992) Polyhydroxyalkanoate production in recombinant *Escherichia coli*. FEMS Microbiol Rev 9(2–4):231–5

Flores N, de Anda R, Flores S et al. (2004) Role of pyruvate oxidase in *Escherichia coli* strains lacking the phosphoenolpyruvate:carbohydrate phosphotransferase system. J Mol Microbiol Biotechnol 8(4):209–21

Fong SS, Burgard AP, Herring CD et al. (2005) *In silico* design and adaptive evolution of *Escherichia coli* for production of lactic acid. Biotechnol Bioeng 91(5):643–8

Francis K, Patel P, Wendt JC et al. (1990) Purification and characterization of two forms of hydrogenase isoenzyme 1 from *Escherichia coli*. J Bacteriol 172(10):5750–7

Frey AD, Bailey JE, Kallio PT (2000) Expression of *Alcaligenes eutrophus* flavohemoprotein and engineered *Vitreoscilla* hemoglobin-reductase fusion protein for improved hypoxic growth of *Escherichia coli*. Appl Environ Microbiol 66(1):98–104

Galkin A, Kulakova L, Yoshimura T et al. (1997) Synthesis of optically active amino acids from alpha-keto acids with *Escherichia coli* cells expressing heterologous genes. Appl Environ Microbiol 63(12):4651–6

Gallagher CE, Cervantes-Cervantes M, Wurtzel ET (2003) Surrogate biochemistry: use of *Escherichia coli* to identify plant cDNAs that impact metabolic engineering of carotenoid accumulation. Appl Microbiol Biotechnol 60(6):713–9

Govantes F, Orjalo AV, Gunsalus RP (2000) Interplay between three global regulatory proteins mediates oxygen regulation of the *Escherichia coli* cytochrome d oxidase (*cydAB*) operon. Mol Microbiol 38(5):1061–73

Guest JR, Angier SJ, Russell GC (1989) Structure, expression, and protein engineering of the pyruvate dehydrogenase complex of *Escherichia coli*. Ann NY Acad Sci 573:76–99

Guest JR, Cole ST, Jeyaseelan K (1981) Organization and expression of the pyruvate dehydrogenase complex genes of *Escherichia coli K12*. J Gen Microbiol 127(1):65–79

Guest JR, Stephens PE (1980) Molecular cloning of the pyruvate dehydrogenase complex genes of *Escherichia coli*. J Gen Microbiol 121(2):277–92

Gunsalus RP, Park SJ (1994) Aerobic-anaerobic gene regulation in *Escherichia coli*: control by the ArcAB and Fnr regulons. Res Microbiol 145(5–6):437–50

Haydon DJ, Quail MA, Guest JR (1993) A mutation causing constitutive synthesis of the pyruvate dehydrogenase complex in *Escherichia coli* is located within the *pdhR* gene. FEBS Lett 336(1):43–7

Hemmi H, Ohnuma S, Nagaoka K et al. (1998) Identification of genes affecting lycopene formation in *Escherichia coli* transformed with carotenoid biosynthetic genes: candidates for early genes in isoprenoid biosynthesis. J Biochem 123(6):1088–96

Hernandez-Montalvo V, Martinez A, Hernandez-Chavez G et al. (2003) Expression of *galP* and *glk* in a *Escherichia coli* PTS mutant restores glucose transport and increases glycolytic flux to fermentation products. Biotechnol Bioeng 83(6):687–94

Hong SH, Lee SY (2001) Metabolic flux analysis for succinic acid production by recombinant *Escherichia coli* with amplified malic enzyme activity. Biotechnol Bioeng 74(2):89–95

Hong SH, Lee SY (2002) Importance of redox balance on the production of succinic acid by metabolically engineered *Escherichia coli*. Appl Microbiol Biotechnol 58(3):286–90

Hoover DM, Ludwig ML (1997) A flavodoxin that is required for enzyme activation: the structure of oxidized flavodoxin from *Escherichia coli* at 1.8 A resolution. Protein Sci 6(12): 2525–37

Hua Q, Joyce AR, Fong SS et al. (2006) Metabolic analysis of adaptive evolution for *in silico*-designed lactate-producing strains. Biotechnol Bioeng 95(5):992–1002

Ingram LO, Aldrich HC, Borges AC et al. (1999) Enteric bacterial catalysts for fuel ethanol production. Biotechnol Prog 15(5):855–66

Ingram LO, Conway T, Clark DP et al. (1987) Genetic engineering of ethanol production in *Escherichia coli*. Appl Environ Microbiol 53(10):2420–5

Jantama K, Haupt MJ, Svoronos SA et al. (2008) Combining metabolic engineering and metabolic evolution to develop nonrecombinant strains of *Escherichia coli* C that produce succinate and malate. Biotechnol Bioeng 99(5):1140–53

Jarboe LR, Grabar TB, Yomano LP et al. (2007) Development of ethanologenic bacteria. Adv Biochem Eng Biotechnol 108:237–61

Jin YS, Stephanopoulos G (2007) Multi-dimensional gene target search for improving lycopene biosynthesis in*Escherichia coli*. Metab Eng 9(4):337–47

Jones SA, Chowdhury FZ, Fabich AJ et al. (2007) Respiration of *Escherichia coli* in the mouse intestine. Infect Immun 75(10):4891–9

Jung YM, Lee JN, Shin HD et al. (2004) Role of *tktA* gene in pentose phosphate pathway on odd-ball biosynthesis of poly-beta-hydroxybutyrate in transformant *Escherichia coli* harboring *phbCAB* operon. J Biosci Bioeng 98(3):224–7

Kabir MM, Shimizu K (2003a) Fermentation characteristics and protein expression patterns in a recombinant *Escherichia coli* mutant lacking phosphoglucose isomerase for poly(3-hydroxybutyrate) production. Appl Microbiol Biotechnol 62(2–3):244–55

Kabir MM, Shimizu K (2003b) Gene expression patterns for metabolic pathway in *pgi* knock-out *Escherichia coli* with and without *phb* genes based on RT-PCR. J Biotechnol 105(1–2): 11–31

Kajiwara S, Fraser PD, Kondo K et al. (1997) Expression of an exogenous isopentenyl diphosphate isomerase gene enhances isoprenoid biosynthesis in *Escherichia coli*. Biochem J 324 (Pt 2):421–6

Kallio PT, Tsai PS, Bailey JE (1996) Expression of *Vitreoscilla hemoglobin* is superior to horse heart myoglobin or yeast flavohemoglobin expression for enhancing *Escherichia coli* growth in a microaerobic bioreactor. Biotechnol Prog 12(6):751–7

Kang MJ, Lee YM, Yoon SH et al. (2005) Identification of genes affecting lycopene accumulation in *Escherichia coli* using a shot-gun method. Biotechnol Bioeng 91(5):636–42

Kaup B, Bringer-Meyer S, Sahm H (2003) Metabolic engineering of *Escherichia coli*: construction of an efficient biocatalyst for D-mannitol formation in a whole-cell biotransformation. Commun Agric Appl Biol Sci 68(2 Pt A):235–40

Kaup B, Bringer-Meyer S, Sahm H (2004) Metabolic engineering of *Escherichia coli*: construction of an efficient biocatalyst for D-mannitol formation in a whole-cell biotransformation. Appl Microbiol Biotechnol 64(3):333–9

Kaup B, Bringer-Meyer S, Sahm H (2005) D: -Mannitol formation from D: -glucose in a whole-cell biotransformation with recombinant *Escherichia coli*. Appl Microbiol Biotechnol 69(4): 397–403

Keenan TM, Nakas JP, Tanenbaum SW (2006) Polyhydroxyalkanoate copolymers from forest biomass. J Ind Microbiol Biotechnol 33(7):616–26

Kim P, Laivenieks M, Vieille C et al. (2004) Effect of overexpression of *Actinobacillus succinogenes* phosphoenolpyruvate carboxykinase on succinate production in *Escherichia coli*. Appl Environ Microbiol 70(2):1238–41

Kim SW, Jung WH, Ryu JM et al. (2008) Identification of an alternative translation initiation site for the *Pantoea ananatis* lycopene cyclase (*crtY*) gene in *E. coli* and its evolutionary conservation. Protein Expr Purif 58(1):23–31

Kim SW, Keasling JD (2001) Metabolic engineering of the nonmevalonate isopentenyl diphosphate synthesis pathway in *Escherichia coli* enhances lycopene production. Biotechnol Bioeng 72(4):408–15

Kim Y, Ingram LO, Shanmugam KT (2007) Construction of an *Escherichia coli* K-12 mutant for homoethanologenic fermentation of glucose or xylose without foreign genes. Appl Environ Microbiol 73(6):1766–71

Knappe J, Blaschkowski HP (1975) Pyruvate formate-lyase from *Escherischia coli* and its activation system. Methods Enzymol 41:508–18

Knappe J, Neugebauer FA, Blaschkowski HP et al. (1984) Post-translational activation introduces a free radical into pyruvate formate-lyase. Proc Natl Acad Sci USA 81(5):1332–5

Knappe J, Sawers G (1990) A radical-chemical route to acetyl-CoA: the anaerobically induced pyruvate formate-lyase system of *Escherichia coli*. FEMS Microbiol Rev 6(4):383–98

Knappe J, Wagner AF (1995) Glycyl free radical in pyruvate formate-lyase: synthesis, structure characteristics, and involvement in catalysis. Methods Enzymol 258:343–62

Kulzer R, Pils T, Kappl R et al. (1998) Reconstitution and characterization of the polynuclear iron-sulfur cluster in pyruvate formate-lyase-activating enzyme. Molecular properties of the holoenzyme form. J Biol Chem 273(9):4897–903

Lara AR, Caspeta L, Gosset G et al. (2008) Utility of an *Escherichia coli* strain engineered in the substrate uptake system for improved culture performance at high glucose and cell concentrations: an alternative to fed-batch cultures. Biotechnol Bioeng 99(4):893–901

Lara AR, Vazquez-Limon C, Gosset G et al. (2006) Engineering *Escherichia coli* to improve culture performance and reduce formation of by-products during recombinant protein production under transient intermittent anaerobic conditions. Biotechnol Bioeng 94(6):1164–75

Laurinavichene TV, Tsygankov AA (2001) H_2 consumption by *Escherichia coli* coupled via hydrogenase 1 or hydrogenase 2 to different terminal electron acceptors. FEMS Microbiol Lett 202(1):121–4

Lee PC, Momen AZ, Mijts BN et al. (2003) Biosynthesis of structurally novel carotenoids in *Escherichia coli*. Chem Biol 10(5):453–62

Lee SJ, Lee DY, Kim TY et al. (2005) Metabolic engineering of *Escherichia coli* for enhanced production of succinic acid, based on genome comparison and *in silico* gene knockout simulation. Appl Environ Microbiol 71(12):7880–7

Lee SY, Yim KS, Chang HN et al. (1994) Construction of plasmids, estimation of plasmid stability, and use of stable plasmids for the production of poly(3-hydroxybutyric acid) by recombinant *Escherichia coli*. J Biotechnol 32(2):203–11

Lim SJ, Jung YM, Shin HD et al. (2002) Amplification of the NADPH-related genes *zwf* and *gnd* for the oddball biosynthesis of PHB in an *E. coli* transformant harboring a cloned *phbCAB* operon. J Biosci Bioeng 93(6):543–9

Lin H, Bennett GN, San KY (2005a) Chemostat culture characterization of *Escherichia coli* mutant strains metabolically engineered for aerobic succinate production: a study of the modified metabolic network based on metabolite profile, enzyme activity, and gene expression profile. Metab Eng 7(5–6):337–52

Lin H, Bennett GN, San KY (2005b) Effect of carbon sources differing in oxidation state and transport route on succinate production in metabolically engineered *Escherichia coli*. J Ind Microbiol Biotechnol 32(3):87–93

Lin H, Bennett GN, San KY (2005c) Fed-batch culture of a metabolically engineered *Escherichia coli* strain designed for high-level succinate production and yield under aerobic conditions. Biotechnol Bioeng 90(6):775–9

Lin H, Bennett GN, San KY (2005d) Genetic reconstruction of the aerobic central metabolism in *Escherichia coli* for the absolute aerobic production of succinate. Biotechnol Bioeng 89(2):148–56

Lin H, San KY, Bennett GN (2005e) Effect of *Sorghum vulgare* phosphoenolpyruvate carboxylase and *Lactococcus lactis* pyruvate carboxylase coexpression on succinate production in mutant strains of *Escherichia coli*. Appl Microbiol Biotechnol 67(4):515–23

Lin H, Vadali RV, Bennett GN et al. (2004) Increasing the acetyl-CoA pool in the presence of overexpressed phosphoenolpyruvate carboxylase or pyruvate carboxylase enhances succinate production in *Escherichia coli*. Biotechnol Prog 20(5):1599–604

Linden H, Misawa N, Chamovitz D et al. (1991) Functional complementation in *Escherichia coli* of different phytoene desaturase genes and analysis of accumulated carotenes. Z Naturforsch [C] 46(11–12):1045–51

Liu X, De Wulf P (2004) Probing the ArcA-P modulon of *Escherichia coli* by whole genome transcriptional analysis and sequence recognition profiling. J Biol Chem 279(13): 12588–97

Lopez de Felipe F, Kleerebezem M, de Vos WM et al. (1998) Cofactor engineering: a novel approach to metabolic engineering in *Lactococcus lactis* by controlled expression of NADH oxidase. J Bacteriol 180(15):3804–8

Maeda T, Sanchez-Torres V, Wood TK (2007a) Enhanced hydrogen production from glucose by metabolically engineered *Escherichia coli*. Appl Microbiol Biotechnol 77(4):879–90

Maeda T, Sanchez-Torres V, Wood TK (2007b) *Escherichia coli* hydrogenase 3 is a reversible enzyme possessing hydrogen uptake and synthesis activities. Appl Microbiol Biotechnol 76(5):1035–42

Martinez I, Zhu J, Lin H, Bennett GN, San KY (2008) Replacing *Escherichia coli* NAD-dependent glyceraldehyde 3-phosphate dehydrogenase (GAPDH) with a NADP-dependent enzyme from *Clostridium acetobutylicum* facilitates NADPH-dependent pathways. Metab Eng 10(6):352–9.

Matthews PD, Wurtzel ET (2000) Metabolic engineering of carotenoid accumulation in *Escherichia coli* by modulation of the isoprenoid precursor pool with expression of deoxyxylulose phosphate synthase. Appl Microbiol Biotechnol 53(4):396–400

Misawa N, Nakagawa M, Kobayashi K et al. (1990) Elucidation of the *Erwinia uredovora* carotenoid biosynthetic pathway by functional analysis of gene products expressed in *Escherichia coli*. J Bacteriol 172(12):6704–12
Misawa N, Shimada H (1997) Metabolic engineering for the production of carotenoids in non-carotenogenic bacteria and yeasts. J Biotechnol 59(3):169–81
Miyake M, Miyamoto C, Schnackenberg J et al. (2000) Phosphotransacetylase as a key factor in biological production of polyhydroxybutyrate. Appl Biochem Biotechnol 84–86: 1039–44
Nanba H, Takaoka Y, Hasegawa J (2003a) Purification and characterization of an alpha-haloketone-resistant formate dehydrogenase from *Thiobacillus* sp. strain KNK65MA, and cloning of the gene. Biosci Biotechnol Biochem 67(10):2145–53
Nanba H, Takaoka Y, Hasegawa J (2003b) Purification and characterization of formate dehydrogenase from *Ancylobacter aquaticus* strain KNK607M, and cloning of the gene. Biosci Biotechnol Biochem 67(4):720–8
Neves AR, Ramos A, Costa H et al. (2002) Effect of different NADH oxidase levels on glucose metabolism by *Lactococcus lactis*: kinetics of intracellular metabolite pools determined by *in vivo* nuclear magnetic resonance. Appl Environ Microbiol 68(12):6332–42
Nnyepi MR, Peng Y, Broderick JB (2007) Inactivation of *E. coli* pyruvate formate-lyase: role of AdhE and small molecules. Arch Biochem Biophys 459(1):1–9
Nomura CT, Taguchi S (2007) PHA synthase engineering toward superbiocatalysts for custom-made biopolymers. Appl Microbiol Biotechnol 73(5):969–79
Ogasawara H, Ishida Y, Yamada K et al. (2007) PdhR (pyruvate dehydrogenase complex regulator) controls the respiratory electron transport system in *Escherichia coli*. J Bacteriol 189(15): 5534–41
Overton TW, Griffiths L, Patel MD et al. (2006) Microarray analysis of gene regulation by oxygen, nitrate, nitrite, FNR, NarL and NarP during anaerobic growth of *Escherichia coli*: new insights into microbial physiology. Biochem Soc Trans 34(Pt 1):104–7
Partridge JD, Sanguinetti G, Dibden DP et al. (2007) Transition of *Escherichia coli* from aerobic to micro-aerobic conditions involves fast and slow reacting regulatory components. J Biol Chem 282(15):11230–7
Patel RN (2000) Microbial/enzymatic synthesis of chiral drug intermediates. Adv Appl Microbiol 47:33–78
Pecher A, Blaschkowski HP, Knappe K et al. (1982) Expression of pyruvate formate-lyase of *Escherichia coli* from the cloned structural gene. Arch Microbiol 132(4):365–71
Peercy BE, Cox SJ, Shalel-Levanon S et al. (2006) A kinetic model of oxygen regulation of cytochrome production in *Escherichia coli*. J Theor Biol 242(3):547–63
Peoples OP, Sinskey AJ (1989) Poly-beta-hydroxybutyrate (PHB) biosynthesis in *Alcaligenes eutrophus* H16. Identification and characterization of the PHB polymerase gene (*phbC*). J Biol Chem 264(26):15298–303
Phue JN, Noronha SB, Hattacharyya R et al. (2005) Glucose metabolism at high density growth of *E. coli* B and *E. coli* K: differences in metabolic pathways are responsible for efficient glucose utilization in *E. coli* B as determined by microarrays and Northern blot analyses. Biotechnol Bioeng 90(7):805–20
Picon A, Teixeira de Mattos MJ, Postma PW (2005) Reducing the glucose uptake rate in *Escherichia coli* affects growth rate but not protein production. Biotechnol Bioeng 90(2): 191–200
Quail MA, Guest JR (1995) Purification, characterization and mode of action of PdhR, the transcriptional repressor of the *pdhR-aceEF-lpd* operon of *Escherichia coli*. Mol Microbiol 15(3):519–29
Reddy SG, Wong KK, Parast CV et al. (1998) Dioxygen inactivation of pyruvate formate-lyase: EPR evidence for the formation of protein-based sulfinyl and peroxyl radicals. Biochemistry 37(2):558–63

Redwood MD, Mikheenko IP, Sargent F et al. (2008) Dissecting the roles of *Escherichia coli* hydrogenases in biohydrogen production. FEMS Microbiol Lett 278(1):48–55

Reed JL, Vo TD, Schilling CH et al. (2003) An expanded genome-scale model of *Escherichia coli* K-12 (iJR904 GSM/GPR). Genome Biol 4(9):R54

Rehm BH (2007) Biogenesis of microbial polyhydroxyalkanoate granules: a platform technology for the production of tailor-made bioparticles. Curr Issues Mol Biol 9(1):41–62

Saito T, Fukui T, Ikeda F et al. (1977) An NADP-linked acetoacetyl CoA reductase from *Zoogloea ramigera*. Arch Microbiol 114(3):211–7

Sakai T, Nakamura N, Umitsuki G et al. (2007) Increased production of pyruvic acid by *Escherichia coli* RNase G mutants in combination with *cra* mutations. Appl Microbiol Biotechnol 76(1):183–92

Salmon K, Hung SP, Mekjian K et al. (2003) Global gene expression profiling in *Escherichia coli* K12. The effects of oxygen availability and FNR. J Biol Chem 278(32):29837–55

Sanchez AM, Bennett GN, San KY (2005a) Effect of different levels of NADH availability on metabolic fluxes of *Escherichia coli* chemostat cultures in defined medium. J Biotechnol 117(4):395–405

Sanchez AM, Bennett GN, San KY (2005b) Efficient succinic acid production from glucose through overexpression of pyruvate carboxylase in an *Escherichia coli* alcohol dehydrogenase and lactate dehydrogenase mutant. Biotechnol Prog 21(2):58–65

Sanchez AM, Bennett GN, San KY (2006) Batch culture characterization and metabolic flux analysis of succinate-producing *Escherichia coli* strains. Metab Eng 8(3):209–26

Sandmann G, Woods WS, Tuveson RW (1990) Identification of carotenoids in *Erwinia herbicola* and in a transformed *Escherichia coli* strain. FEMS Microbiol Lett 59(1–2):77–82

Sauer U, Canonaco F, Heri S et al. (2004) The soluble and membrane-bound transhydrogenases UdhA and PntAB have divergent functions in NADPH metabolism of *Escherichia coli*. J Biol Chem 279(8):6613–9

Sauter M, Sawers RG (1990) Transcriptional analysis of the gene encoding pyruvate formate-lyase-activating enzyme of *Escherichia coli*. Mol Microbiol 4(3):355–63

Sawers G (1999) The aerobic/anaerobic interface. Curr Opin Microbiol 2(2):181–7

Sawers G, Bock A (1988) Anaerobic regulation of pyruvate formate-lyase from *Escherichia coli* K-12. J Bacteriol 170(11):5330–6

Sawers G, Hesslinger C, Muller N et al. (1998) The glycyl radical enzyme TdcE can replace pyruvate formate-lyase in glucose fermentation. J Bacteriol 180(14):3509–16

Sawers G, Watson G (1998) A glycyl radical solution: oxygen-dependent interconversion of pyruvate formate-lyase. Mol Microbiol 29(4):945–54

Schmidt-Dannert C, Umeno D, Arnold FH (2000) Molecular breeding of carotenoid biosynthetic pathways. Nat Biotechnol 18(7):750–3

Schramm G, Zapatka M, Eils R et al. (2007) Using gene expression data and network topology to detect substantial pathways, clusters and switches during oxygen deprivation of *Escherichia coli*. BMC Bioinformatics 8(149):149

Schubert P, Steinbuchel A, Schlegel HG (1988) Cloning of the *Alcaligenes eutrophus* genes for synthesis of poly-beta-hydroxybutyric acid (PHB) and synthesis of PHB in *Escherichia coli*. J Bacteriol 170(12):5837–47

Serres MH, Gopal S, Nahum LA et al. (2001) A functional update of the *Escherichia coli* K-12 genome. Genome Biol 2(9):RESEARCH0035

Shalel-Levanon S, San KY, Bennett GN (2005a) Effect of ArcA and FNR on the expression of genes related to the oxygen regulation and the glycolysis pathway in *Escherichia coli* under microaerobic growth conditions. Biotechnol Bioeng 92(2):147–59

Shalel-Levanon S, San KY, Bennett GN (2005b) Effect of oxygen on the *Escherichia coli* ArcA and FNR regulation systems and metabolic responses. Biotechnol Bioeng 89(5):556–64

Shalel-Levanon S, San KY, Bennett GN (2005c) Effect of oxygen, and ArcA and FNR regulators on the expression of genes related to the electron transfer chain and the TCA cycle in *Escherichia coli*. Metab Eng 7(5–6):364–74

Shi H, Nikawa J, Shimizu K (1999) Effect of modifying metabolic network on poly-3-hydroxybutyrate biosynthesis in recombinant *Escherichia coli*. J Biosci Bioeng 87(5):666–77

Slater S, Gallaher T, Dennis D (1992) Production of poly-(3-hydroxybutyrate-co-3-hydroxyvalerate) in a recombinant *Escherichia coli* strain. Appl Environ Microbiol 58(4): 1089–94

Slater SC, Voige WH, Dennis DE (1988) Cloning and expression in *Escherichia coli* of the *Alcaligenes eutrophus* H16 poly-beta-hydroxybutyrate biosynthetic pathway. J Bacteriol 170(10):4431–6

Slusarczyk H, Felber S, Kula MR et al. (2000) Stabilization of NAD-dependent formate dehydrogenase from *Candida boidinii* by site-directed mutagenesis of cysteine residues. Eur J Biochem 267(5):1280–9

Smolke CD, Martin VJ, Keasling JD (2001) Controlling the metabolic flux through the carotenoid pathway using directed mRNA processing and stabilization. Metab Eng 3(4):313–21

Snoep JL, de Graef MR, Westphal AH et al. (1993) Differences in sensitivity to NADH of purified pyruvate dehydrogenase complexes of *Enterococcus faecalis*, *Lactococcus lactis*, *Azotobacter vinelandii* and *Escherichia coli*: implications for their activity *in vivo*. FEMS Microbiol Lett 114(3):279–83

Song BG, Kim TK, Jung YM et al. (2006) Modulation of *talA* gene in pentose phosphate pathway for overproduction of poly-beta-hydroxybutyrate in transformant *Escherichia coli* harboring *phbCAB* operon. J Biosci Bioeng 102(3):237–40

Steinbuchel A (2005) Non-biodegradable biopolymers from renewable resources: perspectives and impacts. Curr Opin Biotechnol 16(6):607–13

Steinbuchel A, Hein S (2001) Biochemical and molecular basis of microbial synthesis of polyhydroxyalkanoates in microorganisms. Adv Biochem Eng Biotechnol 71:81–123

Stols L, Kulkarni G, Harris BG et al. (1997) Expression of *Ascaris suum* malic enzyme in a mutant *Escherichia coli* allows production of succinic acid from glucose. Appl Biochem Biotechnol 63–65:153–8

Thomas AD, Doelle HW, Westwood AW et al. (1972) Effect of oxygen on several enzymes involved in the aerobic and anaerobic utilization of glucose in *Escherichia coli*. J Bacteriol 112(3):1099–105

Timm A, Steinbuchel A (1992) Cloning and molecular analysis of the poly(3-hydroxyalkanoic acid) gene locus of *Pseudomonas aeruginosa* PAO1. Eur J Biochem 209(1):15–30

Tishkov VI, Popov VO (2006) Protein engineering of formate dehydrogenase. Biomol Eng 23(2–3):89–110

Tomar A, Eiteman MA, Altman E (2003) The effect of acetate pathway mutations on the production of pyruvate in *Escherichia coli*. Appl Microbiol Biotechnol 62(1):76–82

Vadali RV, Fu Y, Bennett GN et al. (2005) Enhanced lycopene productivity by manipulation of carbon flow to isopentenyl diphosphate in *Escherichia coli*. Biotechnol Prog 21(5):1558–61

van Wegen RJ, Lee SY, Middelberg AP (2001) Metabolic and kinetic analysis of poly(3-hydroxybutyrate) production by recombinant *Escherichia coli*. Biotechnol Bioeng 74(1):70–80

Varenne S, Casse F, Chippaux M et al. (1975) A mutant of *Escherichia coli* deficient in pyruvate formate lyase. Mol Gen Genet 141(2):181–4

Vemuri GN, Altman E, Sangurdekar DP et al. (2006) Overflow metabolism in *Escherichia coli* during steady-state growth: transcriptional regulation and effect of the redox ratio. Appl Environ Microbiol 72(5):3653–61

Vemuri GN, Eiteman MA, Altman E (2002) Succinate production in dual-phase *Escherichia coli* fermentations depends on the time of transition from aerobic to anaerobic conditions. J Ind Microbiol Biotechnol 28(6):325–32

Vemuri GN, Minning TA, Altman E et al. (2005) Physiological response of central metabolism in *Escherichia coli* to deletion of pyruvate oxidase and introduction of heterologous pyruvate carboxylase. Biotechnol Bioeng 90(1):64–76

Wagner AF, Schultz S, Bomke J et al. (2001) YfiD of *Escherichia coli* and Y06I of bacteriophage T4 as autonomous glycyl radical cofactors reconstituting the catalytic center of oxygen-fragmented pyruvate formate-lyase. Biochem Biophys Res Commun 285(2): 456–62

Walton AZ, Stewart JD (2002) An efficient Baeyer-Villiger oxidation by engineered *Escherichia coli* cells under non-growing conditions. Biotechnol prog 18(2):262–8.

Wang C, Oh MK, Liao JC (2000) Directed evolution of metabolically engineered *Escherichia coli* for carotenoid production. Biotechnol Prog 16(6):922–6

Wang Q, Wu C, Chen T et al. (2006) Expression of galactose permease and pyruvate carboxylase in *Escherichia coli ptsG* mutant increases the growth rate and succinate yield under anaerobic conditions. Biotechnol Lett 28(2):89–93

Weckbecker A, Hummel W (2004) Improved synthesis of chiral alcohols with *Escherichia coli* cells co-expressing pyridine nucleotide transhydrogenase, NADP+-dependent alcohol dehydrogenase and NAD+-dependent formate dehydrogenase. Biotechnol Lett 26(22):1739–44

Wong MS, Wu S, Causey TB et al. (2008) Reduction of acetate accumulation in *Escherichia coli* cultures for increased recombinant protein production. Metab Eng 10(2):97–108

Yamamoto H, Mitsuhashi K, Kimoto N et al. (2005) Robust NADH-regenerator: improved alpha-haloketone-resistant formate dehydrogenase. Appl Microbiol Biotechnol 67(1):33–9

Yang YT, Aristidou AA, San KY et al. (1999) Metabolic flux analysis of *Escherichia coli* deficient in the acetate production pathway and expressing the *Bacillus subtilis* acetolactate synthase. Metab Eng 1(1):26–34

Yi J, Draths KM, Li K et al. (2003) Altered glucose transport and shikimate pathway product yields in *E. coli*. Biotechnol Prog 19(5):1450–9

Yoon SH, Kim JE, Lee SH et al. (2007a) Engineering the lycopene synthetic pathway in *E. coli* by comparison of the carotenoid genes of *Pantoea agglomerans* and *Pantoea ananatis*. Appl Microbiol Biotechnol 74(1):131–9

Yoon SH, Lee YM, Kim JE et al. (2006) Enhanced lycopene production in *Escherichia coli* engineered to synthesize isopentenyl diphosphate and dimethylallyl diphosphate from mevalonate. Biotechnol Bioeng 94(6):1025–32

Yoon SH, Park HM, Kim JE et al. (2007b) Increased beta-carotene production in recombinant *Escherichia coli* harboring an engineered isoprenoid precursor pathway with mevalonate addition. Biotechnol Prog 23(3):599–605

Yoshida A, Nishimura T, Kawaguchi H et al. (2005) Enhanced hydrogen production from formic acid by formate hydrogen lyase-overexpressing *Escherichia coli* strains. Appl Environ Microbiol 71(11):6762–8

Yoshida A, Nishimura T, Kawaguchi H et al. (2007) Efficient induction of formate hydrogen lyase of aerobically grown *Escherichia coli* in a three-step biohydrogen production process. Appl Microbiol Biotechnol 74(4):754–60

Yun NR, San KY, Bennett GN (2005) Enhancement of lactate and succinate formation in *adhE* or *pta-ackA* mutants of NADH dehydrogenase-deficient *Escherichia coli*. J Appl Microbiol 99(6):1404–12

Zelic B, Bolf N, Vasic-Racki D (2006) Modeling of the pyruvate production with *Escherichia coli*: comparison of mechanistic and neural networks-based models. Bioprocess Biosyst Eng 29(1):39–47

Zelic B, Gostovic S, Vuorilehto K et al. (2004a) Process strategies to enhance pyruvate production with recombinant *Escherichia coli*: from repetitive fed-batch to in situ product recovery with fully integrated electrodialysis. Biotechnol Bioeng 85(6):638–46

Zelic B, Vasic-Racki D, Wandrey C et al. (2004b) Modeling of the pyruvate production with *Escherichia coli* in a fed-batch bioreactor. Bioprocess Biosyst Eng 26(4):249–58

Zhang W, Wong KK, Magliozzo RS et al. (2001) Inactivation of pyruvate formate-lyase by dioxygen: defining the mechanistic interplay of glycine 734 and cysteine 419 by rapid freeze-quench EPR. Biochemistry 40(13):4123–30

Zhou S, Causey TB, Hasona A et al. (2003a) Production of optically pure D-lactic acid in mineral salts medium by metabolically engineered *Escherichia coli* W3110. Appl Environ Microbiol 69(1):399–407

Zhou S, Iverson AG, Grayburn WS (2008) Engineering a native homoethanol pathway in *Escherichia coli* B for ethanol production. Biotechnol Lett 30(2):335–42

Zhou S, Shanmugam KT, Ingram LO (2003b) Functional replacement of the *Escherichia coli* D-(−)-lactate dehydrogenase gene (*ldhA*) with the L-(+)-lactate dehydrogenase gene (*ldhL*) from *Pediococcus acidilactici*. Appl Environ Microbiol 69(4):2237–44

Zhou S, Yomano LP, Shanmugam KT et al. (2005) Fermentation of 10% (w/v) sugar to D: (−)-lactate by engineered *Escherichia coli* B. Biotechnol Lett 27(23–24):1891–6

Zhu J, Shalel-Levanon S, Bennett G et al. (2006) Effect of the global redox sensing/regulation networks on *Escherichia coli* and metabolic flux distribution based on C-13 labeling experiments. Metab Eng 8(6):619–27

Zhu J, Shalel-Levanon S, Bennett G et al. (2007a) The YfiD protein contributes to the pyruvate formate-lyase flux in an *Escherichia coli arcA* mutant strain. Biotechnol Bioeng 97(1):138–43

Zhu Y, Eiteman MA, DeWitt K et al. (2007b) Homolactate fermentation by metabolically engineered *Escherichia coli* strains. Appl Environ Microbiol 73(2):456–64

Chapter 18
Glucose and Acetate Metabolism in *E. coli* – System Level Analysis and Biotechnological Applications in Protein Production Processes

Joseph Shiloach and Ursula Rinas

Contents

Abstract *Escherichia coli* is the main bacterial producer of heterologous proteins. The current production strategies aim at growing the bacteria to high density in order to achieve high levels of desired proteins. The major obstacle for reaching high cell densities with high product titers is the tendency of the bacteria to accumulate acetate during the unrestricted growth on glucose. Moreover, the high demand for precursors and energy required for the biosynthesis of the heterologous protein causes the cells to readjust their anabolic and catabolic reactions which, most often, aggravate the acetate problem. Implementing fed-batch protocols and employing more robust strains, such as *E. coli* B instead of K, can reduce acetate formation. Another approach is to implement metabolic engineering to minimize acetate formation by: (a) turning off genes which directly lead to the formation of acetate, (b) introducing

J. Shiloach (✉)
Biotechnology Unit, NIDDK, NIH Bldg 14A, Room 173, Bethesda, MD 20892, USA
e-mail: ljs@helix.nih.gov

S.Y. Lee (ed.), *Systems Biology and Biotechnology of Escherichia coli*,
DOI 10.1007/978-1-4020-9394-4_18, © Springer Science+Business Media B.V. 2009

genes that channel the carbon flow away from acetate towards other pathways, and (c) by reducing the glucose uptake through deleting or replacing genes of the sugar uptake system. Results show that a more general approach that focuses on global regulators and/or gene sets, encoding multiple pathways will be required to construct a robust strain capable of efficiently executing the production of recombinant proteins at high growth rates without the formation of toxic byproducts such as acetic acid.

18.1 Introduction

Escherichia coli is the major bacterial platform for producing heterologous proteins, which is usually done by growing the recombinant microorganism to high density on glucose as the carbon source. This topic has been the subject of numerous studies since the early 1970s, exploring the limits of bacterial culture density in order to achieve maximum productivity. Research strategies have focused on improving cultivation conditions, process related approaches and manipulation of the bacteria's physiology. The developed growth strategies, together with optimization of media composition, application of fed-batch and dialysis culture techniques have made it possible to grow *E. coli* to cell densities of up to 190 g/L dry cell mass (Shiloach and Fass 2005). High-cell density culture techniques have been successfully employed for large-scale production of recombinant proteins with high yield and high productivities (Choi et al. 2006).

The biosynthesis process exposes the bacteria to metabolic stress which is being reflected in the operation of their Central Carbon Metabolism and is associated with higher acetate production (Dittrich et al. 2005a, Tao et al. 1999). Acetate accumulation is considered an obstacle to enhanced recombinant protein production; it is also considered as one of the factors responsible for the reduced biomass yield in large scale fed-batch cultivation (Enfors et al. 2001, Phue and Shiloach 2005). Research is currently being directed to understand this behavior and the response of various *E. coli* strains to the growth conditions during the cultivation and production process. Few factors, among them, the overloading of the TCA cycle and limitations around the pyruvate node as well as local pockets of oxygen limitations in large-scale cultures, are considered to be the main reason for this phenomenon; therefore, concentrated effort is being directed to prevent the overloading and the related processes and to allow the uninterrupted oxidation of the carbon source. This chapter describes the operation of the Central Carbon Metabolism, possible explanations for the acetate production and ways for its reduction, in particular during the recombinant protein production process. It includes the following sections: a general description of the Central Carbon Metabolism of *E. coli*, a review of acetate production and consumption, a Systems Biology approach to the Central Carbon Metabolism in *E. coli K* (JM109) and *B* (BL21), the effect of recombinant protein production on glucose catabolism, and metabolic engineering approaches to overcome bottlenecks in primary metabolism.

18.2 The Central Carbon Metabolism in *E. coli* – General Description

The Central Carbon Metabolism of *E. coli* in general and specifically the glucose metabolism are well-known, well-studied and well-characterized topics (EcoCyc 2008, EcoSal ASM 2008, KEGG 2008, Nelson and Cox 2003). This metabolism can be described by several interconnected metabolic pathways as seen in Fig. 18.1.

The major pathways are glycolysis (Embden-Meyerhof-Parnas EMP), TCA cycle, glyoxylate shunt, pentose-phosphate pathway, anaplerotic reactions, acetate production, and acetate assimilation.

Glucose assimilation starts with its uptake into the cell via the phosphotransferase system (PTS). D-Glucose is transported by PTS and ultimately enters the cell as glucose-6-phosphate with the concomitant consumption of phosphoenolpyruvate (PEP) and the release of pyruvate. Although the PTS is the dominant transport system it is important to mention that there are alternative high-affinity glucose transport systems (e.g. *mgl*) which are activated at low glucose concentrations (Death

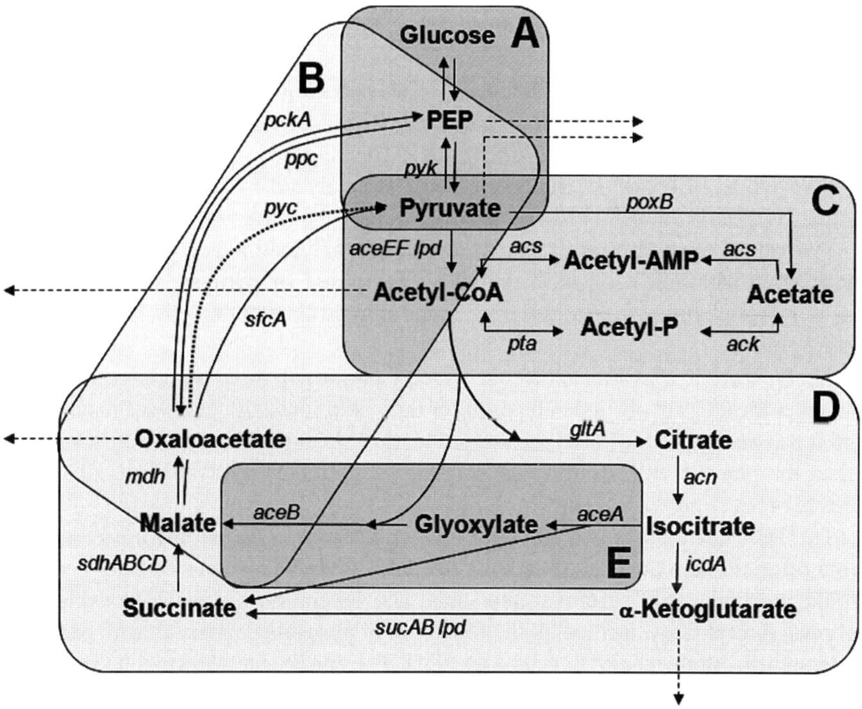

Fig. 18.1 Simplified view of the Central Carbon Metabolism of *E. coli* comprising (**A**) glycolysis and gluconeogenesis, (**B**) anaplerotic reactions, (**C**) acetate formation and assimilation, (**D**) TCA cycle, and **E**. glyoxylate shunt. Arrows with broken lines indicate removel of metabolites for biosynthesis. The arrow with the dotted line indicates an anaplerotic reaction catalysed by pyruvate carboxylase (an enzyme not present in wildtype *E. coli*)

and Ferenci 1994, Franchini and Egli 2006, Wick et al. 2001). Glucose transport can also occur via the galactose-proton-symport system (*galP*)(Chen et al. 1997). In these cases, phosphorylation of glucose is carried out by the cytoplasmic enzyme glucokinase (*glk*) (Meyer et al. 1997)

The glucose-6-phosphate can then be directed into three different routes: it can enter the glycolytic pathway through conversion to fructose-6-phosphate via the phosphoglucose isomerase (*pgi*) reaction, it can enter the oxidative branch of the pentose-phosphate pathway (*zwf*), or it can be converted by phosphoglucomutase (*pgm*) to glucose-1-phosphate for sugar nucleotide synthesis. When entering the glycolytic pathway, the fructose-6-phosphate is converted to fructose-1,6-bisphosphate by 6-phosphofructokinase (*pfkA* and *pfkB*), and then undergoes reversible aldol condensation by fructose bisphosphate aldolase (*fba*) to glyceraldehyde-3-phosphate and dihydroxyacetone-phosphate. The continuation of this pathway is the interconversion of glyceraldehyde-3-phosphate and dihydroxyacetone-phosphate by triosephosphate isomerase (*tpiA*), followed by oxidative phosphorylation of glyceraldhyde-3-phosphate to 1,3-bisphosphoglycerate by glyceraldehyde-3-phosphate dehydrogenase (*gapA*) and the synthesis of ATP by phosphogylcerate kinase (*pgk*) producing 3-phophogylcerate. The two evolutionarily unrelated phosphoglycerate mutases (*gpmA*, *gpmM*) convert the 3-phosphoglycerate to 2-phosphoglycerate. Enolase (*eno*) catalyzes the dehydration of 2-phosphoglycerate to phosphoenolpyruvate (PEP), which contains a high-energy phosphate group that is used both for ATP synthesis and glucose transport by PTS. PEP is converted to pyruvate by two distinct pyruvate kinases (*pykF* and *pykA*). The reverse reaction, the conversion of pyruvate to PEP during gluconeogenesis is catalyzed by phosphoenolpyruvate synthase (*pps*).

Pyruvate is the end-product of glycolysis; it is oxidized to acetyl-CoA and CO_2 by the pyruvate dehydrogenase complex (composed of pyruvate dehydrogenase E1, dihydrolipoamide transacetylase E2, and dihydrolipoamide dehydrogenase E3; *aceEF*, *lpd*).

Acetyl-CoA is a pivotal molecule which can participate is several reactions: it can enter the TCA cycle, it can be used for fatty acids and triglycerides biosynthesis and it can be diverted towards acetate production. Accumulation of acetyl-CoA can affect the glucose utilization by causing accumulation of pyruvate and enhancing acetate production.

The TCA cycle plays two essential roles in the carbon metabolism: it is responsible for the total oxidation of acetyl-CoA, and serves as a source of intermediates for the biosynthesis of several amino acids. The general flow of the TCA cycle is as follows: Acetyl-CoA, formed by the oxidation of pyruvate condenses with oxaloacetate to form citrate by citrate synthase (*gltA*). The citrate is transformed to isocitrate by two genetically distinct aconitases (*acnA* and *acnB*). Next, isocitrate dehydrogenase (*icdA*) performs the oxidative decarboxylation of isocitrate to α-ketoglutarate with generation of NADPH. Following is another oxidative decarboxylation step in which α-ketoglutarate is converted to succinyl-CoA and CO_2 by the α-ketoglutarate dehydrogenase complex (*sucAB*, *lpd*). The succinyl-CoA is converted to succinate by succinyl-CoA synthetase (or succinate thiokinase; *sucCD*). Succinate is oxidized to fumarate by succinate dehydrogenase (*sdhABCD*); the fumarate is reversibly hydrated to malate by three distinct fumarases (*fumA*, *fumB*, *fumC*). In the last reaction of the citric acid cycle, NAD-linked L-malate dehydrogenase (*mdh*) catalyzes the oxidation of malate to oxaloacetate.

The TCA cycle is interconnected to the glyoxylate shunt which is essential for growth on carbon sources such as acetate or fatty acids. This pathway allows the net conversion of acetyl-CoA to metabolic intermediates. In the glyoxylate shunt, isocitrate is cleaved by isocitrate lyase (*aceA*), forming succinate and glyoxylate. Isocitrate lyase competes with the TCA cycle enzyme isocitrate dehydrogenase (*icdA*) for isocitrate. The bifunctional enzyme isocitrate dehydrogenase kinase/phosphatase (*aceK*) regulates the activity of isocitrate dehydrogenase to allow isocitrate lyase to effectively compete for isocitrate. The formed glyoxylate condenses with a second molecule of acetyl-CoA to yield malate in a reaction catalyzed by malate synthase (*aceB*). The malate is subsequently oxidized to oxaloacetate, which can condense with another molecule of acetyl-CoA to start another turn of the TCA cycle. The operation of the TCA cycle can be affected by the removal of the cycle intermediates for biosynthesis of various cell compounds; this can cause accumulation of acetyl-CoA that potentially can affect also the activity of the glycolytic pathway. On the other hand, an active glyoxylate shunt can reduce the accumulation of acetyl-CoA and eliminate interference with both glycolysis and TCA cycle activities.

The role of the anaplerotic reactions is to replace intermediates. These reactions are considered part of the Central Carbon Metabolism and they include: the conversion of PEP to oxaloacetate by PEP carboxylase (PPC shunt, *ppc*), the conversion of oxaloacetate to PEP by PEP carboxykinase (*pck*), the conversion of pyruvate to oxaloacetate by pyruvate carboxylase (not present in wildtype *E. coli*; *pyc*), and the conversion of malate to pyruvate (and *vice versa*) by the malic enzyme (*sfcA*). The glucoenogensis, the conversion of pyruvate to glucose, can also be considered as anaplerotic reaction in which the organism converts excess glucose to glycogen.

Acetate production and assimilation are also part of the Central Carbon Metabolism. Acetate is produced from pyruvate and acetyl-CoA and consumed by conversion back to acetyl CoA. This component of the Central Carbon Metabolism includes the following reactions: acetate production from pyruvate by pyruvate oxidase B (*poxB*), acetate production from acetyl-CoA via acetyl phosphate by phosphotransacetylase (*pta*) and acetate kinase (*ack*), acetate consumption through acetyl-AMP by acetyl CoA synthetase (*acs*), and by the reverse action of acetate kinase (*ack*) and phosphotransacetylase (*pta*).

Lastly, the Central Carbon Metabolism also includes the pentose-phosphate (PP) pathway. The PP pathway serves several metabolic functions which include catabolism of pentoses, glucose, and gluconate, synthesis of pentoses, and providing precursors used in the biosynthesis of lipopolysaccharide, nucleotides, several amino acids and vitamins. This pathway includes two branches: oxidative and nonoxidative. In the oxidative branch, glucose-6-phosphate (G6P) is first oxidized by glucose-6-phosphate dehydrogenase (*zwf*) and then further converted by a series of enzymes to ribulose-5-phosphate (Ru5P) and CO_2. Two molecules of NADP are reduced in the dehydrogenase reactions of this process and can be used for reductive biosynthesis, maintenance of redox balance, and regeneration of oxidative damage. The nonoxidative branch of the pathway comprises reversible reactions that perform the interconversion of the pentose phosphates ribulose-5-phosphate (Ru5P), ribose-5-phosphate (R5P), and xylulose-5-phosphate (Xu5P), and the transfer of either a glycoaldehyde group (transketolase) or a dihydroxyacetone group (transaldolase) among sugar phosphates.

18.3 Acetate Production and Consumption

As was mentioned in the previous section, acetate occupies an important place in the Central Carbon Metabolism of *E. coli*. Acetate accumulation can affect both the bacterial growth and the production of recombinant protein, and serves as an indicator that something went "wrong" in the glucose assimilation process. Acetate accumulation phenomenon has been reviewed comprehensively in the last few years (De Mey et al. 2007, Eiteman and Altman 2006, Shiloach and Fass 2005, Wolfe 2005) and, therefore, it will be described here with an emphasis on how several pathways affect acetate concentration and the possible role of few global controllers.

A genome-scale analysis of the integrated metabolic and transcriptional regulatory networks of *E. coli* shows that the genetic regulatory network responds primarily to the available electron acceptor and to the presence of glucose as the carbon source (Barrett et al. 2005). When carbon flux into the cells exceeds the amphibolic capacity of the central pathways, the flux is diverted to acetate excretion which diminishes the efficiency of carbon conversion to biomass (El-Mansi and Holms 1989). Inverse flux analysis has been used to predict the flux distribution based on the stoichiometries of the reactions in the metabolic network. This approach has been also applied to analyze acetate excretion in aerobic *E. coli* cultures (Delgado and Liao 1997, Farmer and Liao 1997). The results suggest that the anaplerotic pathways, including the reactions catalyzed by PEP carboxylase (*ppc*) and the glyoxylate shunt, are the most likely factors affecting acetate excretion in *E. coli*.

Similar to other metabolites, acetate concentration is the result of production and consumption. The main route for acetate production is from acetyl-CoA through acetyl-phosphate by the two enzymes: phosphotransacetylase (*pta*) and acetatekinase (*ack*). Another minor route for acetate production is directly from pyruvate by pyruvate oxidase B (*poxB*). Although the function of pyruvate oxidase B is not fully understood it is clear that it contributes significantly to aerobic growth efficiency (Abdel-Hamid et al. 2001, Flores et al. 2004a). Any reaction that affects the concentration of acetyl-CoA and pyruvate will, in turn, affect acetate production and hence concentration. Acetyl-CoA concentration is the result of production from glucose through the glycolytic pathway by the conversion of PEP to pyruvate by the reversible enzyme pyruvate kinase (*pyk*), and the conversion of pyruvate to acetyl-CoA by the irreversible reaction catalyzed by the pyruvate dehydrogenase complex (*aceEF*, *lpd*). Acetyl-CoA concentration is also affected by its consumption through the TCA cycle and its consumption for fatty acid biosynthesis. The anaplerotic enzyme PEP carboxylase (*ppc*) converts PEP to oxaloacetate, reducing acetyl-CoA accumulation as a result of higher turnover of the cycle. Anaplerotic reactions can also reduce the acetyl-CoA concentration by lowering the pyruvate concentration through the conversion of PEP back to glucose and glycogen accumulation.

Acetate formation is also affected by the NADH/NAD ratio. Vemuri et al. (2006a) showed that several genes involved in the TCA cycle and respiration are repressed as the glucose consumption rate increases. Deletion of the gene coding for the regulatory protein ArcA (*arcA*) resulted in acetate reduction and increased the biomass yield due to the increased capacities of the TCA cycle and respiratory chain. Acetate formation was completely eliminated by reducing the redox ratio through expression of NADH oxidase (from *Streptococcus pneumonia*) in an *arcA* mutant, even at a very high glucose consumption rate (Vemuri et al. 2006b). NADH and NADPH can be converted into each other through reversible transfer of reducing equivalents between NAD and NADP. The pentose-phosphate pathway and isocitrate dehydrogenase (*icdA*) catalyzed reaction are generally considered as the major sources of the anabolic reductant NADPH which can be converted by the two native *E. coli* transhydrogenases (*pntAB* and *udhA*) into NADH and *vice versa* (Boonstra et al. 1999, Hoffmann et al. 2002, Sauer et al. 2004). Both transhydrogenases have divergent physiological functions: energy-dependent reduction of NADP with NADH by PntAB (Rydstrom 1977, Sauer et al. 2004), and reoxidation of NADPH by UdhA (Boonstra et al. 1999, Hoffmann et al. 2002, Sauer et al. 2004) thus providing *E. coli* primary metabolism with a high flexibility to cope with changing catabolic and anabolic demands.

Acetate assimilation is done by the enzyme acetyl-CoA-synthetase (*acs*) that converts acetate to acetyl-CoA through the intermediate acetyl-AMP. This route is being utilized when acetate is the carbon source and when there is a need to reabsorb the acetate formed when the bacteria grow at high rate, the latter is also known as the "acetate switch"(Wolfe 2005). The components of this switch are phosphotransacetylase (*pta*), acetate kinase (*ack*), and acetyl CoA synthetase (*acs*). This switching behavior is essential for alternating between periods of rapid growth in the presence of abundant nutrients and growth periods where these nutrients are in short supply. Kumari and co-workers (2000) showed that this switch occurs primarily through the induction of *acs*, and that the timing and magnitude of this induction depend, in part, on the direct action of the carbon regulator cyclic AMP receptor protein, synonym: cAMP-catabolite activator protein (*crp* or *cap*) and the aerobic/anaerobic transcriptional regulator (*fnr*). It also depends, probably indirectly, upon the glyoxylate shunt repressor (*iclR*), and its activator the transcriptional regulator of fatty acid metabolism (*fadR*). During aerobic growth of *E. coli* on acetate, phosphotransacetylase (*pta*) and the α-ketoglutarate dehydrogenase complex (*sucAB*, *lpd*) are in direct competition for their common co-factor, HS-CoA. This competition can create a bottleneck at the level of α-ketoglutarate dehydrogenase in the TCA cycle. Addition of pyruvate, glucose or any glycolytic intermediate to acetate-grown cells relieves the bottleneck by reversing the carbon flow through phosphotransacetylase to supply acetyl-phosphate and much-needed HS-CoA (El-Mansi 2005). Growth of *E. coli* on acetate as the sole source of carbon and energy requires operation of the glyoxylate shunt in connection with the expression of the polycistronic *ace* operon (Cortay et al. 1989). Expression of the *aceK* gene is essential for growth on acetate (El-Mansi et al. 1987). The competition at the junction of isocitrate between isocitrate lyase (*aceA*) and isocitrate dehydrogenase (*icdA*) is resolved by the reversible phosphorylation/inactivation of isocitrate dehydrogenase and the operation of the glyoxylate bypass, the expression of which is subject to regulation at the transcriptional and translational levels as well as being dependent on growth rate (El-Mansi et al. 2006). The adaptation to acetate is connected to complex metabolic changes and alterations in gene expression in *E. coli* (Kirkpatrick et al. 2001, Oh et al. 2002, Rosenthal et al. 2008). For example, growth on acetate also induces expression of genes encoding malic enzymes (*maeA*, *sfcA*) and phosphoenolpyruvate synthase (*pps*) while causing repression of glycolytic and glucose phosphotransferase genes (Oh and Liao 2000).

Another mechanism that can affect acetate concentration is the carbon catabolite repression. It allows *E. coli* to alter its metabolism in response to the availability of specific sugar sources. The cAMP-catabolite activator protein (*cap*) complex regulates a number of *E. coli* genes involved in carbon metabolism (Krin et al. 2003). Kao et al. (2005) demonstrated that the gluconeogenic genes in *E. coli* provide a feedback loop to this global regulator in carbon source transition. PTS also plays a role in the carbon catabolite repression; inactivation of PTS components has been applied successfully as a strategy to abolish carbon catabolite repression, resulting in *E. coli* strains that use sugar mixtures more efficiently, such as those obtained from lignocellulosic hydrolysates (Gosset 2005).

A cAMP-independent catabolite repression mechanism found in *E. coli* involves the catabolite repressor/activator (*cra*), which formerly was designated as the fructose repressor (*FruR*), a pleiotropic transcriptional regulatory protein that controls the direction of carbon flux through metabolic pathways (Ramseier et al. 1993). When catabolites bind to Cra, it dissociates from the DNA, causing both catabolite activation and catabolite repression (Saier 1996). Cra controls the expression of genes encoding key enzymes of major pathways of carbon metabolism (Ramseier et al. 1995). Cra exerts a negative effect on the expression of genes encoding glycolytic and Entner-Doudoroff enzymes, while exerting a positive effect on genes encoding the TCA cycle, the glyoxylate shunt and gluconeogenic enzymes (Bledig et al. 1996, Kaga et al. 2002, Negre et al. 1996, Ow et al. 2007).

Based on the above information, several genetic modifications were implemented with an effort to reduce acetate accumulation. These methods are described in more details in the final section of this chapter.

18.4 Systems Biology Approach to the Central Carbon Metabolism in *E. coli* K (JM109) and B (BL21)

E. coli B (BL21) and *E. coli* K (JM109) respond differently to glucose concentration in their growth media, especially when the glucose concentration is 10 grams per liter or more. *E. coli* B is not sensitive to the high glucose concentration, its growth is not affected, and there is very low acetate accumulation. In contrast, *E. coli* K is sensitive to the high glucose concentrations, produces elevated levels of acetate and grows at a slower rate (Shiloach et al. 1996). Investigation of the difference between these two strains can serve as an excellent tool for understanding the regulation and control of the Central Carbon Metabolism when utilizing glucose as sole carbon substrate.

The traditional approach for evaluating and understanding the regulation and the operation of the Central Carbon Metabolism was to concentrate on specific enzymes and genes and sometimes on a specific pathway. More information on the interrelationship between the various pathways will allow a better understanding of the processes controlling glucose utilization, reducing acetate production and improving growth and recombinant protein production.

With the development of new methodologies - especially the high-throughput measurements of DNA, RNA and proteins, and the new mathematical modeling and algorithms - it is possible to examine simultaneously various pathways and to have a sense of the regulation and the operation from a broader perspective. The term for this global approach is Systems Biology (Barrett et al. 2005, Kitano 2002). One of the popular definitions of Systems Biology is the investigation of complex biological processes in a way that aims to understand how individual molecular components combine on a global scale to yield particular structure function relationships and behave in response to specific perturbations. Attempts to utilize global understanding, although in a rather limited way, were implemented long before the term Systems Biology was coined in 2002. The continuing research of the difference between the Central Carbon Metabolism of *E. coli* K and B can serve as an example for implementing the System Biology approach. During the past years several methodologies were implemented to evaluate the relationship between the various pathways of the Central Carbon Metabolism, and the overall response of the system to glucose and acetate concentrations. With the introduction of each new method additional information was obtained and more details became available. Although no powerful mathematical approaches, which currently are part of the Systems Biology, were implemented, better understanding of the metabolism was achieved.

Several factors could be responsible for the different behavior of *E. coli* B compared to *E. coli* K: reduced glucose transport into the cell, increased respiration/O_2 transfer rate, decreased flux from pyruvate to acetate, increased anaplerotic flux from PEP to oxaloacetate, increased flux through the glyoxylate shunt and increased TCA cycle flux. The initial assumption was that perhaps the glyoxylate shunt is fully operational in *E. coli* B and may not be operational or operates at a low rate in *E. coli* K. This assumption had to be proven.

The first attempt to look at the glyoxylate shunt question from a "Systems Biology perspective" was done by metabolic flux analysis together with measuring the concentration and activity of several key metabolites and enzymes, respectively (van de Walle and Shiloach 1998). Metabolic flux analysis (Savinell and Palsson 1992) compares fluxes through specific pathways by using the following assumptions: use known reaction stoichiometries, ignore nonlinearities in kinetics, ignore regulations and assume that the network has been correctly drawn. In this particular case of evaluating the flux through the glyxoylate shunt, the calculations were based on measuring specific glucose uptake rate, specific acetate production rate, growth rate, CO_2 production rate, O_2 uptake rate, cell monomers content, and assuming an ATP/oxygen (P/O) ratio of 1.33 and pseudo steady state concentrations for the intracellular metabolites (van de Walle and Shiloach 1998). The purpose was to determine the flux through the TCA cycle and the glyoxylate shunt; however, because of the singularity of this methodology (Vallino and Stephanopoulos 1990), it was impossible to calculate simultaneously the flux of the two pathways and to receive a direct answer. For *E. coli* B we were able to determine independently the flux through the TCA cycle and the flux through the glyoxylate shunt. But for *E. coli* K only the flux through the TCA cycle could be calculated as this bacterium has only negligible amounts of isocitrate lyase (*aceA*). The results of this initial phase can be summarized as follows: the flux through isocitrate dehydrogenase (*icdA*) was higher in *E. coli* B than in *E. coli* K, isocitrate dehydrogenase was highly active in B and the flux to acetate through the acetate kinase-phosphotransacetylase system (*pta-ack*) was higher in K (van de Walle and Shiloach 1998). In addition, *E. coli* B had a higher internal isocitrate concentration and a lower pyruvate concentration (van de Walle and Shiloach 1998).

The second attempt was to measure simultaneously the flux through the glyoxylate shunt and the TCA cycle using ^{13}C labeled glucose (Noronha et al. 2000). This was done by measuring the distribution of the ^{13}C isotopomers of oxaloacetate and acetyl-CoA. It was concluded that in *E. coli* B, the glyoxylate shunt is active at 22% of the flux through the TCA cycle and is inactive in K. Additionally, in *E. coli* B the flux through the TCA cycle equals the flux through the PPC shunt, while in *E. coli* K the flux of the TCA cycle is only third of the flux through the PPC (Noronha et al. 2000).

The third attempt in utilizing the "Systems Biology perspective" to gain better understanding of the Central Carbon Metabolism was made possible due to the deciphering of the *E. coli* genome and the availability of DNA microarray technology (Blattner et al. 1997, Richmond et al. 1999). By using Northern blots and DNA microarrys, it was possible to simultaneously follow the transcription of genes which are part of several metabolic pathways, and to identify the activated pathways at different growth conditions (Phue et al. 2005, Phue and Shiloach 2004). Although this method allowed the identification of up-regulated and down-regulated genes, it did not provide information on the flux through the various pathways. The results of this study are shown in Fig. 18.2a and b.

In *E. coli* B, the various pathways of the Central Carbon Metabolism are activated whether the glucose concentration is low or high, at both concentrations the tested pathways operate similarly. In contrast, *E. coli* K was responding differently to the various glucose concentrations; its gene activity profile was similar to *E. coli* B only at a low glucose concentration.

The latest step in this effort was done by comparing the transcription level of a group of genes that compose specific metabolic pathways by the semiparametric algorithm using oligo-microarrays (Phue et al. 2007). It was found that as a group,

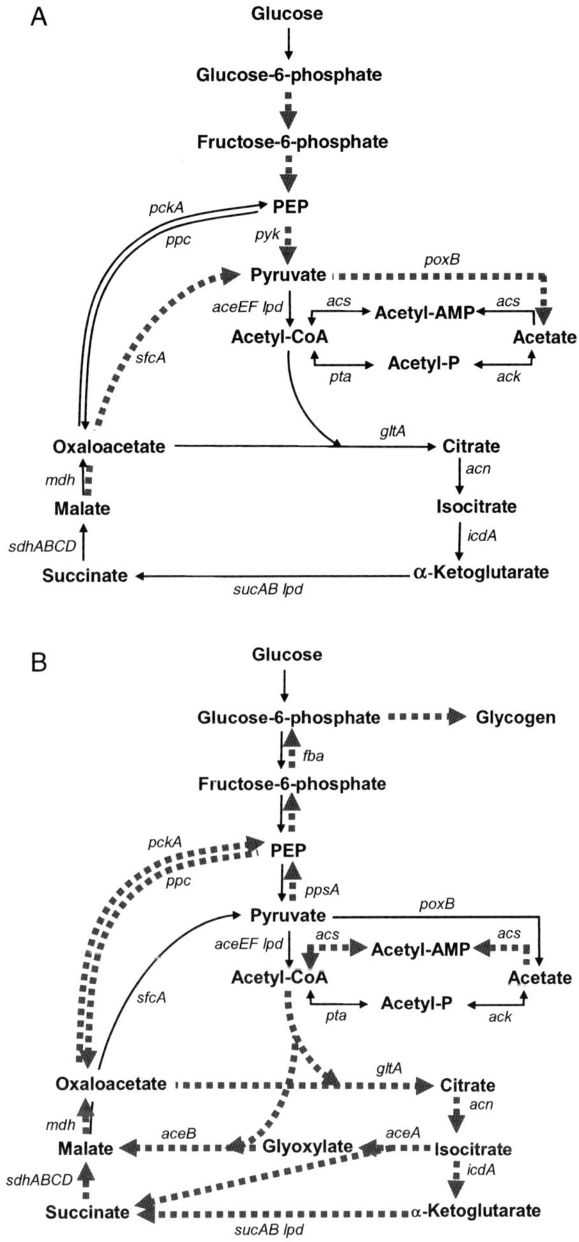

Fig. 18.2 Proposed glucose metabolism during growth of (**a**) *E. coli* K12 (JM101) and (**b**) *E. coli* B (BL21) under glucose excess conditions. The red arrows with broken lines indicate the activated pathways utilized by the different strains

the following pathways were transcribed differently in the two strains: glyoxylate shunt, TCA cycle, fatty acids biosynthesis, gluconeogensis, and anaplerotic pathways. There was no difference between the groups comprising transcription of either glycolysis or the pentose-phosphate pathway genes. This finding confirmed the previous observation that the difference is not the result of a single gene but most likely the effect of one or more global controllers that influence the transcription of complete pathways.

With the information available so far, it is possible to have some explanation why *E. coli* B is producing less acetate when being exposed to high glucose concentration, and why it is utilizing glucose more efficiently than *E. coli* K. But it is impossible to point out why this is happening, why the TCA cycle flux in *E. coli* B is higher than in *E. coli* K, why the glyoxylate shunt is inactive at high glucose concentration in *E. coli* K and why the gluconeogenesis is active in *E. coli* B and not in *E. coli* K. The expectation is that high glucose concentration should activate the glyoxylate shunt; increase the TCA cycle activity, increase the acetate uptake and reduce the acetate concentration. All these actions are observed in *E. coli* B regardless of glucose or acetate concentration; it is puzzling that there is no activation in *E. coli* K, and there is constant activation in *E. coli* B, especially puzzling is the fact that *poxB* is less active in *E. coli* B. Perhaps, additional global analysis of the Central Carbon Metabolism will provide a better explanation. In the meantime, there are numerous efforts to improve *E. coli* K behavior by modifying the Central Carbon Metabolism. These approaches are described in the last section of this chapter.

18.5 Effect of Recombinant Protein Production on Glucose Catabolism

Escherichia coli is still the most prominent bacterial host for recombinant protein production with glucose as the common carbon substrate in recombinant protein production processes. This process can induce a variety of stress reactions in the bacterial host including flux alterations in primary metabolic pathways (Hoffmann and Rinas 2004). Calculations by Stouthamer revealed that protein synthesis is the most energy consuming process of all anabolic activities (Stouthamer 1977, Stouthamer 1980, Stouthamer 1986). According to these estimations, more than 50% of the ATP required for the formation of microbial cells during growth on defined medium with glucose as sole carbon and energy source is used for the polymerization of amino acids into proteins while only 4% is required for the synthesis of amino acids (Stouthamer 1986). Experiments by Anderson and von Meyenburg (1980) suggested that growth of *E. coli* in aerobic cultures under glucose excess conditions is limited by the rates of both respiration and ATP generation through oxidative phosphorylation. Thus, recombinant protein production might be potentially limited by bottlenecks in the energy-generating pathways. Under conditions of glucose excess, part of the glucose is not used for biomass and energy generation through the respiratory chain and proton motive force, rather is diverted towards the formation of overflow metabolites, mainly acetate, causing a reduction in the efficiency of glucose utilization.

Early experiments with genetically modified *E. coli* strains indicated that extracellular accumulation of acetate (Brown 1985, Jensen and Carlsen 1990, Meyer 1984, Shimizu 1988) or other overflow metabolites like glutamate (Rinas 1989) are associated with reduced yields of the recombinant protein produced. The recent developments in the recombinant DNA technology and the widespread utilization of *E. coli* as a microbial protein production factory stimulated research associated with the understanding and solving of the "acetate problem". Although we still do not completely understand the complexity of acetate formation and are far from a non-acetate producing, metabolically balanced and robust *E. coli* designer strain, there has been progress in overcoming this difficulty and improving the recombinant protein production process (see following section). Early hypothesis suggested that metabolic bottlenecks leading to the formation of acetate are localized at the level of TCA cycle activity and in the respiratory chain (Anderson and von Meyenburg 1980, Majewski and Domach 1990). Therefore, acetate formation would be an alternative way for generating ATP, although at reduced efficiency. In fact, acetate formation is observed under conditions of energetic stress/deficiency when the carbon flux into the cells is bigger than the amphibolic capacity of the central pathways, for example, caused by artificially induced futile cycling (Chao and Liao 1994, Patnaik et al. 1992) or at rapid growth in glucose-limited chemostat cultures (Kayser et al. 2005).

It has not only been shown that recombinant protein synthesis is reduced during acetate accumulation, but also that induction of recombinant protein synthesis can lead to enhanced acetate excretion (Akesson et al. 1999, Wittmann et al. 2007). Also reported has been the enhanced pyruvate excretion as a result of recombinant protein synthesis, suggesting an alteration in the pyruvate oxidation pattern (George et al. 1992). An elevated intracellular pyruvate pool, together with enhanced pyruvate excretion was observed during recombinant protein production under glucose excess conditions using a temperature-inducible expression system (Wittmann et al. 2007). These observations suggest that enhanced acetate formation during recombinant protein production results from limitations around the pyruvate node.

Proteomic analyses of inclusion bodies, composed mainly of the recombinant protein product, revealed that dihydrolipoamide dehydrogenase (*lpd*), the common component of the pyruvate and the α-ketoglutarate dehydrogenase complexes, coaggregates during recombinant protein production. This probably leads to additional aggravation of the limitation around the pyruvate node (Rinas et al. 2007) and at the level of TCA cycle activity where α-ketoglutarate dehydrogenase activity is considered as a major bottleneck (El-Mansi 2004, Rinas 1989). Dihydrolipoamide dehydrogenase might be a critical protein since *lpd* knockout mutants of *E. coli* produced significantly more pyruvate and glutamate under aerobiosis (Li et al. 2006). Moreover, *E. coli* strains with deletion of both acetate producing pathways (*ack-pta* and *poxB*) accumulate pyruvate (Dittrich et al. 2005b). Pyruvate excretion in these strains can be prevented by overexpression of genes encoding the pyruvate dehydrogenase complex (Dittrich et al. 2005b) suggesting this complex enzyme as a potential metabolic engineering target for the generation of low acetate producing strains.

As indicated, recombinant protein synthesis driven by strong promoters is a high-energy consuming process potentially limited by bottlenecks in the energy-generating pathways. An example is recombinant protein production using temperature-inducible expression systems. This process caused an immediate drop of the adenylate energy charge, which serves as an indicator of the energetic status of the cells. This occurs at glucose limiting (Hoffmann et al. 2002) as well as at glucose excess growth conditions (Wittmann et al. 2007). Under glucose excess conditions, protein synthesis, driven by the temperature-inducible lambda promoters, caused enhanced excretion of acetate and other byproducts (Wittmann et al. 2007) while protein synthesis under balanced carbon-limited conditions caused redirection of substantially more glucose into the energy-generating respiratory pathway (Hoffmann and Rinas 2001, Schmidt et al. 1999a). Thus, when recombinant protein synthesis is induced under carbon-limiting balanced growth conditions, which do not lead to the formation of acetate, a greater portion of glucose is diverted to carbon dioxide production compared to non-producing conditions (Hoffmann and Rinas 2001, Schmidt et al. 1999b). In balanced fed-batch conditions, about 40–45% of the glucose carbon is converted to carbon dioxide, which increases to 70% after temperature-induced recombinant protein production (Hoffmann and Rinas 2001). During IPTG-induced protein production in balanced carbon limited fed-batch cultures, the flux towards carbon dioxide formation increased from 44–46% of glucose carbon before induction to 50–52% after the onset of recombinant protein production (Schmidt et al. 1999b). An increased respiratory activity upon induction of recombinant protein synthesis has also been noted for other expression systems (Bhattacharya 1997, Lin and Neubauer 2000). The increase in protein synthesis rates upon induction in balanced carbon-limited fed-batch cultures correlated directly with an increase in respiratory activity (Hoffmann and Rinas 2001) together with enhanced glycolytic and TCA cycle activity and reduced pentose-phosphate pathway flux (Luo et al. 2008, Weber et al. 2002). In contrast to the catabolic response in balanced carbon-limited fed-batch cultures, cells reduce TCA cycle activity upon recombinant protein production under glucose excess in batch culture conditions (Wittmann et al. 2007).

Changes in the respiratory activity in response to recombinant protein production are primarily caused by changes on the level of catabolic enzyme activity and not on the amount of catabolic enzymes as the respiratory response is instantaneous (Hoffmann et al. 2002, Schmidt et al. 1999a,b). The cellular response towards recombinant protein production on the level of transcription and translation of genes encoding catabolic enzymes appears to be complex and very specific with respect to the recombinant protein produced and the conditions of induction. General conclusions are difficult to obtain; the most common observations include down-regulation of transcription of genes involved in energy generation, such as TCA cycle, respiration and AcrA-dependent genes (Durrschmid et al. 2008, Haddadin and Harcum 2005, Harcum and Haddadin 2006, Oh and Liao 2000). Proteome analysis indicated both decrease (Wagner et al. 2007) and increase in synthesis rate, or level of proteins (Durrschmid et al. 2008, Hoffmann et al. 2002, Jurgen et al. 2000) encoded by these genes. Contrasting findings, such as decreased transcript levels of TCA cycle and glyoxylate shunt enzymes associated with increased protein levels, have also been reported (Durrschmid et al. 2008).

Global transcriptome analysis of the cellular response towards recombinant protein production indicated that many genes of the glycolytic pathway (e.g. *fba*, *eno*) and PTS (e.g. *ptsG* and *crr*) were downregulated while the gene encoding glucokinase (*glk*) was strongly upregulated (Haddadin and Harcum 2005, Oh and Liao

2000). A strong upregulation of glucokinase in response to recombinant protein production has been noted not only through increased transcription but also through elevated enzyme levels (Arora and Pedersen 1995), indicating a shift in the utilization of the glucose uptake pathway in response to recombinant protein production. The impairment of glucose uptake during recombinant protein production (Lin et al. 2001, Neubauer et al. 2003) might be reflected by the transition from the utilization of the more common PTS towards alternative pathways for supplying overproducing cells with glucose-6-phosphate. On the other hand, reduced synthesis and leakage of periplasmic binding proteins involved in high-affinity glucose uptake (*mglB*) might also contribute to an impairment of glucose uptake under protein production conditions in high-cell density cultures (Rinas and Hoffmann 2004).

18.6 Metabolic Engineering Approaches to Overcome Bottlenecks in Primary Metabolism

The formation of acetic acid is a disturbing side reaction during rapid growth of *E. coli* on glucose (Luli and Strohl 1990). As a result, efforts have been undertaken to reduce the formation of acetic acid either by process control strategies or by metabolic engineering approaches. When implementing process control approaches, the aim is to reduce glucose uptake rate, generally done by limiting the glucose supply through fed-batch culture techniques (Korz et al. 1995, Lee 1996, Shiloach and Fass 2005). This approach has been successfully applied for recombinant protein production in high-cell density fed-batch cultures leading to recombinant protein levels in the range of 5–10 g L^{-1} with *E. coli* strains having tendency towards acetate formation under glucose excess conditions (Hoffmann and Rinas 2004, Schmidt et al. 1999a, Vallejo et al. 2002).

Metabolic engineering efforts have been implemented to generate strains which produce less acetate in protein production processes (for recent review refer to (Eiteman and Altman 2006)). Three major metabolic routes or combinations thereof have been applied to reduce acetate accumulation; (i) knocking out genes that directly lead to the formation of acetate, (ii) introducing genes that lead to redirection of the carbon flow away from glycolysis and acetate formation towards other pathways and metabolites, and (iii) reducing glucose uptake by deleting or replacing genes of the PTS.

Initial approaches focused on mutation or deletion of enzymes that lead to the formation of acetate, in particular blocking the acetate kinase-phosphotransacetylase (*ack-pta*) pathway (Bauer et al. 1990, Hahm et al. 1994). For example, a phosphotransacetylase mutant selected by classical mutagenesis techniques showed improved protein production properties in bioreactor cultures (Bauer et al. 1990). The downregulation of the acetate-generating pathway that includes the enzymes phosphotransacetylase (*pta*) and acetate kinase (*ack*) by using an antisense RNA strategy also improved recombinant protein production (Kim and Cha 2003). Most of these strains have been tested in laboratory-scale shake flask experiments and did not show the robustness required for industrial application. *E. coli* strains that carry single mutations (e.g. *ack*, *pta*, *acs*, *poxB*) do not exhibit the robustness in high-cell density fed-batch cultures compared to the corresponding control strain (Contiero et al. 2000). Inactivation of the *poxB* gene results in slower growth rates and also leads to a reduced carbon conversion efficiency (percentage carbon flux to biomass)(Abdel-Hamid et al. 2001, Li et al. 2007), probably as a result of the activation of energetically less favorable metabolic pathways such as activation of *glk* and repression of PTS genes *ptsG* and *crr* (Li et al. 2007, Vemuri et al. 2005). The deletion of genes that lead to acetate formation (e.g. *ack, pta, poxB*) results in strains that secrete pyruvate (Chang et al. 1999, Diaz-Ricci et al. 1991, Dittrich et al. 2005b, Tomar et al. 2003) and other unusual by-products such as glutamate (Chang et al. 1999) into the culture medium. Taking advantage of this phenomenon, an *E. coli* strain that was engineered for optimal acetate production (Causey et al. 2003), was transformed into an efficient pyruvate producing strain by simply disrupting two genes that lead to acetate formation (*ack*, *poxB*)(Causey et al. 2004). The reduction of pyruvate formation through inactivation of the pyruvate kinase encoding genes (*pykA* and *pykF*) was also considered as a way to reduce acetate formation. The resulting strains metabolized glucose mainly via the PP pathway (Ponce et al. 1998, Siddiquee et al. 2004) and produced less acetate (Ponce 1999, Zhu et al. 2001), but also exhibited reduced growth rates when grown under glucose excess conditions (Ponce 1999, Ponce et al. 1995, 1998, Zhu et al. 2001).

Results obtained by inverse flux analysis suggested that increased flux through anaplerotic pathways (PPC shunt and glyoxylate bypass) should reduce acetate formation (Delgado and Liao 1997, Farmer and Liao 1997). In fact, deregulation of the glyoxylate bypass by disrupting *fadR*, reduced acetate formation without negatively effecting the growth rate (Farmer and Liao 1997, Peng and Shimizu 2006). Increasing the flux through the PPC shunt by overexpressing PEP carboxylase (*ppc*) further decreased acetate formation without impairment of the growth rate (Farmer and Liao 1997). On the other hand, deletion of *ppc* also reduced acetate formation but at the expense of a slower growth rate and a reduced glucose uptake rate (Peng et al. 2004).

Another approach for reducing acetete formation includes the generation of strains overexpressing heterologous genes that encode anaplerotic enzymes that replenish the TCA cycle, for example pyruvate carboxylase (*pyc*). These strains showed better performance in glucose excess batch culture (March et al. 2002) and also revealed reduced acetate production and higher cell yields in controlled chemostat cultures (Vemuri et al. 2005). Directing excess pyruvate away from acetate towards less toxic products through coexpression of other heterologous enzymes (such as acetolactate synthase (*alsS*) from *Bacillus subtilis* which finally leads to the formation of acetoin instread of acetate) also resulted in strains which produced less acetate and performed better as protein producers in batch and fed-batch cultures (Aristidou et al. 1994, 1995). Overexpression of the glucose-6-phosphate dehydrogenase encoding gene (*zwf*), which leads to an increased flux towards the pentose-phosphate pathway (by decreasing the glycolytic flux), resulted in a better performing production strain under carbon excess conditions (Flores et al. 2004b).

In this line, deletion of *zwf* resulted in elevated glycolytic flux and enhanced excretion of acetate and pyruvate (Hua et al. 2003).

Another approach to reduce formation of acetate and to improve protein production under carbon excess conditions involves the reduction of glucose uptake via the PTS (Backlund et al. 2008, Chou et al. 1994, Gosset 2005, Picon et al. 2005, Ponce 1999, Wong et al. 2008). As glucose uptake via the PTS is connected to the generation of pyruvate from PEP, the reduced pyruvate formation might lead to reduced acetate production. The majority of strains with modifications of the PTS produce less acetate, however, they do it at the cost of reduced growth rates (Backlund et al. 2008, Chou et al. 1994, Flores et al. 2002, Picon et al. 2005, Ponce 1999, Wong et al. 2008). Another approach to reduce acetate formation involves the inactivation of the PTS while forcing glucose transport through the galactose-proton symport system composed of the membrane localized galactopermease (*galP*) with subsequent glucose phosphorylation through cytoplasmic glucokinase (*glk*)(De Anda et al. 2006, Flores et al. 2007, Hernandez-Montalvo et al. 2003, Lara et al. 2008). These strains, when carrying multiple copies of *galP* and *glk* genes, exhibit growth rates similar to the PTS wildtype strains, in particular those with an *arcA* background (Flores et al. 2007, Hernandez-Montalvo et al. 2003), but at the same time exhibit increased acetate production rates compared to the PTS wildtype strains (Hernandez-Montalvo et al. 2003). Reducing the glucose uptake rate by fine-tuned expression of *galP* in a PTS mutant strain, reduced acetate formation, but this was associated with a slower growth rate compared with the PTS wildtype strain (Lara et al. 2008). Comparative studies on acetate formation by single global regulator gene knockout mutants (e.g. *arcA*, *arcB*, *cra*, *crp*, *cya*, *fnr*, and *mlc*) also revealed that reduced acetate formation is always connected to reduced glucose uptake and growth rates and *vice versa* (Perrenoud and Sauer 2005). In this way, avoiding utilization of the PTS by replacing glucose with fructose as carbon source also reduced acetate formation and improved protein production, but again at the cost of reduced growth rate (Aristidou et al. 1999).

In summary reducing acetate formation by metabolic engineering or any other means such as reduced glucose feeding or alternative carbon substrates without shifting the carbon flow to other unwanted by-products (e.g. pyruvate) is mainly achieved by reducing the glucose uptake rate concomitant to growth rate reduction.

To successfully generate robust strains producing significantly less acetate in carbon excess conditions it might be necessary to further consider the impact of global regulators, e.g. ArcA (Flores et al. 2007, Vemuri et al. 2006a,b) or FadR (Farmer and Liao 1997, Peng and Shimizu 2006) and the deletion or the introduction of whole gene sets that encode multiple pathways instead of focusing on single genes or pathways. For example, the reduction of the redox ratio in an $arcA^-$ background through expression of an heterologous NADH oxidase eliminated acetate formation even at high glucose consumption rates (Vemuri et al. 2006a). Moreover, eliminating native transcriptional control of a set of TCA cycle enzymes by chromosomal promoter mutation (*sdhCDAB-B0725-sucABCD*), resulted in a strain which produced less acetate and instead directed more glucose to carbon dioxide while maintaining high growth and glucose consumption rates (Veit et al. 2007).

18.7 Concluding Remarks

Enabling recombinant *E. coli* to grow to high density and to produce proteins without severely affecting its metabolism, its growth characteristics, and its protein biosynthesis capabilities is currently an important research and development topic. Comprehensive analysis of the Central Carbon Metabolism, among other metabolic pathways, is on-going in an effort to identify bottlenecks in the metabolism that might affect the growth and production process. So far, this analytical approach has yielded several targets that alleviate some of the growth constraints, especially improving the glucose oxidation process. It is likely that a System Biology approach that takes into account the relationships not only between the metabolic pathways of the Central Carbon Metabolism but also between others metabolic pathways and their relationship with global regulators and other effectors, may improve our understanding of the bacterial behavior under stress, and will result in improving the growth and production process. Although our knowledge of the *E. coli* metabolism and its regulation covers many aspects, at this point, there is not enough information to predict the possible response to different changes and different conditions. The goal is to have a global physiological overview of the *E. coli* metabolism and, accordingly, to construct a robust strain capable of efficiently executing the production of recombinant proteins.

Acknowledgments Funding was provided by the intramural program at the NIDDK, NIH and HZI. The authors thank S. Noronha and S.M.S. Reshamwala for support in the literature survey.

References

Abdel-Hamid AM, Attwood MM, Guest JR (2001) Pyruvate oxidase contributes to the aerobic growth efficiency of *Escherichia coli*. Microbiology 147(Pt 6):1483–98

Akesson M, Karlsson EN, Hagander P et al. (1999) On-line detection of acetate formation in *Escherichia coli* cultures using dissolved oxygen responses to feed transients. Biotechnol Bioeng 64(5):590–8

Andersen KB, von Meyenburg K (1980) Are growth rates of *Escherichia coli* in batch cultures limited by respiration? J Bacteriol 144(1):114–23

Aristidou AA, San KY, Bennett GN (1994) Modification of central metabolic pathway in *Escherichia coli* to reduce acetate accumulation by heterologous expression of the *Bacillus-Subtilis* acetolactate synthase gene. Biotechnol Bioeng 44(8):944–51

Aristidou AA, San KY, Bennett GN (1995) Metabolic engineering of *Escherichia coli* to enhance recombinant protein-production through acetate reduction. Biotechnol Prog 11(4):475–78

Aristidou AA, San KY, Bennett GN (1999) Improvement of biomass yield and recombinant gene expression in *Escherichia coli* by using fructose as the primary carbon source. Biotechnol Prog 15(1):140–45

Arora KK, Pedersen PL (1995) Glucokinase of *Escherichia coli* - induction in response to the stress of overexpressing foreign proteins. Arch Biochem Biophys 319(2):574–78

Backlund E, Markland K, Larsson G (2008) Cell engineering of *Escherichia coli* allows high cell density accumulation without fed-batch process control. Bioprocess Biosyst Eng 31(1):11–20

Barrett CL, Herring CD, Reed JL et al. (2005) The global transcriptional regulatory network for metabolism in *Escherichia coli* exhibits few dominant functional states. Proc Natl Acad Sci USA 102(52):19103–8
Bauer KA, Ben-Bassat A, Dawson M et al. (1990) Improved expression of human interleukin-2 in high-cell-density fermentor cultures of *Escherichia coli* K-12 by a phosphotransacetylase mutant. Appl Environ Microbiol 56(5):1296–302
Bhattacharya SK, Dubey, AK (1997) Effects of dissolved oxygen and oxygen mass transfer on overexpression of target gene in recombinant *E. coli*. Enzyme Microb Technol 20:355–60
Blattner FR, Plunkett G, 3rd, Bloch CA et al. (1997) The complete genome sequence of *Escherichia coli* K-12. Science 277(5331):1453–74
Bledig SA, Ramseier TM, Saier MH, Jr (1996) FruR mediates catabolite activation of pyruvate kinase (*pyk*F) gene expression in *Escherichia coli*. J Bacteriol 178(1):280–3
Boonstra B, French CE, Wainwright I et al. (1999) The *udh*A gene of *Escherichia coli* encodes a soluble pyridine nucleotide transhydrogenase. J Bacteriol 181(3):1030–4
Brown SW, Meyer, HP, Fiechter, A. (1985) Continuous production of human leukocyte interferon with *Escherichia coli* and continuous cell lysis in a two stage chemostat. Appl Microbiol Biotechnol 23:5–9
Causey TB, Shanmugam KT, Yomano LP et al. (2004) Engineering *Escherichia coli* for efficient conversion of glucose to pyruvate. Proc Natl Acad Sci USA 101(8):2235–40
Causey TB, Zhou S, Shanmugam KT et al. (2003) Engineering the metabolism of *Escherichia coli* W3110 for the conversion of sugar to redox-neutral and oxidized products: Homoacetate production. Proc Natl Acad Sci USA 100(3):825–32
Chang DE, Shin S, Rhee JS et al. (1999) Acetate metabolism in a pta mutant of *Escherichia coli* W3110: importance of maintaining acetyl coenzyme A flux for growth and survival. J Bacteriol 181(21):6656–63
Chao YP, Liao JC (1994) Metabolic responses to substrate futile cycling in *Escherichia coli*. J Biol Chem 269(7):5122–6
Chen R, Yap WM, Postma PW et al. (1997) Comparative studies of *Escherichia coli* strains using different glucose uptake systems: Metabolism and energetics. Biotechnol Bioeng 56(5):583–90
Choi JH, Keum KC, Lee SJ (2006) Production of recombinant proteins by high cell density culture of *Escherichia coli*. Chem Eng Sci 61:876–85
Chou CH, Bennett GN, San KY (1994) Effect of modified glucose-uptake using genetic-engineering techniques on high-level recombinant Protein-Production in *Escherichia coli* Dense Cultures. Biotechnol Bioeng 44(8):952–60
Contiero J, Beatty C, Kumari S et al. (2000) Effects of mutations in acetate metabolism on high-cell density growth of *Escherichia coli*. J Ind Microbiol Biotechnol 24(6):421–30
Cortay JC, Bleicher F, Duclos B et al. (1989) Utilization of acetate in *Escherichia coli*: structural organization and differential expression of the ace operon. Biochimie 71(9–10):1043–9
De Anda R, Lara AR, Hernandez V et al. (2006) Replacement of the glucose phosphotransferase transport system by galactose permease reduces acetate accumulation and improves process performance of *Escherichia coli* for recombinant protein production without impairment of growth rate. Metab Eng 8(3):281–90
De Mey M, De Maeseneire S, Soetaert W et al. (2007) Minimizing acetate formation in *E. coli* fermentations. J Ind Microbiol Biotechnol 34(11):689–700
Death A, Ferenci T (1994) Between feast and famine: endogenous inducer synthesis in the adaptation of *Escherichia coli* to growth with limiting carbohydrates. J Bacteriol 176(16):5101–7
Delgado J, Liao JC (1997) Inverse flux analysis for reduction of acetate excretion in *Escherichia coli*. Biotechnol Prog 13(4):361–7
Diaz-Ricci JC, Regan L, Bailey JE (1991) Effect of alteration of the acetic acid synthesis pathway on the fermentation pattern of *Escherichia coli*. Biotechnol Bioeng 38(11):1318–24
Dittrich CR, Bennett GN, San KY (2005a) Characterization of the acetate-producing pathways in *Escherichia coli*. Biotechnol Prog 21(4):1062–7

Dittrich CR, Vadali RV, Bennett GN et al. (2005b) Redistribution of metabolic fluxes in the central aerobic metabolic pathway of *E. coli* mutant strains with deletion of the *ack*A-*pta* and *pox*B pathways for the synthesis of isoamyl acetate. Biotechnol Prog 21(2):627–31

Durrschmid K, Reischer H, Schmidt-Heck W et al. (2008) Monitoring of transcriptome and proteome profiles to investigate the cellular response of *E. coli* towards recombinant protein expression under defined chemostat conditions. J Biotechnol 135(1):34–44

Eiteman MA, Altman E (2006) Overcoming acetate in *Escherichia coli* recombinant protein fermentations. Trends Biotechnol 24(11):530–6

El-Mansi EM, Holms WH (1989) Control of carbon flux to acetate excretion during growth of *Escherichia coli* in batch and continuous cultures. J Gen Microbiol 135(11):2875–83

El-Mansi EM, MacKintosh C, Duncan K et al. (1987) Molecular cloning and overexpression of the glyoxylate bypass operon from *Escherichia coli* ML308. Biochem J 242(3):661–5

El-Mansi M (2004) Flux to acetate and lactate excretions in industrial fermentations: physiological and biochemical implications. J Ind Microbiol Biotechnol 31(7):295–300

El-Mansi M (2005) Free CoA-mediated regulation of intermediary and central metabolism: an hypothesis which accounts for the excretion of alpha-ketoglutarate during aerobic growth of *Escherichia coli* on acetate. Res Microbiol 156(8):874–9

El-Mansi M, Cozzone AJ, Shiloach J et al. (2006) Control of carbon flux through enzymes of central and intermediary metabolism during growth of *Escherichia coli* on acetate. Curr Opin Microbiol 9(2):173–9

Enfors SO, Jahic M, Rozkov A et al. (2001) Physiological responses to mixing in large scale bioreactors. J Biotechnol 85(2):175–85

Farmer WR, Liao JC (1997) Reduction of aerobic acetate production by *Escherichia coli*. Appl Environ Microbiol 63(8):3205–10

Flores N, de Anda R, Flores S et al. (2004a) Role of pyruvate oxidase in *Escherichia coli* strains lacking the phosphoenolpyruvate:carbohydrate phosphotransferase system. J Mol Microbiol Biotechnol 8(4):209–21

Flores N, Leal L, Sigala JC et al. (2007) Growth recovery on glucose under aerobic conditions of an *Escherichia coli* strain carrying a phosphoenolpyruvate:carbohydrate phosphotransferase system deletion by inactivating *arc*A and overexpressing the genes coding for glucokinase and galactose permease. J Mol Microbiol Biotechnol 13(1–3):105–16

Flores S, de Anda-Herrera R, Gosset G et al. (2004b) Growth-rate recovery of *Escherichia coli* cultures carrying a multicopy plasmid, by engineering of the pentose-phosphate pathway. Biotechnol Bioeng 87(4):485–94

Flores S, Gosset G, Flores N et al. (2002) Analysis of carbon metabolism in *Escherichia coli* strains with an inactive phosphotransferase system by (13)C labeling and NMR spectroscopy. Metab Eng 4(2):124–37

Franchini AG, Egli T (2006) Global gene expression in *Escherichia coli* K-12 during short-term and long-term adaptation to glucose-limited continuous culture conditions. Microbiology 152(Pt 7):2111–27

George HA, Powell AL, Dahlgren ME et al. (1992) Physiological-Effects of Tgf-Alpha-Pe40 Expression in Recombinant *Escherichia coli* Jm109. Biotechnol Bioeng 40(3):437–45

Gosset G (2005) Improvement of *Escherichia coli* production strains by modification of the phosphoenolpyruvate:sugar phosphotransferase system. Microb Cell Fact 4(1):14

Haddadin FT, Harcum SW (2005) Transcriptome profiles for high-cell-density recombinant and wild-type *Escherichia coli*. Biotechnol Bioeng 90(2):127–53

Hahm DH, Pan J, Rhee JS (1994) Characterization and evaluation of a *pta* (phosphotransacetylase) negative mutant of *Escherichia coli* HB101 as production host of foreign lipase. Appl Microbiol Biotechnol 42(1):100–7

Harcum SW, Haddadin FT (2006) Global transcriptome response of recombinant *Escherichia coli* to heat-shock and dual heat-shock recombinant protein induction. J Ind Microbiol Biotechnol 33(10):801–14

Hernandez-Montalvo V, Martinez A, Hernandez-Chavez G et al. (2003) Expression of *gal*P and *glk* in a *Escherichia coli* PTS mutant restores glucose transport and increases glycolytic flux to fermentation products. Biotechnol Bioeng 83(6):687–94

Hoffmann F, Rinas U (2001) On-line estimation of the metabolic burden resulting from the synthesis of plasmid-encoded and heat-shock proteins by monitoring respiratory energy generation. Biotechnol Bioeng 76(4):333–40

Hoffmann F, Rinas U (2004) Stress induced by recombinant protein production in *Escherichia coli*. Adv Biochem Eng Biotechnol 89:73–92

Hoffmann F, Weber J, Rinas U (2002) Metabolic adaptation of *Escherichia coli* during temperature-induced recombinant protein production: 1. Readjustment of metabolic enzyme synthesis. Biotechnol Bioeng 80(3):313–9

Hua Q, Yang C, Baba T et al. (2003) Responses of the central metabolism in *Escherichia coli* to phosphoglucose isomerase and glucose-6-phosphate dehydrogenase knockouts. J Bacteriol 185(24):7053–67

Jensen EB,Carlsen S (1990) Production of recombinant human growth hormone in *Escherichia coli*: Expression of different precursors and physiological effects of glucose, acetate, and salts. Biotechnol Bioeng 36(1):1–11

Jurgen B, Lin HY, Riemschneider S et al. (2000) Monitoring of genes that respond to overproduction of an insoluble recombinant protein in *Escherichia coli* glucose-limited fed-batch fermentations. Biotechnol Bioeng 70(2):217–24

Kaga N, Umitsuki G, Clark DP et al. (2002) Extensive overproduction of the AdhE protein by rng mutations depends on mutations in the *cra* gene or in the Cra-box of the *adh*E promoter. Biochem Biophys Res Commun 295(1):92–7

Kao KC, Tran LM, Liao JC (2005) A global regulatory role of gluconeogenic genes in *Escherichia coli* revealed by transcriptome network analysis. J Biol Chem 280(43):36079–87

Kayser A, Weber J, Hecht V et al. (2005) Metabolic flux analysis of *Escherichia coli* in glucose-limited continuous culture. I. Growth-ratedependent metabolic efficiency at steady state. Microbiology-Sgm 151:693–706

Kim JY, Cha HJ (2003) Down-regulation of acetate pathway through antisense strategy in *Escherichia coli*: improved foreign protein production. Biotechnol Bioeng 83(7):841–53

Kirkpatrick C, Maurer LM, Oyelakin NE et al. (2001) Acetate and formate stress: opposite responses in the proteome of *Escherichia coli*. J Bacteriol 183(21):6466–77

Kitano H (2002) Systems biology: a brief overview. Science 295(5560):1662–4

Korz DJ, Rinas U, Hellmuth K et al. (1995) Simple fed-batch technique for high cell density cultivation of *Escherichia coli*. J Biotechnol 39(1):59–65

Krin E, Laurent-Winter C, Bertin PN et al. (2003) Transcription regulation coupling of the divergent *arg*G and *met*Y promoters in *Escherichia coli* K-12. J Bacteriol 185(10):3139–46

Kumari S, Beatty CM, Browning DF et al. (2000) Regulation of acetyl coenzyme A synthetase in *Escherichia coli*. J Bacteriol 182(15):4173–9

Lara AR, Caspeta L, Gosset G et al. (2008) Utility of an *Escherichia coli* strain engineered in the substrate uptake system for improved culture performance at high glucose and cell concentrations: an alternative to fed-batch cultures. Biotechnol Bioeng 99(4):893–901

Lee SY (1996) High cell-density culture of *Escherichia coli*. Trends Biotechnol 14(3):98–105

Li M, Ho PY, Yao S et al. (2006) Effect of *lpd*A gene knockout on the metabolism in *Escherichia coli* based on enzyme activities, intracellular metabolite concentrations and metabolic flux analysis by 13C-labeling experiments. J Biotechnol 122(2):254–66

Li M, Yao SJ, Shimizu K (2007) Effect of *pox*B gene knockout on metabolism in *Escherichia coli* based on growth characteristics and enzyme activities. World J Microbiol Biotechnol 23(4):573–80

Lin HY, Mathiszik B, Xu B et al. (2001) Determination of the maximum specific uptake capacities for glucose and oxygen in glucose-limited fed-batch cultivations of *Escherichia coli*. Biotechnol Bioeng 73(5):347–57

Lin HY, Neubauer P (2000) Influence of controlled glucose oscillations on a fed-batch process of recombinant *Escherichia coli*. J Biotechnol 79(1):27–37

Luli GW, Strohl WR (1990) Comparison of growth, acetate production, and acetate inhibition of *Escherichia coli* strains in batch and fed-batch fermentations. Appl Environ Microbiol 56(4):1004–11

Luo YE, Fan DD, Shang LA et al. (2008) Analysis of metabolic flux in *Escherichia coli* expressing human-like collagen in fed-batch culture. Biotechnol Lett 30(4):637–43

Majewski RA, Domach MM (1990) Simple constrained-optimization view of acetate overflow in *E. coli*. Biotechnol Bioeng 35(7):732–8

March JC, Eiteman MA, Altman E (2002) Expression of an anaplerotic enzyme, pyruvate carboxylase, improves recombinant protein production in *Escherichia coli*. Appl Environ Microbiol 68(11):5620–4

Meyer D, Schneider-Fresenius C, Horlacher R et al. (1997) Molecular characterization of glucokinase from *Escherichia coli* K-12. J Bacteriol 179(4):1298–306

Meyer HP, Leist, C, Fiechter, A (1984) Acetate formation in continuous culture of D1 on defined and complex media. J Biotechnol 1:355–58

Negre D, Bonod-Bidaud C, Geourjon C et al. (1996) Definition of a consensus DNA-binding site for the *Escherichia coli* pleiotropic regulatory protein, FruR. Mol Microbiol 21(2):257–66

Nelson D, Cox M (2003) Lehninger: Principles of Biochemistry. Worth Publishers, New York

Neubauer P, Lin HY, Mathiszik B (2003) Metabolic load of recombinant protein production: inhibition of cellular capacities for glucose uptake and respiration after induction of a heterologous gene in *Escherichia coli*. Biotechnol Bioeng 83(1):53–64

Noronha SB, Yeh HJ, Spande TF et al. (2000) Investigation of the TCA cycle and the glyoxylate shunt in *Escherichia coli* BL21 and JM109 using (13)C-NMR/MS. Biotechnol Bioeng 68(3):316–27

Oh MK, Liao JC (2000) DNA microarray detection of metabolic responses to protein overproduction in *Escherichia coli*. Metab Eng 2(3):201–9

Oh MK, Rohlin L, Kao KC et al. (2002) Global expression profiling of acetate-grown *Escherichia coli*. J Biol Chem 277(15):13175–83

Ow DS, Lee RM, Nissom PM et al. (2007) Inactivating FruR global regulator in plasmid-bearing *Escherichia coli* alters metabolic gene expression and improves growth rate. J Biotechnol 131(3):261–9

Patnaik R, Roof WD, Young RF et al. (1992) Stimulation of glucose catabolism in *Escherichia coli* by a potential futile cycle. J Bacteriol 174(23):7527–32

Peng L, Arauzo-Bravo MJ, Shimizu K (2004) Metabolic flux analysis for a *ppc* mutant *Escherichia coli* based on 13C-labelling experiments together with enzyme activity assays and intracellular metabolite measurements. FEMS Microbiol Lett 235(1):17–23

Peng LF, Shimizu K (2006) Effect of *fad*R gene knockout on the metabolism of *Escherichia col*i based on analyses of protein expressions, enzyme activities and intracellular metabolite concentrations. Enzyme Microb Technol 38(3–4):512–20

Perrenoud A, Sauer U (2005) Impact of global transcriptional regulation by ArcA, ArcB, Cra, Crp, Cya, Fnr, and Mlc on glucose catabolism in *Escherichia coli*. J Bacteriol 187(9):3171–79

Phue JN, Kedem B, Jaluria P et al. (2007) Evaluating microarrays using a serniparametric approach: Application to the central carbon metabolism of *Escherichia coli* BL21 and JM109. Genomics 89(2):300–5

Phue JN, Noronha SB, Bhattacharyya R et al. (2005) Glucose metabolism at high density growth of *E. coli* B and *E. coli* K: Differences in metabolic pathways are responsible for efficient glucose utilization in *E. coli* B as determined by microarrays and northern blot analyses (vol 90, pg 805, 2005). Biotechnol Bioeng 91(5):649–49

Phue JN, Shiloach J (2004) Transcription levels of key metabolic genes are the cause for different glucose utilization pathways in *E. coli* B (BL21) and *E. coli* K (JM109). J Biotechnol 109(1–2):21–30

Phue JN, Shiloach J (2005) Impact of dissolved oxygen concentration on acetate accumulation and physiology of *E. coli* BL21, evaluating transcription levels of key genes at different dissolved oxygen conditions. Metab Eng 7(5–6):353–63

Picon A, de Mattos MJT, Postma PW (2005) Reducing the glucose uptake rate in *Escherichia coli* affects growth rate but not protein production. Biotechnol Bioeng 90(2): 191–200

Ponce E (1999) Effect of growth rate reduction and genetic modifications on acetate accumulation and biomass yields in *Escherichia coli*. J Biosci Bioeng 87(6):775–80

Ponce E, Flores N, Martinez A et al. (1995) Cloning of the 2 Pyruvate-Kinase Isoenzyme Structural Genes from *Escherichia coli* - the Relative Roles of These Enzymes in Pyruvate Biosynthesis. J Bacteriol 177(19):5719–22

Ponce E, Martinez A, Bolivar F et al. (1998) Stimulation of glucose catabolism through the pentose pathway by the absence of the two pyruvate kinase isoenzymes in *Escherichia coli*. Biotechnol Bioeng 58(2–3):292–5

Ramseier TM, Bledig S, Michotey V et al. (1995) The Global Regulatory Protein FruR Modulates the Direction of Carbon Flow in *Escherichia coli*. Mol Microbiol 16(6): 1157–69

Ramseier TM, Negre D, Cortay JC et al. (1993) *In vitro* binding of the pleiotropic transcriptional regulatory protein, FruR, to the *fru, pps, ace, pts* and *icd* operons of *Escherichia coli* and Salmonella typhimurium. J Mol Biol 234(1):28–44

Richmond CS, Glasner JD, Mau R et al. (1999) Genome-wide expression profiling in *Escherichia coli* K-12. Nucleic Acids Res 27(19):3821–35

Rinas U, Hoffmann F (2004) Selective leakage of host-cell proteins during high-cell-density cultivation of recombinant and non-recombinant *Escherichia coli*. Biotechnol Prog 20(3):679–87

Rinas U, Hoffmann F, Betiku E et al. (2007) Inclusion body anatomy and functioning of chaperone-mediated in vivo inclusion body disassembly during high-level recombinant protein production in *Escherichia coli*. J Biotechnol 127(2):244–57

Rinas U, Kracke-Helm, HA, Schügerl, K. (1989) Glucose as a substrate in recombinant strain fermentation technology. By-product formation, degradation and intracellular accumulation of recombinant protein. Appl Microbiol Biotechnol 31:163–7

Rosenthal AZ, Kim Y, Gralla JD (2008) Regulation of transcription by acetate in *Escherichia coli*: *in vivo* and *in vitro* comparisons. Mol Microbiol 68(4):907–17

Rydstrom J (1977) Energy-linked nicotinamide nucleotide transhydrogenases. Biochim Biophys Acta 463(2):155–84

Saier MH, Jr (1996) Cyclic AMP-independent catabolite repression in bacteria. FEMS Microbiol Lett 138(2–3):97–103

Sauer U, Canonaco F, Heri S et al. (2004) The soluble and membrane-bound transhydrogenases UdhA and PntAB have divergent functions in NADPH metabolism of *Escherichia coli*. J Biol Chem 279(8):6613–9

Savinell JM, Palsson BO (1992) Optimal selection of metabolic fluxes for *in vivo* measurement. II. Application to *Escherichia coli* and hybridoma cell metabolism. J Theor Biol 155(2):215–42

Schmidt M, Babu KR, Khanna N et al. (1999a) Temperature-induced production of recombinant human insulin in high-cell density cultures of recombinant *Escherichia coli*. J Biotechnol 68(1):71–83

Schmidt M, Viaplana E, Hoffmann F et al. (1999b) Secretion-dependent proteolysis of heterologous protein by recombinant *Escherichia coli* is connected to an increased activity of the energy-generating dissimilatory pathway. Biotechnol Bioeng 66(1):61–7

Shiloach J, Fass R (2005) Growing *E. coli* to high cell density–a historical perspective on method development. Biotechnol Adv 23(5):345–57

Shiloach J, Kaufman J, Guillard AS et al. (1996) Effect of glucose supply strategy on acetate accumulation, growth, and recombinant protein production by *Escherichia coli* BL21 (lambdaDE3) and *Escherichia coli* JM109. Biotechnol Bioeng 49(4):421–8

Shimizu N, Fukuzono, S, Fujimori, K (1988) Fed-batch cultures of recombinant *Escherichia coli* with inhibitory substance concentration monitoring. J Ferment Technol 66:187–91

Siddiquee KA, Arauzo-Bravo MJ, Shimizu K (2004) Effect of a pyruvate kinase (*pyk*F-gene) knockout mutation on the control of gene expression and metabolic fluxes in *Escherichia coli*. FEMS Microbiol Lett 235(1):25–33

Stouthamer AH (1977) Energetic aspects of the growth of micro-organisms. Symp Soc Gen Microbiol 27:285–315

Stouthamer AH (1980) Energetic regulation of microbial growth. Vierteljahresschrift der Naturforschenden Gesellschaft in Zürich 125:43–60

Stouthamer AH, van Verseveld HW (1986) Stoichiometry of microbial growth. Compr Biotechnol:215–238

Tao H, Bausch C, Richmond C et al. (1999) Functional genomics: expression analysis of *Escherichia coli* growing on minimal and rich media. J Bacteriol 181(20):6425–40

Tomar A, Eiteman MA, Altman E (2003) The effect of acetate pathway mutations on the production of pyruvate in *Escherichia coli*. Appl Microbiol Biotechnol 62(1):76–82

Vallejo LF, Brokelmann M, Marten S et al. (2002) Renaturation and purification of bone morphogenetic protein-2 produced as inclusion bodies in high-cell-density cultures of recombinant *Escherichia coli*. J Biotechnol 94(2):185–94

Vallino J, Stephanopoulos G (1990) Flux determinaion in cellular bioreaction networks: applications to lysine fermentations. In: Todd P, et al. (ed) Frontiers in bioprocessing, CRC Press, Boca Raton, FL

van de Walle M, Shiloach J (1998) Proposed mechanism of acetate accumulation in two recombinant *Escherichia coli* strains during high density fermentation. Biotechnol Bioeng 57(1):71–8

Veit A, Polen T, Wendisch VF (2007) Global gene expression analysis of glucose overflow metabolism in *Escherichia coli* and reduction of aerobic acetate formation. Appl Microbiol Biotechnol 74(2):406–21

Vemuri GN, Altman E, Sangurdekar DP et al. (2006a) Overflow metabolism in *Escherichia coli* during steady-state growth: transcriptional regulation and effect of the redox ratio. Appl Environ Microbiol 72(5):3653–61

Vemuri GN, Eiteman MA, Altman E (2006b) Increased recombinant protein production in *Escherichia coli* strains with overexpressed water-forming NADH oxidase and a deleted ArcA regulatory protein. Biotechnol Bioeng 94(3):538–42

Vemuri GN, Minning TA, Altman E et al. (2005) Physiological response of central metabolism in *Escherichia coli* to deletion of pyruvate oxidase and introduction of heterologous pyruvate carboxylase. Biotechnol Bioeng 90(1):64–76

Wagner S, Baars L, Ytterberg AJ et al. (2007) Consequences of membrane protein overexpression in *Escherichia coli*. Mol Cell Proteomics 6(9):1527–50

Weber J, Hoffmann F, Rinas U (2002) Metabolic adaptation of *Escherichia coli* during temperature-induced recombinant protein production: 2. Redirection of metabolic fluxes. Biotechnol Bioeng 80(3):320–30

Wick LM, Quadroni M, Egli T (2001) Short- and long-term changes in proteome composition and kinetic properties in a culture of *Escherichia coli* during transition from glucose-excess to glucose-limited growth conditions in continuous culture and vice versa. Environ Microbiol 3(9):588–99

Wittmann C, Weber J, Betiku E et al. (2007) Response of fluxome and metabolome to temperature-induced recombinant protein synthesis in *Escherichia coli*. J Biotechnol 132(4):375–84

Wolfe AJ (2005) The acetate switch. Microbiol Mol Biol Rev 69(1):12–50

Wong MS, Wu S, Causey TB et al. (2008) Reduction of acetate accumulation in *Escherichia coli* cultures for increased recombinant protein production. Metab Eng 10(2):97–108

Zhu T, Phalakornkule C, Koepsel RR et al. (2001) Cell growth and by-product formation in a pyruvate kinase mutant of *E. coli*. Biotechnol Prog 17(4):624–8

Chapter 19
Performance Characteristics for Sensors and Circuits Used to Program *E. coli*

Jeffrey J. Tabor, Eli S. Groban, and Christopher A. Voigt

Contents

Abstract The behavior of *E. coli* can be reprogrammed by the introduction of foreign segments of DNA. Three classes of genetic parts, termed sensors, circuits and actuators comprise the DNA programs. Sensors are gene products which allow the cell to detect physical or chemical information in its environment. Genetic engineers can use sensors directly from nature, modify them in some manner, or design them *de novo* to control cellular processes with extracellular or intracellular signals. Genetic circuits act to process information from sensors in order to dictate the behavior of the cell. They can be designed with combinations of "off the shelf" regulatory parts such as transcription factors and promoters, or in some cases can be used "as is" from nature. Finally, genetic circuits govern the expression of actuators, genes whose products perform some physical function to alter the state or the environment within which the cell exists. Using recent DNA synthesis and assembly technologies, genetic sensors, circuits and actuators can be combined to create programs that

C.A. Voigt (✉)
Department of Pharmaceutical Chemistry, University of California, San Francisco, CA 94143-2280 USA; Biophysics Program, San Francisco, CA 94158 USA
e-mail: cavoigt@picasso.ucsf.edu

S.Y. Lee (ed.), *Systems Biology and Biotechnology of Escherichia coli*,
DOI 10.1007/978-1-4020-9394-4_19, © Springer Science+Business Media B.V. 2009

command cells to perform a series of tasks. This approach will transform the way that genetic engineers approach problems in biotechnology. This review covers the construction of genetic sensors and circuits for use in *E. coli*, as well as genetic methods to perturb their performance features.

19.1 Introduction

To program novel behaviors into *E. coli*, handfuls of genetic parts, or segments of DNA with defined functions, are introduced into the cell. In the background, thousands of regulatory and metabolic reactions operate simultaneously and in direct physical contact with the heterologous parts. The engineered components can operate as insulated modules or can be functionally integrated with the preexisting networks of the host cell. Despite what would appear to be long odds, surprisingly complex behaviors with medical, industrial or academic relevance can be achieved.

In this chapter, we will discuss some of the principles which guide the programming of *E. coli*. We define biological programs as strings of genetic parts encoded on segments of DNA which are introduced to the cell on plasmid vectors or integrated into the genome. The designed DNA fragments carry three classes of parts which we will refer to as *sensors*, *circuits* and *actuators* (Voigt 2006). Each of these functions is encoded on a piece of DNA. When combined they create a genetic program that provides a set of instructions that the cell can read and execute. Though the sensor/circuit/actuator construction paradigm can be applied to program any number of genetically tractable organisms (Drubin et al. 2007, Greber and Fussenegger 2007, Sia et al. 2007), this chapter will be limited to a discussion of *E. coli* where much of the foundational work has been accomplished.

Sensors transmit information to genetic circuits. *Genetic circuits* are groups of regulatory molecules which control gene expression to program the cellular response to sensory inputs. Genetic circuits are ubiquitous in the genomes of natural organisms and the characterization of their input-output ranges and dynamic and steady-state responses, or *performance features*, can inform the construction of synthetic analogs with defined properties. In some cases the entire DNA segment encoding a natural circuit can be used "out of the box", or as found in nature, simply being connected to user defined sensors and actuators. Synthetic genetic circuits are built by designing a piece of DNA which carries a series of regulatory parts which interact in a defined manner.

Genetic circuits drive *actuators* which act to change the state or behavior of the host cell or its environment. Actuators range from simple reporters like Green Fluorescent Protein (GFP) to entire organelles. The programming of reliable and sophisticated behaviors in *E. coli* will require actuator expression and function to be tightly governed by environmental, physiological or metabolic signals which are transmitted through genetic circuits via sensors.

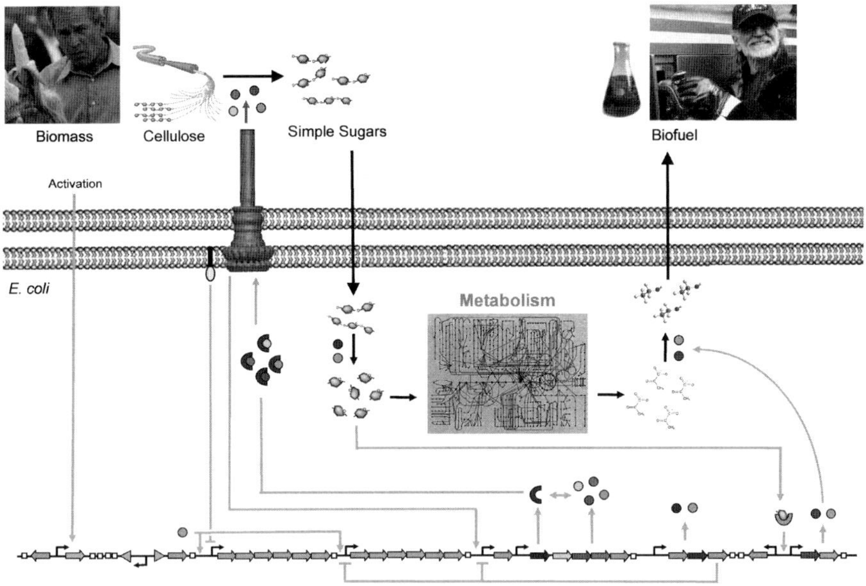

Fig. 19.1 A hypothetical *E. coli* program to convert biomass to liquid fuel. Complex plant material requires that multiple enzymes be exported in a timed sequence. The enzymes need to be exported from the cell, in this case using a type III secretion system imported from Salmonella. The build up of simple sugars induces a pathway to break them down into glucose and covert a metabolic product into a fuel. This is an example of integrated bioprocessing, where multiple steps of a manufacturing process are programmed into a single organism. This requires the combination of sensors, circuits, and actuators to control and respond to a sequence of events

Programs written from sensors circuits and actuators can coordinate sophisticated multistep behaviors with applications in biotechnology (Fig. 19.1). This type of integrated bioprocessing includes, for example, sensing, integrating and responding to media conditions or cell growth stages or densities within a fermenter for optimized yields of an industrially relevant natural product.

Historically limited to piecemeal stitching of naturally occurring DNA fragments, modern DNA synthesis and assembly methods allow the arbitrary connection of sensors, circuits and actuators. Very large (genome scale) biological programs can now be written *in silico* and constructed commercially (Endy 2008, Gibson et al. 2008). The reprogramming of genomes will enable streamlining of the cell through the wholesale addition, deletion or modification of regulatory and metabolic pathways. This will in turn increase the stability, efficiency and productivity (Posfai et al. 2006) of engineered cellular processes.

19.2 Sensors

Genetic sensors typically receive information from the extracellular environment or internal cell state, which is then transmitted to gene regulatory networks. Environmental sensing in *E. coli* largely comprises three strategies: classical regulation, two-component sensing and riboregulation. We will discuss some of the best studied and most widely engineered examples of these sensors throughout this section.

Sensors can receive myriad physical and chemical inputs including small or macromolecules, pH, temperature, light and even signals from other cells. This chapter will focus only on small molecule signals which are the most widely used inputs for engineering *E. coli*.

19.2.1 Classical Regulation

Classical regulation is the control of promoter activity by ligand binding proteins (Fig. 19.2A). The sensor is a cytoplasmic transcription factor which receives an environmental signal by directly binding to a small molecule ligand. Ligand binding triggers a conformational rearrangement which results in increased or decreased affinity of the transcription factor for cognate DNA operator sequences. The sensory output can be transmitted in two ways, by activation or the relief of repression. Activation typically occurs by transcription factor-mediated recruitment of the RNA polymerase complex at the promoter while repression occurs by its occlusion (Wagner 2000).

Classical transcription factors are the most widely used sensors for programming *E. coli*. This is due to the simplicity of their components, their rapid output (strong transcriptional responses occur on the order of 1 minute (Guzman et al. 1995)), the ease with which their input and output specificities can be re-engineered, and the availability of their inducer compounds (Wagner 2000). Here, common strategies are outlined for re-engineering the specifities and performance features of classically regulated transcription factors. Throughout this section we will focus on a particularly well elaborated example, the tetracycline responsive TetR protein.

19.2.1.1 Re-engineering Classically Regulated Sensors

The steady-state quantitative relationships between the concentration of input signal and output gene expression, or transfer functions (Canton et al. 2008, Weiss et al. 1999, Yokobayashi et al. 2002), have been characterized for many classically regulated systems. The features of transfer functions arise from the rate of occupation of promoters by transcription factors and RNA polymerases at different input concentrations (Bintu et al. 2005b). The transfer function of a circuit can be measured by linking it to a sensor, varying the amount of input and measuring the output with a reporter gene (Fig. 19.3). Transfer functions are useful in the design of cellular behaviors because they define the minimal and maximal amount of sensory input which generate circuit responses, the magnitude of induction at any given input concentration and the sensitivity of the circuit to input (Bintu et al. 2005a).

The dynamic range of induction, or magnitude of output in the fully activated (ON) state divided by that of the inactive (OFF) state, is a critical feature of any sensor. In many cases, a large dynamic range of induction is desirable because it more clearly differentiates the absence and presence of an environmental input. Increased

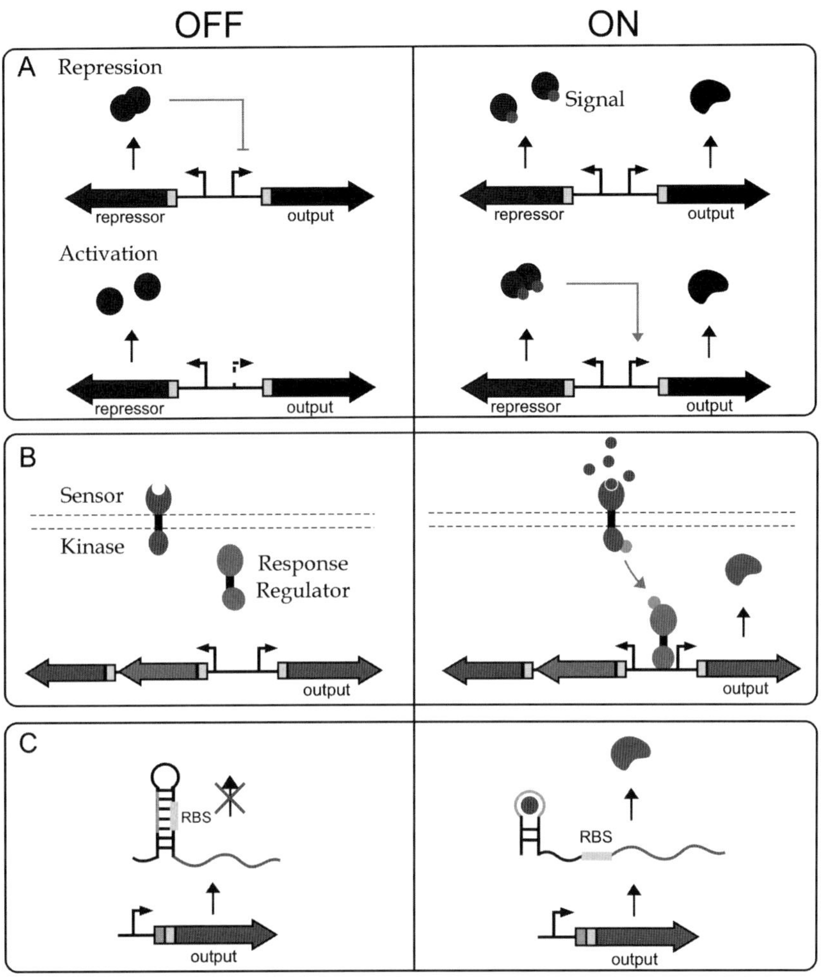

Fig. 19.2 (**A**) Classical transcriptional regulation. In classical systems a cytoplasmic transcription factor protein regulates the target genes in response to the presence of input ligand (*grey dots*). Classical regulation can occur in two forms, repression or activation. With repression, the transcription factor binds to the promoter in the absence of ligand (*left*) and undergoes a conformational change upon ligand binding which causes it to dissociate from the DNA, activating transcription (*right*). With activation, the transcription factor does not associate with the promoter in the absence of ligand (*dashed*), but does in its presence, increasing the rate of transcription. (**B**) Two-component sensing. A membrane associated sensor-kinase protein associates with an extracellular ligand at its sensor domain, which drives a structural change in its cytoplasmic kinase domain. This triggers autophosphorylation of the cytoplasmic domain. The phosphate group (*light grey dot*) is then transferred to the receiver domain of a cytoplasmic, diffusible response regulator protein. When phosphorylated, the response regulator changes conformation and binds to its cognate operator sites near promoters, activating or repressing gene expression. (**C**) An engineered riboswitch. A constitutive promoter drives the expression of a gene with an engineered RNA hairpin occluding its ribosome binding site (RBS, *grey*) and blocking translation. The hairpin also carries an aptamer sequence upstream of the RBS, which can bind to a cognate ligand (*large dot*), triggering a structural rearrangement which liberates the RBS for productive translation. Adapted from Topp and Gallivan, 2007

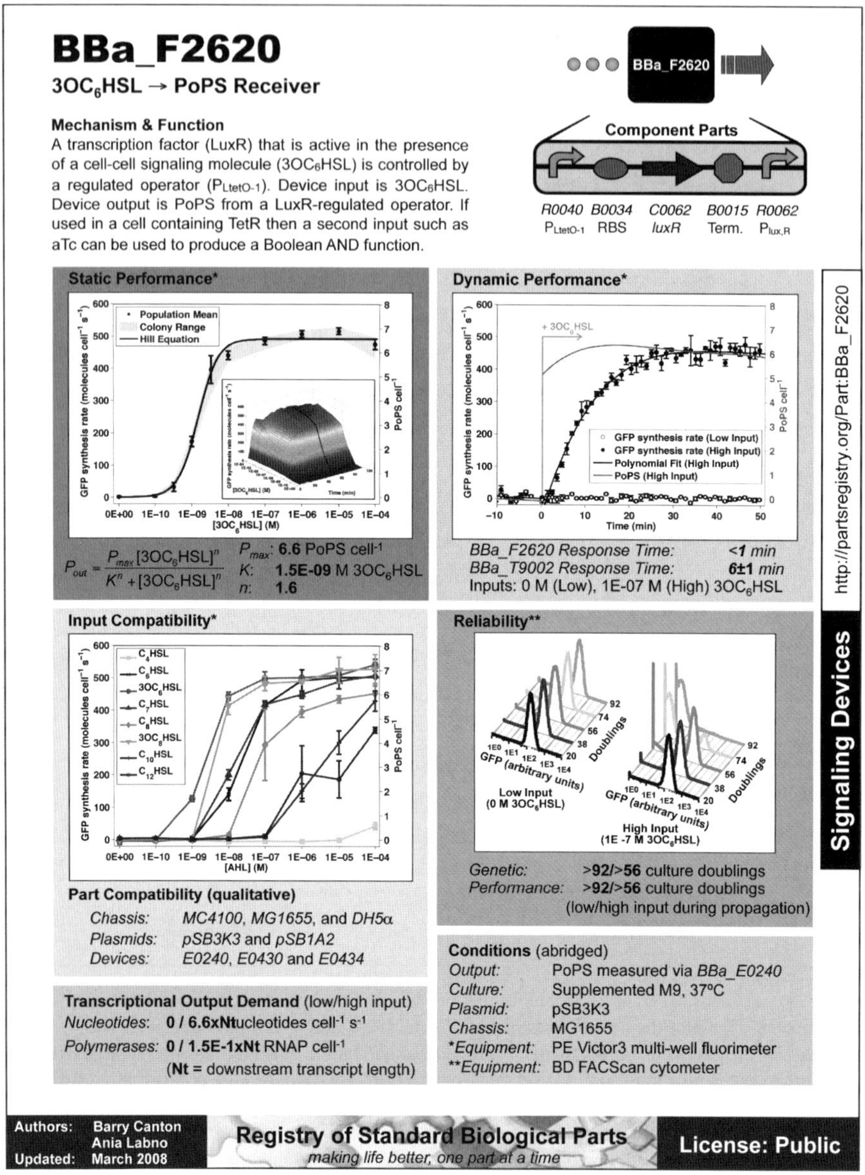

Fig. 19.3 Performance Feature Specification Sheet (from Canton et al. 2008)

dynamic ranges can be achieved by increasing the transcription rate of the ON state, decreasing the transcription rate of the OFF state, or both. The ON state can most easily be increased by strengthening the −35 and −10 RNA polymerase recognition sequences while the repressed state can be lowered by changing the configuration of operator sites around the promoter (Cox et al. 2007, Lutz and Bujard 1997,

Lutz et al. 2001). The *sensitivity*, or rate of increase in transcriptional output as a function of ligand concentration (Fig. 19.5) is largely proportional to the cooperativity of binding of the transcription factor at the promoter. We will discuss strategies for programming cooperativity in Section 19.3.

19.2.1.2 Increasing Dynamic Range

The dynamic range of classical transcription factor systems can be increased by changing the architecture of the output promoter. Traditionally this been accomplished by the addition, deletion or reorganization of the operator sites (Brosius et al. 1985, de Boer et al. 1983, Guzman et al. 1995, Lutz and Bujard 1997). In this section we will discuss efforts specific to the TetR protein.

TetR has been used as the basis for engineering a more tightly repressed and strongly inducible sensing system. To accomplish this two high affinity operator sites were added to an otherwise strong promoter. TetR was then constitutively expressed to repress the promoter in the absence of the input ligand. The system showed virtually no expression in the OFF state, was sensitive to very low levels of input and showed a ~5000-fold dynamic range of induction (Lutz and Bujard 1997). The performance features of the re-engineered system were all marked improvements over the naturally occurring version from which it was derived (de la Torre et al. 1984, Kleckner et al. 1978) and as a result TetR has become one of the most widely used sensors for programming *E. coli*.

19.2.1.3 Changing Operator Specificity

Novel transcription factor:promoter pairs can also be derived from natural systems. The introduction of two mutations within the TetR operator sequence can reduce the affinity to levels insufficient for *in vivo* repression. Rational redesign of DNA binding domains or directed evolution can then be used to re-establish the affinity of the transcription factor for the mutant operators. Indeed, such methods have generated novel transcription factor:promoter pairs based on the TetR (Helbl and Hillen 1998, Helbl et al. 1998) LacR and lambda Cro (Backes et al. 1997) systems as well. Importantly, these novel specificities can be generated with very small numbers of amino acid substitutions in the transcription factors, allowing the rapid generation and screening of many new orthogonal sets in the cellular context. Similar strategies are likely to be amenable to virtually any classically regulated promoter system in *E. coli*.

19.2.1.4 Changing the Input Ligand

The input specificities of classically regulated systems can also be reprogrammed. This is typically accomplished by randomly mutating amino acid residues around the ligand binding pocket and screening variants in functional assays *in vivo* (Collins et al. 2005, 2006, Hawkins et al. 2007). We will discuss efforts to reprogram the ligand specificity of the TetR protein in this section.

TetR has been evolved to recognize an alternate ligand with strong preference over the natural ligand (Henssler et al. 2004). Importantly, the novel inducer is not recognized by the wild type TetR protein, a feature which gives rise to two orthogonal sensors. The combination of novel input and output specificities has the potential to generate completely orthogonal sensing systems which can be used in parallel with one another. Indeed, two TetR variants which sensed different ligands and activated different promoters were recently introduced into the same *E. coli* cell to control the expression of two separate genes (Kamionka et al. 2004). This work demonstrates the value of classically regulated sensing systems as a platform for the construction of genetic control elements with broad applications in biological design.

19.2.2 Two-Component Sensing

A common strategy for environmental sensing in bacteria is a process known as two-component sensing. The canonical two-component system consists of a membrane-bound sensor protein that receives an environmental signal at an extracellular sensory domain and passes the information to a cytoplasmic response regulator protein (Fig. 19.2B). This occurs via the transfer of a phosphate moiety from the cytoplasmic kinase domain of the sensor protein to the receiver domain of the response regulator protein, which can then bind to DNA operator sites at a DNA binding domain to activate or repress gene expression (Hoch and Silhavy 1995).

These sensors are slower to respond than their classically regulated counterparts. For example, the well studied EnvZ/OmpR system of *E. coli* reaches half maximal response to the presence of an input signal in about 5 minutes but requires much longer (on the order of 1 hour) to reach steady-state (Batchelor and Goulian 2006). This happens despite the fact that the phosphotransfer event occurs on a seconds time scale at most (Laub et al. 2007).

The re-engineering of two-component systems has been aided by the modularity of the protein structure. Modular systems are those that are composed of multiple interchangeable subcomponents, or modules. In two-component systems, the extracellular sensory domain of the sensor kinase protein can be replaced by the sensor module from a similar protein. Likewise, the kinase domain of a given sensor protein can be swapped with another to change its specificity for a response regulator (Fig. 19.4). Similar to the classically regulated systems, the specificity of the sensor kinase for its input signal can be altered by computational design methods.

19.2.2.1 Domain Swapping

Sensors can easily be rewired to new outputs by domain swapping. This involves fusing non-cognate sensor and kinase domains at a splice site in a linker region. Most two-component engineering efforts to date have been based on domain swapping, a design process by which chimeric proteins are built from the subdomains of two or more pre-existing proteins (Fig. 19.4). This type of engineering allows

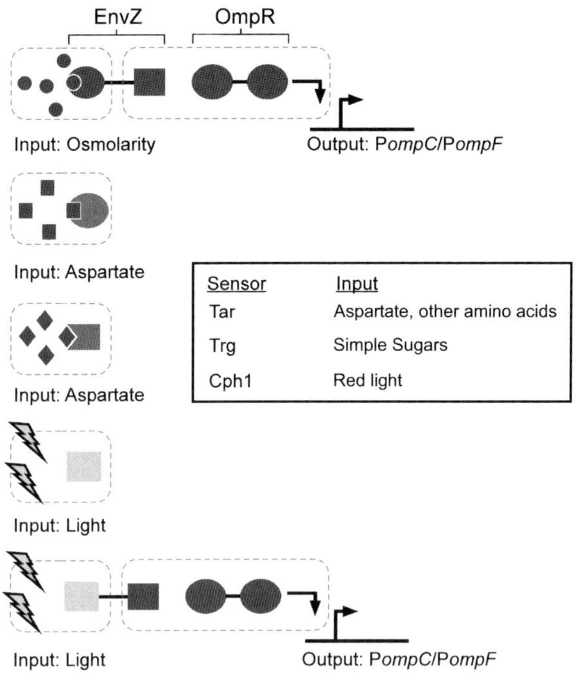

Fig. 19.4 Domain Swapping. The sensory domain of EnvZ receives inputs and transfers information to the response regulator OmpR through the kinase domain. OmpR then activates the expression of the output gene from a target promoter. Other sensory domains can replace the naturally occurring EnvZ sensory domain to create chimeric sensor proteins which activate output promoters in response to different inputs

the sensing pathway to be rewired such that, for example, the output promoter will respond to a completely different input ligand.

The early discovery of a convenient module boundary (Utsumi et al. 1989) made the osmo-responsive EnvZ/OmpR two-component system of *E. coli* a favorite target for many engineering efforts (Baumgartner et al. 1994, Levskaya et al. 2005, Looger et al. 2003). In the natural configuration, the sensor kinase EnvZ phosphorylates the response regulator OmpR in response to changes in osmolarity. Phosphorylated OmpR then binds to operator sites at a promoter, activating or repressing gene expression (Aiba et al. 1989, Aiba and Mizuno 1990, Forst et al. 1989). In the pioneering domain swapping effort, Inoyue and co-workers fused the cytoplasmic domain of EnvZ with the sensory domain of the transmembrane aspartate receptor (TAR), thus rewiring the EnvZ/OmpR pathway to be activated by the amino acid aspartate (Utsumi et al. 1989, Fig. 19.4).

The sensory domain of the chemoreceptor protein Trg has similarly been fused to the cytoplasmic domain of EnvZ (Baumgartner et al. 1994). The Trg sensory domain interacts with periplasmic sugar binding proteins only when they are bound to their ligands to direct *E. coli* chemotaxis. The hybrid Trg-EnvZ protein allowed control of the EnvZ/OmpR pathway with the unnatural ligand ribose via the ribose binding protein (RBP) (Baumgartner et al. 1994).

The sensory domain of a other sensor kinases have also been used to control a chemotactic signaling. NarX is a histidine kinase which senses nitrate and nitrite (Williams and Stewart 1997). Replacement of the sensory domain of the TAR protein with the sensory domain of the NarX kinase has programmed *E. coli* to chemotax away from extracellular nitrate and nitrite (Ward et al. 2002).

In 2005, the osmosensing domain of EnvZ was replaced with a light sensing domain from the *Synechocystis* phytochrome protein Cph1 to program *E. coli* to respond to light (Levskaya et al. 2005). This also required the introduction of a two gene metabolic pathway to produce the chromophore PCB, which binds to the engineered sensor kinase (Gambetta and Lagarias 2001). A confluent lawn of the engineered *E. coli* could then be used as a high resolution film capable of directly converting a two-dimensional light input pattern into a pigment output pattern.

19.2.2.2 Redesigning Ligand Specificities

Other efforts have used computational methods to redesign of periplasmic sugar binding protiens to sense ligands as varied as trinitrotoluene (TNT), L-Lactate, (Looger et al. 2003) and Zn^{2+} (Dwyer et al. 2003) for control of gene expression through the Trg-EnvZ/OmpR pathway. Unlike domain swapping strategies, these studies required detailed knowledge of the three-dimensional structure of the parental proteins. The structural information guided the authors to consider between 5 and 17 amino acids residues as candidates for mutation, and the computational searches typically yielded small lists of candidate protein sequences which were directly amenable to experimental evaluation.

19.2.2.3 Designing the Histidine Kinase-Response Regulator Interface

There are at least 32 natural two-component systems in *E. coli*, all of which have similar structures at the sensor/response regulator interface (Ulrich et al. 2005). To maintain the fidelity of signal transmission through any one of these pathways the sensors and response regulators have evolved a great deal of pairwise molecular specificity (Skerker et al. 2005). Knowledge of the specificity determinants of the histidine kinase-response regulator interactions could allow rewiring of input-output relationships.

Bioinformatic algorithms have recently been used to elucidate regions of the histidine kinase proteins responsible for response regulator specificity. This information enabled the rewiring of two-component pathways by mutating sensor/response regulator interaction domains. The substitution of as few as three amino acid resides within a cytoplasmic subdomain of EnvZ reprogrammed its specificity away from OmpR to numerous other response regulators (Skerker et al. 2008). The ability to redesign protein/protein interfaces adds a valuable degree of freedom which will greatly increase the number of possible alternative two-component signaling pathways that can be constructed in *E. coli*.

19.2.3 Riboregulators

RNA molecules can sense inputs, often through interactions with small or macromolecular ligands, and transmit the information to control gene expression. This typically occurs via the formation of a ligand binding pocket within the regulatory

RNA (riboregulator) which triggers an overall change in its secondary structure. These structural rearrangements can hide or liberate regulatory domains which can then modulate gene expression *in cis* or *in trans* (on the same or another gene). To date, 16 *E. coli* genes have been shown to be subject to cis-acting regulation by ligand binding riboregulators termed riboswitches (Barrick and Breaker 2007).

Bacterial riboswitch sensors convert ligand binding into a change in the transcription or translation rate of the mRNA within which they are embedded (Winkler and Breaker 2003). Though not as widely utilized as their protein counterparts, the structural and functional simplicity of RNA makes it a very attractive platform for the engineering of sensing in bacteria (Isaacs et al. 2006). This is because secondary structure, which governs much of the overall shape and function of RNA, can be computationally predicted with good accuracy (Mathews et al. 1999) and experimentally verified much more rapidly than can three-dimensional protein structures (Soukup and Breaker 1999c). This allows realistic *in silico* design of riboregulators *de novo*, a monumentally difficult task in the protein world.

19.2.3.1 Reprogramming

Riboregulation is also compelling because simple base pairing rules and robust directed evolution methods allow the construction of many orthogonal regulators based on a single parent structure (Bayer and Smolke 2005, Isaacs et al. 2004, Jose et al. 2001, Koizumi et al. 1999, Soukup and Breaker 1999a, Soukup and Breaker 1999b, Soukup et al. 2001, Tang and Breaker 1997). The modular structure of riboregulators also allows them to be introduced into many different genes and even ported between vastly different organisms with surprising ease (Yen et al. 2004). Moreover, unlike in two-component engineering, the sensory domains of riboregulators need not bear any structural or evolutionary relationship to the regulatory domains to which they are fused (Bayer and Smolke 2005, Buskirk et al. 2004, Jose et al. 2001, Soukup and Breaker 1999b).

As a concise demonstration of the design advantages of riboregulators, a riboswitch was recently designed *de novo* to reprogram *E. coli* chemotaxis (Topp and Gallivan 2007). In this work an antisense RNA domain was engineered to base pair with and occlude a ribosome binding site (RBS) upstream of the open reading frame of a chemotaxis-dependent gene, inhibiting translation and subsequently chemotaxis (Fig. 19.2C). A ligand binding (aptamer) domain for the small molecule theophylline was included within the riboregulator such that when theophylline was present, a local base pairing rearrangement occurred which liberated the ribosome binding site, allowing translation. In this way, the engineered riboswitch guided *E. coli* to swim up a gradient of a chemical that does not normally function as an attractant. Though domain swapping and directed evolution have enabled the rewiring of chemotaxis at the protein level as well (Derr et al. 2006, Ward et al. 2002), the benefits of riboregulation are manifest in this example as high throughput efforts have allowed rapid increases in the dynamic range of induction in response to ligand (Lynch et al. 2007, Topp and Gallivan 2008).

19.2.4 Cell-Cell Communication

Cells also have the ability to sense the presence of other cells in the environment. In bacteria this often occurs through a process known as quorum sensing (Miller and Bassler 2001). In short, cells produce membrane-diffusible signals which diffuse into other cells and function as ligands for classical transcription factors. This type of sensing can drive coordinated decision making in cell communities, which enables more sophisticated behaviors.

Cell-cell communication sensors have been used in *E. coli* to control the density of a bacterial population (You et al. 2004), coordinate the timing and magnitude of gene expression between two different cell types (Brenner et al. 2007), drive multicellular pattern formation (Basu et al. 2004, 2005), coordinate the invasion of a malignant mammalian cell (Anderson et al. 2006) or even create a synthetic ecosystem (Balagadde et al. 2008). Each of these circuits was constructed from the Lux-type quorum sensing circuit of *V. fischeri*. A full review of the engineering applications of this type of cell-cell communication system is reviewed elsewhere (Salis et al. 2009).

19.3 Circuits: Processing Sensory Information

Genetic circuits, or networks of interacting regulatory molecules, can integrate one or more sensory inputs into logical and dynamic genetic outputs (Hasty et al. 2002, Kaern et al. 2003, Wall et al. 2004). Circuits have previously been constructed in *E. coli* which generate memory (Atkinson et al. 2003, Gardner et al. 2000), oscillations (Atkinson et al. 2003, Elowitz and Leibler 2000) or pulses (Basu et al. 2004) of gene expression. Other circuits have been designed to function as logic gates, capable of integrating information from multiple sensors to produce a single output (Anderson et al. 2007, Guet et al. 2002, Yokobayashi et al. 2002). Genetic circuits can also coordinate cell-cell communication and community-level decision making (Balagadde et al. 2008, Basu et al. 2005, Brenner et al. 2007, You et al. 2004) This section provides an overview of the performance features and engineering considerations for some of the best characterized and most useful genetic circuit motifs.

19.3.1 Classical Regulation

The simplest genetic circuits are the classical ligand-inducible transcription systems described in Section 19.2.1. In these simple circuits, the presence of input signal positively influences the transcription of an output gene. The transfer function of classically regulated circuits is important because it describes the level of gene expression out of the circuit in response to a given concentration of input signal. This is important when linking multiple circuits in series, because if the output of one circuit is not quantitatively matched with the input of another, then information transfer through the system breaks down. It is of particular interest to discuss the performance features of classically regulated circuits here as they constitute the foundation of many more complex circuit designs.

19.3.1.1 Simple Promoters

In classically regulated circuits the output abundance typically varies as a positive sigmoidal function of the input concentration (Bintu et al. 2005a) (Fig. 19.5A). This relationship arises because there are two input ranges where the system is non-responsive and one input range under which it is. At low input levels, well below the K_D of the transcription factor for the ligand, there is virtually no change in output. As the input ligand concentration approaches the K_D of the transcription factor, there is a monotonic increase in output protein abundance proportional to input.

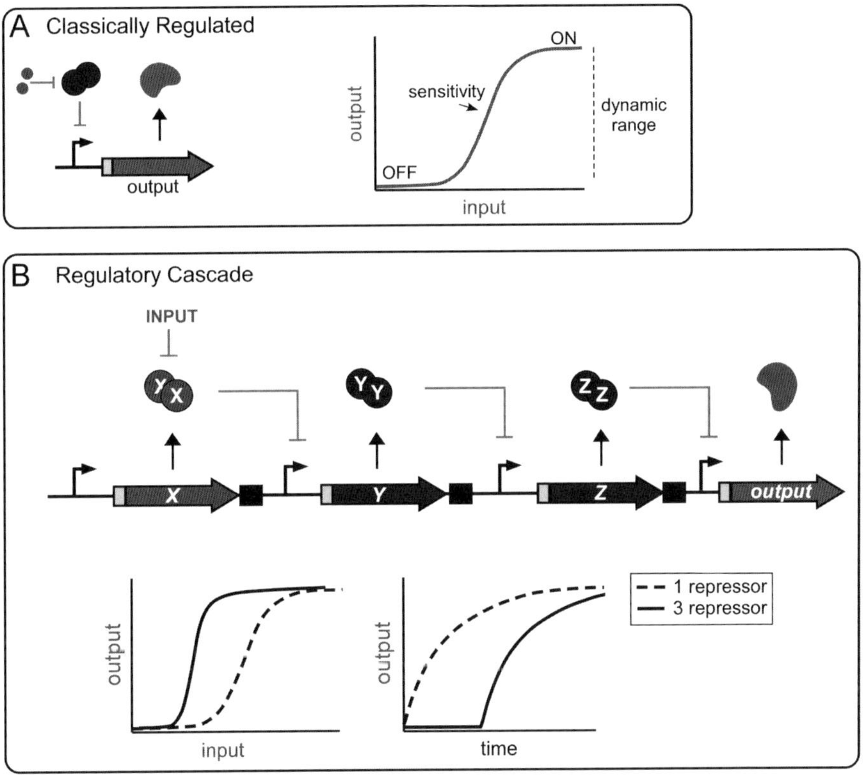

Fig. 19.5 (**A**) Transfer function. Classically regulated promoters typically show sigmoidal response profiles to the concentration of inducer. In the range of inducer (small dot) concentrations well below the K_D of the transcription factor, output changes little changes in input. In the responsive region the concentration of output increases steadily and continuously as a function of input concentration. At inducer concentrations well above the KD of the transcription factor there is no further increase in output. (B) Regulatory cascade. An input signal inactivates repressor protein X, resulting in the derepression of repressor Y. Upon accumulation of Y, repressor Z is repressed and its levels decline, increasing the concentration of the output. (*Lower left*) Cascading results in ultrasensitivity and lower sensing thresholds. A single repressor version of the above circuit (*dashed line*) shows a standard sigmoidal response. The 3 repressor cascade amplifies signal, reducing the absolute concentration of inducer required to activate the circuit and increasing the sensitivity of the response. (*Bottom right*) Cascading generates lags in response time. The single repressor circuit (*dashed line*) responds instantaneously to introduction of inducer while the 3 repressor cascade generates a significant latency in the response

The sensitivity (Wall et al. 2004), or slope of the response curve, in this range is largely determined by the cooperativity of binding of the transcription factor at the promoter of interest. Cooperativity refers to an effect where the affinity of a transcription factor for its DNA operator site increases as a consequence of a previous binding event by another transcription factor at a nearby operator (Bintu et al. 2005a, Ptashne and Gann 2002). This is often the result of protein-protein interaction domains which drive multimerization of the transcription factors on the DNA. Finally, as ligand concentrations increase well above the relevant K_D, the pool of transcription factors or relevant DNA operators become saturated and the output does not increase with further increases in input (Fig. 19.5A).

Certain features of the transfer function can be altered by changing the number and type of operator sites near the output promoter in a classically regulated system. For example the sensitivity, or log-log slope of the input-output function in the responsive range, is less than or equal to 1 for promoters with a single operator site. This is true whether the system is regulated by an activator or repressor (Bintu et al. 2005a,b). The addition of a second operator which enables cooperative binding can significantly increase sensitivity, typically ~2–4 fold (Bintu et al. 2005a,b). DNA looping can also be used to increase the sensitivity of the response (Vilar and Leibler 2003).

In activator systems, if binding is not cooperative, the sensitivity of the response remains the same with the introduction of a second operator, but the dynamic range of induction increases multiplicatively. In repressor based systems, additional operators which do not result in cooperative binding can still increase the sensitivity because the presence of a repressor at any the first site can significantly occlude the RNA polymerase, inherently facilitating binding of a repressor at the second site (Bintu et al. 2005a).

Continuous Response

Classically regulated transcriptional systems have the property of continuous responsivity. Continuous response means that the abundance of the output gene product in a single cell scales proportionally to the concentration of input signal in the environment. This allows the homogenous "fine-tuning" of output expression levels across an entire population. The fine-tuning of expression also allows the control of protein variance between individual cells, which has been shown to naturally decrease as protein abundance increase (Bar-Even et al. 2006). As we will discuss in Section 19.3.2.2, many natural genetic circuits lack continuous responsivity and some have even been intentionally modified to acquire it.

Speed of Response

An important performance feature of any circuit is the rise time, or time required after the addition of an input for the output to reach 50% of its steady-state value. This value, which has been measured for several systems in *E. coli* is approximately 1 cell cycle (45–135 minutes in these studies) (Mangan et al. 2006, Rosenfeld et al. 2002). The time required for an *E. coli* cell to fully respond to an environmental stimulus via the classical mode of regulation is therefore greater than the time required for it to produce a complete copy of itself. The response time of classically regulated circuits can be increased by adding protease tags (Andersen et al. 1998) to speed degradation of regulatory proteins. The slow response times of classically regulated circuits will be compared with those of more complex circuits below.

19.3.1.2 Complex and Biphasic Promoters

Promoters bearing multiple operator sites which activate and repress gene expression can result in non-monotonic behavior in response to a monotonic increase in input signal. For example, the P_{RM} promoter of phage λ has three operator sites for the transcription factor CI. CI initially binds at two high affinity sites and has an activating effect on promoter output. When CI reaches higher concentrations, however, it binds to a low affinity site and functions as a repressor. A circuit wherein CI is expressed proportionally to an input can therefore result in an output which is OFF at both low and high input and ON only at intermediate inputs (Michalowski et al. 2004).

The operator for the AraC activator has been added to the *E. coli* lac promoter to generate a two-tiered activation response (Lutz and Bujard 1997). In this design, transcription increases proportional to the concentration of the first input IPTG but saturates at an intermediate level. This response is solely a function of promoter derepression. When provided saturating IPTG, the promoter can then undergo a second tier of activation proportional to the concentration of the activator arabinose. This occurs as a result of AraC mediated recruitment of RNA polymerase at the derepressed promoter. Many mutants of this promoter have also been constructed which offer different performance features as well (Lutz et al. 2001).

19.3.1.3 Regulatory Cascades

Multiple classically regulated circuits can be linked in series such that the output of one circuit serves as the input to the next (Fig. 19.5B). Cascades can be used to temporally order the expression of many different output genes in response to a single input stimulus (Kalir et al. 2001), allow cells to respond to increasingly small amounts of input (Hooshangi et al. 2005) and filter out transient or noisy input signals.

There are several inherent trade-offs in the use of regulatory cascades. For example, inducer sensitivity and signal amplification can be increased with the number of regulatory steps, but this occurs at a cost to response time. Moreover, lengthening can oftentimes require the redesign of upstream elements to ensure that the output ranges of the existing segment are matched to the input ranges over which the downstream segment can respond (Basu et al. 2004, Hooshangi et al. 2005, Yokobayashi et al. 2002).

Signal Amplification and Ultrasensitivity

To directly measure the performance features of genetic cascades, Weiss and coworkers constructed several synthetic genetic circuits that systematically increased the length of a cascade. This included circuits with 1, 2, and 3 repressors connected in series (Fig. 19.5B). As repressors were added to the cascade, the authors observed that the output reached half-maximal response at lower inducer concentrations; about 40% lower inducer per repressor added. Signal amplification allows cells to respond to inputs which are present in the environment at concentrations below the limit of detection of the natural sensory apparatus.

As with other circuit designs that we have discussed, regulatory cascades can increase sensitivity to the input (Hooshangi et al. 2005, Pedraza and van Oudenaarden 2005) (Fig. 19.5B). In the Weiss example, the range of inducer concentrations required to generate a full response decreased approximately 5-fold upon the addition of the second repressor and 8-fold upon addition of the third. Moreover, a mathematical model indicated that sensitivity would continue to increase as more than three repressors were added to the chain (Hooshangi et al. 2005).

Activation Delays

The relaying of an input signal through a multi-step regulatory cascade results in a temporal lag in response (Fig. 19.5B). Whereas a single repressor showed near immediate response and reached a steady-state output at two hours, the two repressor system took greater than six hours to reach steady-state (Hooshangi et al. 2005). The addition of the third repressor delayed signal transmission dramatically. This circuit showed no response to inputs at times less than two hours, and took 10 hours to reach steady-state. Furthermore, the model showed that with every two additional repressors added the rise time would continue to increase two-fold.

Cascade-Mediated Control of Complex Cellular Processes

The expression of many genes can be temporally ordered if they are regulated by cascades. The *E. coli* flagellum is encoded within 14 operons which contain its structural and regulatory genes. Upon induction, each operon is activated in an order commensurate with the sequence of assembly of the proteins which make up the flagellar apparatus (Kalir et al. 2001). The regulators in this cascade are able to activate each of their target operons in sequence with minutes long lag times in between. This highly regulated sequence of events is probably encoded at the DNA level by variable operator sequences at each promoter for which the regulators have slightly different binding affinities (Kalir et al. 2001). In this scenario, free floating cytoplasmic regulator proteins will occupy stronger operator sites before occupying any given lower affinity operator, allowing rank ordering of gene expression.

Quantitative measurements of gene expression in this system allowed the development of a rigorous computational model which could then be used to make predictive perturbations to circuit behavior (Kalir and Alon 2004). Similarly detailed measurements of the regulatory interactions and their effect on gene expression will be invaluable in the troubleshooting, manipulation and optimization of forward engineered systems as well. Though synthetic biology is far from reliably designing structures as complex as the flagellum, one can envision many smaller scale applications where cascades could be used to time orders of expression in complex processes. For example, timed protein expression could facilitate the stepwise biosynthesis of novel antibiotics (Pfeifer et al. 2001), boost drug production (Keasling 2008) convert agricultural waste into fuel (Service 2007) or even coordinate the expression of existing complex cellular machines (Temme et al. 2008).

19.3.2 Feedback and Feed Forward Regulation

Linking the output of a classically regulated circuit back to its input or forward through intermediate regulators can dramatically alter its dynamic and steady-state properties. In this section we review the most common natural and engineered feedback and feed forward circuits, focusing on the impact of overall architecture and key parameters on circuit performance.

19.3.2.1 Negative Feedback

Negative feedback occurs when the output of a given circuit represses its own production (Fig. 19.6). Circuits controlled by negative feedback have unique response characteristics which are critical for certain biological design applications. Though negative feedback can be implemented as an inhibitory step at any point between production and decay of a gene product this section focuses on transcriptional feedback, which has been widely employed in the construction of synthetic circuits.

Response Accelerators

The response times of negative feedback circuits are markedly reduced compared to their analogous classically regulated counterparts (Savageau 1974). Using engineered variants of the *tet* system, Alon and coworkers experimentally demonstrated

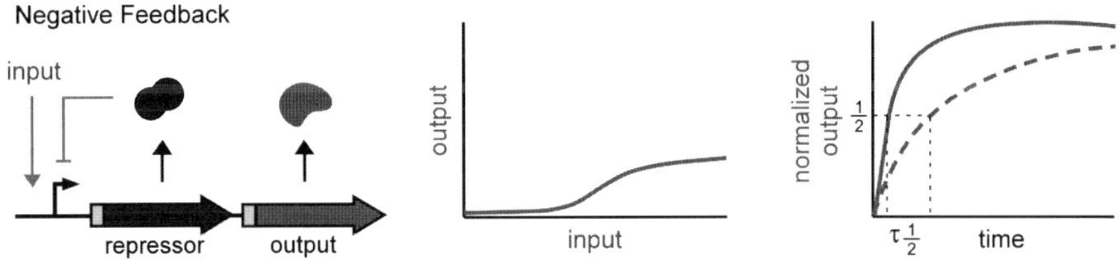

Fig. 19.6 Negative Transcriptional Feedback. A repressor protein is encoded under the control of the promoter which it regulates. The shape of the input/output curve is the same as in Fig. 19.5A above, but the system reaches equal or less output at any given concentration of input. The rise time (t(1/2)), or time required for the circuit to reach 50% of its steady-state output is significantly decreased in negative feedback (*solid line*) as compared to classically regulated (*dashed line*) systems

a reduction in rise time from over two hours to 15 minutes upon the introduction of negative feedback (Rosenfeld et al. 2002) (Fig. 19.6). The acceleration of the response is proportional to the strength of repression, a parameter which can be engineered by altering the number, strength or location of operator sites (Basu et al. 2004, Cox et al. 2007). Acceleration also increases with the cooperativity of binding of the repressor protein to the promoter (Rosenfeld et al. 2002, Savageau 1974). This term can be changed by the addition or removal of operator sites (Bintu et al. 2005a) or by the selection of repressor proteins with different oligomerization properties (Ninfa and Mayo 2004).

Though the negative feedback component reduces response time it also reduces the steady-state output of a circuit (Bashor et al. 2008, Rosenfeld et al. 2002). The rise time acceleration in negative feedback circuits occurs because shortly after induction the promoter is not repressed. Only after the accumulation of repressor does the activity of the promoter decrease to steady-state. This is in contrast to the classically regulated promoter which is active at a high level at all times after induction, resulting in a higher steady-state output which takes more time to achieve. The negative feedback circuit architecture is only useful, therefore, if the circuit output is operational at reduced steady-state expression levels.

Noise Buffering

Stochastic fluctuations, or noise, in gene expression is inevitable in genetic circuits and can reduce the fidelity of signal transmission and cellular decision making (McAdams and Arkin 1997). Moreover, as the number of components in an engineered circuit increases, the effects of noise in any one component can be compounded (Hooshangi et al. 2005, Pedraza and van Oudenaarden 2005).

It has long been recognized that negative feedback circuit architecture can reduce noise in output gene expression (El-Samad and Khammash 2006, Savageau 1974). To experimentally validate this effect, Becskei and Serrano constructed a synthetic circuit wherein a repressor protein inhibited its own transcription in *E. coli* (Becskei and Serrano 2000). Negative feedback reduced noise, measured as the coefficient of variation in protein expression across a population of cells, up to 70% over a circuit without feedback. Moreover, and as predicted (Savageau 1974), the magnitude of noise buffering was proportional to the strength of feedback.

The reason that negative feedback circuits buffer fluctuations is intuitive. In classically regulated transcriptional systems, fluctuations in any step of protein expression (transcription, mRNA decay, translation, etc.) are amplified by subsequent steps and cause variation in protein abundances between individual cells. In negative feedback circuits, fluctuations that cause increases in the output protein concentration are quickly dampened by increased repression while fluctuations that cause the output levels to decrease reduce repressor abundances and increase transcription rates. The result is that the system returns to steady state more rapidly after random fluctuations.

There is a caveat to the use of negative feedback as a safeguard against noise in engineered circuits. Though noise decreases proportional to feedback strength over a large range of protein abundances (Becskei and Serrano 2000, Thattai and van Oudenaarden 2001), noise can actually increase if the strength of negative feedback becomes too strong (Shahrezaei et al. 2008). This is due to a phenomenon known as the "small number effect" where the impact of intrinsic fluctuations in chemical reactions increases rapidly as the concentration of reactants becomes very small (Bar-Even et al. 2006, Kaern et al. 2005). That is, at smaller protein concentrations each random protein production or decay event has a larger impact on the mean concentration. This highlights the general biological design principle that increasing the number of proteins in a cell reduces noise in protein abundance (Bar-Even et al. 2006).

19.3.2.2 Positive Feedback

Positive feedback occurs when the output of a circuit activates its own production (Fig. 19.7A). Circuits with positive feedback can have many features which are valuable in the engineering of more robust, decisive cellular behaviors including ultrasensitivity, bistability, hysteresis and memory (Fig. 19.7B–D). This section

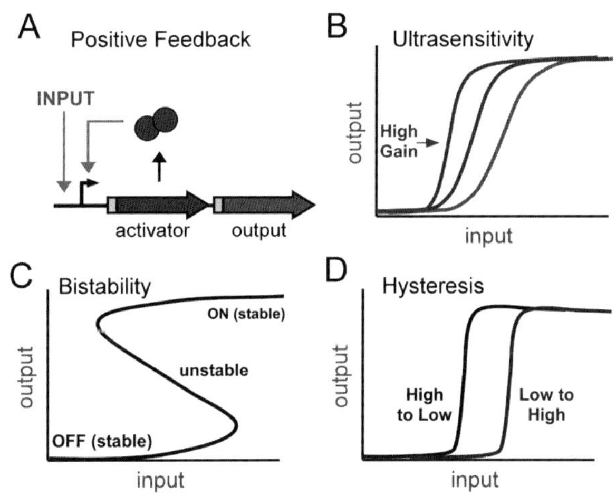

Fig. 19.7 Positive transcriptional feedback. (**A**) A self activating protein is expressed under control of the input. (**B**) Positive feedback circuits with lower kinetic orders of transcription factor binding and cooperativity result in ultrasensitive responses to inducer. Sensitivity is measured as the slope of the relationship between output and input. This increases from the classically regulated system (*right most line*) to 2 positive feedback systems with increasing kinetic constants of activation (*left most lines*). (**C**) Positive feedback circuits with very high kinetic orders of activation can achieve bistability. In these circuits, cells rest at low output levels or high output levels but never at intermediate output levels. (**D**) Hysteresis. When starting at low inducer concentrations and moving higher the circuit requires some amount of inducer to switch ON. When starting in the ON state and reducing the concentration of inducer available to the circuit, the switch happens at a significantly lower concentration

describes the performance features of positive feedback loops and how they can be changed by modifying the underlying circuit parameters.

Response Delays

In contrast to negative feedback circuits which accelerate response times, positive feedback circuits are thought to slow the rise to steady-state. Though it has not been measured in a well-controlled experimental setting, the magnitude of the rise time delay is predicted to be proportional to the strength of the positive feedback step (Savageau 1974). For a transcriptional circuit, feedback strength is governed by the binding affinity of the output transcription factor for its DNA operators, the mode by which the transcription factor interacts with RNA polymerase and its cooperativity (Bintu et al. 2005a, Ninfa and Mayo 2004).

The most direct strategy for manipulating the magnitude of delay in a positive feedback circuit is to vary the DNA operator sites at the promoter to which the activator binds. This can be done by varying the number, spacing and sequence of the operators. Single nucleotide mutations within operator sites can significantly reduce the affinity of a transcription factor for its operator (Basu et al. 2004, Falcon and Matthews 2000, Frank et al. 1997, Takeda et al. 1989). Increasing or decreasing the spacing between multiple operators can affect both binding affinity (Chen and Kadner 2000) and cooperativity of binding (Smith and Sauer 1995).

The introduction of positive feedback increases the steady-state output level of a classical transcriptional circuit. To compensate, one can decrease the production or increase the decay rate of the circuit output. For example, weakening the strength of the self-activating promoter or adding a degradation tag to transcription factor would reduce the steady-state and serve to more closely match the expression levels of a the two circuits.

Ultrasensitivity

It has been demonstrated that regulatory systems with positive feedback are more sensitive to inducer than systems without feedback (Fig. 19.7B) (Savageau 1974). Positive feedback has since been experimentally verified to impart ultrasensitivity in both natural and engineered circuits (Ferrell and Machleder 1998, Bashor et al. 2008). Ultrasensitivity occurs in positive feedback circuits where the strength of the feedback is not so large that the system loses the ability to occupy intermediate output states. The level of ultrasensitivity can be controlled by manipulating the strength of feedback. This can be achieved by changing the stability of the activator protein or its cooperativity or binding affinity at the promoter.

Bistability

Positive feedback can also create a bistable switch (Ferrell 2002). Bistable circuits can occupy only one of two states, canonically an OFF and an ON state, in response to a continuous range of input concentrations (Fig. 19.7C). This can be very useful in circuit design and will be discussed further in Section 19.3.3. It is challenging to design bistable circuits based on positive feedback (Ajo-Franklin et al. 2007) because if either of the states is quantitatively off balance with the other, the system will only be able to occupy one state (Ferrell 2002). For example, leaky transcription of the positive feedback element is often sufficient to trip the switch and keep the circuit in a monostable ON state under all conditions.

A bistable switch based on positive transcriptional feedback has been constructed in *E. coli* (Isaacs et al. 2003). This circuit was composed of a temperature-sensitive transcriptional activator expressed under the control of the promoter which it activated. High kinetic constants of dimerization and transcriptional activation provided the non-linear responsivity required for bistability. At permissive temperatures, leaky transcription tripped the feedback switch driving all cells in the population to reach a stable ON state. At destabilizing temperatures, a lack of activator accumulation kept all cells in the OFF state. At intermediate temperatures the population bifurcated such that individual cells occupied either the ON or OFF state. This digital response occurred because intermediate protein expression states in any cell are unstable and small fluctuations are amplified to drive cells to quickly settle in either of the stable states (Ferrell 2002, Isaacs et al. 2003).

Bistable circuits have a unique property in that they can achieve different steady-state output responses under identical input conditions depending on their history (Fig. 19.7D) (Ferrell 2002, Ninfa and Mayo 2004). That is, if the circuit begins in the OFF state it requires a greater input concentration to switch than if it began in the ON state. This characteristic, known as hysteresis, is useful in the engineering of robust cellular decision making. This is because hysteresis makes circuits with switch-like behaviors less sensitive to fluctuations in input signal near the switch point.

Ninfa and coworkers designed a transcriptional positive feedback circuit with a dominant repressor protein to construct a bistable switch in *E. coli* (Atkinson et al. 2003). In the absence of inducer, the repressor inactivated the feedback loop and the switch was OFF. At activating concentrations of inducer the circuit rapidly switched to the ON state. If the circuit had previously been exposed to high levels of inducer, however, it switched ON at $\sim 70\%$ lower inducer concentrations. Two key circuit parameters drove this system to exhibit hysteresis. First there was very high sensitivity within the switching range making intermediate expression states unstable. Second, the dynamic range of induction was large, on the order of 20-fold. These are the two most critical design requirements in the construction of positive feedback circuits with hysteretic properties (Angeli et al. 2004, Ferrell 2002, Ninfa and Mayo 2004).

Controlling Feedback Saturation

In a bistable switch, the magnitude of output gene expression in the ON state is determined by the protein production and decay parameters of the circuit. The level of gene expression in an activated bistable switch can therefore not be fine tuned. Because the steady-state output level is often an important design consideration in genetic engineering applications, we will discuss several strategies for controlling the magnitude of the ON state, or point of feedback saturation here.

In a simple positive feedback circuit, where an activator protein drives its own promoter, the steady-state output of the fully activated circuit is determined by the maximal rate of production and decay rate of the protein. In the synthetic circuit constructed by Collins and coworkers, the per cell output of the fully activated switch decreased continuously as the activator protein was destabilized (Isaacs et al. 2003). It is likely though that other circuit parameters such as promoter or RBS strength, or mRNA stability could be modified to achieve a similar result.

Eliminating Bistability to Generate a Continuous Response

Sugar inducible systems like *lac* and *ara* are the most widely used elements for engineered genetic control in *E. coli*. They are bistable because sugar-mediated transcriptional activation increases the rate of sugar uptake from the environment, generating a positive feedback loop. For many engineering applications this bistability is undesirable. Bistability creates discontinuous jumps in output as inducer is added. This hampers the freedom of the genetic engineer to set the circuit at intermediate output phenotypes. Moreover, at intermediate inducer concentrations the population can bifurcate such that some cells occupy the OFF state, some the ON state and none occupy an intermediate state. In many applications in biotechnology it is beneficial for all cells in a population to behave identically. Fortunately, the bistable feedback circuits which nature provides can be modified for continuous input-output control and population homogeneity.

Several groups have shown that by expressing sugar uptake genes constitutively the positive feedback loops can be broken and bistability eliminated, allowing continuous induction over a large range of inducer (Khlebnikov et al. 2001, Khlebnikov and Keasling 2002, Khlebnikov et al. 2000, Morgan-Kiss et al. 2002). The deletion of the sugar catabolic genes from the host also aids in the homogeneity of the response (Morgan-Kiss et al. 2002).

19.3.2.3 Feed Forward Loops

A common genetic circuit in *E. coli* is the feed forward loop (FFL), where an input is split into two pathways, which then reconverge on an output (Milo et al. 2002, Shen-Orr et al. 2002). In its simplest form, an FFL consists of two regulatory genes (canonically X and Y) and one output gene (Z). Feed forward architecture results when X regulates the production of Y and both in turn regulate the production of Z (Fig. 19.8).

There are two major classes of FFLs. In the first class, termed coherent FFLs, the sign of the regulatory interaction remains the same all the way through the circuit. That is X regulates Y and Z in the same manner that Y regulates Z. Coherent FFLs therefore regulate outputs similarly to single transcription factors, but introduce several quantitative performance differences. In the second class of FFLs, termed incoherent FFLs, the regulatory effect changes after the circuit splits, resulting in

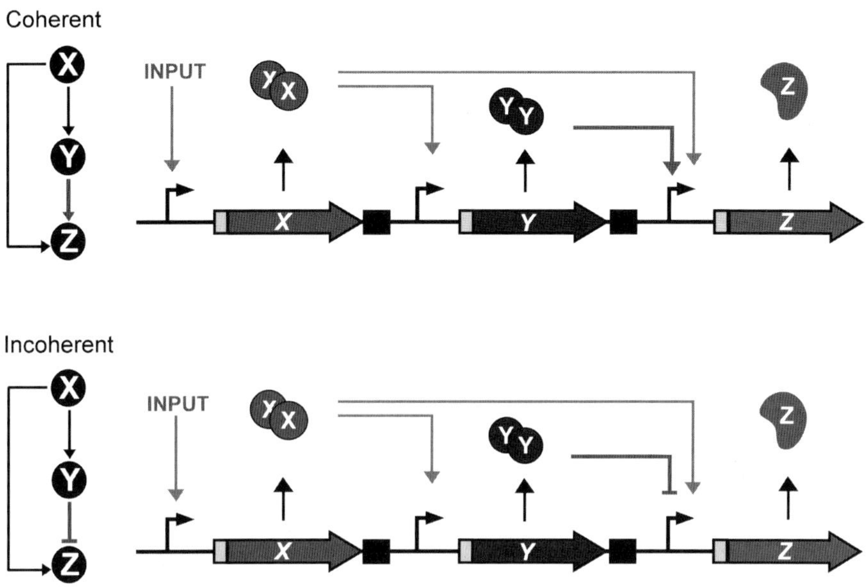

Fig. 19.8 Feed Forward Loops (FFLs). FFLs are genetic circuits composed of three proteins, X, Y and Z. X and Y are transcription factors which regulate the expression of Z. The "feed-forward" connectivity refers to the fact that X also regulates Y. Coherent FFLs result when the regulatory relationship between X and Z is the same as that between X and Y. Incoherent FFLs arise when these two relationships are opposite

opposing regulation at the output. As we will discuss below, this type of circuit can result in interesting dynamic behaviors such as overshoots or pulses of gene expression.

Coherent FFLs: Activation Delays

A FFL is coherent if the regulatory effect of X on Z is the same as the effect of X on Z through Y (Fig. 19.8). Coherent FFLs have been shown to act as sign-sensitive delays in *E. coli* signal processing networks (Kalir et al. 2005, Mangan and Alon 2003, Mangan et al. 2003). "Sign-sensitive" refers to the fact that the circuits generate a lag in the transcriptional response to either the introduction or removal of an environmental signal, but not both. Activation delays can function as noise filters in that they prevent the circuit from responding to transient pulses of signals. Coherent FFLs are useful tools then for the engineering of sense-response behaviors in which the cell must parse sustained signal from input noise in the environment.

The basis of the delay in this type of FFL is intuitive. Z depends on the presence of X and Y for expression. Though the presence of input signal immediately activates X, Y cannot be expressed until X first accumulates. From that point, Y must then accumulate to a concentration sufficient to activate Z. Indeed, increasing the basal expression level of Y decreases the length of the delay (Mangan et al. 2003).

The basal expression rate of an activator protein in a synthetic circuit is simple to tune with promoters or RBSs of different strengths, for example.

The natural arabinose responsive circuit of *E. coli* is a coherent FFL. This is not true, however, for the minimized pBAD circuit from which one of the natural regulators has been removed (Guzman et al. 1995). The natural arabinose FFL circuit generates a delay in the activation of transcription after induction (Mangan et al. 2003). In *E. coli*, the absence of glucose increases intracellular cyclic adenosine monophosphate (cAMP) levels which activate the transcription factor CRP (X). CRP activates the expression of the *araC* (Y) gene, the product of which is a transcription factor whose function is dependent upon arabinose. The output araBAD (Z) promoter functions as a logical AND gate, requiring the presence of cAMP:CRP and arabinose:AraC for productive transcription. This FFL results in a ~0.2 cell cycle or 10–20 minute delay in activation of the Z promoter after the onset of inducing conditions. The delay is shown to be sign sensitive as the removal of the stimulus does not result in a delayed inactivation response as compared to a simple AND gate promoter without a feed forward connectivity between the two transcription factors.

Deactivation Delays

The sign sensitivity of a FFL mediated delay can be changed by changing the activation logic of the Z promoter from AND to OR (Mangan and Alon 2003). Alon and coworkers proofed this concept by demonstrating that part of the *E. coli* flagellar apparatus is expressed under the control of a coherent FFL in which Z is expressed as a SUM function of X and Y (Kalir et al. 2005). SUM is a modified OR where the influence of X and Y on Z output is additive. Moreover, SUM is a simple operation to engineer in *E. coli*. SUM can be achieved by simply placing two different promoters in series. In this configuration, the first or second promoter can drive expression of the output gene, and if both are active, the rate of production of mRNA is greater.

In the flagellar example, X activates Y and the two transcription factors additively activate the operons that produce the flagellar motor (Kalir et al. 2005). If the input signal is removed and X is transiently inactivated, the circuit prolongs flagellar expression because Y levels linger. The authors show that the delay occurs under a wide range of circuit parameters, and that manipulation of the kinetic parameters of regulation can alter the length of the delay (Kalir and Alon 2004). A similar effect was shown for the Salmonella SPI-1 Type III Secretion System, which contains both a feed forward and split positive feedback loop (Temme et al. 2008).

Incoherent FFLs

An incoherent FFL consists of a circuit where X activates Y and Z but Y represses Z (Fig. 19.8). There are over 100 examples of incoherent FFLs in the *E. coli* genome (Mangan et al. 2006). This circuit generates several interesting and unique dynamical outputs such as pulses of gene expression and time-derivative sensing (Basu et al. 2004). In this section we will discuss the performance features of incoherent FFLs in *E. coli*, the effect of key molecular parameters on their function, and their application in the construction of some of the most sophisticated synthetic cellular behaviors to date.

ON Accelerators

Because X first activates and then indirectly represses the expression of Z, incoherent FFLs result in "overshoot dynamics" in the expression of Z (Mangan and Alon 2003). This means that Z temporarily reaches abundances greater than the final steady-state. Also, because the strength of a partially repressible promoter driving Z must be stronger than that of a non-repressible promoter capable of generating the same steady-state, the rise time of the output Z is necessarily increased in an incoherent FFL of this form (Mangan and Alon 2003). This property is similar to the accelerated response of negative feedback loops as described above. In a natural example, Alon and coworkers have demonstrated that the incoherent FFL in the galactose utilization network of *E. coli* results in a 1.75-fold overshoot of the steady-state output and an approximately 3-fold acceleration in rise time (Mangan et al. 2006).

In incoherent FFL circuits, important performance features such as the magnitude of response acceleration, the steady-state output and the size of the overshoot are particularly sensitive to the parameters associated with the repressor Y. In general, the higher the expression level of Y and the greater its repressive effects, the greater the acceleration of the circuit (Mangan and Alon 2003).

Pulse Generators

A pulse generator is a genetic circuit capable of activating and then completely repressing output gene expression in response to the addition of an input. Incoherent FFLs can generate pulses of gene expression if the repression of Z by Y is very strong. In 2004, Weiss and coworkers constructed a synthetic incoherent FFL in *E. coli*. In their design X was the transcription factor LuxR which is activated by the membrane permeable quorum sensing compound AI-1, Y was the strong transcriptional repressor CI and Z was the reporter gene *gfp*.

Because the circuit was constructed *de novo*, the authors could easily investigate the effects of genetic parameters such as the rate of synthesis of Y, and the strength of repression Z by Y. The authors noted that if either of these two parameters was too great, the circuit could never be activated by inducer (Basu et al. 2004). Under a range of permissive kinetic parameters, however, the circuit showed robust pulse generation after addition of inducer. The true pulse of gene expression occurred because the Y protein CI is a very strong repressor of its target promoter, capable of bringing output expression back to zero.

Critical pulse features such as amplitude and duration could be controlled by varying the kinetic parameters of the Y protein or the rate or concentration at which inducer was added. Specifically, the stronger the RBS or the affinity for the output promoter the shorter and smaller the resulting pulse. Furthermore, at fixed Y kinetic parameters, the pulse amplitude varied proportionally to both the absolute concentration and the rate of increase of inducer. This synthetic circuit is an elegant demonstration of the level of behavioral sophistication that can be designed *de novo* and optimized to the specifications of the engineer.

19.3.2.4 Dynamic Circuits

Several genetic circuits have been engineered which drive dynamic responses. A striking example is the three protein transcriptional ring oscillator known as the "repressilator" (Fig. 19.9A) (Elowitz and Leibler 2000). In this circuit, protein A represses protein B, protein B represses protein C and protein C represses protein A. Oscillations occur because the addition of an input signal can cause one of the proteins, say A, to become abundant and repress the next protein in line (B). Because B is repressed, C begins to rise in abundance and can then in turn repress A. This process continues until A rises again, and in this manner the circuit encodes genetic oscillations. The repressilator was capable of generating three to four oscillations in a given cell, but showed a notable lack of uniformity across the population.

In another example, Ninfa and coworkers constructed a two-component transcriptional oscillator in which a transcription factor first activates itself and then activates its own repressor (Fig. 19.9A) (Atkinson et al. 2003). In this circuit an input triggers the activator to initially accumulate. After some time the activator is repressed by the accumulating repressor. As activator levels subsequently fall, so do repressor levels, triggering another round of activator accumulation. This circuit drove dampened oscillations over four periods, which spanned almost 60 hours.

A circuit based on cell-cell communication has been constructed to program population level oscillations in *E. coli* (Balagadde et al. 2005). In this design a gene

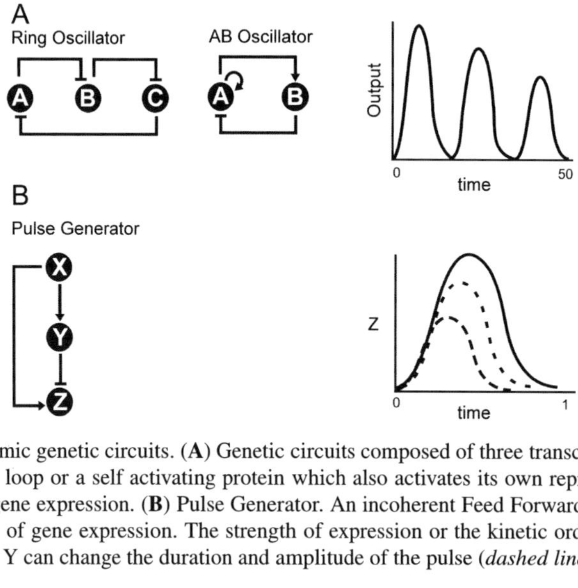

Fig. 19.9 Dynamic genetic circuits. (**A**) Genetic circuits composed of three transcriptional repressors in a closed loop or a self activating protein which also activates its own repressor can cause oscillations in gene expression. (**B**) Pulse Generator. An incoherent Feed Forward Loop produces temporal pulses of gene expression. The strength of expression or the kinetic order of repression of the repressor Y can change the duration and amplitude of the pulse (*dashed lines*)

which triggered cell death was expressed under the control of a quorum sensing circuit. The circuit was OFF at low cell densities but switched ON at high density. Microscopic monitoring demonstrated that *E. coli* expressing this circuit regularly oscillated in density from 1 to 3 cells per picoliter of media with a period of about 20 hours.

As discussed in Section 19.3.2.2, Weiss and coworkers also constructed a dynamic circuit capable of generating a temporal pulse of gene expression in response to a single, step introduction of input signal (Fig. 19.9B) (Basu et al. 2004). The amplitude and duration of the pulse could be programmed by changing the strength or production rate of the repressor in the circuit. Moreover, because the circuit input was a membrane diffusible quorum sensing compound, a cell could be triggered to pulse by production of the activator in a nearby cell.

19.3.3 Switches and Logic

Genetic switches are circuits which rapidly transition between discreet states in response to input. Logical devices are circuits which interpret the states of multiple switches to produce a single, unified output. Switches and logic are useful because they aid the programming of desirable IF/THEN behaviors in *E. coli*. Genetic logic is carried out by circuits which can be rationally designed or combinatorially screened.

Extensibility, or the ease with which a switch or logic device can be connected to a different input or output is a desirable trait in switches and logic devices. Extensibility requires knowledge of the transfer functions of the parts. For example, the output range of a given switch or switches must be matched to the input range of a given logic device in order for signal transmission to proceed properly through the circuit. If expression in the OFF state of a switch is leaky and it is interpreted as ON by the downstream logic gate, then the circuit will not properly respond to input signals. If the transfer functions of switches and logic gates are well documented, however, they can be used "off the shelf" and connected to other well characterized parts.

NOT Gate

One of the most useful and frequently constructed genetic logic operations is the Boolean NOT gate. Commonly referred to as an inverter, the NOT gate inverts the sign of the regulatory relationship between the input and output of the circuit. In the simplest system, this is accomplished by the introduction of a transcriptional repressor between the input and output (Fig. 19.10A). An input signal which would otherwise activate expression of the output therefore inactivates it via the activation of a repressor. Besides inverting the input/output logic, NOT gates are also known to increase input sensitivity (Hooshangi et al. 2005, Karig and Weiss 2005, Pedraza and van Oudenaarden 2005) and lower sensing thresholds (Karig and Weiss 2005).

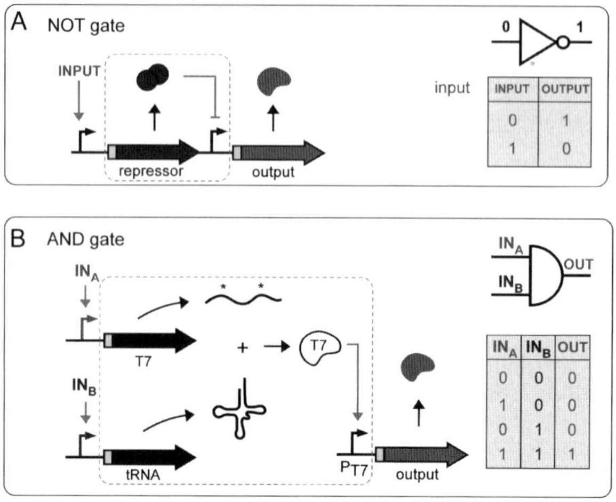

Fig. 19.10 Transcriptional Logic (**A**) NOT gate. Also known as a genetic inverter, the NOT gate encodes a repressor under the control of the environmental input. The repressor inactivates expression from an otherwise active output promoter. The inverter device (*dashed box*) comprising the repressor protein and the output promoter is an independent module which can be placed between any input promoter and output gene. The logic of the NOT circuit (*upper right*) is shown in the truth table (*bottom right*). (**B**) AND gate (*dashed box*) comprises an untranslatable T7 RNA polymerase mRNA bearing two stop codons (*asterisks*) in the open reading frame and a suppressor tRNA which incorporates amino acids at the stop codons to allow productive translation. Only when both halves are transcribed is T7 RNAP produced and does the output promoter become active. Each half of the AND gate can be driven by any inducible promoter, activated by its cognate input signal. Adapted from Anderson et al., 2007

Many genetic circuits containing NOT gates fail to function properly when constructed. This often occurs because basal expression of the repressor in the absence of input can be sufficient to inhibit the output promoter, constitutively trapping the inverter in the OFF state. The abundance of the repressor protein can then be reduced to match the relevant sensitivity of the output promoter. This can be accomplished by weakening the RBS on the repressor mRNA, weakening the operator sites at the output promoter (Hooshangi et al. 2005, Weiss 2001, Yokobayashi et al. 2002) or randomly mutating the repressor to reduce its strength (Yokobayashi et al. 2002)

Switches and Memory

Memory is required for many sophisticated functions in electronic systems and is also ubiquitous in molecular biology, forming the basis for the burgeoning field of epigenetics. One popular biological design goal which relies on memory is to construct cells that can count how old they are or how many times they and their ancestors have been exposed to some signal over a long period of time. Memory can be implemented as an extreme form of hysteresis in circuits with strong positive feedback. In such systems, previous exposure to high input signal triggers a circuit to remain active even when the signal goes to zero (Ferrell 2002).

In 2000, Collins and coworkers constructed a memory switch in *E. coli* (Gardner et al. 2000). The switch was comprised of two cross-inhibiting transcriptional repressors. If repressor A was expressed, it repressed B and the switch was OFF. If an input was added which inactivated A, B accumulated and in turn, repressed A. This turned the switch ON. This switch generated stable memory over at least 22 hours, allowing a cell many generations away from the ancestor which had received the signal to maintain a stable response. This switch required proper matching of the transfer functions of its two subcomponents. If the expression level of one repressor was too strong in the OFF state the system became monostable. This required the screening of several combinations of promoters and RBSs of different strengths.

Arkin and coworkers have also constructed a memory device based on a permanent genetic rearrangement event. This circuit makes use of the recombinase encoded by the *fimE* gene to flip an improperly oriented promoter into alignment with an output gene (Ham et al. 2006). The DNA reorganization event is permanent, resulting in stable long-term circuit memory. Moreover, because the *fimE* gene can be expressed as the output of any sensor, the *fimE* switch is modular and can potentially generate memory of any input stimulus capable of activating gene expression. An advantage of this circuit is that it produces virtually no basal expression when the promoter is in the opposite orientation from the gene it controls.

AND Gate

The logical AND operation, where the presence of two inputs (A and B) are required to activate output expression, is a useful concept for biological design and can be applied to the construction of many more sophisticated logical operations. The most parsimonious strategy for the construction of a genetic AND gate involves two interdependent genetic components which, when expressed simultaneously can initiate a downstream gene expression step. Such a system was recently implemented at the transcriptional level in *E. coli* (Anderson et al. 2007). In this setup, inducible promoter A drives the expression of an mRNA encoding the T7 RNA polymerase (RNAP) gene. The mRNA is non-functional, however, as two specific stop codons are introduced into the coding sequence. Inducible promoter B drives the expression of a tRNA which encodes an amino acid at those stop codons, rescuing translation of the RNAP. The circuit output is a promoter which is only transcribed by T7 RNAP protein such that it requires the presence of the two inputs A and B (Fig. 19.10B). Importantly, this system was designed to be modular such that any two inducible promoters could be used to drive the AND gate. This modularity allowed the circuit to integrate signals from four different promoters and drive two separate output genes.

In the initial circuit design, the two components of the AND were not properly matched. The basal, or leaky expression rate of the T7 mRNA was significantly high that the circuit produced positive output in the presence of only one input. To reduce leaky expression, the authors randomly mutated the RBS preceding the T7 open reading frame and screened a library for variants dependent upon both inputs for activation. A majority of the variants in the library showed significant dependence on both inputs, indicating that the design was quite robust to variable expression levels. When the promoter driving the T7 mRNA was replaced, however, the new RBS failed to generate enough mRNA to activate the AND gate even when the promoter was fully active. To restore functionality an RBS library was again screened and again produced a viable circuit.

Other Logic

To construct other types of genetic logic, Leibler and coworkers randomly connected five promoters to three classical transcription factors which either activated or repressed them. Two ligands were chosen as inputs and one of the transcription factors was chosen to repress an output reporter gene. Several switch-like logical responses including NAND, NOR and NOT IF arose repeatedly from the circuit library (Guet et al. 2002). Interestingly, circuits with similar connectivities, or profiles of regulatory contacts between components, were capable of generating different logical responses while networks with different connectivities were capable of generate the same logic. Many of the constructed circuits also produced intermediate or "fuzzy" logic.

A large number of intermediate logical operations were also observed in a related study wherein four different transcription factor binding sites were randomly placed in three locations around a single promoter (Cox et al. 2007). This combinatorial approach revealed that activator sites function most effectively when placed directly upstream of the −35 site and function poorly if at all when placed downstream of it. Repressor sites are more tolerant to different locations but are most effective when placed between the −35 and −10 sites. These efforts demonstrate the power of screening random combinations of regulators to achieve a desired logical operation.

19.4 Actuators: Interfacing Cells with the Environment

A fundamental motivation for programming cells is that they have the ability to modify the chemistry and biology of their surrounding environments. Actuators are defined as gene products which carry out any type of cellular process or behavior from an enzyme capable of synthesizing drugs or fuels to the synthesis and control of entire organelles and molecular machines. This section is meant to only briefly outline some of the things that *E. coli* can do.

State Reporters

State reporters are molecules whose only function is to be observed or measured. When linked to genetic circuits, reporters can provide a "print-out" of information coming in from cellular sensors and circuits. In biosensing applications the acquisition of information about the presence, absence, concentration or temporal profile of an input signal in the environment or the cell is itself the goal of the system. Reporters can also provide a physical read out of the solution of computations performed by genetic circuits. The most common reporters are proteins such as β-galactosidase or Green Fluorescent Protein (GFP), the abundances of which can easily be measured by standard molecular biological techniques.

Metabolic Engineering

Metabolic engineering involves the expression of enzymes which divert cellular metabolites into alternative pathways to produce desired output products (Lee and Papoutsakis 1999). The enzymes used in metabolic engineering are therefore actuators which can be expressed as the outputs of genetic circuits. A typical metabolic design might employ sensors which detect the presence of upstream metabolites to time the expression of the biosynthetic enzymes which act upon them.

One application of metabolic engineering is the production of liquid fuels (Jarboe et al. 2007, Keasling 2008, Mielenz 2001, Service 2007). To this end, Liao and coworkers recently re-engineered *E. coli* amino acid metabolism for the production of branched chain alcohols, compounds which have desirable fuel properties (Atsumi et al. 2008). This required the expression of one of several two-enzyme clusters which converted intermediate metabolites from amino acid biosynthetic pathways to the various alcohols. Endogenous amino acid metabolic genes could also be overexpressed as complementary actuators to increase flux through the pathways and bump up fuel yields.

Metabolic actuators can be used to reprogram *E. coli* for the production of therapeutic compounds as well (Pfeifer et al. 2001, Swartz 2001, Zhang et al. 2006). For example, Keasling and coworkers have introduced a large number of non-native isoprenoid biosynthetic enzymes into *E. coli* to efficiently convert the ubiquitous metabolite acetyl-CoA into artemisinic acid, a direct precursor to the potent and otherwise prohibitively expensive anti-malarial compound artemisinin. Optimization of enzyme expression levels and compensatory engineering to eliminate toxic byproducts has resulted in profound improvements in yield, approximately 1 million fold increase in a 4 years span (Keasling 2008). These efforts are likely to reduce the cost of artemisinin orders of magnitude, to prices compatible with its utilization in many underdeveloped countries with high malarial death rates.

Most metabolic engineering efforts to date have expressed the actuators under the control of classically regulated circuits. These have been chosen for their simplicity and the continuous fine-grained control that they offer over enzyme expression levels. The construction of more sophisticated sensor-circuit-actuator systems should facilitate the design of increasingly ambitious microbial factories and help to optimize yields.

Organelle Transfer

Clusters of genes encoding entire organelles can also be used as actuators. Historically, the ability to manipulate such large fragments of DNA has required the presence of fortuitous restriction sites in the natural organelle sequences or specialized polymerase chain reaction (PCR) based methods. Improved DNA synthesis technologies now allow the *de novo* fabrication of organelle scale fragments.

In the initial demonstration of organelle transfer, 11 genes responsible for the synthesis of cytoplasmic gas vesicles in *B. megaterium* were moved into *E. coli* (Li and Cannon 1998). Expression of this gene cluster from a classically regulated circuit on a standard expression plasmid resulted in the formation of functional vesicles capable of significantly increasing the buoyancy of *E. coli* in aqueous media.

Similar strategies have resulted in the transfer of the fully functional nitrogen fixation (*nif*) (Dixon et al. 1976) and O antigen lipopolysaccharide (Bastin et al. 1991) enzyme clusters from *Klebsiella* and enteropathogenic *E. coli*, as well as the Type III protein secretion organelle from *Salmonella* (Wilson et al. 2007) and the cryptic Type II organelle from *E. coli* itself (Francetic et al. 2000). These efforts used unmodified, contiguous genomic DNA fragments which were recombined into plasmids and introduced into *E. coli* "as is". These strategies therefore relied on expression from the natural promoters and RBSs of the relevant genes, and necessarily introduced the possibility of regulation by undefined control elements. The utility of organelle actuators will undoubtedly benefit from control through sensors and circuits.

Building Genetic Programs

Sensors, circuits and actuators are true modular engineering components only when they can easily and arbitrarily be linked together. Several methodologies have recently been developed which allow the combination of multiple stretches of DNA without the need for inherent restriction sequences. One example is a universal, iterative cloning method for the assembly of standardized "Biobrick" parts (Shetty et al. 2008). In this method, a DNA part is computationally designed to internally lack several specific restriction sites. These restriction sites are then added to the upstream and downstream regions of the part and used as universal handles for the iterative, arbitrary connection of any two Biobricks. This standardized strategy increases the efficiency and ease with which any two parts can be combined (composability).

A PCR-based strategy termed SLIC has recently been developed for the "one-pot" assembly of up to 10 unrelated stretches of DNA in a specific order (Li and Elledge 2007). This method uses oligonucleotide primers to add specific linker sequences to any piece of DNA which then guide the order of assembly. The benefits to this approach are that no specific restriction sites need be avoided in the internal sequence of any part and that more than two parts can be combined in one step. Other advanced assembly strategies based on large scale oligonucleotide synthesis and polymerase chain reaction (PCR) assembly have allowed the construction of complete viral (Cello et al. 2002, Smith et al. 2003, Tumpey et al. 2005) and even bacterial (Gibson et al. 2008) genomes from computationally designed DNA information.

Standardization and assembly technologies are already helping eliminate barriers between the design and physical construction of DNA (Endy 2005), a process which has been the historical rate limiting step in genetic engineering. A true leap in biological design will occur when these technologies become more widely available and less expensive, allowing true modular assembly of sensors, circuits and actuators. In an early watershed example, Collins and coworkers linked a DNA damage sensor to a bistable genetic switch to drive an actuator which triggered biofilm formation in *E. coli*. In this bottom up programming effort, the *E. coli* could stably and strongly switch ON biofilm formation phenotype in the presence of DNA damaging environmental inputs such as UV light or a chemical mutagen (Kobayashi et al. 2004).

Finally, when genetic parts are linked together in a design, their quantitative input/output properties must be properly matched (Yokobayashi et al. 2002). As discussed in Section 19.3.3 above, if the OFF state of a sensor is sufficiently leaky to activate a downstream genetic circuit, the circuit will not be capable of receiving signaling information from the sensor. There are many strategies for matching the transfer functions of multiple parts, but until universal metrics of genetic activity can be established (Endy 2005, Canton et al. 2008) there will always be a significant troubleshooting component in the assembly of functional systems.

19.5 Conclusions

The vast molecular genetic literature on *E. coli* has made it the subject of choice for many early efforts in synthetic biology. Five decades of work have given genetic engineers a rich repository of parts, often sensors and actuators, which can be taken out of their natural context and used for new, user-defined purposes. More recent efforts have established useful circuit design principles that have further pushed the level of sophistication of behaviors that can be designed.

Complementing the scientific contributions, DNA synthesis and sequencing technologies have become increasingly high throughput and less expensive in the past few years. Further advances will bolster biological design by allowing researchers to bypass the arduous process of physically constructing designed DNA sequences. In the end, *E. coli* synthetic biology serves two major purposes. It enables the goal-oriented engineering of strains which can carry out novel functions of medical, industrial or academic interest and it serves as a bottom-up complement to top-down systems approaches for the elucidation of the molecular principles which govern cellular behavior.

References

Ajo-Franklin CM, Drubin DA, Eskin JA et al. (2007) Rational design of memory in eukaryotic cells. Genes Dev 21(18):2271–6

Andersen JB, Sternberg C, Poulsen LK et al. (1998) New unstable variants of green fluorescent protein for studies of transient gene expression in bacteria. Appl Environ Microbiol 64(6):2240–6

Anderson JC, Clarke EJ, Arkin AP et al. (2006) Environmentally controlled invasion of cancer cells by engineered bacteria. J Mol Biol 355(4):619–27

Anderson JC, Voigt CA, Arkin AP (2007) Environmental signal integration by a modular AND gate. Mol Syst Biol 3:133

Angeli D, Ferrell JE, Jr., Sontag ED (2004) Detection of multistability, bifurcations, and hysteresis in a large class of biological positive-feedback systems. Proc Natl Acad Sci USA 101(7): 1822–7

Atkinson MR, Savageau MA, Myers JT et al. (2003) Development of genetic circuitry exhibiting toggle switch or oscillatory behavior in *Escherichia coli*. Cell 113(5):597–607

Atsumi S, Hanai T, Liao JC (2008) Non-fermentative pathways for synthesis of branched-chain higher alcohols as biofuels. Nature 451(7174):86–9

Backes H, Berens C, Helbl V et al. (1997) Combinations of the alpha-helix-turn-alpha-helix motif of TetR with respective residues from LacI or 434Cro: DNA recognition, inducer binding, and urea-dependent denaturation. Biochemistry 36(18):5311–22

Balagadde FK, Song H, Ozaki J et al. (2008) A synthetic *Escherichia coli* predator-prey ecosystem. Mol Syst Biol 4:187

Balagadde FK, You L, Hansen CL et al. (2005) Long-term monitoring of bacteria undergoing programmed population control in a microchemostat. Science 309(5731):137–40

Bar-Even A, Paulsson J, Maheshri N et al. (2006) Noise in protein expression scales with natural protein abundance. Nat Genet 38(6):636–43

Bashor CJ, Helman NC, Yan S et al. (2008) Using engineered scaffold interactions to reshape MAP kinase pathway signaling dynamics. Science 319(5869):1539–43

Bastin DA, Romana LK, Reeves PR (1991) Molecular cloning and expression in *Escherichia coli* K-12 of the rfb gene cluster determining the O antigen of an *E. coli* O111 strain. Mol Microbiol 5(9):2223–31

Basu S, Gerchman Y, Collins CH et al. (2005) A synthetic multicellular system for programmed pattern formation. Nature 434(7037):1130–4

Basu S, Mehreja R, Thiberge S et al. (2004) Spatiotemporal control of gene expression with pulse-generating networks. Proc Natl Acad Sci U S A 101(17):6355–60

Batchelor E, Goulian M (2006) Imaging OmpR localization in *Escherichia coli*. Mol Microbiol 59(6):1767–78

Baumgartner JW, Kim C, Brissette RE et al. (1994) Transmembrane signalling by a hybrid protein: communication from the domain of chemoreceptor Trg that recognizes sugar-binding proteins to the kinase/phosphatase domain of osmosensor EnvZ. J Bacteriol 176(4):1157–63

Bayer TS, Smolke CD (2005) Programmable ligand-controlled riboregulators of eukaryotic gene expression. Nat Biotechnol 23(3):337–43

Becskei A, Serrano L (2000) Engineering stability in gene networks by autoregulation. Nature 405(6786):590–3

Bintu L, Buchler NE, Garcia HG et al. (2005a) Transcriptional regulation by the numbers: applications. Current Opinion in Genetics & Development 15(2):125–135

Bintu L, Buchler NE, Garcia HG et al. (2005b) Transcriptional regulation by the numbers: models. Current Opinion in Genetics & Development 15(2):116–124

Brenner K, Karig DK, Weiss R et al. (2007) Engineered bidirectional communication mediates a consensus in a microbial biofilm consortium. Proc Natl Acad Sci USA 104(44):17300–4

Brosius J, Erfle M, Storella J (1985) Spacing of the −10 and −35 regions in the tac promoter. Effect on its in vivo activity. J Biol Chem 260(6):3539–41

Buskirk AR, Landrigan A, Liu DR (2004) Engineering a ligand-dependent RNA transcriptional activator. Chem Biol 11(8):1157–63

Canton B, Labno A, Endy D (2008) Refinement and standardization of synthetic biological parts and devices. Nat Biotechnol 26(7):787–93

Cello J, Paul AV, Wimmer E (2002) Chemical synthesis of poliovirus cDNA: generation of infectious virus in the absence of natural template. Science 297(5583):1016–8

Chen Q, Kadner RJ (2000) Effect of altered spacing between uhpT promoter elements on transcription activation. J Bacteriol 182(16):4430–6

Collins CH, Arnold FH, Leadbetter JR (2005) Directed evolution of *Vibrio fischeri* LuxR for increased sensitivity to a broad spectrum of acyl-homoserine lactones. Mol Microbiol 55(3): 712–23

Collins CH, Leadbetter JR, Arnold FH (2006) Dual selection enhances the signaling specificity of a variant of the quorum-sensing transcriptional activator LuxR. Nat Biotechnol 24(6): 708–12

Cox RS, 3rd, Surette MG, Elowitz MB (2007) Programming gene expression with combinatorial promoters. Mol Syst Biol 3:145

de Boer PA, Crossley RE, Rothfield LI (1983) Proc Natl Acad Sci USA 80:21–25

de la Torre JC, Ortin J, Domingo E et al. (1984) Plasmid vectors based on Tn10 DNA: gene expression regulated by tetracycline. Plasmid 12(2):103–10

Derr P, Boder E, Goulian M (2006) Changing the specificity of a bacterial chemoreceptor. J Mol Biol 355(5):923–32

Dixon R, Cannon F, Kondorosi A (1976) Construction of a P plasmid carrying nitrogen fixation genes from *Klebsiella pneumoniae*. Nature 260(5548):268–71

Drubin DA, Way JC, Silver PA (2007) Designing biological systems. Genes Dev 21(3):242–54

Dwyer MA, Looger LL, Hellinga HW (2003) Computational design of a Zn2+ receptor that controls bacterial gene expression. Proc Natl Acad Sci USA 100(20):11255–60

El-Samad H, Khammash M (2006) Regulated degradation is a mechanism for suppressing stochastic fluctuations in gene regulatory networks. Biophys J 90(10):3749–3761

Elowitz MB, Leibler S (2000) A synthetic oscillatory network of transcriptional regulators. Nature 403(6767):335–8

Endy D (2005) Foundations for engineering biology. Nature 438(7067):449–53

Endy D (2008) Genomics. Reconstruction of the genomes. Science 319(5867):1196–7

Falcon CM, Matthews KS (2000) Operator DNA sequence variation enhances high affinity binding by hinge helix mutants of lactose repressor protein. Biochemistry 39(36):11074–83

Ferrell JE, Jr. (2002) Self-perpetuating states in signal transduction: positive feedback, double-negative feedback and bistability. Curr Opin Cell Biol 14(2):140–8

Ferrell JE, Jr., Machleder EM (1998) The biochemical basis of an all-or-none cell fate switch in *Xenopus* oocytes. Science 280(5365):895–8

Francetic O, Belin D, Badaut C et al. (2000) Expression of the endogenous type II secretion pathway in *Escherichia coli* leads to chitinase secretion. Embo J 19(24):6697–703

Frank DE, Saecker RM, Bond JP et al. (1997) Thermodynamics of the interactions of lac repressor with variants of the symmetric lac operator: effects of converting a consensus site to a non-specific site. J Mol Biol 267(5):1186–206

Gambetta GA, Lagarias JC (2001) Genetic engineering of phytochrome biosynthesis in bacteria. Proc Natl Acad Sci USA 98(19):10566–71

Gardner TS, Cantor CR, Collins JJ (2000) Construction of a genetic toggle switch in *Escherichia coli*. Nature 403(6767):339–42

Gibson DG, Benders GA, Andrews-Pfannkoch C et al. (2008) Complete chemical synthesis, assembly, and cloning of a *Mycoplasma genitalium* genome. Science 319(5867):1215–20

Greber D, Fussenegger M (2007) Mammalian synthetic biology: engineering of sophisticated gene networks. J Biotechnol 130(4):329–45

Guet CC, Elowitz MB, Hsing W et al. (2002) Combinatorial synthesis of genetic networks. Science 296(5572):1466–70

Guzman LM, Belin D, Carson MJ et al. (1995) Tight regulation, modulation, and high-level expression by vectors containing the arabinose PBAD promoter. J Bacteriol 177(14):4121–30

Ham TS, Lee SK, Keasling JD et al. (2006) A tightly regulated inducible expression system utilizing the fim inversion recombination switch. Biotechnol Bioeng 94(1):1–4

Hasty J, McMillen D, Collins JJ (2002) Engineered gene circuits. Nature 420(6912):224–30

Hawkins AC, Arnold FH, Stuermer R et al. (2007) Directed evolution of Vibrio fischeri LuxR for improved response to butanoyl-homoserine lactone. Appl Environ Microbiol 73(18):5775–81

Helbl V, Hillen W (1998) Stepwise selection of TetR variants recognizing tet operator 4C with high affinity and specificity. J Mol Biol 276(2):313–8

Helbl V, Tiebel B, Hillen W (1998) Stepwise selection of TetR variants recognizing tet operator 6C with high affinity and specificity. J Mol Biol 276(2):319–24

Henssler EM, Scholz O, Lochner S et al. (2004) Structure-based design of Tet repressor to optimize a new inducer specificity. Biochemistry 43(29):9512–8
Hoch J, Silhavy T (1995) Two Component Signal Transduction. Washington, DC. ASM Press
Hooshangi S, Thiberge S, Weiss R (2005) Ultrasensitivity and noise propagation in a synthetic transcriptional cascade. Proc Natl Acad Sci USA 102(10):3581–6
Isaacs FJ, Dwyer DJ, Collins JJ (2006) RNA synthetic biology. Nat Biotechnol 24(5):545–54
Isaacs FJ, Dwyer DJ, Ding C et al. (2004) Engineered riboregulators enable post-transcriptional control of gene expression. Nat Biotechnol 22(7):841–7
Isaacs FJ, Hasty J, Cantor CR et al. (2003) Prediction and measurement of an autoregulatory genetic module. Proc Natl Acad Sci USA 100(13):7714–9
Jarboe LR, Grabar TB, Yomano LP et al. (2007) Development of ethanologenic bacteria. Adv Biochem Eng Biotechnol 108:237–61
Jose AM, Soukup GA, Breaker RR (2001) Cooperative binding of effectors by an allosteric ribozyme. Nucleic Acids Res 29(7):1631–7
Kaern M, Blake WJ, Collins JJ (2003) The engineering of gene regulatory networks. Annu Rev Biomed Eng 5:179–206
Kaern M, Elston TC, Blake WJ et al. (2005) Stochasticity in gene expression: from theories to phenotypes. Nat Rev Genet 6(6):451–64
Kalir S, Alon U (2004) Using a quantitative blueprint to reprogram the dynamics of the flagella gene network. Cell 117(6):713–20
Kalir S, Mangan S, Alon U (2005) A coherent feed-forward loop with a SUM input function prolongs flagella expression in *Escherichia coli*. Mol Syst Biol 1:2005 0006
Kalir S, McClure J, Pabbaraju K et al. (2001) Ordering genes in a flagella pathway by analysis of expression kinetics from living bacteria. Science 292(5524):2080–3
Kamionka A, Sehnal M, Scholz O et al. (2004) Independent regulation of two genes in *Escherichia coli* by tetracyclines and Tet repressor variants. J Bacteriol 186(13):4399–401
Karig D, Weiss R (2005) Signal-amplifying genetic circuit enables in vivo observation of weak promoter activation in the Rhl quorum sensing system. Biotechnol Bioeng 89(6):709–18
Keasling JD (2008) Synthetic biology for synthetic chemistry. ACS Chem Biol 3(1):64–76
Khlebnikov A, Datsenko KA, Skaug T et al. (2001) Homogeneous expression of the P(BAD) promoter in *Escherichia coli* by constitutive expression of the low-affinity high-capacity AraE transporter. Microbiology 147(Pt 12):3241–7
Khlebnikov A, Keasling JD (2002) Effect of lacY expression on homogeneity of induction from the P(tac) and P(trc) promoters by natural and synthetic inducers. Biotechnol Prog 18(3):672–4
Khlebnikov A, Risa O, Skaug T et al. (2000) Regulatable arabinose-inducible gene expression system with consistent control in all cells of a culture. J Bacteriol 182(24):7029–34
Kleckner N, Barker DF, Ross DG et al. (1978) Properties of the translocatable tetracycline-resistance element Tn10 in *Escherichia coli* and bacteriophage lambda. Genetics 90(3):427–61
Kobayashi H, Kaern M, Araki M et al. (2004) Programmable cells: interfacing natural and engineered gene networks. Proc Natl Acad Sci USA 101(22):8414–9
Koizumi M, Soukup GA, Kerr JN et al. (1999) Allosteric selection of ribozymes that respond to the second messengers cGMP and cAMP. Nat Struct Biol 6(11):1062–71
Laub MT, Biondi EG, Skerker JM (2007) Phosphotransfer profiling: systematic mapping of two-component signal transduction pathways and phosphorelays. Methods Enzymol 423:531–48
Lee SY, Papoutsakis ET (1999) Metabolic Engineering. Marcel Dekker, New York
Levskaya A, Chevalier AA, Tabor JJ et al. (2005) Synthetic biology: engineering *Escherichia coli* to see light. Nature 438(7067):441–2
Li MZ, Elledge SJ (2007) Harnessing homologous recombination in vitro to generate recombinant DNA via SLIC. Nat Methods 4(3):251–6
Li N, Cannon MC (1998) Gas vesicle genes identified in *Bacillus megaterium* and functional expression in *Escherichia coli*. J Bacteriol 180(9):2450–8
Looger LL, Dwyer MA, Smith JJ et al. (2003) Computational design of receptor and sensor proteins with novel functions. Nature 423(6936):185–90

Lutz R, Bujard H (1997) Independent and tight regulation of transcriptional units in *Escherichia coli* via the LacR/O, the TetR/O and AraC/I1-I2 regulatory elements. Nucleic Acids Res 25(6):1203–10

Lutz R, Lozinski T, Ellinger T et al. (2001) Dissecting the functional program of *Escherichia coli* promoters: the combined mode of action of Lac repressor and AraC activator. Nucleic Acids Res 29(18):3873–81

Lynch SA, Desai SK, Sajja HK et al. (2007) A high-throughput screen for synthetic riboswitches reveals mechanistic insights into their function. Chem Biol 14(2):173–84

Mangan S, Alon U (2003) Structure and function of the feed-forward loop network motif. Proc Natl Acad Sci USA 100(21):11980–5

Mangan S, Itzkovitz S, Zaslaver A et al. (2006) The incoherent feed-forward loop accelerates the response-time of the gal system of *Escherichia coli*. J Mol Biol 356(5):1073–81

Mangan S, Zaslaver A, Alon U (2003) The coherent feedforward loop serves as a sign-sensitive delay element in transcription networks. J Mol Biol 334(2):197–204

Mathews DH, Sabina J, Zuker M et al. (1999) Expanded sequence dependence of thermodynamic parameters improves prediction of RNA secondary structure. J Mol Biol 288(5):911–40

McAdams HH, Arkin A (1997) Stochastic mechanisms in gene expression. Proc Natl Acad Sci USA 94(3):814–9

Michalowski CB, Short MD, Little JW (2004) Sequence tolerance of the phage lambda PRM promoter: implications for evolution of gene regulatory circuitry. J Bacteriol 186(23):7988–99

Mielenz JR (2001) Ethanol production from biomass: technology and commercialization status. Curr Opin Microbiol 4(3):324–9

Miller MB, Bassler BL (2001) Quorum sensing in bacteria. Annu Rev Microbiol 55:165–99

Milo R, Shen-Orr S, Itzkovitz S et al. (2002) Network motifs: simple building blocks of complex networks. Science 298(5594):824–7

Morgan-Kiss RM, Wadler C, Cronan JE, Jr (2002) Long-term and homogeneous regulation of the *Escherichia coli* araBAD promoter by use of a lactose transporter of relaxed specificity. Proc Natl Acad Sci USA 99(11):7373–7

Ninfa AJ, Mayo AE (2004) Hysteresis vs. graded responses: the connections make all the difference. Sci STKE 2004(232):pe20

Pedraza JM, van Oudenaarden A (2005) Noise propagation in gene networks. Science 307(5717):1965–9

Pfeifer BA, Admiraal SJ, Gramajo H et al. (2001) Biosynthesis of complex polyketides in a metabolically engineered strain of *E-coli*. Science 291(5509):1790–1792

Posfai G, Plunkett G, 3rd, Feher T et al. (2006) Emergent properties of reduced-genome *Escherichia coli*. Science 312(5776):1044–6

Ptashne M, Gann A (2002) Genes & Signals. Cold Spring Harbor Laboratory Press, New York, 2002

Rosenfeld N, Elowitz MB, Alon U (2002) Negative autoregulation speeds the response times of transcription networks. J Mol Biol 323(5):785–93

Salis H, Tamsir A, Voigt CA (2009) Engineering bacterial sensors and signals. Bacterial sensing and Signaling (in press, 2009)

Savageau MA (1974) Comparison of classical and autogenous systems of regulation in inducible operons. Nature 252(5484):546–9

Service RF (2007) Cellulosic ethanol. Biofuel researchers prepare to reap a new harvest. Science 315(5818):1488–91

Shahrezaei V, Ollivier JF, Swain PS (2008) Colored extrinsic fluctuations and stochastic gene expression. Mol Syst Biol 4:196

Shen-Orr SS, Milo R, Mangan S et al. (2002) Network motifs in the transcriptional regulation network of *Escherichia coli*. Nat Genet 31(1):64–8

Shetty RP, Endy D, Knight TF, Jr (2008) Engineering BioBrick vectors from BioBrick parts. J Biol Eng 2(1):5

Sia SK, Gillette BM, Yang GJ (2007) Synthetic tissue biology: tissue engineering meets synthetic biology. Birth Defects Res C Embryo Today 81(4):354–61

Skerker JM, Perchuk BS, Siryaporn A et al. (2008) Rewiring the specificity of two-component signal transduction systems. Cell 133(6):1043–54
Skerker JM, Prasol MS, Perchuk BS et al. (2005) Two-component signal transduction pathways regulating growth and cell cycle progression in a bacterium: a system-level analysis. PLoS Biol 3(10):e334
Smith HO, Hutchison CA, 3rd, Pfannkoch C et al. (2003) Generating a synthetic genome by whole genome assembly: phiX174 bacteriophage from synthetic oligonucleotides. Proc Natl Acad Sci USA 100(26):15440–5
Smith TL, Sauer RT (1995) P22 Arc repressor: role of cooperativity in repression and binding to operators with altered half-site spacing. J Mol Biol 249(4):729–42
Soukup GA, Breaker RR (1999a) Design of allosteric hammerhead ribozymes activated by ligand-induced structure stabilization. Structure 7(7):783–91
Soukup GA, Breaker RR (1999b) Engineering precision RNA molecular switches. Proc Natl Acad Sci USA 96(7):3584–9
Soukup GA, Breaker RR (1999c) Relationship between internucleotide linkage geometry and the stability of RNA. Rna 5(10):1308–25
Soukup GA, DeRose EC, Koizumi M et al. (2001) Generating new ligand-binding RNAs by affinity maturation and disintegration of allosteric ribozymes. Rna 7(4):524–36
Swartz JR (2001) Advances in *Escherichia coli* production of therapeutic proteins. Curr Opin Biotechnol 12(2):195–201
Takeda Y, Sarai A, Rivera VM (1989) Analysis of the sequence-specific interactions between Cro repressor and operator DNA by systematic base substitution experiments. Proc Natl Acad Sci USA 86(2):439–43
Tang J, Breaker RR (1997) Rational design of allosteric ribozymes. Chem Biol 4(6):453–9
Temme K, Salis H, Tullman-Ercek D et al. (2008) Induction and relaxation dynamics of the regulatory network controlling the type III secretion system encoded within *Salmonella* pathogenicity island 1. J Mol Biol 377(1):47–61
Thattai M, van Oudenaarden A (2001) Intrinsic noise in gene regulatory networks. Proc Natl Acad Sci USA 98(15):8614–9
Topp S, Gallivan JP (2008) Random walks to synthetic riboswitches–a high-throughput selection based on cell motility. Chembiochem 9(2):210–3
Tumpey TM, Basler CF, Aguilar PV et al. (2005) Characterization of the reconstructed 1918 Spanish influenza pandemic virus. Science 310(5745):77–80
Ulrich LE, Koonin EV, Zhulin IB (2005) One-component systems dominate signal transduction in prokaryotes. Trends Microbiol 13(2):52–6
Utsumi R, Brissette RE, Rampersaud A et al. (1989) Activation of bacterial porin gene expression by a chimeric signal transducer in response to aspartate. Science 245(4923):1246–9
Vilar JM, Leibler S (2003) DNA looping and physical constraints on transcription regulation. J Mol Biol 331(5):981–9
Voigt CA (2006) Genetic parts to program bacteria. Curr Opin Biotechnol 17(5):548–57
Wagner R (2000) Transcription regulation in prokaryotes. Oxford University Press, Oxford, New York
Wall ME, Hlavacek WS, Savageau MA (2004) Design of gene circuits: lessons from bacteria. Nat Rev Genet 5(1):34–42
Ward SM, Delgado A, Gunsalus RP et al. (2002) A NarX-Tar chimera mediates repellent chemotaxis to nitrate and nitrite. Mol Microbiol 44(3):709–19
Weiss R (2001) Cellular Computation and Communications Using Engineered Genetic Regulatory Networks, Massachussetts Institute of Technology
Weiss R, Homsy GE, Knight TF, Jr (1999) Towards in vivo digital circuits. DIMACS Workshop on Evolution as Computation 1:1–18
Williams SB, Stewart V (1997) Discrimination between structurally related ligands nitrate and nitrite controls autokinase activity of the NarX transmembrane signal transducer of *Escherichia coli* K-12. Mol Microbiol 26(5):911–25

Wilson JW, Coleman C, Nickerson CA (2007) Cloning and transfer of the *Salmonella* pathogenicity island 2 type III secretion system for studies of a range of gram-negative genera. Appl Environ Microbiol 73(18):5911–8

Winkler WC, Breaker RR (2003) Genetic control by metabolite-binding riboswitches. Chembiochem 4(10):1024–32

Yen L, Svendsen J, Lee JS et al. (2004) Exogenous control of mammalian gene expression through modulation of RNA self-cleavage. Nature 431(7007):471–6

Yokobayashi Y, Weiss R, Arnold FH (2002) Directed evolution of a genetic circuit. Proc Natl Acad Sci USA 99(26):16587–91

You L, Cox RS, 3rd, Weiss R et al. (2004) Programmed population control by cell-cell communication and regulated killing. Nature 428(6985):868–71

Zhang W, Ames BD, Tsai SC et al. (2006) Engineered biosynthesis of a novel amidated polyketide, using the malonamyl-specific initiation module from the oxytetracycline polyketide synthase. Appl Environ Microbiol 72(4):2573–80

Chapter 20
Systems Metabolic Engineering of *E. coli*

Sang Yup Lee and Jin Hwan Park

Contents

Abstract Metabolic engineering can be defined as purposeful modification of metabolic pathways and other cellular network to achieve desired cellular phenotype and performance. Rational metabolic engineering developed a couple of decades ago changed the way strains have been developed, which had traditionally been performed by random mutagenesis and selection. Now, we are observing another paradigm shift towards systems-level metabolic engineering, powered by the methods and tools developed in the discipline of systems biology. Systems metabolic engineering allows whole-cell-wide metabolic engineering based on the findings of systems biological studies including omics and computational analyses. Not only the metabolic network, but also gene regulatory and signaling networks can be engineered to develop an optimal strain. Also, it is important to consider fermentation and downstream processes during the upstream strain development. In this chapter, the general strategies of systems metabolic engineering are reviewed with relevant examples recently reported.

S.Y. Lee (✉)
Metabolic and Biomolecular Engineering National Research Laboratory, Department of Chemical and Biomolecular Engineering (BK21 program), Department of Bio and Brain Engineering, Department of Biological Sciences, and Bioinformatics Research Center, Center for Systems and Synthetic Biotechnology, Institute for the BioCentury, BioProcess Engineering Research Center, KAIST, 335 Gwahangno, Yuseong–gu, Daejeon 305-701, Korea
e-mail: leesy@kaist.ac.kr

S.Y. Lee (ed.), *Systems Biology and Biotechnology of Escherichia coli*,
DOI 10.1007/978-1-4020-9394-4_20,

20.1 Systems Biology and Biotechnology

Advances in genomics and functional genomics opened up new possibilities of better understanding of biological systems as a whole. Systems biology emerged towards this goal is also advancing rapidly by integrating mathematical and computational analyses with biological experiments. The so called 'omics' technologies including genomics, transcriptomics, proteomics, metabolomics and fluxomics are providing us with unprecedentedly large amounts of data which allow data-driven discovery of biological phenomena. Unlike classical molecular biology, which deals with individual cellular components, systems biology looks at the cells at the systems-level. The genome-wide high-throughput analysis by using various systems biological tools has been proved to be beneficial for elucidating the cell physiology from global point of view. This benefit can thus be employed to study biotechnologically important microorganisms as well, which generated a new research field called systems biotechnology (Lee et al. 2005b). In the field of metabolic engineering, the emergence of systems biology has greatly contributed to the increasing number of successful applications by utilizing the expanding number of systems biological tools available (Stephanopoulos 2007, Tyo et al. 2007). Application of systems biology and biotechnology to real bioprocess development has been recently reviewed (Lee et al. 2005b, Park et al. 2008).

Escherichia coli remains one of the most widely used microorganisms, due to the extensive knowledge on its metabolism, well-established omics and molecular biological techniques, fermentation techniques including high cell density cultivation technique, and availability of the complete genome sequence. Thus, *E. coli* is an important model organism for systems biology, which can be extended to other organisms. Also, systems-level engineering of *E. coli* based on genome-wide high-throughput data and computational analyses has emerged as an important strategy for developing bioprocesses in industry. Recently, several notable achievements were reported, which demonstrated the successful application of systems biology and biotechnology to developing superior *E. coli* strains (Lee et al. 2007, Park et al. 2007, Wang et al. 2006). In this chapter, the overall strategy for systematic engineering of *E. coli* is introduced and its industrial importance is discussed with relevant examples.

20.2 Systems Metabolic Engineering Roadmap

Because systems biology is a relatively new field, guidelines for the construction of superior industrial strains are still being refined and perfected. These guidelines are continuously updated to satisfy the increasing demands of metabolic and biochemical engineers. In this regard, here we present a roadmap of systems metabolic engineering based on currently available various omics and computational tools, which will guide us to develop improved strains suitable for industrial-scale production of target bioproducts starting from the wild-type strain (Fig. 20.1). The details for each of those tools are covered in the following chapters of this book.

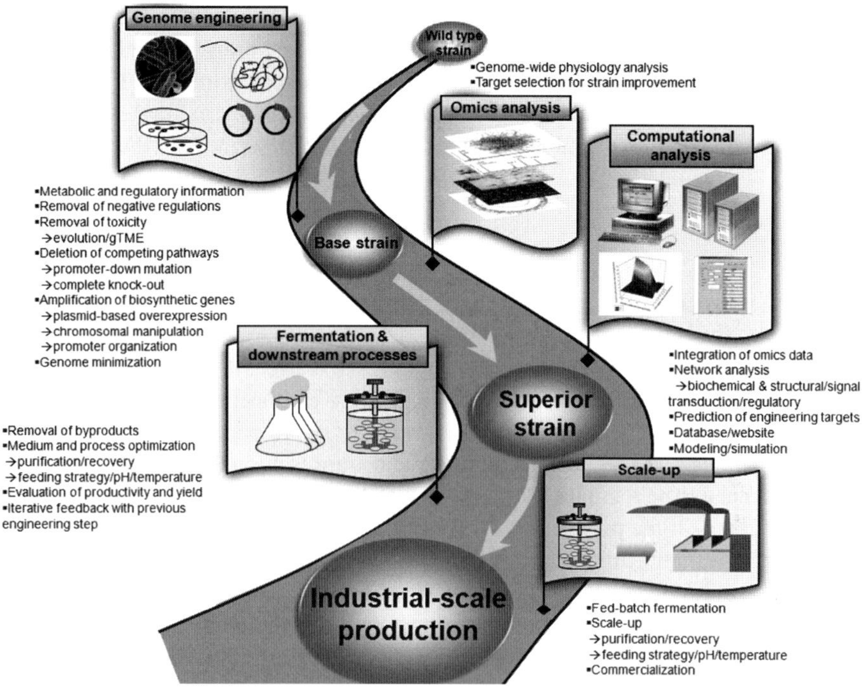

Fig. 20.1 Systems metabolic engineering roadmap starting from a wild-type strain to the final strain suitable for industrial-scale production. Factors to be considered in each step are listed

In the first step, a so-called base strain is constructed by intuitive rational metabolic engineering of a wild-type strain using conventional strategies of strain improvement. Some of the general strategies that can be employed are removal of negative regulations and competing pathways by site-specific genome engineering, and the amplification of rate-controlling (limiting) biosynthetic pathways by plasmid-based or chromosomally integrated gene overexpression.

Next, additional metabolic engineering target genes are selected by genome-wide omics analysis using the microorganism's genome, transcriptome, proteome, fluxome and metabolome data. Notably, integration of these omics analyses and computational simulation results might be able to provide additional information on the cellular status at various hierarchical levels, thereby allowing us to identify new targets suitable for further improvement of the strain. Various methods of computational analysis, including construction of biological networks and databases and subsequent data mining and simulations, are becoming essential tools for the efficient integration of these data.

After this initial engineering of the cellular metabolism to construct the base strain and subsequent strain improvement based on multi-level genome-wide high-throughput analysis, the engineered strain can undergo lab-scale fermentation, through which various fermentation performance data including the product yield, productivities and byproduct formation are evaluated. The resulting data are used to refine the strain design by repeating the above procedure in a feedback manner and can be iterated until the satisfactory performance is observed. All these procedures can also be repeated for the development of industrial-scale processes. The details of each step and relevant examples are described in the following section.

20.2.1 Genome Engineering

Before stepwise metabolic engineering, all things that affect the biosynthesis of target bioproducts need to be manipulated by genome engineering based on genome sequence data and biochemical characterization studies. Important factors that affect target biosynthesis are feedback inhibition and transcriptional attenuation regulation, which need to be removed by site-specific genome engineering (Lee et al. 2007, Park et al. 2007). If the product is toxic to the cell, it is not possible to overproduce this product without killing the cell unless toxicity is removed. Several strategies have been employed to reduce toxicity, which include the manipulation of genes related with transporter if the product is secreted into the medium, adaptive evolution, which enables cells to increase tolerance against the toxic product through continuous exposure (Guimaraes et al. 2008), and global transcription machinery engineering (gTME) which uses error prone PCR to engineer transcription factors affecting expression of multiple genes, the consequence of which results in product tolerance (Alper and Stephanopoulos 2007).

Once the above problems are solved, it is often necessary to amplify the rate-controlling biosynthetic pathways and to delete competing pathways by overexpressing the corresponding genes and knocking out the corresponding genes, respectively. To amplify the relevant genes, two methods can be considered, plasmid-based overexpression or chromosomal integration, depending on the desired expression level of the corresponding genes and stability (Lee et al. 2007). Deletion of competing pathways can also be carried out by two different methods, complete gene knock-out or attenuation mutation. Complete deletion of competing pathways might result in an auxotrophic strain for certain nutrients (Lee et al. 2007, Park et al. 2007), thus requiring supplementation of these nutrients. If this becomes a problem, the competing pathway can be attenuated instead of being completely knocked-out (Lee et al. 2007).

Another important modification carried out by genome engineering that can be considered is to have the minimal set of genes needed to sustain bacterial life, which potentially redirect more cellular resources to the desired product. It has been reported that an *E. coli* strain with genome reduction up to 15% still maintains good growth profiles and protein production (Posfai et al. 2006). However, genome deletion needs to be carefully performed as deletion of a large part of genome might cause unwanted physiological changes during actual industrial-scale fermentation including cell lysis, byproducts formation, and strain instability.

20.2.2 Omics Analysis

For further characterization of cellular status and identification of additional targets to be engineered, various omics analyses can be utilized (Table 20.1). For example, the transcriptome, proteome, and nucleotide sequences were compared between the parent strain *E. coli* W3110 and the rationally engineered L-threonine-overproducing base *E. coli* strain TF5015 to understand regulatory mechanisms of L-threonine production and the physiological changes in the base strain (Lee et al. 2003). As a result, genes involved in the glyoxylate shunt, the tricarboxylic acid cycle, and amino acid biosynthesis, were identified to be significantly upregulated, whereas ribosomal protein genes were found to be downregulated. Furthermore, two important mutations in the *thrA* and *ilvA* genes were identified as essential for the overproduction of L-threonine.

Identification of new engineering target genes based on transcriptome profiling has been reported for the enhanced production of recombinant protein. Transcriptome profiles before and after induction during the high cell-density culture revealed several target genes down-regulated (Choi et al. 2003). Overexpression of the *prsA* (encoding phosphoribosyl pyrophosphate synthetase) and *glpF* (glycerol transporter) genes, which were selected among the down-regulated genes, allowed a significant increase in IFG-I_f production (from 1.8 to 4.3 g/L).

Another recent report suggests that transcriptome analysis is effective for the initial screening of target genes for molecular breeding. A xylitol-producing strain was constructed by inserting genes of NADPH-dependent D-xylose reductase and D-xylose permease into the *E. coli* chromosome. Analysis of the recombinant strain's transcriptome under xylitol-producing and non-producing conditions revealed that xylitol production down-regulated 56 genes, among which the *yhbC* gene was selected as a target gene to be engineered. Deletion of the *yhbC* gene resulted in a 2.7-fold increase in xylitol production (Hibi et al. 2007).

Proteome analysis can also be a powerful tool for the systems-level analysis and engineering of strains. For example, the importance of Eda (2-keto-3-deoxy-6 phosphogluconate aldolase) in poly(3-hydroxybutyrate) production in an engineered *E. coli* was identified by comparative proteome profiling (Han et al. 2001). In another study, increased expression of the 30S ribosomal protein S1 in poly(3-hydroxybutyrate) producing *E. coli* lacking the *pgi* gene was also identified by comparative proteome profiling (Kabir and Shimizu 2003).

Also, proteome profiles of *E. coli* in response to the overproduction of human leptin, a serine rich protein (Han et al. 2001). It was found that the levels of enzymes involved in the biosynthesis of serine family amino acids significantly decreased. Based on this result, the *cysK* gene (encoding cysteine synthase A) was overexpressed, resulting in a 4-fold increase in the specific leptin productivity. This strategy was extended successfully to enhance production of another serine-rich protein.

In another example, overexpression of the *ppsA* gene (encoding phage shock protein A) was suggested as a good strategy based on proteome profiling for increasing the yield of soluble antibody by 50% (Aldor et al. 2005). These results show that

Table 20.1 Successful applications of systems biological tools for the characterization and improvement of *E. coli*

Systems biology tools	Product	Approach	Results	Refs
Transcriptome	xylitol	Comparative analysis of producing and nonproducing conditions for identifying factors suppressing NADPH supply and deleting them	2.7-fold increase in xylitol production	Hibi et al. 2007
Transcriptome	IGF-I fusion protein	Comparative analysis of wild-type and recombinant strains for identifying gene targets and amplifying them	139% increase in IGF-I_f production	Choi et al. 2003
Proteome	Humanized antibody	Comparative analysis of control and production strains for identifying and amplifying gene targets	50% increase in the yield of soluble antibody	Aldor et al. 2005
Proteome	Human leptin	Comparative analysis of wild-type and recombinant strains for understanding genotypic characteristics	4-fold increase in the specific leptin productivity, and improved production of another serine-rich protein	Han et al. 2003
Proteome	Poly(3-hydroxybutyrate)	Comparative analysis of wild-type and recombinant strains for understanding genotypic characteristics	The importance of Eda (2-keto-3-deoxy-6-phosphogluconate aldolase) in poly(3-hydroxybutyrate) production by engineered *E. coli* was identified	Han et al. 2001
Proteome	Poly(3-hydroxybutyrate)	Comparative analysis of wild-type and recombinant strains for identifying and amplifying gene targets	Increased expression of 30S ribosomal protein S1 in poly(3-hydroxybutyrate) producing *E. coli* lacking pgi was also identified	Kabir and Shimizu 2003

Table 20.1 (continued)

Systems biology tools	Product	Approach	Results	Refs
Transcriptome and proteome	L-threonine	Comparative analysis of wild-type and mutant strains to elucidate underlying mechanism for overproduction of L-threonine	Identification of significant mutations in *thrA* and *ilvA* for overproduction of L-threonine	Lee et al. 2003
In silico simulation	Succinic acid	Metabolic flux analysis to identify gene targets to be modified	Production of a high yield of 1.29 mol succinic acid per mol glucose	Wang et al. 2006
Genome and *in silico* simulation	Lycopene	Combination of gene knockout targets identified from metabolic flux analysis and genome-wide transposon method	Maximum production of lycopene in a mutant strain by 8.5-fold higher than recombinant *E. coli* K-12 wild-type and a 2-fold higher than an engineered parental strain.	Alper et al. 2005
Genome and *in silico* simulation	Succinic acid	Comparative genome analysis of *E. coli* and *Mannheimia succiniciproducens* combined with metabolic flux analysis to identify candidate genes to be manipulated for overproducing succinic acid in *E. coli*	More than 7-fold increase in succinic acid production	Lee et al. 2005
Genome, transcriptome and *in silico* simulation	L-valine	Systematic development of L-valine producing *E. coli*	Production of a high yield of 0.378 g L-valine per g glucose	Park et al. 2007
Genome, transcriptome and *in silico* simulation	L-threonine	Systematic development of L-threonine producing *E. coli*	Production of a high yield of 0.393 g L-threonine per g glucose by batch culture, and 82.4 g/L L-threonine by fed-batch culture	Lee et al. 2007

proteomics can be successfully used to identify target genes to be engineered towards enhanced bioproducts formation.

20.2.3 Computational Analysis

As systems metabolic engineering is practiced with omics analysis as discussed above, computational analysis plays a critical role in integrating the data and information extracted (Fig. 20.1). Accordingly, many computational frameworks are actively being developed and applied. In particular, the genome-scale stoichiometric modeling and simulation by metabolic flux analysis have become one of the most representative computational methods applicable in systems metabolic engineering.

Genome-scale stoichiometric modeling and optimization-based flux analysis techniques have become popular because of their relative simplicity and ability to predict the flux distributions under various genotypic and environmental conditions (Kim et al. 2008). The details on the construction and applications of genome-scale metabolic models are described in a chapter by Prof. Palsson in this book. Brifely, the number of genome-scale metabolic models is continuously increasing. Also, the size and coverage of these models are also expanding to become more realistic models as in the case of *E. coli* (Edwards and Palsson 2000, Feist et al. 2007, Reed et al. 2003), *Helicobacter pylori* (Schilling et al. 2002, Thiele et al. 2005), *Haemophilus influenza* (Edwards and Palsson 1999, Schilling and Palsson 2000) and *Mannheimia succiniciproducens* (Hong et al. 2004, Kim et al. 2007). This expansion enables more precise simulation of the metabolic characteristics, and thus enables identification of target genes for obtaining improved phenotypes. Recently, the development of so far the most comprehensive metabolic model of *E. coli* has been reported (Feist et al. 2007). This model composed of 2077 reactions and 1039 metabolites describes the metabolism of *E. coli* in much more expanded manner than the previous models in terms of the number of genes (1260 genes) incorporated in the model.

Notable improvement in the production of various chemicals in *E. coli* was achieved by engineering the target genes identified by the genome-scale metabolic simulations (Alper et al. 2005a,b, Lee et al. 2007, Park et al. 2007, Wang et al. 2006). Different methods of simulation have also been developed for other applications such as identifying indispensability of genes and understanding adaptive evolution (Fong and Palsson 2004, Reed et al. 2003). The true power of genome-scale metabolic simulation lies in its ability to systematically predict flux distributions in diverse situations, in particular, to predict the effects of genotypic alterations, thus saving time and effort of performing actual experiments, which are numerous if one is to consider all possible scenarios.

20.2.4 Downstream Process

Strain needs to be further improved by considering the factors identified during the actual midstream and downstream industrial processes. Omics and computational analyses of the production strain under actual fermentation condition might provide additional engineering targets that are essential to eliminate any unforeseen fermentation problems, such as unexpectedly high concentration of byproducts that hinder cell growth and production. As an example, computational analysis has been employed to identify pathways responsible for the increased acetate production during the high cell density fed-batch fermentation of L-threonine producing *E. coli* strain. From this analysis, the production strain was further engineered to recycle the byproducts to biomass formation and/or energy production, resulting in a 20.4% increase in a volumetric productivity (Lee et al. 2007).

The presence of any byproduct together with the desired product increases the cost of recovery and purification. As cells naturally produce both desired and undesired bioproducts, it is necessary to remove pathways that lead to the formation of undesirable byproducts. For example, *E. coli* B strain produces only ethanol when biosynthetic pathways for other byproducts including lactate, formate, and succinate were removed (Zhou et al. 2008). In another example, removal of pathways leading to the formation of L-isoleucine, pantothenate, and L-leucine increased the flux towards a desired product L-valine (Park et al. 2007). These examples emphasize the importance of considering midstream and downstream processes during the strain development for the overall optimization of the bioprocess.

20.3 Successful Applications of Systems Metabolic Engineering

Recently, several excellent examples of systems metabolic engineering for the construction of superior strains of *E. coli* have been reported. These examples are summarized in Table 20.1. Development of 100% rationally designed *E. coli* strain overproducing L-valine (Park et al. 2007) is one of the examples that follow the systems metabolic engineering roadmap described above. First, a base strain was constructed by removing all the known negative regulations and competing pathways that hinder the L-valine production, and amplifying the local biosynthetic pathways. Next, transcriptome analysis and computational gene knock-out simulation were employed to identify more target genes engineered. At the end, a superior *E. coli* strain producing L-valine with a high yield of 0.378 g L-valine per g glucose was successfully developed from the initial wild-type strain producing 'zero' g L-valine per g glucose.

Another good example that follows the roadmap is the development of L-threonine overproducing *E. coli* strain (Fig. 20.2b). The L-threonine overproducing strain was developed by following almost the same procedure as that for developing L-valine producing strain (Lee et al. 2007). Additionally, the desired expression level of target genes selected by transcriptome analysis was determined using a computation method called *in silico* flux response analysis. They went one step further to optimize the downstream process by removing the severe problem of acetate accumulation observed during the fed-batch fermentation, by *in silico* flux response analysis. The final engineered strain was able to produce L-threonine with a high yield of 0.393 g L-threonine per g glucose by batch culture. This strain was capable of producing 82.4 g/L L-threonine by fed-batch culture.

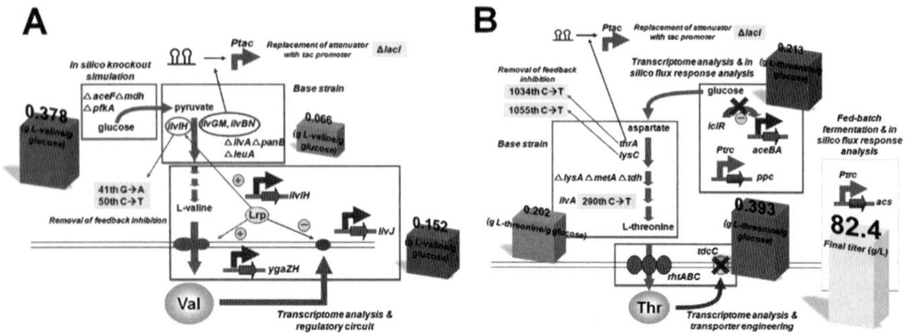

Fig. 20.2 Schematic presentation of stepwise strain improvement for the production of L-valine (**a**) and L-threonine (**b**). Successive engineering along the roadmap (Fig. 20.1) gradually increases the product yield. Results from each experiment indicated by box are reflected in the block near the corresponding box. Values of 0.066 and 0.202 represent the yields of L-valine and L-threonine, respectively, obtained with the corresponding base strains, in which negative regulations were removed and the rate-controlling biosynthetic enzymes were amplified. Values of 0.152 and 0.213 indicate the yields of L-valine and L-threonine achieved by additional engineering based on the transcriptome analysis for L-valine and the transcriptome analysis combined with *in silico* flux response analysis for L-threonine, respectively. Values of 0.378 and 0.393 are the final yields achieved for L-valine and L-threonine after *in silico* knock-out simulation for L-valine and transcriptome-based transporter engineering for L-threonine. For L-threonine, fed-batch fermentation was additionally carried out, during which accumulation of significant amounts of byproduct (acetic acid) was observed. *In silico* flux response analysis was performed again to engineer the acetate uptake system. The final concentration of L-threonine achieved by fed-batch fermentation of the final engineered strain is 82.4 g/L. The gray boxes in the left side of each figure represent mutations introduced into the genome to remove feedback inhibition and *lacI* gene. The 'X' marks for L-threonine indicate deletion of the *iclR* and *tdcC* genes to remove negative regulation by IclR and to block L-threonine uptake, respectively. Ptac indicates the replacement of transcriptional attenuator with the *tac* promoter. Ptrc indicates the replacement of native promoter with the *trc* promoter. The thick leading to L-valine or L-threonine arrows indicate increased fluxes by directly overexpressing the corresponding genes or by knocking out the genes suggested by *in silico* simulation. The lines around 'Lrp' indicate the global regulation by Lrp. The plus (+) and minus (−) symbols indicate activation and repression of corresponding regulation by Lrp, respectively. These figures were re-drawn from Park et al. (2007) and Lee et al. (2007)

In another example, systems metabolic engineering approach was taken to develop succinic acid overproducing *E. coli* strain through comparative genome scanning and metabolic flux analysis with a natural succinic acid overproducer *M. succiniciproducens* (Lee et al. 2005a). Comparative genome analysis of *E. coli* and *M. succiniciproducens* predicted five candidate genes to be manipulated for overproducing succinic acid in *E. coli*. Then, metabolic flux analysis was carried out to find an optimal combination of the selected genes that provide the strain with the maximum biomass and succinic acid production capability upon their knock-out. This resulted in the selection of genes which enhanced the succinic acid production by more than 7-fold (Lee et al. 2005a). In a different study, an *E. coli* strain capable of producing succinic acid with a yield of 1.29 mol succinic acid per mol glucose was developed by engineering the strain based on genome-scale metabolic flux analysis (Wang et al. 2006). Likewise, metabolic flux analysis and genome-wide transposon library search were performed for the development of *E. coli* strain overproducing lycopene (Alper et al. 2005b). Although knock-out of gene targets predicted by metabolic flux analysis allowed increased production of lycopene, the amount produced was below the stoichiometrically maximal value. This limit was overcome by ideally combining gene targets with transposon method, some of which led to 8.5-fold higher production of lycopene. These examples demonstrate that systems metabolic engineering is a powerful and essential strategy for developing improved strains for the production of various bioproducts.

20.4 Future Perspectives

The importance of systems biology in strain development has just begun to be validated with a series of successful reports. By combining multiple omics data and integrating them with computational analysis, systems biology provides us with valuable information on cellular physiological status, and consequently new strategies for strain improvement. However, it is true that we are not yet capable of truly integrating these omics data for understanding the biological systems as a whole. Research on inter-relating the omics data at various hierarchical levels is needed. One of the most successful applications of computational analysis in systems metabolic engineering is genome-scale metabolic flux analysis as described above. Much effort is being exerted to refine the genome-scale metabolic models and simulation methods. For the latter, various algorithms to predict gene knock-out and amplification targets are being developed. Also, better constraints to be applied during the genome-scale metabolic simulations are being developed for more accurate prediction of metabolic flux distributions. Genome-scale metabolic models thus far employed are rather limited as they lack regulatory constraints. Thus, complex regulatory circuits controlling overall cellular metabolism needs to be deciphered to understand the regulatory mechanisms governing the metabolism (Akesson et al. 2004). Besides stoichiometric modeling and simulation, research on developing other methods such as dynamic modeling, data mining and machine learning methods will also be more actively performed to integrate gene regulatory and signaling networks together with metabolic network (Bonneau et al. 2007, Friedman 2004). Furthermore, one should not give up using so-called random approach yet. Even though it will be always nice to develop 100% genotypically-defined strain, our ability to develop a satisfactory strain might be limiting in many cases. Evolution of strains under rational selective pressure, engineering regulatory proteins, and even random mutagenesis can be combined with systems metabolic engineering to further improve the strain performance.

In this chapter, systems metabolic engineering strategies for the development of strains were presented together with successful examples of developing superior *E. coli* strains capable of overproducing various bioproducts. It is expected that more fine-tuned strategies of systems metabolic engineering will be developed as our knowledge on cellular metabolism and regulation advances and more experimental and computational tools for systems biological studies are developed. It is believed that more successful examples of systems metabolic engineering will appear in the near future.

Acknowledgments This work was supported by the Korea-Australia Collaborative Research Project on the Development of Sucrose-Based Bioprocess Platform (N02071165) from the Korean Ministry of Knowledge Economy. Also, this work was supported by the Korean Systems Biology Program from the Ministry of Education, Science and Technology through the Korea Science and Engineering Foundation (No. M10309020000-03B5002-00000). Further supports by LG Chem Chair Professorship and Microsoft are appreciated.

References

Akesson M, Forster J, Nielsen J (2004) Integration of gene expression data into genome-scale metabolic models. Metab Eng 6(4):285–93

Aldor IS, Krawitz DC, Forrest W et al. (2005) Proteomic profiling of recombinant *Escherichia coli* in high-cell-density fermentations for improved production of an antibody fragment biopharmaceutical. Appl Environ Microbiol 71(4):1717–28

Alper H, Jin YS, Moxley JF et al. (2005a) Identifying gene targets for the metabolic engineering of lycopene biosynthesis in *Escherichia coli*. Metab Eng 7(3):155–64

Alper H, Miyaoku K, Stephanopoulos G (2005b) Construction of lycopene-overproducing *E. coli* strains by combining systematic and combinatorial gene knockout targets. Nat Biotechnol 23(5):612–616

Alper H, Stephanopoulos G (2007) Global transcription machinery engineering: a new approach for improving cellular phenotype. Metab Eng 9(3):258–67

Bonneau R, Facciotti MT, Reiss DJ et al. (2007) A predictive model for transcriptional control of physiology in a free living cell. Cell 131(7):1354–65

Choi JH, Lee SJ, Lee SJ et al. (2003) Enhanced production of insulin-like growth factor I fusion protein in *Escherichia coli* by coexpression of the down-regulated genes identified by transcriptome profiling. Appl Environ Microbiol 69(8):4737–42

Edwards JS, Palsson BO (1999) Systems properties of the *Haemophilus influenzae* Rd metabolic genotype. J Biol Chem 274(25):17410–6

Edwards JS, Palsson BO (2000) The *Escherichia coli* MG1655 *in silico* metabolic genotype: its definition, characteristics, and capabilities. Proc Natl Acad Sci USA 97(10):5528–33

Feist AM, Henry CS, Reed JL et al. (2007) A genome-scale metabolic reconstruction for *Escherichia coli* K-12 MG1655 that accounts for 1260 ORFs and thermodynamic information. Mol Syst Biol 3:121

Fong SS, Palsson BO (2004) Metabolic gene-deletion strains of *Escherichia coli* evolve to computationally predicted growth phenotypes. Nat Genet 36(10):1056–8

Friedman N (2004) Inferring cellular networks using probabilistic graphical models. Science 303(5659):799–805

Guimaraes PM, Francois J, Parrou JL et al. (2008) Adaptive evolution of a lactose-consuming *Saccharomyces cerevisiae* recombinant. Appl Environ Microbiol 74(6):1748–56

Han MJ, Yoon SS, Lee SY (2001) Proteome analysis of metabolically engineered *Escherichia coli* producing Poly(3-hydroxybutyrate). J Bacteriol 183(1):301–8

Hibi M, Yukitomo H, Ito M et al. (2007) Improvement of NADPH-dependent bioconversion by transcriptome-based molecular breeding. Appl Environ Microbiol 73(23):7657–63

Hong SH, Kim JS, Lee SY et al. (2004) The genome sequence of the capnophilic rumen bacterium *Mannheimia succiniciproducens*. Nat Biotechnol 22(10):1275–81

Kabir MM, Shimizu K (2003) Fermentation characteristics and protein expression patterns in a recombinant *Escherichia coli* mutant lacking phosphoglucose isomerase for poly(3-hydroxybutyrate) production. Appl Microbiol Biotechnol 62(2–3):244–55

Kim HU, Kim TY, Lee SY (2008) Metabolic flux analysis and metabolic engineering of microorganisms. Mol Biosyst 4(2):113–20
Kim TY, Kim HU, Park JM et al. (2007) Genome-scale analysis of *Mannheimia succiniciproducens* metabolism. Biotechnol Bioeng 97(4):657–671
Lee JH, Lee DE, Lee BU et al. (2003) Global analyses of transcriptomes and proteomes of a parent strain and an L-threonine-overproducing mutant strain. J Bacteriol 185(18):5442–51
Lee KH, Park JH, Kim TY et al. (2007) Systems metabolic engineering of *Escherichia coli* for L-threonine production. Mol Syst Biol 3:149
Lee SJ, Lee DY, Kim TY et al. (2005a) Metabolic engineering of *Escherichia coli* for enhanced production of succinic acid, based on genome comparison and *in silico* gene knockout simulation. Appl Environ Microbiol 71(12):7880–7
Lee SY, Lee DY, Kim TY (2005b) Systems biotechnology for strain improvement. Trends Biotechnol 23(7):349–58
Park JH, Lee KH, Kim TY et al. (2007) Metabolic engineering of *Escherichia coli* for the production of L-valine based on transcriptome analysis and *in silico* gene knockout simulation. Proc Natl Acad Sci USA 104(19):7797–802
Park JH, Lee SY, Kim TY et al. (2008) Application of systems biology for process development. Trends Biotechnol 26(8):404–412
Posfai G, Plunkett G, 3rd, Feher T et al. (2006) Emergent properties of reduced-genome *Escherichia coli*. Science 312(5776):1044–6
Reed JL, Vo TD, Schilling CH et al. (2003) An expanded genome-scale model of *Escherichia coli* K-12 (iJR904 GSM/GPR). Genome Biol 4(9):R54
Schilling CH, Covert MW, Famili I et al. (2002) Genome-scale metabolic model of *Helicobacter pylori* 26695. J Bacteriol 184(16):4582–93
Schilling CH, Palsson BO (2000) Assessment of the metabolic capabilities of *Haemophilus influenzae* Rd through a genome-scale pathway analysis. J Theor Biol 203(3):249–83
Stephanopoulos G (2007) Challenges in engineering microbes for biofuels production. Science 315(5813):801–4
Thiele I, Vo TD, Price ND et al. (2005) Expanded metabolic reconstruction of *Helicobacter pylori* (iIT341 GSM/GPR): an *in silico* genome-scale characterization of single- and double-deletion mutants. J Bacteriol 187(16):5818–30
Tyo KE, Alper HS, Stephanopoulos GN (2007) Expanding the metabolic engineering toolbox: more options to engineer cells. Trends Biotechnol 25(3):132–7
Wang Q, Chen X, Yang Y et al. (2006) Genome-scale *in silico* aided metabolic analysis and flux comparisons of *Escherichia coli* to improve succinate production. Appl Microbiol Biotechnol 73(4):887–94
Zhou S, Iverson AG, Grayburn WS (2008) Engineering a native homoethanol pathway in *Escherichia coli* B for ethanol production. Biotechnol Lett 30(2):335–42

Index